Radon and Its Decay Products

A C S   S Y M P O S I U M   S E R I E S **331**

Radon and Its Decay Products

Occurrence, Properties, and Health Effects

Philip K. Hopke, EDITOR

University of Illinois

Developed from a symposium sponsored by
the Divisions of Environmental Chemistry and
Nuclear Chemistry and Technology
at the 191st Meeting
of the American Chemical Society,
New York, New York,
April 13–18, 1986

American Chemical Society, Washington, DC 1987

Library of Congress Cataloging-in-Publication Data

Radon and its decay products.
(ACS symposium series, ISSN 0097-6156; 331)

"Developed from a symposium sponsored by the
Divisions of Environmental Chemistry and Nuclear
Chemistry and Technology at the 191st Meeting of the
American Chemical Society, New York, New York,
April 13-18, 1986."

Includes bibliographies and index.

1. Radon—Toxicology—Congresses. 2. Radon—
Physiological effect—Congresses. 3. Radon—
Environmental aspects—Congresses. 4. Radon—
Decay—Congresses. 5. Air—Pollution, Indoor—
Measurement—Congresses.

I. Hopke, Philip K., 1944-   . II. American
Chemical Society. Division of Environmental
Chemistry. III. American Chemical Society. Division of
Nuclear Chemistry and Technology.

RA1247.R33R33   1987   363.7'392   86-32042
ISBN 0-8412-1015-2

ACS Symposium Series

M. Joan Comstock, *Series Editor*

1987 Advisory Board

Foreword

The ACS SYMPOSIUM SERIES was founded in 1974 to provide a medium for publishing symposia quickly in book form. The format of the Series parallels that of the continuing ADVANCES IN CHEMISTRY SERIES except that, in order to save time, the papers are not typeset but are reproduced as they are submitted by the authors in camera-ready form. (The papers in this book were prepared with a different reference style from the style usually used in the ACS SYMPOSIUM SERIES.) Papers are reviewed under the supervision of the Editors with the assistance of the Series Advisory Board and are selected to maintain the integrity of the symposia; however, verbatim reproductions of previously published papers are not accepted. Both reviews and reports of research are acceptable, because symposia may embrace both types of presentation.

Contents

Preface

RADON AND ITS DECAY PRODUCTS have been studied for a long time; now the results of these studies are visible and important to a wide audience. The discovery in the past two years of very high concentrations of radon in houses in the eastern United States has increased the public's awareness of the radon problem. Radon is no longer a hazard only to uranium miners and to people living in houses built with or on contaminated materials. The health risks of radon are substantial compared with other natural radioactivity. The realization of the potential for widespread effects of radon and its decay products on lung cancer rates has been growing in the radon research community for almost a decade. Indoor air surveys in Canada and Sweden in the late 1970s as well as other, more limited, sampling efforts showed that higher radon levels existed in indoor air than had been expected.

Radon and its decay products were a major focus of meetings held in Capri, Italy, in 1983 and in Maastricht, the Netherlands, in 1985. These meetings reported on ongoing national surveys in European countries, on laboratory and field studies on the properties of radon decay products, and on the models for relating the airborne radioactivity concentrations to the human lung.

A major meeting in the United States was developed to bring together researchers from around the world to review the current state of knowledge and to discuss problems that still needed to be addressed. This book, based on that symposium, presents new and important results of the work of researchers active in the field.

As organizer of that symposium and editor of this volume, I would like to thank the authors who agreed to present their work in this forum and the reviewers who helped to improve the quality of the presentations. In particular, I wish to thank Victoria Corkery, without whose invaluable assistance this volume would not have been possible.

PHILIP K. HOPKE
University of Illinois
Urbana, IL 61801

October 17, 1986

Chapter 1

Radon and Its Decay Products: An Overview

Philip K. Hopke

Departments of Civil Engineering and Nuclear Engineering and Institute for Environmental Studies, University of Illinois, Urbana, IL 61801

This volume contains the reports presented at a
symposium held as part of the 191th National Meeting
of the American Chemical Society. The chapters present
recent findings on a number of aspects of the indoor
radon problem. In this introductory chapter, an overview of these reports is presented that highlights
some of these results and suggests areas where
uncertainties still remain.

During the past two years, indoor radon and its decay products have
come to be recognized as a major potential threat to public health.
Although radon studies have been on-going for many years, it is only
with the discovery of very high levels of radon in the Reading Prong
area of eastern Pennsylvania, New Jersey, and New York that public
attention has focused on the question of indoor radon, its sources,
its properties, its health implications and how it may be controlled.
This interest has also manifested itself in a number of technical
meetings to present the current state of knowledge and to discuss
the critical questions regarding the behavior and effects of radon
and its decay products (Bosnjakovic et al., 1985; APCA, 1986).
In order to focus on more of the basic research problems related
to radon, a symposium was organized in conjunction with the 191th
National Meeting of the American Chemical Society. This volume
presents most of the reports given at that symposium. There are
five major groups of reports; occurrence, measurement methods,
physical and chemical properties of radon and its decay products,
health effects, and mitigation of radon levels.

Occurrence

During the past decade, it has been recognized that indoor radon
levels do not arise only or even primarily from building materials.
This conclusion is in contrast to the previous understanding as

expressed by the United Nations Scientific Committee on the Effects
of Atomic Radiation (UNSCEAR, 1977). It has become certain that
although unusual building materials such as alum shale concrete or
phosphogypsum can be significant contributors, ordinary houses
made of ordinary materials can have extraordinary levels of radon.
In most cases, the radon infiltrates from the soil into the building
under the influence of a pressure differential caused by the
structure itself. The amount of radon that results is dependent
on the amount of radon in the soil gas, the ability of the soil
gas to move through the soil, and the characteristics of the building
that provides the driving force and entry pathways.

Initial efforts at predicting areas where houses may be at
higher risk have focused on searching for areas with above average
levels of soil radium. However, the report by Sextro et al. (1987)
shows that houses on soils with below average radium concentrations
can have high indoor radon if the soil is highly permeable and the
house extracts a large fraction of its makeup air from the soil.
Thus, it is important to simultaneously consider the combination
of soil and building characteristics that govern the amount of
movement of radon-laden soil gas.

Hess et al. (1987) show that for areas where there is high radon
in the potable water supply, water can represent a significant source
of indoor radon. However, there are ways of removing radon from
water including aeration or carbon adsorption.

A number of countries have now completed or are finishing
surveys of their housing stock for indoor radon levels. The results
of surveys from Ireland, United Kingdom, Norway, Sweden, Finland,
and Japan are presented here and Magno and Guimond (1987) describe
the survey planned for the United States. The results reinforce
the difficulty in predicting the concentration of radon in any
specific dwelling. However, there are some geographical patterns
that do emerge that relate to the soil radium content. In those
cases, it is useful to make additional local area surveys such as
in Cornwall in the United Kingdom (Cliff et al., 1987).

In all of these indoor radon surveys positively skewed,
approximately log-normal distributions are obtained. It should be
noted that the local area survey may yield a different distribution
from the national and thus estimates of numbers of high level
houses in susceptible areas can be substantially underestimated
from an aggregated national distribution.

Measurement Methods

Several papers present reviews of measurement methods or improvements
in existing methods. Yamashita et al. (1987) present the description
of a portable liquid scintillation system that can be used for thoron
(Rn-220) as well as radon (Rn-222) in water samples. Thoron measurements have not been made for houses where radon in water may be a
significant source. Such an instrument could be useful in making
such determinations as well as in studying geochemical problems as
described in this report. A review of measurement methods by Shimo
et al. (1987) and of development and calibration of track-etch
detectors (Yonehara et al., 1987) are also included. Samuelsson

(1987) provides a review and critique of the closed cannister
method for emanation rate determinations.

Properties

The physico-chemical properties of radon and its decay products are
presented in a series of reports primarily focusing on the decay
products. However, Stein (1987) presents a review of his pioneering
studies of radon chemistry and the reactions of radon with strong
oxidizing agents. Although radon is not chemically active in indoor
air, it is interesting to note that radon is not an "inert" gas.

There are a series of papers that focus on the behavior of the
radon decay products and their interactions with the indoor
atmosphere. Previous studies (Goldstein and Hopke, 1983) have
elucidated the mechanisms of neutralization of the Po-218 ionic
species in air. Wilkening (1987) reviews the physics of small
ions in the air. It now appears that the initially formed polonium
ion is rapidly neutralized, but can become associated with other ions
present. Reports by Jonassen (1984) and Jonassen and McLaughlin
(1985) suggest that only 5 to 10% of the decay products are
associated with highly mobile ions and that much of the activity
is on large particles that have a bipolar charge distribution.
Bandi et al. (1987) present the results of a series of calculations
on the attachment of the decay products to the existing particles
for the various models of the attachment process.

However, these results may need to be modified by the findings
of Porstendorfer et al. (1987), Vanmarcke et al. (1987), and Knutson
et al. (1985) that the "unattached" Po-218 containing molecules are
actually part of an ultrafine mode (0.7 - 2.0 nm) in the activity
size distribution. Thus, they are not free molecules and will move
with a reduced diffusion coefficient based on the size of these
ultrafine particles.

Raes (1987) presents the results of additional cluster formation
modeling studies based on the methods he presented earlier (Raes,
1985; Raes and Janssens, 1985). These results would suggest that
the ultrafine mode is the result of ion-based cluster formation.
Chu et al. (1987) present results of laboratory studies of the
formation of an ultrafine aerosol by converting SO_2 to sulfuric acid
using measurement methods described by Holub and Knutson (1987) and
Kulju et al. (1987). It was found that the size of the resulting
activity distributions is dependent on the SO_2 concentration. The
role of humidity is still unclear and more studies are needed, but
it appears that both future theoretical models and laboratory
studies will be extremely fruitful in elucidating the behavior of
Po-218 from shortly after its formation until its incorporation
into the existing accumulation mode aerosol.

Health Effects

There are several groups of chapters discussing the health effects
of radon and its decay products. In Pohl-Ruling et al. (1987) and
Reubel et al. (1987), the direct effect of ionizing radiation on
cell behavior are presented. Such studies help to elucidate the
relationship between the radiation dose and the response at the

cellular level. In order to deposit dose to the cells of the
bronchial epithelium, there must be deposition of the particle-
attached activity. The past dosimetric estimates have used the
theoretical model of Davies (1973) as corrected by Ingham (1975)
for calculating particle deposition from a given aerosol size
distribution. In the chapter by B.S. Cohen (1987), it is shown that
these deposition velocity estimates are low by a factor of 2 down
to particle sizes of 0.04 μm. Using radiolabeled monodisperse
particles, the deposition of particles in the various generation
of the bronchii have been measured. These results will be very
important in the development of new dose estimates because of the
increase in activity deposition that has been found.

Harley and Cohen (1987) present some such revised dose estimates
based on these deposition results. James (1987) reviews the cells
at risk and the important link between dose and response. Both
suggest that more work is needed to improve the models for deposition
of particles, amount and extent of radiometric dose, and the
potential responses to that dose.

Steinhausler (1987) and Martell (1987) review the dosimetric
models and related model studies. Their view is that there are still
very large uncertainties in the existing data and in the extrapolation from the exposure and response data for underground miners
and experimental animals to the health effects of the radon progeny
levels to which the general public is exposed. B.L. Cohen (1987)
describes his work to relate radon measurements with lung cancer
rates for various geographical areas to test the concept of a dose
threshold.

An important part of the dose question is the relationship
between radon concentration, decay product concentrations and
unattached fraction. The airborne particle concentration is a
critical factor in determining the retention of decay products in
the air. Thus, as the particle concentration increases, the
equilibrium factor between the radon and decay products, F,
increases. Various values of F have been reported ranging from
0.30 as an average value for German houses (Urban et al., 1985)
while Stranden (1987) uses 0.50 for Norwegian homes and McLaughlin
(1987) has reported individual values as high as 0.80. Although
this increased decay product concentration increases the exposure,
the higher particle concentration also leads to a lower "unattached"
fraction, f. This fraction is credited in the dose models with
yielding a much higher dose to the bronchii per unit airborne
radioactivity concentration relative to the "attached" activity.
Thus, Vanmarcke et al. (1987) shows that over the range of values
they have measured, the dose is relatively constant per unit radon
concentration according to the James-Birshall model (NEA, 1982).
This inverse relationship between equilibrium factor and "unattached"
fraction and their relationship to the resulting dose is important
in considering how to most efficiently and effectively monitor for
exposure. This inverse relationship suggests that it is sufficient
to determine the radon concentration. However, it is not clear
how precisely this relationship holds and if the dose models are
sufficiently accurate to fully support the use of only radon
measurements to estimate population exposure and dose.

Mitigation Methods

This volume does not have mitigation as a major focal point. This aspect of the radon issue was a large part of the APCA Specialty Conference on Indoor Radon (APCA, 1986). However, there were presentations on housing construction practice based on the current understanding of modes of radon ingress that could limit radon problems (Ericson and Schmied, 1987), on electrostatic filtration for removing particles and associated decay products (Jonassen, 1987), and on adsorption of radon on activated carbon as an alternative control strategy (Bocanegra and Hopke, 1987).

It is clear that some proper planning in the design phase of new houses can lead to less likelihood of indoor radon problems and make provision for lower cost modifications later to mitigate against such problems. The experience in Sweden is that houses with low radon concentrations can be constructed on high risk soils by proper design and construction practices.

As indicated above, there is a relationship between particle concentration, equilibrium factor and the amount of highly mobile radioactive particles. Removal of the accumulation mode particles may decrease the decay product exposure, but increase the dose because of the high effectiveness of the "unattached" activity in dose deposition. Thus, air cleaning may not succeed in lower risk unless both factors are taken into account. Jonassen explores electrostatic filtration in this context. Finally, design considerations are presented for a possible alternative control system using activated carbon in an alternating bed system.

Conclusions

This volume represents a collection of papers that provide a considerable amount of recent results and reflect the current level of scientific understanding of radon related problems. However, with the increased public interest and the resulting increased scientific study, it can be expected that there will be many important new findings and our knowledge of the nature and extent of the indoor radon problem will be greatly expanded in the next few years.

Literature Cited

APCA, Indoor Radon, Air Pollution Control Association, Pittsburgh, PA (1986).

Bandi, F., A. Khan, and C.R. Phillips, Effects of Aerosol Poly-dispersity on Theoretical Calculation of Unattached Fractions of Radon Progeny, this volume (1987).

Bocanegra, R. and P.K. Hopke, The Feasibility of Using Activated Charcoal for Indoor Radon Control, this volume (1987).

Bosnjakovic, B., P.H. van Dijkum, M.C. O'Riordan, and J. Sinnaeve, eds., Proceedings of Exposure to Enhanced Natural Radiation and Its Regulatory Implications, Sci. Total Environ., Volume 45 (1985)

Ericson, S.-O. and H. Schmied, Modified Design in New Construction Prevents Infiltration of Soil Gas Carrying Radon, this volume (1987).

Chu, K.D., P.K. Hopke, E.O. Knutson, K.W. Tu, and R.F. Holub, The Induction of an Ultrafine Aerosol by Radon Radiolysis, this volume (1987).

Cliff, K.D., A.D. Wrixon, B.M.R. Green, and J.C.H. Miles, Radon and Its Decay-Product Concentrations in UK Dwellings, this volume (1987).

Cohen, B.S., Deposition of Ultrafine Particles in the Human Tracheo-bronchial Tree: A Determinant of the Dose from Radon Daughters, this volume (1987).

Cohen, B.L., Surveys of Radon Levels in U.S. Homes as a Test of the Linear-No Treshold Dose-Response Relationship for Radiation Carcino-genesis, this volume (1987).

Davies, C.N., Diffusion and Sedimentation of Aerosol Particles from Poiseuille Flow in Pipes, Aerosol Sci. 4:317-328 (1973).

Goldstein, S.D. and P.K. Hopke, Environmental Neutralization of Po-218, Environ. Sci. Technol. 19:146-150 (1985).

Harley, N.H. and B.S. Cohen, Updating Radon Daughter Bronchial Dosimetry, this volume (1987).

Hess, C.T., J.K. Korsah, and C.J. Einloth, Radon in Houses Due to Radon in Potable Water, this volume (1987).

Holub, R.F. and E.O. Knutson, Measurement of ^{218}Po Diffusion Coefficient Spectra Using Multiple Wire Screens, this volume (1987).

Ingham, D.B., Diffusion of Aerosols from a Stream Flowing Through a Cylindrical Tube, Aerosol Sci. 6:125-132 (1975).

James, A.C., A Reconsideration of Cells at Risk and Other Key Factors in Radon Daughter Dosimetry, this volume (1987).

Jonassen, N., Electrial Properties of Radon Daughters, presented to the International Conference on Occupational Radiation Safety and Mining, Toronto, Canada (1984).

Jonassen, N. and J.P. McLaughlin, The Reduction of Indoor Air Concentrations of Radon Daughters Without the Use of Ventilation, Sci. Total Environ. 45:485-492 (1985).

Jonassen, N., The Effect of Filtration and Exposure to Electric Fields on Airborne Radon Progeny, this volume (1987).

Knutson, E.O., A.C. George, L. Hinchliffe, and R. Sextro, Single Screen and Screen Diffusion Battery Method for Measuring Radon Progeny Size Distributions, 1–500 nm, presented to the 1985 Annual Meeting of the American Association for Aerosol Research, Albuquerque, NM, November (1985).

Kulju, L.M., K.D. Chu, and P.K. Hopke, The Development of a Mobility Analyzer for Studying the Neutralization and Particle Producing Phenomena Related to Radon Progeny, this volume (1987).

Magno, P.J. and R.J. Guimond, Assessing Exposure to Radon in the United States: An EPA Perspective, this volume (1987).

Martell, E.A., Critique of Current Lung Dosimetry Models for Radon Progeny Exposure, this volume (1987).

McLaughlin, J.P., Population Doses from Radon Decay Products in Ireland, this volume (1987).

Nuclear Energy Agency (NEA), Dosimetry Aspects of Exposure to Radon and Thoron Daughters, report by a group of experts, Paris, OECD (1982).

Pohl-Ruling, J., P. Fischer, and E. Pohl, The Effect of Radon and Decay Products on Peripheral Blood Chromosomes, this volume (1987).

Porstendorfer, J., A. Reineking, and K.H. Becker, Free Fractions, Attachment Rates and Plateout Rates of Radon Daughters in Houses, this volume (1987).

Raes, F., Description of the Properties of Unattached ^{218}Po and ^{212}Pb Particles by Means of the Classical Theory of Cluster Formation, Health Phys. 49:1177–1187 (1985).

Raes, F. A. Jannsens, and H. Vanmarcke, Modeling Size Distributions of Radon Decay Products in Realistic Environments, this volume (1987).

Raes, F. and Janssens, A., Ion-Induced Aerosol Formation in a H_2O-H_2SO_4 System – I. Extension of the Classical Theory and Search for Experimental Evidence, J. Aerosol Sci. 16:217–227 (1985).

Reubel, B., C. Atzmuller, F. Steinhausler, and W. Huber, Biophysical Effects on Human Lung Cells Due to Radon Exposure, this volume (1987).

Samuelsson, C., A Critical Assessment of Radon-222 Exhalation Measurements Using the Closed-Can Method, this volume (1987).

Sextro, R.G., B.A. Moed, W.W. Nazaroff, K.L. Revzan, and A.V. Nero, Investigations of Soil as a Source of Indoor Radon, this volume (1987).

Shimo, M., T. Iida, and Y. Ikebe, Intercomparison of Different Instruments for Measurement of Radon Concentration in Air, this volume (1987).

Stein, L., Chemical Properties of Radon, this volume (1987).

Steinhausler, F., On the Validity of Risk Assessments for Radon Daughters Induced Lung Cancer, this volume (1987).

Stranden, E., ^{222}Rn in Norwegian Dwellings, this volume (1987).

United Nations Scientific Committee on the Effects of Atomic Radiation (UNSCEAR), Sources and Effects of Ionizing Radiation, UNIPUB, New York (1977).

Urban, M., A. Wicke, and H. Keifer, Bestimmung der Strahlenbelastung der Bevolkerung durch Radon und dessen kurzlebige Zerfallsprodukte in Wohnhausern und im Freien, Report No. KIK 3805, Kernforschungs-zentrum Karlsruhe GmbH, Karlsruhe, West Germany, September 1985.

Vanmarcke, H., A. Janssens, F. Raes, A. Poffijn, P. Berkvens, and R. Van Dingenen, On the Behavior of Radon Daughters in the Domestic Environment and Its Effect on fht Effective Dose Equivalent, this volume (1987).

Wilkening, M., Effect of Radon on Some Electrical Properties of Indoor Air, this volume (1987).

Yamashita, K., H. Yoshikawa, M. Yanaga, K. Endo, and H. Nakahara, Determination of ^{220}Rn and ^{222}Rn Concentrations in Fumarolic Gases, this volume (1987).

Yonehara, H., H. Kimura, M. Sakanoue, E. Iwata, S. Kobayashi, K. Fujimoto, T. Aoyama, and T. Sugahara, Improvement in the Measurement of Radon Concentrations by a Bare Track Detector, this volume (1987).

RECEIVED October 27, 1986

OCCURRENCE

Chapter 2

Investigations of Soil as a Source of Indoor Radon

R. G. Sextro, B. A. Moed, W. W. Nazaroff[1], K. L. Revzan, and A. V. Nero

Indoor Environment Program, Lawrence Berkeley Laboratory, University of California, Berkeley, CA 94720

The predominant source of indoor radon in most single-family housing in the U.S. is the soil adjacent to the house substructure. We have examined factors influencing the production and transport of radon in soil and into buildings. A number of important parameters have been identified and their effect on radon production and migration assessed, including radium concentration, moisture content, air permeability, and grain size distribution of soils. The potential regional variations in parameters affecting radon have been evaluated by examining geographic data, including surface radium concentrations and general soil data. We have also investigated factors influencing radon migration into individual dwellings. Coupling between the building shell and the surrounding soil has been demonstrated experimentally, and pressure-field mapping and soil permeability measurements have been carried out.

Soil is now recognized as a significant, if not predominant, source of radon in the indoor environment. This is especially true in those dwellings with elevated indoor radon concentrations. (In this paper, radon refers to ^{222}Rn, and radium to ^{226}Ra. Although some of the discussion applies to ^{220}Rn as well, its presence in indoor air is usually limited by its short half-life of 55 s.) Several recent studies of radon entry into homes have been done, including investigations or discussions of the influence of pressure differentials across the building shell, and the role of soil as a source of indoor radon (Nazaroff, et al., 1985a; Nazaroff and Doyle, 1985; Nero and Nazaroff, 1984; DSMA, 1983). There are a number of reasons to investigate both

[1]Current address: Environmental Engineering Science, California Institute of Technology, Pasadena, CA 91125

the characteristics of soils and the interaction with the building structure. The intrinsic properties of the soil, combined
with the details of the building structure and its operation, are
important determinants of the radon entry rate into the structure. In addition, both the house and the soil are affected by
changes in ambient environmental conditions.

This paper summarizes our recent investigations of soil as a
source of radon. These investigations have a two-fold purpose,
and attack the problem from two distinct but complementary
approaches. The first effort is to assemble data at a geographical scale, appropriate for understanding the general characteristics of soils. This approach provides information on the
appropriate scales and ranges of important soil variables, and
how these might vary within and among regions. Another important
aspect of this approach is to derive values of parameters that
may be used in a predictive technique for identifying areas where
high indoor radon levels are most likely (Nazaroff et al. 1986;
Sextro, 1985).

The second approach is to investigate both experimentally and
theoretically the influence of soils at a localized level, on the
scale of an individual house, for example. Production and transport of radon in soils and migration of radon into houses depend
upon local soil conditions, and results of these studies will
improve our basic understanding of the physical processes
involved. This in turn will aid in the systematic development of
more effective entry identification and mitigation techniques to
prevent or reduce radon entry into dwellings.

Factors Influencing Production and Migration of Radon in Soils

General Description. The parameters that influence radon production and migration through soils and into buildings are illustrated schematically in Figure 1. The ranges of these various
parameters, and in some cases their nominal values, are presented
in Table I. Ra-226, the immediate parent of radon, is part of the
^{238}U decay series, and is found in all soil materials.

As shown in Table I, radium content of surface soils not associated with U mining or milling varies by slightly more than an
order of magnitude, with a typical value of $\sim$40 Bq kg^{-1}. In
areas near U mining and milling, a much larger range of soil
radium concentrations has been found, e.g. (Powers et al. 1980).
Soil is a mixture of solid materials, air and, usually, water and
organic matter. The radium content of soil often reflects that
of the rocks from which the solid materials are derived by physical and chemical activity. The observed ranges are from $\sim$ 0 to
20 Bq kg^{-1} for ultrabasic rocks (dunite) to 1 to 1835 Bq kg^{-1} for
igneous metamorphic rocks (gneiss) (Wollenberg, 1984). While
these ranges are broader than those for measured for soils, the
mean values for rocks, excluding alkali rocks, is consistent with
the means observed for soils.

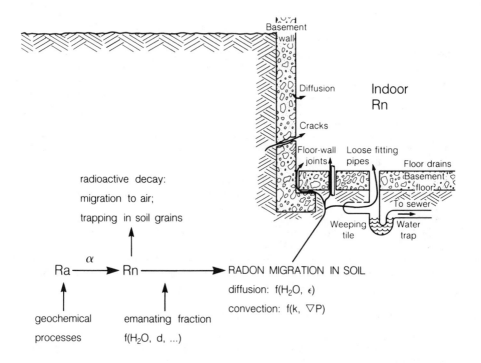

Figure 1. Schematic illustration of factors influencing the production and migration of radon in soils and into buildings. Geochemical processes affect the radium concentration in the soil. The emanating fraction is principally dependent upon soil moisture (H_2O) and the size distribution of the soil grains (d). Diffusion of radon through the soil is affected primarily by soil porosity (ϵ) and moisture content, while convective flow of radon-bearing soil gas depends mainly upon the air permeability (k) of the soil and the pressure gradient (∇P) established by the building.

Table I. General Soil Characteristics

Parameter	Range	Typical Value	Note
Radium in surface soils, A_{Ra}	10 - 200 Bq kg^{-1}	40 Bq kg^{-1}	1
Emanating fraction, r	0.05 - 0.7	0.25	2
Porosity, ϵ	0.35 - 0.65	0.5	3
Radon in soil gas, C_∞	(0.7 - 22) $\times 10^4$ Bq m^{-3}	2.7 $\times 10^4$ Bq m^{-3}	4
Diffusion coefficient, D_e	(0.7 - 7) $\times 10^{-6}$ m^{2}s^{-1}	2 $\times 10^{-6}$ m^{2}s^{-1}	5
Permeability, k	10^{-16} - 10^{-7} m^2	—	—
Differential pressure, ΔP	0 - ca. 10 Pa	3 Pa	6

Notes:

1. Myrick et al. 1983

2. Arithmetic mean from Sisigina, 1974; and Barretto 1973.

3. Mid-point of range

4. Radon concentration in soil gas at depths well below the surface can be estimated from:

$$C_\infty = A_{Ra} \ r \ \rho_s \ \frac{1 - \epsilon}{\epsilon},$$

using typical values for A_{Ra}, ϵ and r from this table and a soil grain density, $\rho_s = 2.65 \times 10^3$ kg m^{-3}.

5. Nero and Nazaroff, 1984.

6. Nazaroff et al. 1986.

The alpha decay of radium yields a recoiling radon atom, with a kinetic energy of 86 keV. The distance the recoil travels depends upon the composition and density of the materials it encounters. Typical ranges are 60 μm in air, 75 nm in water and 35 nm in crystalline material. There are several possible end points for the recoil. The radon atom may stop within the crystalline material in which it started, or become embedded in an adjacent soil grain. The diffusion of radon out of this material is extremely slow, although there is some evidence that diffusion back through the channel of material altered by the passage of the recoil occurs more quickly (Tanner, 1980). Another possibility is for the radon to terminate its recoil in water in the soil pore space. The recoil range in water is much smaller than typical pore dimensions and diffusion from water into air in the pore space is relatively rapid. Thus water, as a very effective recoil absorber, is an important variable in determining the amount of radon present in the (intergranular) pores in the soil.

The ratio of mobile radon to the total radon produced is referred to as the emanating fraction or emanation coefficient. The range of observed emanation coefficients is indicated in Table I. The effect of moisture on the emanation coefficient has been noted by several authors (Thamer et al. 1981; Strong and Levins, 1982; Stranden et al. 1984). The emanation coefficient for dry materials has been observed to be about a factor of four less than at water saturation, with the emanation coefficient changing rapidly as water is first added to dry materials; the coefficient then increases slowly as more water is added to the system. However, most studies have been done with uranium ore tailings and the response of soils may differ.

The resulting radon concentrations in soil gas have been observed to range from 7×10^3 to more than 200×10^3 Bq m^{-3} with a typical concentration in the range of $20 - 40 \times 10^3$ Bq m^{-3} (Nero and Nazaroff, 1984). This radon may then migrate through the soil and into houses by molecular diffusion and by convective flow. These two processes are generally described by Fick's law and Darcy's law, respectively. In the former case, the diffusional flux is driven by the concentration gradient. Darcy's law relates the volumetric flux density (fluid velocity per unit cross-sectional area) to the pressure gradient in the pores of the medium and its permeability. While a detailed derivation and discussion of the underlying mathematics is beyond the scope of this paper, results of a recent formulation of the problem (Nazaroff et al. 1986) are presented here in order to illustrate the range and significance of the important parameters.

A general transport equation describing the rate of change of the radon activity concentration in the pore space results from combining the effects of diffusion and convection:

$$\frac{\partial I_{Rn}}{\partial t} = D_e \, \nabla^2 I_{Rn} \, \frac{k}{\epsilon \, \mu} \, \vec{\nabla}P \cdot \vec{\nabla}I_{Rn} + G - \lambda_{Rn} \, I_{Rn}, \qquad (1)$$

where I_{Rn} is the activity concentration of radon, D_e is the interstitial diffusion coefficient (which relates the gradient of the interstitial concentration to the flux density across the pore area), k is the intrinsic permeability, ϵ the soil porosity, μ the dynamic viscosity of air, P the pressure, G is the radon generation rate in the soil pores, and λ_{Rn} is the radon decay constant. The first term on the right hand side of the equation accounts for diffusion, the second element describes pressure-driven convective flow, and the remaining terms account for production and radioactive decay of mobile radon in the soil pore space.

This formulation is based on several assumptions, including one of small soil moisture content (meaning that negligible radon is dissolved in liquid in the pore space). Other simplifying assumptions are that the soil is isotropic and homogeneous with respect to diffusion coefficient, air permeability, porosity, emanating fraction, radium content, and bulk density. Further, that Darcy's law describes the flow velocity, and, for the range of pressures of interest here, air is treated as an incompressible fluid.

This equation cannot be solved analytically, except for the simpler geometries (Clements and Wilkening, 1974; Schery, et al., 1984). Some authors have used numerical modeling approaches for investigating radon migration (DSMA 1985). For our purposes here, it will suffice to demonstrate the ranges over which diffusive or convective flow are significant or may be neglected. One way of elucidating this is to use dimensional analysis. Equation (1) may be converted to dimensionless form by multiplying and dividing the variables by combinations of characteristic time λ_{Rn}^{-1}, length L and pressure difference P_o. The resulting dimensionless equation is

$$\frac{1}{N_r} \frac{\partial I_{Rn}^*}{\partial t^*} = \frac{1}{Pe_p} \nabla^{*2} I_{Rn}^* + \vec{\nabla}^* P^* \cdot \vec{\nabla}^* I_{Rn}^* + \frac{1}{N_r} (G^* - I_{Rn}^*), \quad (2)$$

where the asterisks denote dimensionless quantities, and $Pe_p = k \, \epsilon \, P_o \, (\mu D_e)^{-1}$, and $N_r = P_o \, k \, (\mu \, \epsilon \, L^2 \, \lambda_{Rn})^{-1}$.

The dimensionless group Pe_p is essentially the ratio of the rate of convective transport to the rate of diffusive transport. Similarly, N_r describes the relative importance of radioactive decay to convective flow as a method of removing radon from the soil pores. In the case of $Pe_p \gg 1$, diffusion can be neglected and the first term in equation (1) drops out. If in addition $N_r \gg 1$, then radioactive decay can be neglected as a removal term. If $Pe_p \ll 1$, then diffusive radon migration dominates, and the second term in equation (1) can be neglected.

Using the definition of Pe_p and N_r and the nominal values from Table I for the parameters of interest, the conditions under which any of these dimensional considerations apply can be evaluated. For example, $Pe_p = 1$ when the air permeability is 2.3 $\times 10^{-11}$ m^2. For soils with larger permeabilities, such that $Pe_p \gg 1$, diffusion can be neglected as a transport process. Similarly, N_r depends directly on k, and on L^{-2}, where L is a characteristic dimension (length) over which transport is considered. Choosing 3 m as the length L, $N_r = 1$ when k is 5 $\times 10^{-11}$ m^2. Thus for high permeability soils in the vicinity of a house, radioactive decay can be ignored compared to convective flow as a means of removing radon from the soil pores.

Representative soil categories and their associated air permeabilities are given in Table II. In the examples discussed above, the transition region where Pe_p is about 1 occurs when k is on the order of 10^{-11} m^2, which corresponds to medium to fine sand soils. For soils with larger air permeabilities, such as

sands and gravels, convective flow is the dominant radon transport process, while for low permeability soils such as uniform
silts and clays, molecular diffusion is the more significant process.

Table II. Soil Permeabilities

Soil Type	Permeability (m^2)
Clay	10^{-16}
Sandy Clay	5×10^{-15}
Silt	5×10^{-14}
Sandy Silt and Gravel	5×10^{-13}
Fine Sand	5×10^{-12}
Medium Sand	10^{-10}
Coarse Sand	5×10^{-10}
Gravel	10^{-8}

The range of permeability values presented in Table II reveal
another important aspect of the role of air permeability of
soils. The permeabilities range over 8 orders of magnitude in the
extreme cases of gravels to clays. Even in what may be more typical soil categories, the range of permeabilities is still 4 to 5
orders of magnitude. By comparison, the values for other parameters, shown in Table I, have a much smaller range, and consequently local or regional differences in them may not be as significant in determining radon entry rates.

Geographical Data on Soils. As noted earlier, a geographically-
based investigation of soils has the important possibility of providing data that can be used for estimating the potential of high
indoor radon concentrations at a regional scale. There is evidence that such regional variations exist. Nero et al. (1985)
found that the geometric means of indoor radon concentrations
measured in 22 different geographical areas across the U.S. to
vary by an order-of-magnitude. In fact, these geometric mean
values are distributed approximately lognormally with a geometric
standard deviation close to 2. Among the soil characteristics
listed in Table I, geographic data on two parameters, radium concentration and air permeability, have been investigated and are
discussed here in more detail.

Data on the geographic distribution of surficial radium concentrations were acquired by the National Airborne Radiometric
Reconnaissance (NARR) survey, part of the National Uranium
Resource Evaluation (NURE) program conducted by the U.S. Department of Energy in the mid-1970s. The data were originally collected and tabulated by 1° by 2° quadrangle map area, and the data
cover approximately 450 out of a total of 474 such quadrangles

that cover the contiguous 48 states. Magnetic tapes from these aerial measurements contain data on the surficial concentrations of ^{40}K, ^{238}U, and ^{232}Th recorded continuously along the flight line. The primary flight lines, oriented east-west, were typically spaced 3 miles apart for states in the western U.S., and 6 miles in the eastern U.S; however, for all or parts of some quadrangles, the flight-line spacing was 0.25 miles. These tapes have been processed at LBL and the data volume reduced by averaging the data collected along a flight line over 0.0160° of longitude, yielding a grid of data elements with a spacing of approximately 1 mile by 0.25, 3 or 6 miles, depending upon the flight-line spacing.

Data on surface radium concentrations has been compiled for 394 of the 474 quadrangles covering the conterminous 48 states. The resulting distribution is illustrated in Figure 2, where the parameters shown for the distribution are calculated from the binned data. This distribution has a GM of 25 Bq kg^{-1} and geometric standard deviation (GSD) of 1.75. Based on this distribution, approximately 20% of the surface radium concentration data are above 40 Bq kg^{-1} and 0.7% are above 100 Bq kg^{-1}.

Surface radium concentrations in soils can also vary widely within a region, as shown in Figure 3 for two areas selected to represent those quadrangles with geometric mean radium concentrations < 20 Bq kg^{-1} (Figure 3a), and those areas with concentrations > 40 Bq kg^{-1} (Figure 3b). As can be seen from the figure, the differences in the area radium concentration distributions are quite large. Less than 0.5% of the radium concentration data from the Chico Quadrangle are above 40 Bq kg^{-1}, and less than 0.01% are above 60 Bq kg^{-1}. In contrast, almost 65% of the surface radium concentration data from the San Jose Quadrangle have values greater than 40 Bq kg^{-1} with 0.5% above 200 Bq kg^{-1} and 0.05% greater than 300 Bq kg^{-1}.

In addition to these frequency distributions, maps of various types have been produced to depict the spatial distribution of the surface radium concentrations. One example is shown in Figure 4, which represents data from the aerial survey of the Newark Quadrangle, covering part of eastern Pennsylvania and northern New Jersey. The data in this figure are plotted on an approximately six mile by six mile grid. Thus each point represents an average of the data collected along six miles of the flight line. The frequency distribution of surface radium concentrations for this quadrangle area has a GM of 35 Bq kg^{-1}, a GSD of 1.45 and an arithmetic mean of 37 Bq kg^{-1}. From these distribution parameters, one can estimate that approximately 7% of the surface radium concentration data are greater than 60 Bq kg^{-1} and less than 0.2% of the data are above 100 Bq kg^{-1}.

While these data and the resulting maps are useful in illustrating both the trends and variability in surface radium concentrations, detailed field examination of surface radium concentrations in selected areas is needed to determine the reliability of

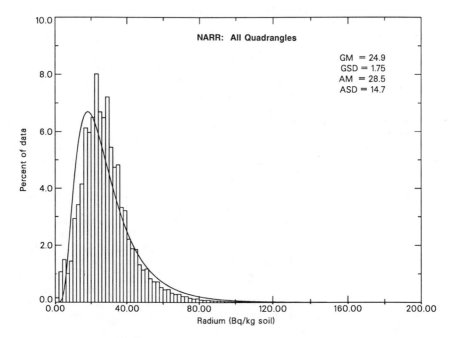

Figure 2.   Distribution of the surface radium concentration
data from the National Airborne Radiometric Reconnaissance survey for 394 1° by 2° quadrangles covering most of the contiguous 48 states. The distribution parameters are calculated from
the data and the lognormal distribution based on the geometric
mean, and standard deviation from the data is shown as a solid
curve.

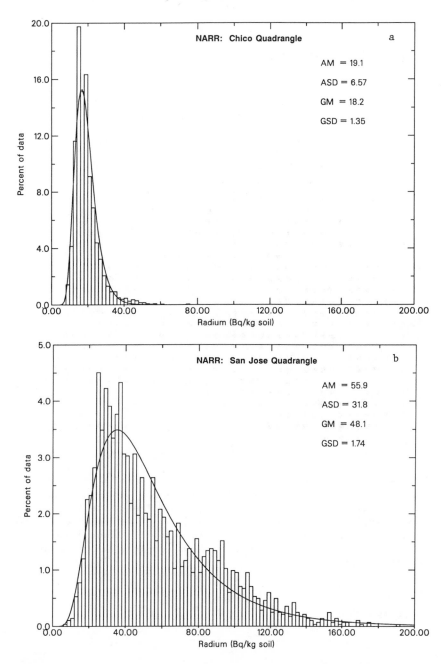

Figure 3. Distribution of surface radium concentrations for a quadrangle (a) with a GM < 20 Bq kg^{-1} and for a quadrangle (b) with a GM > 40 Bq kg^{-1}. The curves represent lognormal distributions based on the distribution parameters calculated from the data.

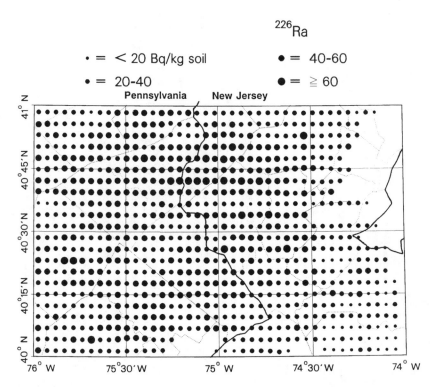

Figure 4. The spatial distribution of inferred surface radium concentrations for the Newark Quadrangle, covering eastern Pennsylvania and northern New Jersey. The state boundaries are shown as solid lines, while the county boundaries are indicated by broken lines. The spacing between flight lines is 6 miles, and each point represents data averaged along six miles of each flight line.

the data. Such a field validation study has been conducted for comparison with NARR aerial data obtained in the Spokane, WA, area. One mile segments of two different flight lines were investigated; each traversed granitic outcrops and the aerial data for those flight-line segments showed considerable variation. In general, a correspondence was observed between the radium concentrations based on the aerial data and the in-situ measurements of radium content and radon flux from soil (Moed et al., 1984).

As noted in Table I, average surface radium concentrations appear to vary by about a factor of 20. This can also be seen from the distributions from the NARR data. Soil permeabilities, on the other hand, have much larger variations, and thus, in principle, may have a greater influence on the spatial variations in average indoor radon concentrations that have been observed. As with the case of surface radium concentrations, the spatial variability of air permeabilities of soils is an important element in developing a predictive capability.

Information on surface soils is available from a number of sources, including surface soil maps compiled by the U.S. Geological Survey and the geological surveys of various states. At the present time, the coverage of such maps is not complete, nor has any systematic data on air permeability of soils been compiled. However reports issued by the Soil Conservation Service (SCS) of the U.S. Department of Agriculture contain information on most soils on a county-by-county basis. While no direct air permeability information is contained in these reports, the data and descriptive material contained there may be useful in estimating air permeabilities.

We have used these indicators to estimate permeability values for surface soils in Spokane County, WA. A general soil map of this county is shown in Figure 5. The spatial extent of two soil associations, the Naff-Larkin-Freeman association (area 1) and the Garrison-Marble-Springdale association (area 2), are shaded in the figure, and detailed estimates of air permeability for these two association have been made, as described briefly below. More information is given in Nazaroff et al. (1986).

The Naff-Larkin-Freeman association contains fine- to medium-textured soils, moderately-well to well-drained. The Garrison-Marble-Springdale soils, on the other hand, are gravelly and sandy (coarse-textured) and are "somewhat-excessively" to "excessively" drained (USDA, 1968). In addition to the soil textures, water permeabilities are described qualitatively for these soils, leading to the estimated range of intrinsic permeabilities shown in Table III for these two soil associations.

There are a number of uncertainties associated with estimating air permeabilities in this way. The soil associations described by the SCS on the county maps generally consist of one or more major soil series and may also include one or more minor soil series. The fraction of the mapped area occupied by these minor soil series, and the degree of similarity of physical

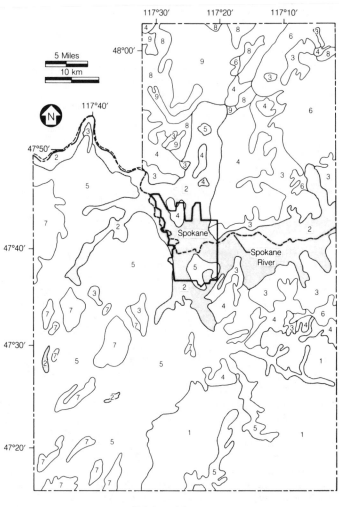

Figure 5. General soil map from the Soil Conservation Service for Spokane County, WA. Two soil associations discussed in the text are shaded. The Spokane River is indicated by a dashed line and the approximate boundary for the city of Spokane is shown by a heavy border.

properties between the major and minor soils will control the uncertainties associated with the derivation of soil characteristics for a given geographical area. In the case of the two soil associations discussed here, 77 to 80% of the respective mapped areas are actually occupied by the major soil series.

Table III. Permeabilities of selected soils in Spokane County, WA

| Soil Association | Permeability (m^2) | |
	Derived from SCS Information	Measured by LBL (Range)
Naff-Larkin-Freeman	10^{-14} to 10^{-12}	—
Garrison-Marble-Springdale	10^{-12} to 10^{-10}	$\sim 2 \times 10^{-11}$ $(0.3 - 60) \times 10^{-11}$

Detailed Site-specific Studies. Experimental measurements of parameters affecting soil gas transport and entry into houses is very limited. One recent study of two homes with basements has demonstrated coupling between the building shell and the surrounding soil and measured convective flow through the soil (Nazaroff et al. 1985b). Utilizing an array of soil probes surrounding each house and by depressurizing the basement using a blower door, the resulting negative pressure field in the soil was mapped. The effect of the enhanced basement depressurization was seen at distances up to 5 m from the basement wall for one house and up to 3 m in the other. Soil gas flows, measured utilizing SF_6 as a tracer gas, were also determined. In one case net migration velocities of greater than 1 m h^{-1} were observed with a basement depressurization of 30 Pa at an injection point 1.5 m from the basement wall.

A similar set of experiments were conducted as part of detailed radon investigations in one house in the Spokane (WA) area. The pressure-field in the surrounding soil was mapped using techniques described in Nazaroff et al. (1985b). Soil permeabilities were measured in-situ utilizing an air permeameter similar to that discussed in DSMA, (1983). An array of 30 soil probes was placed in the soil surrounding the house, as can be seen from Figure 6. The probe depths ranged from 0.5 to 1 m deep; the depths were usually limited by the ability to penetrate the soil which contained a large number of rocks of varying sizes scattered throughout the soil column. The deepest probes, with soil depths between 0.8 and 1.0 m, are indicated in the figure with squares, while probes with depths between 0.5 and 0.7 m are

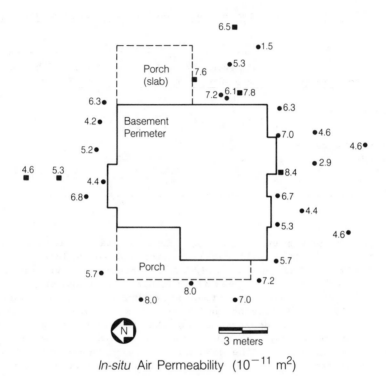

In-situ Air Permeability $(10^{-11}\ m^2)$

Figure 6. Plan view of a study house in Spokane County, WA. Probe locations with depths between 0.8 and 1 m are indicated by squares while those probes with depths between 0.5 and 0.7 m are shown as circles. The numbers indicate the in-situ air permeability measured at each probe location.

designated by circles. The measured air permeabilities for each location are shown in Figure 6 in units of 10^{-11} m^2. As can be seen, with the exception of one probe on the eastern side of the house measuring 1.5×10^{-11} m^2, the measured permeabilities are quite uniform, within $\pm$ 50% of the average value of 5.7×10^{-11} m^2. (This particular probe struck a sub-surface rock upon emplacement, and when extracted after the experments, was found to have a bent tip. This may have impaired the probe operation.)

The pressure-field mapping was done by depressurizing the basement using a blower door installed in a door between the basement and the outside. The blower was operated to produce a pressure of 34 Pa below atmospheric pressure. This pressure differential, although greater than the ~ 10 Pa values measured across the building shell in the wintertime, provided a stable baseline that was not affected significantly by variable wind loading on the building shell during the course of the experiments. All negative pressure values in the soil probes were measured with atmospheric pressure as a reference, using a differential pressure transducer described previously (Nazaroff et al. 1985b). When the blower door was turned off the soil probe pressures returned to zero, indicating that the observed pressure field in the soil was directly related to basement depressurization.

The induced pressure field in the soil surrounding the house is shown in Figure 7. There are two significant features exhibited in these data. First, in contrast with the soil permeability measurements, the pressure field is not spatially uniform; rather, marked differences in coupling between the building shell and the surrounding soil can be seen along the building perimeter. In addition, the depressurization observed in the furthest probe points on the eastern and northern side of the house demonstrate the spatial extent of the pressure field. These two sampling locations are greater than 5 m away from the building shell. Second the pressure-field map for this house shows a number of localized areas where the negative pressure is significant. Visual inspection of the interior basement wall usually resulted in finding a hole or unsealed pipe penetration through the stone basement wall opposite the area where significant soil depressurization was observed.

This house was also part of a study of remedial measures for the reduction of indoor radon concentrations (Turk et al. 1986), and based on pre-mitigation tests, almost 20% of the air infiltration in this house was estimated to be soil gas. Thus, the effective radon entry rate for this building shell is one-to-two orders of magnitude larger than that estimated for houses in the earlier study (Nazaroff, et al. 1985b).

In addition to the house described above, air permeability of soils was measured at several other house sites in the Spokane, WA, vicinity as part of a study of remedial measures for radon (Turk et al. 1986). Soil probes 1-cm-diameter and ranging between 1 and 1.5 m long were placed in the soil at two locations

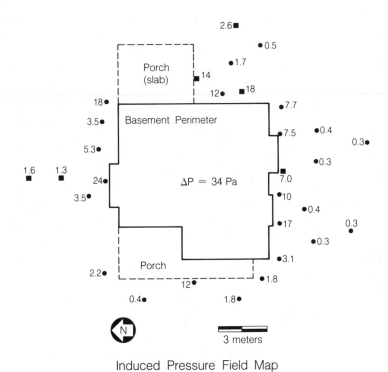

Induced Pressure Field Map
(− ΔP in Pa)

Figure 7.   Pressure field map for the same house as shown in
Figure 6.   The probe depth code is the same as for Figure 6.
The numbers show the pressure decrease at each probe measured
with respect to atmospheric pressure induced by a pressure
difference in the basement of -34 Pa.

at each house site. As it turned out, all sampling locations were in the Garrison-Marble-Springdale soil association shown in Figure 5. The average of the 11 measurements is indicated in Table III, along with the range of the observations. The agreement between the experimental data and the air permeability estimates for this soil association, derived from the SCS data and descriptions, is reasonably good.

Summary and Conclusions

Factors influencing the production and migration of radon in soils have been examined, and various sources of geographic data have been discussed. Two significant soil characteristics include air permeability and, less importantly, radium concentration. While there are, at present, few opportunities to compare the larger-scale data with on-site field measurements, those comparisons that have been made for both surface radium concentrations and air permeability of soils show a reasonable correspondence. Further comparisons between the aerial radiometric data and surface measurements are needed. Additional work and experience with SCS information on soils will improve the confidence in the permeability estimates, as will comparisons between the estimated permeabilities and actual air permeability measurements performed in the field.

The detailed site-specific experiments have helped demonstrate qualitatively the influence of building depressurization on soil gas movement and radon entry. These studies, along with the use of certain remedial techniques for reduction of indoor radon concentrations that rely on diversion of soil gas flow or alteration of the pressure differences across the building shell, have corroborated the hypothesis that radon production and convective transport through soils is a significant, if not predominant, source of radon in those houses with elevated indoor levels. A comparison of the quantitative results of these studies indicates that the specific characteristics of a house and building site are quite important, and that further experimental studies in dwellings with other substructure types and with other soil characteristics are necessary in developing a more general understanding of radon entry into structures.

Acknowledgments

This paper has benefitted from reviews by H. Wallman and D. Grimsrud of LBL. This work was supported by the Assistant Secretary for Conservation and Renewable Energy, Office of Building and Community Systems, Building Systems Division, and by the Director, Office of Energy Research, Office of Health and Environmental Research, Human Health and Assessments Division and Pollutant Characterization and Safety Research Division of the U.S. Department of Energy (DOE) under Contract No. DE-AC03-76SF00098. It was also supported by the U.S. Environmental Protection Agency

(EPA) through Interagency Agreement DW89930801-01-0 with DOE. Although the research described here is partially supported by the EPA, it has not been subjected to EPA review and therefore does not necessarily reflect the views of EPA, and no official endorsement by EPA should be inferred.

Literature Cited

Barretto, P.M.C., Emanation Characteristics of Terrestrial and Lunar Materials and the Radon-222 Loss Effect on the Uranium-Lead System Discordance, Ph.D. thesis, Rice University, Houston (1973).

Clements, W.E., and Wilkening, M.H., Atmospheric pressure effects on ^{222}Rn transport across the earth-air interface, J. Geophys. Res. 79:5025-5029 (1974).

DSMA Atcon, Ltd., Review of Existing Instrumentation and Evaluation of Possibilities for Research and Development of Instrumentation to Determine Future Levels of Radon at a Proposed Building Site, Report INFO-0096, Atomic Energy Control Board, Ottawa, Canada (1983).

DSMA Atcon, Ltd., A Computer Study of Soil Gas Movement into Buildings, Report 1389/1333, Department of Health and Welfare, Ottawa, Canada (1985).

Moed, B.A., Nazaroff, W.W., Nero, A.V., Schwehr, M.B., and Van Heuvelen, A., Identifying Areas with Potential for High Indoor Radon Levels: Analysis of the National Airborne Radiometric Reconnaissance for California and the Pacific Northwest, Report LBL-16955, Lawrence Berkeley Laboratory, Berkeley CA (1984).

Myrick, T.E., Berven, B.A., and Haywood, F.F., Determination of Concentrations of Selected Radionuclides in Surface Soil in the U.S., Health Phys. 45:631-642 (1983).

Nazaroff, W.W., and Doyle, S.M., Radon Entry into Houses Having a Crawl Space, Health Phys. 48:265-281 (1985).

Nazaroff, W.W., Feustel, H., Nero, A.V., Revzan, K.L., Grimsrud, D.T., Essling, M.A., and Toohey, R.E., Radon Transport into a Detached One-story House with a Basement, Atmos. Environ. 19:31-46 (1985a).

Nazaroff, W.W., Lewis, S.R., Doyle, S.M., Moed, B.A., and Nero, A.V., Experiments on Pollutant Transport from Soil into Residential Buildings by Pressure-Driven Air Flow, Report LBL-18374, Lawrence Berkeley Laboratory, Berkeley CA (1985b), submitted to Environ. Sci. Technol.

Nazaroff, W.W., Moed, B.A., Sextro, R.G., Revzan, K.L., Nero, A.V., Factors Influencing Soil as a Source of Indoor Radon: A Framework for Geographically Assessing Radon Source Potentials, Report LBL-20645, Lawrence Berkeley Laboratory, Berkeley CA (1986).

Nero, A.V., and Nazaroff, W.W., 1984, Characterizing the Source of Radon Indoors, Rad. Prot. Dos. 7:23-39 (1984).

Nero, A.V., Schwehr, M.B., Nazaroff, W.W., and Revzan, K.L., Distribution of Airborne [222]Rn Concentrations in U.S. Homes, Report LBL-18274, Lawrence Berkeley Laboratory, Berkeley CA (revised 1985), to be published in Science.

Powers, R.P., Turnage, N.E., and Kanipe, L.G., Determination of Radium-226 in Environmental Samples, in Proc. Natural Radiation Environment III,
eds.) pp. 640-660, U.S. Dept. of Commerce, National Technical Information Service, Springfield VA, (1980).

Schery, S.D., Gaeddert, D.H., and Wilkening, M.H., Factors Affecting Exhalation of Radon from a Gravelly Sandy Loam, J. Geophys. Res. 89, 7299-7309 (1984).

Sextro, R.G., Understanding the Origin of Radon Indoors--Building a Predictive Capability, Report LBL-20210, Lawrence Berkeley Laboratory, Berkeley CA (1985), submitted to Atmos. Environ.

Sisigina, T.I., Assessment of Radon Emanation from the Surface of Extensive Territories, in Nuclear Meteorology pp. 239-244, Israeli Program of Scientific Translations, Jerusalem, (1974).

Stranden, E., Kolstad, A.K., and Lind, B., The Influence of Moisture and Temperature on Radon Exhalation, Rad. Prot. Dos. 7:55-58 (1984).

Strong, K.P., and Levins, D.M., Effect of Moisture Content on Radon Emanation from Uranium Ore and Tailings, Health Phys. 42:27-32 (1982).

Tanner, A.B., Radon Migration in the Ground: a Supplementary Review, in Proc. Natural Radiation Environment III, Conf-780422, (Gesell, T.F., and Lowder, W.M., eds.) pp. 5-56, U.S. Dept. of Commerce, National Technical Information Service, Springfield VA, (1980).

Thamer, B.J., Nielson, K.K., and Felthauser, K., The Effects of Moisture on Radon Emanation Including the Effects on Diffusion, Report BuMines OFR 184-82, PB83-136358, U.S. Dept. of Commerce, National Technical Information Service, Springfield VA (1981)

Turk, B.H., Prill, R.J., Fisk, W.J., Grimsrud, D.T., Moed, B.A., Sextro, R.G., Radon and Remedial Action in Spokane River Valley Residences, Report LBL-21399, Lawrence Berkeley Laboratory, Berkeley CA (1986).

U.S. Department of Agriculture (USDA), Soil Survey, Spokane County, Washington
tion Service, Washington, D.C., (1968).

Wollenberg, H.A., Naturally Occurring Radioelements and Terrestrial Gamma-ray Exposure Rates: An Assessment Based on Recent Geochemical Data, Report LBL-18714, Lawrence Berkeley Laboratory, Berkeley CA (1984).

RECEIVED August 20, 1986

Chapter 3

Radon in Houses Due to Radon in Potable Water

C. T. Hess, J. K. Korsah[1], and C. J. Einloth

Department of Physics, University of Maine, Orono, ME 04469

Radon concentration in the air of 10 houses has been
measured as a function of water use and meteorological
parameters such as barometric pressure, wind velocity
and direction, indoor and outdoor temperature, and
rainfall. Results of calibrations and data collected
in winter, spring, fall, and summer are given for
selected houses. Average values of radon concentration
in air are from 0.80 to 77 pCi/l. Water use average
ranges from 70 to 240 gal/day. Average potential
alpha energy concentrations in these houses range
from 0.02 to 1.6 working levels. The radon level due
to water use ranges from 0 to 36% of the house radon
from soil and water combined. The radon level change
due to use of a filter on the water supply shows a
60% reduction in radon in the house. Conclusions
are that water radon can be a major fraction of the
radon in houses. The ratio of airborne radon
concentration due to water use to the radon
concentration in water is $4.5 \times 10^{-5} - 13 \times 10^{-5}$.

Although much research effort has been spent in the past several
years on the problem of radon in groundwater and household air,
radon continues to be an indoor air pollution problem in several
areas in the United States and elsewhere. High levels of radon and
its daughters in uranium mines have been associated with an increase
of the incidence of lung cancer among mine workers and, assuming that
the relationship between cancer incidence and exposure is linear, the
relatively low average concentrations in U.S. homes could still be
causing several thousand lung cancer cases per year. Although much
work has been done in terms of relating radon concentrations to
parameters such as ventilation rate (Nazaroff, 1983) and water use
(Hess et al., 1982; Hess et al., 1983) more work is still needed to
adequately characterize the way in which ^{222}Rn varies in homes as a
basis for developing effective control measures.

[1]Current address: West Africa Computer Science Institute, P.O. Box 4725RS, Monrovia,
Liberia, West Africa

0097-6156/87/0331-0030$06.00/0

Research shows a strong correlation between radon source magnitude and air exchange rate. A theoretical model relating the radon source magnitude (or input rate) to the indoor concentration is given by Bruno (Bruno, 1983) and others. However, such a model does not take into account important meteorological parameters such as barometric pressure, indoor and outdoor temperature differences etc., as well as other parameters such as water use, all of which can contribute to the radon concentration levels. For example, experiment has shown that even in cases of minimum ventilation (such as a house with closed doors and windows), changes in radon concentrations can occur with barometric pressure due to infiltration (self airing) through the pores of the surrounding walls and through gaps around closed doors and windows (Steinhausler, 1975; Stranden et al., 1979). Moreover, while some research shows that soil gas transport is the major contribution to indoor radon levels, other research suggests that domestic water can be a major source of radon in a significant number of houses, especially in New England (Hess et al., 1982; Bruno, 1983). More quantitative analyses need to be done in this area in terms of characterizing the relative contribution of various parameters to radon concentrations in houses. In particular, the build-up of radon levels in an energy efficient ('tight') house due to changes in meteorological parameters and water use needs to be quantitatively described to more accurately estimate the relative importance of these parameters to other parameters such as ventilation rate.

1. We have designed and built a microprocessor-based Data Acquisition System for radon research that is currently being used for field studies. Among other things the system will be used to monitor radon levels in conjunction with other meteorological data in a controlled space over relatively long periods of time.
2. We have selected 10 houses for measurement. Parameters such as house age, house building material, and radon in water levels were considered.
3. We have collected the data from these houses over week intervals for a variety of seasons.

Instrument Hardware Design

The heart of the Data Acquisition System is the HP-85 (Hewlett-Packard, 1982) microcomputer which has been interfaced to a Wrenn Counter (Wrenn et al., 1975) using a Ludlum model 2200 scaler rate meter (Ludlum Measurements, Inc., 1982), a Rainwise weather station (Rainwise, Inc., 1984), and a water meter. These various components are interfaced to the HP-85 via a peripheral interface circuit as shown in the block diagram of Figure 1. For the interface to function properly, the General Purpose I/O (GPIO) has been configured such that two of its ports (A and B) form a 16-bit input port while ports C and D form a 16-bit output port.

Under normal operating conditions the system is designed to acquire data from the peripherals at times determined by the preset time on the scaler rate meter. When the rate meter finishes its time interval the interface is signaled to acquire the stored

counts from the rate meter in byte parallel, bit serial format into
its Random Access Memory (RAM) for later processing storage on tape.
After time out the stored counts in the rate meter are outputted
asynchronously in BCD format at a rate of about 1 KHz. This rate is
a little too high to output directly to the computer since the BASIC
program used is relatively slow, hence the use of the RAM which acts
as a buffer for the data. The weather station data and the water
use data (in gallons) are also acquired, processed and stored on
tape. When data processing and storage are completed, the rate
meter is initialized again to count for the preset time. System
design is such that data can be continuously stored at 10 minute
intervals for one week. If the sampling time is doubled, data can
be stored continuously for two weeks, and so on. Although the HP-85
is not provided with a battery backup, the rest of the instrumentation is. Thus, once the instrumentation has been set up in the
field, it does not have to be reinitiated in case of a power
failure. As soon as power is resumed, the supervisor program will
automatically reload into computer memory and run. Normal data
acquisition will then resume. Data acquisition on tape will simply
resume where it left off.

The weather station is fitted with various sensors and is
capable of monitoring the following parameters: Time, indoor
temperature, outdoor temperature, barometric pressure, wind
direction, wind speed, rainfall and humidity. The water meter used
in our demonstration is a Neptune (Neptune Measurement Company, 1984)
impulse switch which develops an electrical impulse for every ounce
of water flow.

System Software Design

Basically the supervisory program is designed to acquire data from
all the peripherals at fixed intervals and store them on tape after
initial processing (other programs are later used for more involved
analysis of the data). Other features of the supervisory program
include the option of printing any data stored on file, continuing
new data storage from the last entry on tape, or erasing any
unwanted data on tape before data acquisition. Default parameters
are included in the software to avoid the need for supervision of
the system. Upon power start up, the software allows the system to
perform such initial chores as ensuring that the weather station and
interface power are on and alerting the user to set the time at the
weather station if necessary.

Radon concentration was calibrated using a U.S. E.P.A. radium
solution standard in a 50 gallon steel drum. Buildup times were
used to evaluate the tightness of the drum and to evaluate the
calibration factor for the Wrenn-Ludlum detector, which was
calculated to be 1.85 cpm/pCi/l (See Figure 2). By fitting the time
constant for buildup to the data obtained from the drum, a time
constant with leakage from the drum as well as radioactive decay is
determined. Then the concentration is corrected for the leakage
rate from the drum, and a calibration factor is formed using the
maximum concentration measured and the counts per minute obtained
experimentally. The curve labeled theoretical is found using the
half-life for radon decay only and shows a slower buildup for a
perfectly sealed barrel.

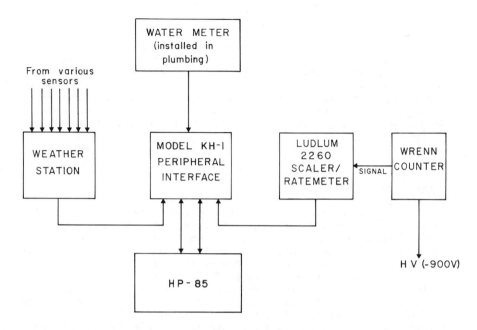

Figure 1. Block diagram of instruments and connections.

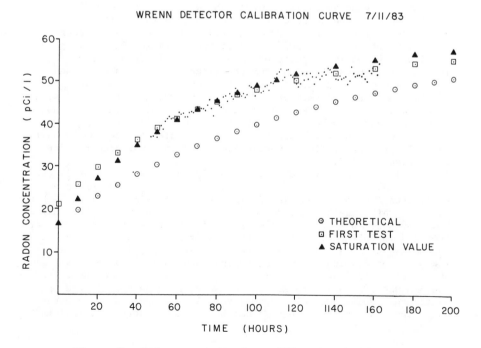

Figure 2. Radon concentration calibration in barrel.

House Selection

The 10 houses were selected for a variety of house types. These
factors are given in Table I. The house code numbers are listed
on the left side in column 1. The building material and foundation
material and size are given in columns 3, 4, and 5. Building volume
parameters, floor number, room number, ceiling height and area of
house are given in columns 6, 7, 8, and 9. Heating system, number
of stoves, fuel type, furnace type and building tightness are shown
in columns 10, 11, 12, and 13. Geological parameters such as rock-
soil type and water radon concentration are given in columns 14 and
15.

Plots of Data Collected

After transferring the data collected by the HP-85 to the IBM 3030
we plotted the results for House 3090 using a Calcomp plotter.
These results represent data collected on 12 April 1984. Figure 3a
shows the outside temperature versus time of day. Note the increase
of temperature in the afternoon time period. Figure 3b shows inside
temperature for the same 24 hour period. Figure 3c shows the
barometric pressure for 24 hours. Note the slowly increasing
pressure. Humidity is shown in Figure 3d. It shows a slow decrease.
Figure 3f shows the wind direction versus time of day. The predominant wind directions were 280° and 135°. This indicates a wind
direction shift of almost 140° (nearly a reversal of wind direction).
North is taken as 0.0° or 360°. The wind speed is shown in Figure
3g. Average wind speed is about 3 mph, which is considered to be
light wind. Figure 3e shows the water consumption in the house.
Peaks occur at 9:00 and 10:00 and 13:00. The water has a radon
concentration of 55,300 pCi/l. Radon concentration in air is given
in Figure 3h. A large pulse of radon is visible from 9:00 to 12:00.
These peaks represent radon from the water supply uses discussed
below. Radon concentration has a stable slow varying portion at
60 cts. and a sudden peak at 180 cts./.5hr. Using these results, a
quantitative test is possible for the dilution of radon in the house.
These data can also be examined to find the ventilation time for the
house which is about 2 hours for the radon pulse to decline by half.
Table II presents various average radon and dose measurements for
each house studied.

Table III shows the average radon in the air when water is used
on a typical day. The average is taken from the periods of water use
of greater than 5 gallons in 10 minutes. These time intervals begin
with the water use and extend to 2 hours after the water use. All
radon measurements in the air during this water use time interval are
averaged in Column 3. The non-water-use times are averaged in
Column 2. Columns 2 and 3 have the average picocuries per liter
taken in the 10 minute time intervals. Houses 3016, 3045, 3090,
3046, 3014, and 3094 show an increase of radon in air during water
use. House 3045 has very high radon which depends very little on
water use. House 3003 also showed very little dependence on water
use. The dose due to water use can be found by subtracting column 2

Table I. Characteristics of Houses Studied

| Number | Age | Building Material | | | Building | | | Area |
		House	Found.	Rock	Floors	Rooms	Ceil.	Sq. Ft.
3003	15	Wood	Full	--	2	3	8	1440
3014	5	Wood	Part	4100	3.5	3	8	2300
3016	7	Wood	Part	gr. ledge	3	3	8	3224
3045	5	Wood	None	bsmt.flr.	2	7	8	1152
3046	14	Granite	Full	some	2	1	9	1800
3047	17	Wood	Full	--	2	6	8	1296
3090	12	Wood	Full	400	3	9	9	2800
3094	204	Wood	Full	60	3	7	8	1100
3097	6	Wood	Full	15	2	4	8	867
3017	---	Wood	Gravel	--	3	5	--	----

| Number | # Stove | Heating | | Tightness | Geology | |
		Fuel	Furnace		Soil	Water
3003	2	w/o	Forced Hot Air	A	G	---
3014	1	Solar	-----	T	G	25,900 (1,900 Filter on)
3016	1	Wood	----	T	G	38,100
3045	2	Wood	Rad.	T	G	29,800
3046	1	w/o	Forced Hot Air	A	G	6,600
3047	1	w/o	Base	A	G	33,700
3090	1	w/o	Base	A	G	56,900
3094	1	Wood	Base	T	G	53,700
3097	1	Wood	Base	D	G	----
3017	1	Wood	Base	T	G	----

Age column is given in years, rock column is portion of rock used in a foundation, fireplace or wall in square feet, height of ceiling in feet. Area is in square feet. Blanks in furnace column indicate no use of furnace, tightness A is average, T is tight, D is drafty, as stated by the homeowner, soil column shows soil from granite bedrock in a foundation, fireplace or wall. Water column is radon concentration in pCi/l.

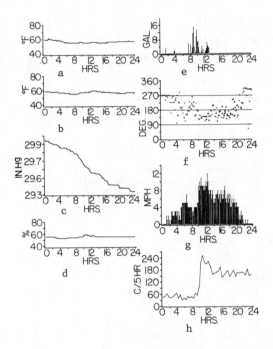

Figure 3a, Outside temperature; 3b, Inside temperature; 3c, Barometric Pressure; 3d, Humidity; 3e, Water consumption; 3f, Wind direction; 3g, Wind speed; 3h, Radon concentrations all versus time of day.

Table II.   Houses Studied with Location, Collection Dates Averaged
Results Average Radon in Air, Average Dose due to Radon,
Average Radon in Water and Average Water Usage

House #	Location	Collection Dates	Avg. pCi/l in Air	Avg. Dose WLM/month	Avg. pCi/l in Water	Avg. H₂0 Usage (gal/day)
3003	Searsport	10/11/83-10/20/83 Fall	2.48	0.052	28,800	----
3003	Searsport	7/27/84-8/03/84 Summer	1.34	0.028	34,200	230
3014	Mount Desert Island	11/01/83-11/10/83 Fall	2.10	0.044	2,200	----
3014	Mount Desert Island	05/03/84-05/14/84 Spring	2.33 1.23	0.065 0.034	25,300 1,900	130 132
3016	Mount Desert Island	10/20/83-11/01/83 Fall	9.94	0.210	41,800	----
3016	Mount Desert Island	04/24/84-05/03/84 Spring	1.91	0.053	37,360	119
3017	Mount Desert Island	06/27/84-07/05/84 Summer	0.85	0.024	14,400	128
3045	Dedham	03/01/84-03/11/84 Winter	58.9	1.636	27,800	190
3045*	Dedham	12/04/84-12/12/84 Fall	45.4	0.961	25,800	136
3047	Dedham	02/16/84-03/01/84 Winter	37.4	0.792	33,300	259
3049	Dedham	03/21/84-03/29/84 Spring	4.26	0.090	6,500	70
3090	Ellsworth	04/12/84-04/22/84 Spring	4.55	0.096	55,300	238
3094	Ellsworth	04/04/84-04/12/84 Spring	1.81	0.038	33,100	160
3097	East Eddington	07/16/84-07/24/84 Summer	1.06	0.022	38,000	166

*New Floor in Greenhouse.

from column 3 and multiplying by 0.0211 (wl/pCi/l). This factor
assumes 720 hours per month of occupancy which is 4.23 times greater
than the miner occupancy and 0.5 equilibrium for the daughters
relative to the radon. In many cases the water radon is smaller
than the soil gas radon. This method assumes that all radon is
rapidly lost and rapidly produced in time intervals smaller than 2
hours. Slowly produced radon from drains or toilet tanks will be
incorrectly attributed to the soil gas by this subtraction method.
Only filter methods shown in the following section can correctly
assess the slow radon component. The transfer coefficient column
is the ratio of the average air radon concentration due to water
use divided by the water radon concentration.

Table III. Houses Studied and Separation of Radon Counts

HOUSE #	(NON-WATER) AVG. pCi/l	AVG. pCi/l DUE TO WATER	DOSE WLM/month	TRANSFER COEFFICIENT
3016	1.12	1.76	.0135	4.2×10^{-5}
3045	64.20	62.30	-----	----------
3017	0.76	0.83	.0015	5.7×10^{-5}
3003	1.10	0.77	-----	----------
3090	5.79	7.26	.0311	1.3×10^{-4}
3046	4.40	5.15	.0158	9.3×10^{-5}
3014	1.57	2.23	.0139	8.7×10^{-5}
3094	1.11	1.68	.0120	5.0×10^{-5}

Filtration Test Results

The radon levels of House 3014 were measured before and after the
use of a radon filter with one cubic foot of granular activated
carbon in a water softener tank. This filter reduced the water
radon from 25,400 pCi/l to 1,900 pCi/l. Detailed analysis of the
radon in air and water use for this house is shown in Figures 4 and
5. Figure 4 shows the radon produced with the filter bypassed and
Figure 5 shows the radon in air and water use with the data analyzed
for average radon before and after filtration by averaging the radon
in air. The average value for filter bypassed is 2.39 pCi/l for days
1 and 2, and 2.67 pCi/l for days 5, 6, and 7. The radon in air for
the days with the filter connected (days 8, 9, 10) is 1.00 pCi/l.
This corresponds to a dose reduction in the house of more than a
factor of 2. It also shows that more than half of the household
radon is caused by the radon from water. The 0.76 pCi/l is the
radon caused by soil gas and building materials. These results are
summarized in Table IV.

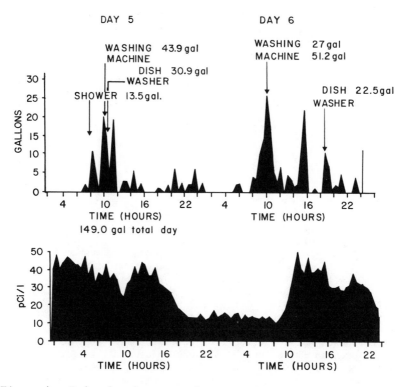

Figure 4. Radon in air compared to water use versus time of day.

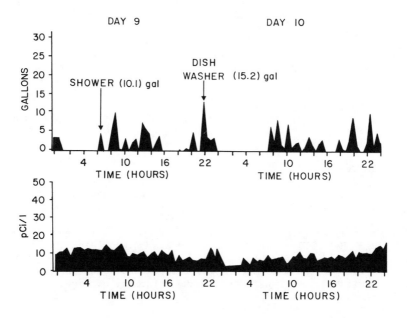

Figure 5. Radon in air compared to water use versus time of day.

Table IV. House 3014 with Change in Radon in Air Due to Change
in Water Radon Caused by Use of a Water Radon Filter

DAY	FILTER	WATER RADON (pCi/1)	AIR RADON (pCi/1)	DOSE WLM/month	RATIO AIR/WATER
1,2	off	25,373	1.81	0.0381	4.1×10^{-5}
3,4	off	25,373	HOUSE VACANT	------	----------
5,6,7	off	25,373	2.02	0.0426	4.9×10^{-5}
8,9,10	on	1,863	0.76	0.0160	----------

Conclusions

The instrument for measurement of radon with weather parameters and
water use was designed, calibrated and used in the field for 1.5
years with excellent results. The power surge and power interruptions were the only limitations on the use of the instrument. The
system software was written mostly during the design phase of the
project. However, modifications were made in the software to
permit better data collection during power interruptions. Battery
backup was needed to keep the data during interruptions of our one
week studies.

Ten houses were selected from past studies. Four of the houses
were measured in two different seasons to assess the effect of
weather changes. Houses were selected to exemplify both old and
recent construction and to include wood, oil, and solar heat.
Plots of the data are included for a representative house on one
day. Indoor-outdoor temperatures, humidity, barometric pressure,
wind direction and speed, water consumption and radon concentration
are shown versus time of day. The data were transmitted in serial
form and required data conditioning before use in the IBM computer.
Average values of radon concentration ranged from .8 to 77 pCi/1.
Water use was averaged and found to vary from 70 to 240 gallons/day.
The calculated dose for radon exposure assuming 100% occupancy at
.5 equilibrium for daughters ranged from 0.02 to 1.6 working level
months/month. Radon concentration in air was also monitored in a
house with a radon in water filter. This filter reduced the radon
levels in water from 25,000 to 1,900 pCi/1 and reduced the air radon
level from 2.8 pCi/1 to 1.0 pCi/1.

Acknowledgments

The research on which this report is based was financed in part by
the United States Department of the Interior, Geological Survey,
through the Maine Land and Water Resources Center, and by a grant
from Central Maine Power Company, Maine Yankee Atomic Power Company,
and Bangor Hydro-Electric Company. We also acknowledge the typing
and corrections of Patricia L. Heal and Ruby Ackert. Contents of
this publication do not necessarily reflect the views and policies
of the United States Department of the Interior, nor does mention
of trade names or commercial products constitute their endorsement
by the United States Government.

Literature Cited

Bruno, R. C., Sources of Indoor Radon in Houses: A Review, Journal of the Air Pollution Control Association 33:105-115 (1983).

Hewlett Packard, GPIO Interface Owners Manual Series 80 HP829401, 1000 N.E. Circle Blvd., Corvallis, OR 97330 (1980).

Hess, C. T., C. V. Weiffenbach and S. A. Norton, Environmental Radon and Cancer Correlations in Maine, Health Phys. 45:339-348 (1983).

Hess, C. T., C. V. Weiffenbach and S. A. Norton, Variations of Airborne and Waterborne Rn-222 in Houses in Maine, Environment International 8:59-66 (1982).

Ludlum Measurements, Inc., Instruction Manual, Model 2200 Portable Scaler Ratemeter, 501 Oak Street, Box 248, Sweetwater, Texas 79556 (1982).

Nazaroff, W. W., F. J. Offermann and A. W. Robb, Automated System for Measuring Air-Exchange Rate and Radon Concentration in Houses, Health Phys. 45:525-537 (1983).

Neptune Measurement Company, Emerald Road, Greenwood, South Carolina 29646 (1984).

Rainwise, Inc., Box 475, Bar Harbor, Maine 04609 (1984).

Steinhausler, F., Long Term Measurements of ^{222}Rn, ^{220}Rn, ^{214}Pb, and ^{212}Pb Concentrations in the Air of Private and Public Buildings and their Dependence on Meteorological Parameters, Health Phys. 29:705-713 (1985).

Stranden, E., L. Berteig and F. Ugletveit, A Study on Radon in Dwellings, Health Phys. 36:413-421 (1979).

Wrenn, M. E., H. Spitz and N. Cohen, Design of a Continuous Digital Output Environmental Radon Monitor, IEEE Trans. on Nuclear Science NS-22:645-647 (1975).

RECEIVED August 4, 1986

Chapter 4

Measurements of Radon Concentrations in Residential Buildings in the Eastern United States

Andreas C. George and Lawrence E. Hinchliffe

Environmental Measurements Laboratory, U.S. Department of Energy, New York, NY 10014

As part of a program to develop and test radon survey techniques, passive activated carbon samplers were used to measure radon concentrations in 380 buildings in six states in the eastern United States. Measurements were made in the basement and living areas of each residential building, and in some work locations of several plant buildings during summer and winter. The activated carbon samplers performed well in these tests, and the logistics via U. S. mail were satisfactory. The lowest concentrations of radon were found in: Long Island, NY, Luzerne County, PA, and South Carolina. The remainder of the test areas showed a substantial number of buildings to have radon levels above the Environmental Protection Agency's and the National Council on Radiation Protection and Measurements's guideline action levels. The arithmetic mean concentrations of radon in the living areas in the different regions ranged from 1.1 pCi/1 to 9.3 pCi/1 (0.0055 to 0.0465 WL) during the winter months. The dose equivalent rate to the bronchial epithelial cells of the occupants of these buildings ranged from 1.4 to 10.7 rem/yr. Although only one area lies on the Reading Prong, which extends through New Jersey, several of the regions can be considered hot areas or anomalous regions with similar problems as the Prong. The estimated risk from the observed radon distributions is such that a substantial fraction of the northeastern United States population is exposed to levels high enough to warrant immediate attention.

In recent years, there has been growing concern about homes with high indoor radon concentration levels and a potential for increased risk of lung cancer among occupants of these buildings. Residential buildings measured in the past two years, frequently had radon concentrations 1 to 2 orders of magnitude greater than average. These high concentration levels suggest that it is necessary to obtain exposure data from a wide variety of buildings and geographical locations throughout the United States in order to better define and understand the radiation exposure of the general population to indoor radon and radon progeny.

The first comprehensive measurements in the United States of indoor radon and radon progeny and particle size distributions were made by the Environmental Measurements Laboratory (EML) in the late 1970's at places associated with the sites declared excess by Atomic Energy Commission. Because of the concern expressed by local government officials and private citizens about possible elevated radon levels in nearby residences due to transport and infiltration of radon from contaminated sites, the Department of Energy (DOE) sponsored studies to determine the concentration levels in uncontaminated homes situated near these sites. The measurements were made in buildings in selected locations considered background areas in the New York City metropolitan area (George and Breslin, 1980) and at suspect high level areas in the vicinity of the Middlesex Sampling Plant in Middlesex, NJ, at the Lake Ontario Ordinance Works Facility in Lewiston, NY, and at the Canonsburg Industrial Park in Canonsburg, PA (George and Eng, 1981). Annual mean concentrations of radon or radon progeny were estimated in each building by 1 or 2 week-long measurements repeated several times over a period of 2 yr (1976-1978 or 1978-1980). Passive integrating radon monitors (George and Breslin, 1977) and active integrating working level monitors (Guggenheim et al., 1979) developed by EML were used throughout the monitoring period. As the demand for radon measurements increased, the need for simpler and less costly monitoring devices grew proportionately, requiring the development of instruments and methods applicable to broader field studies.

Measurements of radon progeny, which are the more difficult to make, can be replaced with measurements of radon to provide information on the upper limit for the potential alpha energy exposure from radon progeny. Therefore, simple passive radon monitors for integrated measurements of 1 to 12 months (Alter and Fleischer, 1981) or as little as 3 to 4 days (George, 1984) are developed. From previous studies we found that a single measurement of the average concentration of radon over periods of days, repeated in different seasons, is adequate to assess the exposure and radiation dose of the occupants of buildings (George and Breslin, 1980; George, 1984). This approach was undertaken by EML and others to develop and successfully test passive activated carbon monitors (George, 1984; Cohen and Cohen, 1983) suitable for integrated measurements of environmental levels of radon over a period of a few days (3 to 7 days). The advantages of this new device are that it is simple, reusable, maintenance-free, completely passive, no transfer of sample is required, and it is also readily handled in the mail. The first use of this technology was to make measurements of radon in selected homes near Philadelphia, Pennsylvania and in homes near Damascus, MD (George et al., 1984). In this survey of 72 residential buildings, in the two localities, homes with high concentrations of radon were identified. At about the same time, from newspaper articles published in certain areas in the northeastern United States, radon was beginning to be considered a possible indoor air pollutant and home owners were alerted to the

possible health effects from radon exposure in certain situations.
     As part of the radon program at EML to develop or improve and
field test radon monitors, a modified activated carbon device
(Warner, 1986) was developed to obtain higher measurement
sensitivity. As a result, we have surveyed 380 buildings in six
states in the eastern United States. The purpose of the
measurements reported in this paper was to test the feasibility of
the new version of the passive activated carbon device and to obtain
data on indoor radon levels in different geographical locations.

Location of Buildings

The 380 buildings surveyed were located in six states (shown in
Figure 1) and consisted primarily of single family residences with
the exception of several plant buildings operated by E.I. Dupont for
the U. S. Department of Energy. The plant locations were surveyed
with the cooperation of the Health Physics Group at the Savannah
River Plant in Aiken, South Carolina who also arranged for the radon
measurements to be taken in 41 residential buildings in 15 towns
within a radius of 40 km from the plant.
     In the Washington, D.C. area, in cooperation with television
station WJLA-TV, we measured 52 residential buildings, which were
located in 9 towns in northern Virginia and in 22 towns in Maryland.
Three residential buildings in Washington, D.C. were also included.
     Through the cooperation of WNEP-TV16, serving the Wilkes-Barre/
Scranton area in Pennsylvania, measurements for radon concentration
levels were made in 42 residential buildings in 34 towns in northern
Pennsylvania. In an alternate site in Luzerne County Pennsylvania,
we measured radon concentrations in 41 residential buildings in 22
towns. This request came from the Commission on Economic Opportunity, a non profit social service agency that winterized the homes
of low income families.
     We surveyed 20 homes for radon in 14 towns in New Jersey within
a radius of 25 km from Chester, New Jersey in cooperation with Bell
Communication Research, Inc. In a second geographical region
encompassing 5 counties centered around Morris County, we measured
radon concentrations in 50 residential buildings scattered in 24
towns through the cooperation of a local insurance company.
     In New York State, we surveyed three geographical regions, one
in Long Island, one in the Albany region and the third in the
Syracuse area. In Long Island we measured radon concentration
levels in 38 residential buildings in 23 towns in conjunction with
measurements of indoor gamma radiation exposures. In the Albany
area through the cooperation of the New York State Energy Research
and Development Authority (ERDA)., we measured radon concentration in
29 buildings in 16 towns within a radius of 20 km from Albany. The
third region, in the vicinity of Syracuse, was surveyed for radon
through a request from an engineering company specializing in
building energy conservation. We measured 26 buildings in 6 towns
centered around Syracuse.
     From a questionnaire that the occupants of the buildings
returned to us along with the exposed detectors, we were able to
determine building characteristics. Most of the buildings were
single-family residences with full basements. Suprisingly, many

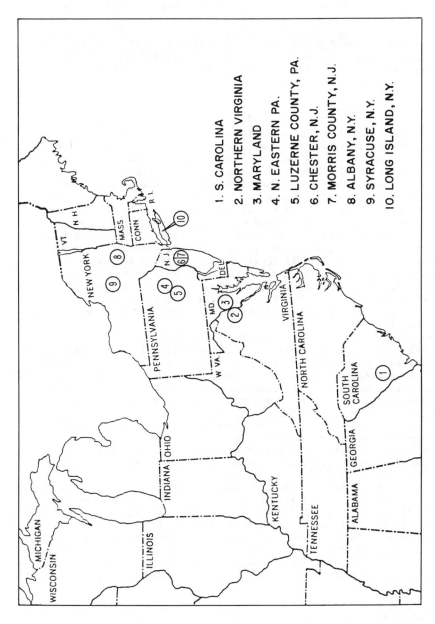

Figure 1.   Geographical distribution of residences in six states.

1. S. CAROLINA
2. NORTHERN VIRGINIA
3. MARYLAND
4. N. EASTERN PA.
5. LUZERNE COUNTY, PA.
6. CHESTER, N.J.
7. MORRIS COUNTY, N.J.
8. ALBANY, N.Y.
9. SYRACUSE, N.Y.
10. LONG ISLAND, N.Y.

basements were finished and some people spend a substantial part
of their time there. The buildings were of wood-frame or brick
construction with heating systems of all types. Domestic water
supplies were mostly from public sources, but a substantial number
of homes used well water located on site.

The buildings ranged in age from <1 yr old to >100 yr old.
Newer homes seemed to be built with energy conserving measures in
mind, although some owners of older homes (>15 years old), indicated
that they added some form of insulation to render them air tight for
energy conservation purposes.

Monitoring Procedures

The estimation of annual mean concentration of radon was determined
in each building by making measurements during the winter (heating
season) and during the summer. The regions we surveyed generally
have a six month period in which homes are maintained at maximum
closed conditions from October to April. Summer weather conditions
generally prevail from May through September.

Measurements for radon were made twice in each building during
the two seasons. One detector was usually exposed in a location in
the home where occupants were likely to spend most of their time.
We denoted that location as living area. A second detector was
placed in the basement if there was one, because basements are more
often the source of radon in a building. Since living habits change
with more frequent use of basements as playrooms, machine shops or
even as bedrooms, we thought it prudent to measure the highest
potential radon concentration levels to which residents of buildings
might be exposed.

The detector used to measure indoor radon was the latest
version of the passive activated carbon device developed at EML
(George, 1984; Warner, 1986), which consists of a thin-walled
aluminum canister with a screen cover to expose 80 g of carbon to
the test atmosphere. Although not as physically rugged as earlier
models, properly packed this monitoring device was as successful in
conducting the surveys through the mail.

After the end of the 4-day exposure, the detectors were
returned to EML for analysis. The amount of radon adsorbed on the
carbon device was determined by counting the gamma rays of radon
progeny in equilibrium with radon. The concentrations of radon in
the buildings were determined from the radioactivity in the device
and the calibration factor, obtained in a radon chamber, that takes
into consideration the length of exposure and a correction for the
amount of water vapor adsorbed during the exposure. The lower limit
of detection with this technique is 0.2 pCi/l for a measurement
period of 4 days when the test sample is counted for 10 min, 4 days
after the end of exposure. More than 90% of the radon monitoring
devices were analyzed successfully. Most of the unsuccessful
measurements were due to delays or losses caused by the participants.

Results and Discussion

The results of all measurements are summarized in Tables I through
IX, where we present the range of values and the arithmetic means

Table I. Radon Concentrations and Radiation Dose Rate Equivalent to Epithelial Cells of the Occupants of Residential and Plant Buildings In and Near the Savannah River Plant

	Radon Concen. (pCi/l)					% Equal to or Greater Than		Dose Rate (rem/yr)
	Min.	Max.	Mean	GM	GSD	4 pCi/l	8 pCi/l	
Living Area								
Winter	0.3	12	1.48	0.95	2.0	2	0	(4.1)
Summer	0.1	2.6	0.85	0.63	1.9	0	0	
Basement								
Winter	0.8	5.6	2.7	2.3	1.9	17	2	(9.6)
Plant Buildings								
Summer	0.1	0.8	0.37	0.29	3.0	0	0	(1.3)

Note:

Number of buildings = 41 private residences and 41 locations in the plant.

Table II. Radon Concentrations and Radiation Dose Rate Equivalent to
Epithelial Cells of the Occupants of Residential Buildings in
Northern Virginia and Maryland

	Radon Concen. (pCi/1)					% Equal to or Greater Than		Dose Rate (rem/yr)
	Min.	Max.	Mean	GM	GSD	4 pCi/1	8 pCi/1	
Living Area								
Winter	0.3	28.4	4.3	2.1	3.6	31	15	(10.7)
Summer	0.3	718	1.8	1.2	2.3	7	0	
Basement								
Winter	0.5	59	7.6	4.2	3.6	51	30	(22.1)
Summer	0.6	21	4.9	3.2	2.8	39	18	

Note:

Number of buildings = 41 private residences and 41 locations in the plant.

Building ages were from 2 to 200 years.

Percent of buildings <15 years old = 50.

Table III. Radon Concentrations and Radiation Dose Rate Equivalent to Epithelial Cells of the Occupants of Residential Buildings in Northeastern Pennsylvania

	Radon Concen. (pCi/1)					% Equal to or Greater Than		Dose Rate (rem/yr)
	Min.	Max.	Mean	GM	GSD	4 pCi/1	8 pCi/1	
Living Area								
Winter	0.3	35.7	3.9	1.90	2.9	20	12	
Summer	0.07	7.1	1.4	0.93	2.7	6	0	(9.4)
Basement								
Winter	0.63	56.7	6.2	3.5	2.9	45	22	
Summer	0.38	16.8	4.3	2.7	3.2	27	18	

Note:

Number of buildings = 42.

Building ages were from 1 to 160 years.

Percent of buildings <15 years old = 50.

Table IV. Radon Concentrations and Radiation Dose Rate Equivalent to Epithelial Cells of the Occupants of Residential Buidlings In Luzerne County In Pennsylvania

	Radon Concen. (pCi/1)					% Equal to or Greater Than		Dose Rate (rem/yr)
	Min.	Max.	Mean	GM	GSD	4 pCi/1	8 pCi/1	
Living Area								
Winter	0.15	9.1	1.4	0.95	2.3	4	2	(3.6)
Summer	0.0.5	4.08	0.63	0.35	2.8	0	0	
Basement								
Winter	0.48	17.8	3.9	2.7	2.0	28	6	
Summer	0.24	21.9	2.1	0.90	2.6	6	3	

Note:

Number of buildings = 41.

Building ages were from 3 to 175 years.

Percent of buildings <15 old = 24.

Table V. Radon Concentrations and Radiation Dose Rate Equivalent to Epithelial Cells of the Occupants of Residential Buildings In and Around Chester, New Jersey

	Radon Concen. (pCi/1)					% Equal to or Greater Than		Dose Rate (rem/yr)
	Min.	Max.	Mean	GM	GSD	4 pCi/1	8 pCi/1	
Living Area								
Winter	0.2	145*	2.3	1.3	4.2	22	10	(6.6)
Summer	0.2	10	1.4	0.5	3.3	4	0	
Basement								
Winter	0.6	40	7.9	3.0	4.0	42	24	(26.0)
Summer	0.5	30	6.8	2.1	3.8	32	16	

Note:

Number of buildings = 20.

Building ages were from 8 to 100 years.

Percent of buildings <15 years old = 10.

* This value not included in calculating the arithmetic mean.

Table VI. Radon Concentrations and Radiation Dose Rate Equivalent to
Epithelial Cells of the Occupants of Residential Buildings in Hunterdon,
Morris, Warren, Somerset and Sussex Counties in New Jersey

	Radon Concen. (pCi/1)					% Equal to or Greater Than		Dose Rate (rem/yr)
	Min.	Max.	Mean	GM	GSD	4 pCi/1	8 pCi/1	
Living Area								
Winter	0.2	76	4.3	1.5	3.3	21	9	(9.4)
Summer	0.2	8.5	1.0	0.6	2.7	3	1	
Basement								
Winter	0.3	140	9.3	3.5	3.1	44	24	(25.7)
Summer	0.3	22	5.2	3.0	3.1	40	19	

Note:

Number of buildings = 50.

Building ages were from 1 to 65 years.

Percent of buildings <15 years old = 70.

Table VII. Radon Concentrations and Radiation Dose Rate Equivalent to Epithelial Cells of the Occupants of Residential Buildings in the Vicinity of Albany, New York

	Radon Concen. (pCi/1)					% Equal to or Greater Than		Dose Rate (rem/yr)
	Min.	Max.	Mean	GM	GSD	4 pCi/1	8 pCi/1	
Living Area								
Winter	0.3	23	2.9	1.2	3.5	17	7	(7.6)
Summer	0.3	4.8	1.4	1.0	2.8	8	0	
Basement								
Winter	0.5	221*	6.1	1.7	4.8	30	16	(22.0)
Summer	0.6	53	6.3	1.7	4.8	30	16	

Note:

Number of buildings = 29

Building ages were from 0.5 to 155 years.

Percent of buildings <15 years old = 41.

* This value not included in calculating the arithmetic mean.

Table VIII. Radon Concentrations and Radiation Dose Rate Equivalent to
Epithelial Cells of the Occupants of Residential Buildings in the Vicinity of
Syracuse, New York

	Radon Concen. (pCi/1)					% Equal to or Greater Than		Dose Rate (rem/yr)
	Min.	Max.	Mean	GM	GSD	4 pCi/1	8 pCi/1	
Living Area								
Winter	0.5	27	4.0	1.9	3.4	27	12	(8.5)
Summer	0.2	1.9	0.8	0.6	2.2	0	0	
Basement								
Winter	0.6	60	8.2	3.0	4.0	42	24	(22.0)
Summer	0.6	13	4.2	2.8	3.1	38	18	

Note:

Number of buildings = 26.

Building ages were from 0.5 to 150 years.

Percent of buildings <15 years old = 12.

Table IX. Radon Concentrations and Radiation Dose Rate Equivalent to Epithelial Cells of the Occupants of Residential Buildings in Long Island, New York

	Radon Concen. (pCi/1)					% Equal to or Greater Than		Dose Rate (rem/yr)
	Min.	Max.	Mean	GM	GSD	4 pCi/1	8 pCi/1	
Living Area								
Winter	0.05	1.6	0.55	0.39	2.7	0	0	
Summer	0.09	1.0	0.28	0.22	2.4	0	0	(1.4)
Basement								
Winter	0.50	2.8	1.14	1.1	1.4	0	0	
Summer	0.04	3.2	1.04	0.9	1.7	0	0	(3.8)

Note:

Number of buildings = 39

Building ages were from 2 to 100 years.

Percent of buildings <15 years old = 23.

for the living areas and basements in each geographical region. The geometric means (GM) and the geometric standard deviations (GSD) were calculated for each region from the log normal distribution plots. The average of the arithmetic mean values obtained from winter and summer measurements was used to calculate the radiation dose equivalent rate to the epithelial cells of the lungs. The average radon concentration from the arithmetic means measured in the living area and basement as pCi/l was converted to working level (WL), by assuming that the ratio between radon and its decay products, known as the working level ratio (WL-R) was 50%. One WL is the quantity of alpha energy delivered from a mixture of 100 pCi each of ^{218}Po, ^{214}Bi and ^{214}Po in 1 liter of air = 1.3 x 10^5 MeV. WL-R = 100% x WL/radon concentrations (pCi/l). The dose conversion factor of 0.7 rad/working level month (WLM) (Harley and Pasternack, 1982) was used to calculate the mean absorbed dose to the epithelial cells and a quality factor (QF) of 20 was applied to convert the absorbed dose to dose equivalent rate. For example, from the average value of (WL) obtained from the arithmetic mean radon concentrations measured in the living area during winter and summer in South Carolina (Table I), the calculated dose equivalent rate is 4.1 rem/yr, e.g.,

$$(1.48 + 0.85)/2 = 1.16 \text{ pCi/l.}$$

At 50% WL-R, WL = .0058, and

$$\frac{.0058 \text{ WL} \times 730 \text{ hr/month} \times 12 \text{ month/yr}}{173 \text{ hr/working month}} = 0.29 \text{ WLM/yr}$$

and

$$0.29 \text{ WLM/yr} \times 0.7 \text{ rad/WLM} \times 20 \text{ (QF)} = 4.1 \text{ rem/yr.}$$

Because housewives and children may spend all of their time indoors, we used an occupancy factor of 100%. For other occupants this may drop to about 80% and the dose equivalent rate will drop accordingly.

The geometric mean radon concentrations in the living areas are usually about one-half of those in the basement during the winter months when homes are normally closed up. During the summer, radon levels in the living areas are one-third or less of those in the basements. The data in Figure 2 show that while there is little seasonal difference in radon levels for the basement, there are pronounced effects for living areas that are subject to changes in ventilation conditions.

The distribution of radon concentrations in basements and living areas during the winter months for different geographical areas are shown in Figures 3 and 4, respectively. Figure 3 shows the radon distributions in the basements of buildings in the six regions along with that measured in 4984 homes located on the

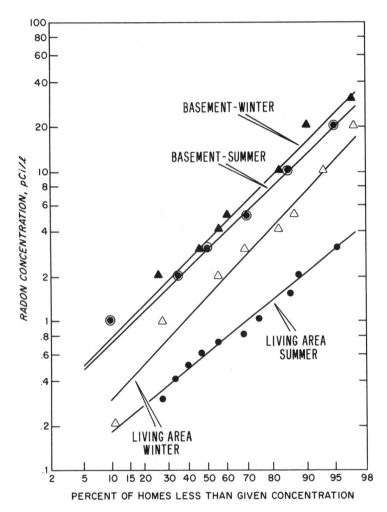

Figure 2.   Distribution of radon concentrations in residential buildings in Morris County, New Jersey.

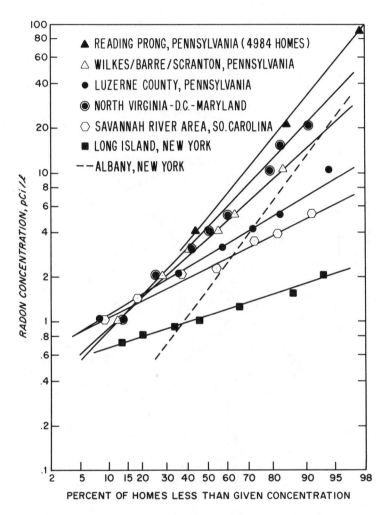

Figure 3. Distribution of radon concentrations in the basements of buildings during winter in different geographical locations.

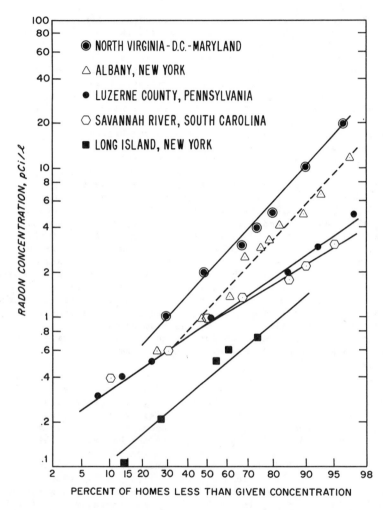

Figure 4. Distribution of radon concentrations in the living areas of buildings during winter in different geographical locations.

Reading Prong in Pennsylvania. The distributions of the radon
measurements from the Chester, NJ region and the nearby region
centered around Morris County, NJ and that at Syracuse, NY are
similar to those measured in Northern Virginia and Maryland and the
Wilkes-Barre/Scranton area and are not plotted in Figure 3 to avoid
overcrowding.

The lowest levels of radon were measured in Long Island, New
York. Somewhat higher levels were measured in South Carolina and in
Luzerne county in Pennsylvania. These three regions show the
smallest geometric standard deviations which suggests some similarities within regions such as building construction and geology. The
measurements also show similar concentrations to those obtained in
an earlier study (George and Breslin, 1980) from which the National
Council on Radiation Protection and Measurements (NCRP) (NCRP, 1984)
used the mean concentration (0.8 pCi/1 or 0.004 WL in the living
areas) to calculate the exposure rate in the United States. In
general, the arithmetic mean concentrations of radon during the
winter months ranged from 1.1 pCi/1 to 9.3 pCi/1.

The percentage of homes having radon concentrations above 4
pCi/1 (the EPA suggested guideline) ranged from 0% to 51%. The
corresponding percentage of homes having radon above 8 pCi/1 (the
NCRP recommended guideline) ranged from 0% to 30%. For comparison,
among the 4984 measured homes on the Reading Prong 55% and 37%
exceed the 4 pCi/1 and 8 pCi/1 guidelines, respectively (NCRP,
1984). With the exception of the three regions in Long Island,
South Carolina and Luzerne County in Pennsylvania, the other
geographical locations can be characterized as regions with sources
high in radon and with the potential of high indoor radon
exposures. It is important to note that most of the areas measured
in this study (except for Chester, NJ and some homes centered around
Morris County, NJ) lie outside of the Reading Prong, an area
identified as a high radon region extending from Reading,
Pennsylvania through New Jersey and up to near the border of New
York and Connecticut.

Figure 4 shows the distributions of radon concentrations in
living areas during the winter. The distribution curve for the
Syracuse, New York area (not shown to avoid figure overcrowding) is
very similar to that from northern Virginia-Maryland. Those from
Morris County, and Chester, NJ lie between those from northern
Virginia-Maryland and Albany, NY. As in the basements, the living
areas in Long Island homes also have the lowest radon concentration
levels. Residences in South Carolina and in Luzerne County, PA have
the same arithmetic mean concentrations (1.4 pCi/1), whereas the
mean concentrations in the homes in the remaining regions ranged
from 2.3 pCi/1 to 4.3 pCi/1 with an average concentration of 3.6
pCi/1.

To test if there are statistically significant differences
among different parameters such as geographical areas, location in
the home, and seasons, the data were analyzed using the Mann-Whitney
non-parametric test (Hollander and Wolf, 1973). The test verfies if
there are significant radon concentration differences between two
parameters at the 95% level of confidence. General conclusions from
the analysis of the data shows living areas and basements were
distinct in all locations and seasons. Most living areas were
different from winter to summer, while most basements were not.

Differences between geographical locations have been found in living areas in the summer showing Long Island and northern Virginia-Maryland to be distinct from most other locations. During the winter, living areas in Long Island, South Carolina and in Luzerne County, PA were found to differ from all other locations. Comparing homes of different ages, we found no correlation between the age of the building and radon concentration levels. It seems other variables such as regional geology and ventilation have a larger effect on indoor radon.

Summer measurements in basements seem to show two groups: Long Island and Luzerne County, having low concentrations of radon, differ substantially from all other locations. Measurements in South Carolina, the southernmost region, lie somewhere in between, being indistinguishable from all other locations. Winter measurements in basements again show Long Island to differ substantially from all other locations, with very little difference among the rest of the areas.

Conclusions

The radon concentration data for 380 buildings, located in six states, indicate that there is almost a 10-fold spread of annual dose equivalent to the lungs (1.4–10.7 rem) of building occupants who spend their entire time in the living area. The corresponding dose rates in the basements of these buildings ranged from 3.8 to 26 rem. The results of this study also indicate higher radon concentrations than were observed in other regions within some of the same states (George and Breslin, 1980; George and Eng, 1981). Radon levels varied from one region to another because of probable differences in: source strength, energy conservation measures, and building technology. A surprising finding is that residential buildings in the northeastern United States, located outside the Reading Prong, have radon levels approaching those of the Prong. This indicates that there are anomalous regions on a wider scale, suggesting that a National Survey is needed to address this problem. Knowledge of the national distribution is essential in finding and examining hot areas, formulating strategy for controlling excessive concentrations and making a reliable estimate of the average population exposure and risk. The results of our study show that the estimated risk from the observed radon distributions in living areas is such that a substantial fraction of the general population may be exposed to levels high enough to warrant immediate attention. Where basements are used for ordinary habitation, another fraction of the population is exposed to levels as high or higher than the occupational limit of 4 WLM/yr.

Another important finding is that simple passive radon monitoring devices, used for short integrating periods, can be very useful in acquiring the data necessary to undertake a national survey for indoor radon.

Literature Cited

Alter, H. W., and R. L. Fleischer, Passive Integrating Radon Monitor in Environmental Monitoring, <u>Health Phys.</u> 40:693 (1981).

Cohen, B. L., and E. S. Cohen, Theory and Practice of Radon
Monitoring with Charcoal Adsorption, Health Phys. 45:501 (1983).

George, A. C., and A. J. Breslin, in: Workshop on Methods for
Measuring Radiation In and Around Uranium Mills, (E. O.   Harward,
ed.) Vol. 3, No. 9, pp. 105-115, Atomic Industrial Forum, Inc.
(1977).

George, A. C. and Breslin, A. J. 'The Distribution of Ambient Radon
and Radon Daughters in Residential Buildings in the New Jersey - New
York Area'. Natural Radiation Environmental III, Vol. 2, CONF-
780422, Technical Information Center, U. S. Department of Energy,
Springfield, VA (1980).

George, A. C., and J. Eng, Indoor Radon Measurements in New Jersey,
New York and Pennsylvania, Health Phys. 45:397 (1981).

George, A. C., Passive Integrated Measurement of Indoor Radon Using
Activated Carbon, Health Phys. 46:867 (1984).

George, A. C., M. Duncan, and H. Franklin, Measurements of Radon in
Residential Buildings in Maryland and Pennsylvania, USA, Rad. Prot.
Dosimetry 7:291 (1984).

Guggenheim, F. S., A. C. George, R. T. Graveson, and A. J. Breslin,
A Time Integrating Environmental Radon Daughter Monitor, Health
Phys. 36:452 (1979).

Harley, N., and B. C. Pasternack, Environmental Radon Daughter Alpha
Dose Factors in a Five-Lobed Human Lung, Health Phys. 42:789
(1982).

Hollander, M., and D. A. Wolf, Non Parametric Statistical Methods,
John Wiley and Sons, Inc., New York, NY (1973).

NCRP, Exposures from the Uranium Series with Emphasis on Radon and
Its Daughters, National Council on Radiation Protection and
Measurements Report No. 77 (1984).

Warner, M. D., Analytical Chem. 58:44A-47A (1986).

RECEIVED August 20, 1986

Chapter 5

Assessing Exposure to Radon in the United States: Perspective of the Environmental Protection Agency

Paul J. Magno and Richard J. Guimond

Office of Radiation Programs, U.S. Environmental Protection Agency, Washington, DC 20460

Exposure to radon and its decay products inside homes is now recognized as a significant public health problem. To better define the scope and magnitude of this problem, the Environmental Protection Agency (EPA) has developed a national radon exposure assessment program. This program consists of a national radon survey to determine the national frequency distribution of radon levels in residential structures, a States Assistance Program to identify high risk radon areas, and a Data Quality Program to assure reliable and consistent radon and radon decay product measurement results for both individual and governmental agencies.

Exposure to radon and its decay products inside homes is now recognized as a significant public health problem. Estimates of the number of lung cancers in the U.S. population caused by indoor radon range (Nero, 1983; Harley, 1984; NCRP, 1984) from 5,000 to 20,000 per year with some individuals being exposed to radon levels which may cause lifetime lung cancer risks in excess of 50 percent. To better define the scope and magnitude of this problem the Environmental Protection Agency (EPA) is developing a program to assess radon exposures on a nationwide basis. This national radon assessment program has the following objectives:

1. Determination of the frequency distribution of radon levels in residential structures on a nationwide basis through a national survey.
2. Identification of high risk areas through a program to assist States in conducting surveys to identify these areas, and through the development of predictive geological models.
3. Establishment of data quality programs which will result in consistent and reliable radon and radon decay products measurements for both individuals and governmental agencies.

National Survey of Indoor Radon Levels

EPA is planning a national survey of radon levels in homes. The objectives of this study are to determine the frequency distribution of radon levels in residential structures on a nationwide basis and to investigate factors which affect these levels. This study is needed to obtain a more accurate estimate of the average radon level in homes and to provide reliable data on the number of homes exceeding various radon levels. Such information will provide a better understanding of the magnitude of the public health problem associated with indoor radon levels. In addition this information will establish the base line level against which results of other surveys and indoor radon measurements can be compared.

A relatively large number of indoor radon measurements have already been made, particularly during the last 6-12 months. However, most of these measurements are not useful in defining the national frequency distribution of indoor radon levels because there are a number of uncertainties and limitations associated with these measurements. These uncertainties include the different purposes for the measurements, the widely varying sample designs, and the many different sample collection and measurement procedures used.

The only information presently available on the national frequency distribution of indoor radon levels is a 1984 analysis by Nero at the Lawrence Berkley Laboratory (Nero et al., 1984). Using data from about 500 houses, Nero developed a frequency distribution of radon levels in U.S. single family houses. This distribution is characterized by a geometric mean of 0.9 pCi/L and a geometric standard deviation of 2.8.

However, since the data used in this study are subject to the limitations and uncertainties cited above, the results of this analysis represent only a very rough approximation of the national frequency distribution of indoor radon levels. EPA's national survey will seek to more accurately characterize this distribution through use of a larger sample size, a statistically based survey design, and consistent, quality assured sample collection and measurement procedures.

Sample Population

The target population for a survey is that set of elements about which the data will be used to make inferences. The target population for the national radon survey will be all occupied residential housing units in the United States which are used as primary residences. This will include both single and multi-family housing units. The survey will not include schools, workplaces, group quarters (dormitories, institutions, rooming houses, etc.), seasonal units, or unoccupied units.

Sample Design

To permit inferences about the target population, probability sampling methods will be used in designing the survey. Sample collection and interview procedures being considered include face to face interviews using field study staff, random digit dialing telephone interviews or some combination of these procedures. The face to face procedure is the most likely method at the present time and will be described here to illustrate the manner in which the probability sample will be selected.

This survey method would use a three stage sample design. The first stage or the primary sample units are all the counties in the United States. A random sample of counties would be selected with probability proportional to the number of housing units in the county. The second stage units are census blocks or enumeration districts within the selected counties and they would also be randomly selected with probability proportional to the number of housing units. The third stage units are individual housing units systematically selected from the second stage sampling units.

The survey will be designed to allow determination of the frequency distribution of radon levels by both housing type and a combination of geographical/geological regions. Housing types will involve two strata, single family units and multi-family units. Geographical/geological regions will involve several strata in the range of 6 to 8. In addition, information obtained from a questionnaire used in the survey will be used to analyze other factors which may be related to indoor radon levels. The main emphasis on this type of information will be related to housing characteristics, for example, the housing unit's substructure, i.e., basement, crawl space, slab on grade, etc.

Sample Size

The sample size of the survey will depend primarily on the need to accurately estimate the upper percentiles of the national distribution and the number of subgroup populations for which frequency distributions are needed and the level of accuracy needed for these distributions (Cox et el., 1985). For example, the sample size needed to determine the radon level exceeded by 1% of the housing units in the U.S. with a relative standard error (RSE) of 10% is about 20,000 units. If, however, the radon level exceeded by 0.1% of the housing unit is desired, this would require a sample size of 200,000 housing units. These sample sizes could be reduced to 5,000 and 50,0000 units respectively, if a RSE of 20% were considered adequate. A design effect of 2 was used in estimating these sample sizes. Table 1 shows the sample sizes as a function of both percentile of interest and precision of the estimates.

These estimates are for a national distribution and will
increase depending upon the number of subpopulation groups
for which frequency distribution data is desired. Determining
the sample size, therefore, requires decisions regarding the
percentile of interest and level of accuracy needed, and
balancing these against the costs of the required sample.

Table I: Number of Sampling Units for Various
           Population Percentages

Population Percentages (P)	Sample Size (No. of Units)		
	5% RSE	10% RSE	20% RSE
0.1	800,000	200,000	50,000
1	80,000	20,000	5,000
5	15,000	4,000	1,000

P = Percentage of housing units exceeding known radon
      concentration
RSE = relative standard error
Design Effect = 2.0

Measurement Procedures

Average radon concentrations in the selected housing units
will be measured over a 12-month period using alpha-track
detectors. These detectors will be placed in general living
areas of the houses, usually a living room, family room, or
bedroom. Duplicate detectors will be deployed in 5 to 10
percent of the houses for aulity control purposes.

Schedule

The projected schedule for the various stages of the national
survey are shown below:

Complete Design	October 1986
Complete Pilot Test	March 1987
Initiate National Survey	July 1987
Complete National Survey	April 1989
Complete Data Analysis	October 1989

Identification of High Risk Areas

The main goal of EPA's radon program is the reduction of lung
cancer risks from radon exposure through the use of preventive
or mitigative techniques with emphasis on houses with highly
elevated risk levels. However, before such techniques can be
applied, areas/houses with highly elevated radon levels need to
be identified. Therefore, an important part of EPA's radon
assessment program is the identification of areas with highly
elevated indoor radon levels. Once an area with high levels
has been identified, more detailed surveys can be conducted or
individuals can be notified of the need to have the radon level
measured in their houses
     EPA is approaching this problem in two ways: (1) by
encouraging and providing assistance for State surveys to
identify high risk areas and (2) through the development of
predictive methodology to identify such areas.
     As part of its radon assessment program, EPA is providing
the following assistance to States in conducting radon surveys:
(1) radon analyses using charcoal canisters.
(2) assistance in survey design and sample selection.
(3) assistance in data analysis, management and
     interpretation.
EPA is working with the U.S. Geological Survey (USGS) to
develop methods using geological information for predicting
areas which have a high potential for elevated indoor radon
levels. Under an interagency agreement with EPA, the USGS is
selecting and characterizing up to 20 areas according to their
geological and (surficial) soil characteristics related to
radon availability. Factors considered in characterizing and
selecting these areas include: (1) uranium and radium content
of bedrock, (2) grain size, porosity and permeability of
surficial soils, (3) moisture content of soils, and (4)
fractures in bedrock.
     The distribution of radon levels in houses located in
these 20 areas will be measured either through EPA/State
cooperative surveys or as a part of EPA's national survey. If
correlations can be shown to exist between geological
characteristics and indoor radon levels, then this information
can be used to design surveys to more efficiently identify
homes with high indoor radon levels.

Quality Assurance

Numerous radon and radon decay measurements in houses are now
being made by a large number of private and governmental
organizations. In order to assure valid and consistent
measurements, it is important that proven methods be used
following standardized procedures. To address this need, EPA
issued "Interim Indoor Radon and Radon Decay Product
Measurement Protocols" and established a Radon/Radon Progeny
Measurement Proficiency program.

Measurement Protocols

In February 1986, EPA issued a document (Ronca-Battista et al., 1986) titled "Interim Radon and Radon Decay Product Measurement Protocols," describing seven methods for measuring radon and its decay products in houses. The methods addressed are those that have been evaluated by EPA and found to be satisfactory; other methods may be added as they are reviewed by EPA. In addition, portions of the document may be revised as new information and data becomes available.

This document provides procedures for measuring radon concentrations with continuous radon monitors, charcoal canisters, alpha-track detectors, and grab radon techniques. It also provides procedures for measuring radon decay product concentrations with continuous working level monitors, radon progeny integrating sampling units (RPISU), and grab radon decay product methods. The procedures provide information needed by an experienced user to calibrate, deploy, and operate the devices or instruments as well as to develop an adequate quality assurance program. Specifications for the location of the measurement, the house conditions during the measurement, and minimum requirements for quality control are included in each procedure.

EPA is also planning to issue a second protocol document on how to apply these protocols for different purposes. This document will provide guidance on where in the house the measurement should be made and the appropriate sampling times for each method which will vary depending upon the purpose of the measurement.

Measurement Proficiency Program

The Environmental Protection Agency has established a Radon/Radon Progeny Measurement Proficiency (RMP) program (Singletary et al., 1986) to assist State agencies and the public by providing a list of laboratories capable of measuring radon and/or radon progeny with results of known quality. The RMP program offers organizations making the measurements an opportunity to test and demonstrate their radon and/or radon progeny measurement proficiency on a quarterly basis. Participants will submit one or more types of detectors to be exposed to a known concentration in a radon calibration chamber. The detectors will be returned to the participants who will analyze them using standard procedures without knowing the exposure level. Measurement proficiency will be determined by comparing detector analyses to known values. After each test, participants will receive the Radon/Radon Progeny Analytical Proficiency Report which summarizes each laboratory's performance and provides means and standard deviations. Each laboratory's analyses will be identified by unique letters and numbers known only to the program coordinator(s) and the participating laboratory. A separate report titled "Radon/Radon Progeny Cumulative Proficiency Report" will be distributed to an appropriate agency in each State and the requesting public. This report will list laboratories, by name only, which have met the minimum guidelines set for the RMP Program.

Entry into the program is by application. Only organizations using methods for which EPA has issued measurement protocols may apply. Passive detectors may be submitted for radon chamber exposures by mail. Active devices must be placed in the chamber by a representative of the participating organization. Measurements will be made in quadruplicate and the results averaged. The first measurement proficiency test was conducted March 24 to April 3, 1986. Participants failing the first test will be provided a second test during the first week of May.

Organizations whose measurement results are with ± 25% of the known exposure level will be considered to have met the proficiency test requirements.

Literature Cited

Cox, B.G., DeWitt, D.S., Whitmore, R.W., Singletary, H.M. Howard, C.E., and Starner, K.K., A Survey Design for a National Study of Indoor Radon Concentrations in Residential Dwellings, RTI/3080/10-01F, Research Triangle Institute, Research Triangle Park, North Carolina (November 1985).

Harley, N.H., Radon and Lung Cancer in Mines and Homes, New England Journal of Medicine, 310:1525-1526 (1984).

National Council on Radiation Protection and Measurements, Exposures from the Uranium Series with Emphasis on Radon and its Daughters, NCRP Report No. 77, National Council of Radiation Protection and Measurements, Bethesda, Maryland (March 1984).

Nero, A.V., Indoor Radiation Exposures from ^{222}Rn and its Daughters: A view of the Issue, Health Physics, 45:273-288 (1983).

Nero, A.V., Schwehr, M.B., Nazaroff, W.W., and Revzan, K.L., Distribution of Airborne ^{222}Radon Concentrations in U.S. Homes", LBL 18274, Lawrence Berkley Laboratory, Berkley, California (November 1984).

Ronca-Battista, M., Magno, P.J., Windham, S., and Sensentaffar, E., Interim Indoor Radon and Radon Decay Product Measurement Protocols, EPA 520/1-8604, U.S. Environmental Protection Agency, Washington, D.C. (April 1986).

Singletary, H.M., Starner, K., and Howard, E., Implementation Strategy for The Radon/Radon Progeny Measurement Proficiency Evaluation and Qualtiy Assurance Program, EPA 5201.-86-03, U.S. Environmental Protection Agency, Washington, D.C. (February 1986).

RECEIVED October 27, 1986

Chapter 6

Radon-222 in Norwegian Dwellings

Erling Stranden

National Institute of Radiation Hygiene, P.O. Box 55, N-1345 Østerås, Norway

Results of Rn-222 measurements in 1500 dwellings in 75
municipalities in Norway are reported. The study was
conducted to assess geographical variations in Rn-222
concentrations and to assess the relative importance
of the different sources. The population average indoor
Rn-222 concentration is assessed to be 80-100 Bq^{-3}.
The equilibrium factor was studied in 58 dwellings and
a factor of 0.5 was found to be representative for
Norwegian dwellings. About 1 % of Norwegian dwellings
are expected to have Rn-222 concentrations exceeding
800 Bqm^{-3} (400 Bqm^{-3} Rn-222 progeny). The highest values
occur in alum shale and granite areas, and bedrock and
subsoil are the dominating radon sources. In a pilot
investigation, the radon exposure in dwellings was
correlated against lung cancer incidence and smoking
habits data from the Norwegian Cancer Registry. Even
though the exposure data are uncertain, the data
strongly suggest that Rn-222 progeny exposure in
dwellings may be a significant contributor to lung
cancer incidence among the general population. The
results support the "relative risk model" and suggest
that between 10 and 30 % of lung cancers in the
Norwegian population are initiated by Rn-222 exposure
in dwellings.

In recent years several studies have shown that the indoor concentrations of Rn-222 and its progeny are surprisingly high in certain
countries and areas. In most countries, Rn-222 progeny in indoor
air constitute a major fraction of the population dose from natural
sources of ionizing radiation, and in some countries Rn-222 progeny
exposure in dwellings is the dominating mode of population exposure
to ionizing radiation. (Castrén et al, 1984, Swedjemark and
Mjønes, 1984.)

0097-6156/87/0331-0070$06.00/0

ICRP (1984) have recently published recommendations on the limitation of exposure of the general public to natural radiation, and the World Health Organization, WHO, is preparing guide lines for Rn-222 progeny in indoor air. These guide lines and recommendations make it necessary to conduct surveys on Rn-222 in indoor air to assess the distribution of concentrations in existing houses and to assess geographical variations to be able to define areas where further investigations are needed. In order to limit Rn-222 progeny exposure in future houses, knowledge of the main Rn-222 sources and their distribution is needed.

In this paper, the main results of a study of Rn-222 and progeny in Norwegian dwellings are summarized. The study was primarly conducted to assess the regional variations in indoor Rn-222 concentrations and to study the importance of the different Rn-222 sources. The results of the survey have also been used to discuss possible health effects and the feasibility of epidemiological studies.

Material and Methods

To assess the regional variations of Rn-222 in Norwegian dwellings, a simple and practical sampling procedure was chosen. The local health authorities in 75 municipalities were asked to select about 20 houses in their municipality, and dosemeters were sent by mail to the local health inspectors who in their turn contacted the householders.

All types of houses were included in the sampling, except blocks of flats. In the country as a whole, less than 20 % of the dwellings are in blocks of flats. In our survey, Oslo, the capital of Norway, is not included and in the remaining part of the population, only about 10 % is living in blocks of flats.

In total, 1500 houses were investigated. In each dwelling, measurements were performed in the living room and in one of the bedrooms. All measurements were performed during the heating season.

The method of measurements is based on a combination of activated charcoal and TLD, and the integration time used is 5-7 days. Details on the principle of the dosimeter and its calibration have been published elsewhere. (Stranden et al, 1983.)

In areas where the highest indoor Rn-222 concentrations were measured, the source term was studied at the site. Houses with high concentration were inspected visually, especially the basement, and the heating and ventilation system was discussed with the householders. Field measurements of Rn-222 exhalation rates were perfored near the houses and at furture building sites in the municipality. The method for the exhalation rate measurements is based on the same activated charcoal/TLD combinations as the dosimeter for measurements of indoor Rn-222. Details on the method is published elsewhere. (Stranden et al, 1985.)

In connection with our field measurements, samples of geological materials were collected for analyses in the laboratory. Activity concentrations were measured by standard gamma spectrocopy with a 90-cm^3 Ge(Li) detector and a Canberra Model 8100 multichannel analyser. Details on calibration and procedure were published earlier (Stranden, 1985). Rn-222 exhalation measurements were also performed

on the geological samples by enclosing them in a closed container
and measuring the growth of Rn-222 in the container air. In some
of the samples, the influence of moisture and temperature on
exhalation rate was studied. (Stranden, et al, 1984: Stranden,
et al, 1984a.)

In municipalities where the water supplies are from deep-
bored wells, samples of household water were collected for Rn-222
analyses. The water samples were analysed by liquid scintillation
counting. The method for sampling and sample preparation were
adopted from Partridge et al (1979).

Rn-222 in Norwegian Dwellings

Indoor concentrations

In Fig. 1 the cumulative frequency of the measured mean values for
induvidual houses is plotted on a log-normal scale. The aritmetric
mean value in our measurements is 160 Bq/m^3. Areas with high concentrations are overrepresented in this distribution, (as seen from
the figure) and by population weighing the distribution for the
municipalities, a population weighted average of 110 Bq/m^3 in the
heating season is obtained.

Seasonal and Spatial Variations

In 47 houses, Rn-222 measurements were performed by Track Etch
detectors from Terradex Inc. during two periods, November-May and
June-October. From these measurements, the normal seasonal
variations were assessed. In Fig. 2 a plot of the cumulative
distribution of mean Rn-222 concentrations during summer and
winter are shown. From this figure, the mean concentration during
summer is about 50 % of the mean concentration during the winter
months.

In most cases, the Rn-222 concentration is lower in the bedroom than in the living room. (The average ratio was 0.91 with a
standard deviation of 0.15.) There might be several reasons for
this. Better airing of the bedroom than of the living room and
the fact that most traditional single family houses in Norway have
the bedroom in the first floor above ground and the living room in
the floor at ground level are probably the most feasible
explanations.

Variations with Age of the House

For each house measured, a questionaire was filled in by the householder. Among other things, the householder was asked to give the
year of construction for the house. In Fig. 3, the mean Rn-222
concentration is plotted against the age of the house. In this
plot, alum shale areas are excluded. As seen from this figure,
we did not find any systematic variations with the age of the house.

Equilibrium Factor and Rn-222-progeny

In 58 dwellings, the equilibrium factor was measured by grab

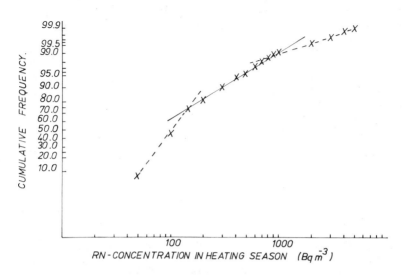

Fig.1. Cumulative frequency distribution of all measurements during the heating season.

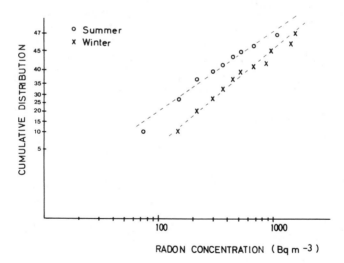

Fig.2. Cumulative frequency distribution of measurements in 47 houses during summer and winter.

sampling of Rn-222 and progeny. The equilibrium equivalent Rn-222 concentration (EEC) was determined by the Kusnetz method (Kusnetz, 1956) and the Rn-222 concentration by the scintillation counting on Lucas flasks. Fig.4, shows the distribution of the equilibrium factor, F, defined as $F = EEC/C_{Rn}$, where C_{Rn} is the Rn-222 concentration in Bq/m^3. The mean value of the equilibrium factor in this study was 0.5. The individual values for F may not be representative for the individual houses, but it is belived that the mean value is usable for average dose estimates.

In an earlier paper, results of measurements of Rn-222 (thoron progeny) have been discussed. In normal areas, the concentration of Rn-220 progeny is usally lower than 2-3 Bq/m^3. In some houses in a Th-rich area, the Rn-220 progeny concentration was higher than 10 Bqm^{-3}. (Stranden, 1984).

Sources

Soil and Bedrock

The highest Rn-222 concentration in indoor air were found in alum shale and granite areas. We also found high indoor Rn-222 concentrations in one area where the activity concentrations in the ground was "normal". These houses were on very porous ground (glacial esker) and the permeability of the ground was probably thus high.

In Table I results of Rn-226 activity measurements on geological samples are shown together with measurements on Rn exhalation rates from the samples. The exhalation rates varies considerably with the moisture content of material. The exhalation rate is low for dry samples and when the moisture content increases, the exhalation rate increases until it reaches a plateau. When the moisture content increases further, a rapid increase in radon exhalation occur. When the saturation level of moisture is reached, the exhalation rate drops dramatically. The exhalation rates given in Table I are obtained by assuming that the most probable moisture content is whithin the plateau of exhalation rate/moisture curve. (Stranden et al, 1984, Stranden et al, 1984a).

Table I. Activity concentrations and exhalation rates from samples of geological materials

Material	No of samples	Activity concentration of Ra-226 ($Bq\ kg^{-1}$)	Exhalation rates from samples + ($Bq\ h^{-1}\ kg^{-1}$)
Alum shale	6	900 – 5600	3 – 10
Alum shale-soil	6	700 – 1800	3 – 5
Granite	4	70 – 270	0.2 – 0.6

+ Based on studies on the influence of moisture.
(Stranden et al, 1984, Stranden et al, 1984a)

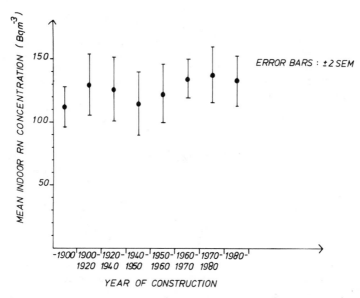

Fig. 3. Rn-222 concentration versus year of construction.

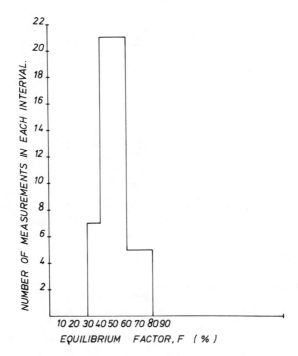

Fig.4. Distribution of equilibrium factor measured in 58 dwellings.

        In 85 sites, the exhalation from the ground was studied.
70 of these measurements were performed near where indoor Rn-222
concentrations were measured the previous winter, and 15 measurements were performed on future building ground. Exhalation measurements were also performed in "normal" areas where no enhanced indoor
Rn-222 concentrations had been found. In Table II the main results
of these measurements are shown.

Table II.   Indoor Rn-222 concentration and exhalation rates from the
            ground surface in different areas

Type of area	Rn-222 in indoor air (Bq m^{-3})	Rn-222 exhalation rate at ground surface (Bq h^{-1} m^{-2})
Alum shale	30 + − 5300	20 + − 10 000
Granite	30 − 800	120 − 800
Esker	100 − 3000	100 − 300
"Normal" ++	20 − 200	< 20 − 150

+   Where the lowest values occur, the ground water is higher than
     the alum shale layer.

++   Mostly clay and soil.

        In the granite and alum shale areas, exhalation measurements
above the ground were well correlated with the indoor air measurements. In the esker area, the exhalation rates were normal, and the
high indoor Rn-222 concentrations here are believed to be caused by
convective flow due to underpresure in the houses. (Stranden et al,
1985).

Household Water

In Norway, most municipalities have water supplies from surface water.
However, a few municipalities have large water works based on
drilled wells, and in some areas, small waterworks, supplying a few
households each, are relatively common. In order to study the household water as a Rn source, water samples from tap water from 58 wells
were analysed. Some of these wells were supplying whole municipalities
and some of them only one or two households. In Table III, the distribution of Rn-222 concentration in the wells studied is shown together
with the number of people supplied by the wells.

        The highest concentration we found was 7000 kBq m^{-3}.

        Annanmäki (1978) studied the transfer of Rn from water to air.
By applying occupancy factors for different areas in the house, he
found that 1 Bq m^{-3} in household water would result in 1.4x10^{-4} Bqm^{-3}
in breathing air. By applying this transfer factor, the highest Rn-222
concentrations in air due to well water is 1000 Bq m^{-3}.

As seen from Table III most of the larger water works have low Rn concentrations. Only 6 wells with a total of 30 users had Rn concentrations above 1500 kBq m^{-3}.

Table III. Rn-222 concentrations in household water from drilled wells

Rn-222 concentration in water (kBq m^{-3})	Number of wells	Number of users
< 100	22	63 000
101 - 500	18	8 800
501 - 1000	9	860
1001 - 1500	3	60
> 1500	6	30

Building Materials

In a previous paper (Stranden 1979) presented activity concentration measurements in the main Norwegian building materials. We have also studied Rn exhalation from samples of building materials and from walls (Stranden and Berteig 1980; Stranden 1983). Rn exhalation is strongly dependent on the moisture content of the building material, and the Rn exhalation may vary by a factor 20 or more due to differences in moisture content (Stranden et al 1984, Stranden et al 1984a). For mature materials, the variation in moisture will not be very large, and by assuming that the materials are mature, a relatively good estimate of the Rn exhalation rate may be obtained. In Table IV, the data from our previous studies are summarized.

Table IV. Activity concentrations and exhalation rates from main building materials

Building material	Ra-226 activity concentration (Bq kg^{-1}) Range and mean	Exhalation rate per unit mass and unit Ra-226 activity concentration $(Bqh^{-1}kg^{-1})/(Bqkg^{-1})$	Estimated range of exhalation rate from typical walls $(Bqh^{-1} m^{-2})$ ++
Concrete	4 - 130 (28)	$1 - 3 . 10^{-3}$	5 - 50
Clay brick	34 - 90 (60)	$1 - 5 . 10^{-4}$	2 - 10
LECA +	37 - 81 (52)	$1 - 2 . 10^{-3}$	10 - 20

+   Light Expended Clay Aggreate

++ Based on Ra-226 activity concentration, typical wall thickness and exhalation rate measurements from samples and walls. (Stranden and Berteig, 1980; Stranden 1983, Stranden et al, 1984; Stranden et al, 1984a.)

Possible Health Effects

Occupational studies have demonstrated an enhanced frequency of
bronchial cancer among various groups of Rn-222 progeny exposed
underground miners, especially uranium miners. (Archer et al, 1976:
Kunz et al 1978: Kunz et al 1975: Lundin et al, 1971;Müller et al,
1983; Sevc and Placek, 1973; Seve et al, 1976; Whittmore and
Mc Millan 1983.)
     However, so far, only a few small-scale epidemiological
surveys have been carried out or started to study the possible
influence of the exposure to Rn-222 progeny in dwellings on the
etiology of lung cancer in the general population, and the evidence
of these findings is inconclusive.
     By reviewing all available epidemiological findings, Jacobi
and Paretzke (1985) concluded that a proportional hazard model will
be best fitted to the epidemiological data. The findings suggest
that a significant fraction of the lung cancer cases in the population can be correlated to exposure to Rn-222 progeny in dwellings
and the relative fractions of radiogenic lung cancer rates might be
nearly the same for smokers and non-smokers, and for men and women.
     The Cancer Registry of Norway was started in 1952, and all new
cases of cancer in Norway have ever since been subject to compulsory
reporting. The data can be broken down on sex, age and residence.
In the first few years, there was a known underreporting, bur reliable
cancer incidence data from 1955 and up to date are available (Cancer
Registry, 1978).
     In 1964/65, a large scale smoking habit investigation was
performed by the Cancer Registry (Zeiner-Henriksen 1976). These
data may be broken down on sex, age and residence, and the amount
of tobacco (cigarettes, pipe, combined) can be assessed. Follow-up
investigations have been carried out, and it is thus possible to
assess the trends in smoking habits in the years following 1965.
     The relative risk model may be expressed in a simple form:

$$R = Ro \ (\ 1 + \alpha \ C) \qquad\qquad (1)$$

where: R = lung cancer incidence rate in the population
         Ro = lung cancer risk when there is
              no Rn progeny
         α = relative risk factor for Rn progeny exposure
         C = Rn progeny concentration

     By assuming that cigarette smoking and Rn in dwellings are the
main causes for lung cancer in the general population, we performed
the following simple two- variable correlation anaslyses:
     1) Age adjusted lung cancer incidence in the period
        1966-1985 was correlated against smoking habits
        (expressed as average number of cigarettes smoked
        per day in the grown population) for men and women.
     2) The ratio, actual incidence rate to incidence rate
        predicted by the correlation between lung cancer and
        smoking habits was correlated against Rn concentration.
        (Mean Rn concentrations in the population groups under
        study varied from 80 Bqm^{-3} to 180 Bqm^{-3}.)

In this way it was possible to test the relative risk model. The results of this correlation analyses may be summarized as follows:

a) The ratio used in 2) above increased by increasing radon concentration along the same line for both men and women.

b) By combining data from this ratio for men and women (6 points, each point based on about 4×10^6 person years), and normalizing the ratio to unity for zero Rn concentration, a relative tisk factor of 0.001-0.003 per Bqm^{-3} of Rn (life risk and life time exposure) was found at the 95 % confidence level. This corresponds to a relative risk factor of 0.002-0.06 per Bqm^{-3} for Rn progeny. This is in the same range as the values predicted from the relative risk model of Jacobi and Paretzke (1985), and the results suggest that it is possible that between 10 and 30 % of the lung cancers in Norway (mean Rn-concentration 100 Bqm^{-3}) is caused by Rn progeny in dwellings.

Due to lack of man-power and resources, the sampling used in our study was not truly random. And the results should only be interpreted as indicative.

Taking the possible health effects caused by inhaled Rn-222 progeny in the general public into account, better risk estimates based on epidemiological studies on the general population are urgently needed.

To be able to perform such large-scale epidemiological studies several conditions should be fulfilled: Reliable lung cancer incidence data for the population under study, large regional variations in indoor Rn-222 progeny concentrations, low mobility in the population, reliable data on the building stock and last but not least: Reliable data on smoking habits among different parts of the population.

Furthermore, the exposure data should be obtained by stratified random sampling of houses based on population registers and data on building stock.

In 1980, a complete census was conducted in Norway including data on the building stock. The census was conducted by the Central Bureau of Statistics, and the data on building stock are available in a suitable form. (Norwegian Central Bureau of Statistics, personal communication.)

There are large variations in Rn-222 concentrations among relatively large population groups in Norway. This, together with the excellent data on cancer incidence and smoking habits,makes it feasible to conduct a large-scale retrospective multi-variable correlation analyses on Rn-222 progeny exposure in Norwegian dwellings. Such a study should however be based on stratified random sampling. The data on building stock and person registers make it possible to develop a stratified random sampling scheme covering most of the municipalities in Norway.

Discussion and Conclusions

From the measured indoor Rn-222 concentration during the heating season, measurements of equilibrium factors and assessments of the seasonal variations, it is possible to assess population averaged indoor concentrations of Rn-222 progeny in Norwegian dwellings.

It is also possible to assess the distribution of concentrations in the building stock. In Table V, the main results of such assessments are shown. This table suggests that the mean indoor concentration of Rn-222 progeny in Norwegian dwellings is high compared to most countries. Furthermore, in several areas, exceptionally high levels have been measured in individual houses, and about 3 % of the dwellings in Norway are expected to have Rn-222 progeny concentrations above 200 Bq m^{-3}. This is suggested as an action level by ICRP (1984). About 1 % of the dwellings are expected to have Rn-222 progeny concentrations above 400 Bq m^{-3}.

Table V.   Assessment of mean value and distribution of Rn-222 progeny in Norwegian dwellings

Mean value Bq m^{-3}	Expected percentage of Norwegian dwellings having indoor air concentrations above the given levels (Bq m^{-3} of Rn-222 progeny)		
	> 100	> 200	> 400
40 – 50	10	3	1

From the investigation on the source term, the ground is the most importance radon source, and when the Rn-222 concentrations are much higher than normal, the ground is always the main reason. In Table VI, we have used the data on the source term to assess the contribution from the different sources in typical Norwegian dwellings. The contribution from household water is assessed by transfer factors found by Annanmäki (1978) and the contribution from building materials have been assessed by using data on exhalation measurements, common use of the material and an air exchange rate of 0.5 hr^{-1}.

Table VI.   Estimates of the source strength of the main Rn-222 sources

Source	Contribution to indoor air Rn-222 concentration (Bq m^{-3})	
	Range	Most probable
Ground	10 – 5500	100
Building materials	5 – 50	10
Household water	0 – 1000	0

The highest indoor Rn-226 concentrations occur in areas with alum shale in the ground. Measurements in these areas yield a good agreement between exhalation rates and indoor Rn-222 concentrations. Even in granite areas, high indoor Rn-222 concentrations may occur, but no extreme values were found in such areas. These findings are also confirmed by the field studies.

Even though Ra-226 concentrations in the bedrock and soil are
"normal, the Rn influx from the bedrock may be high if the permeability of the material under the house is high, as in esker areas.
In this case, exhalation measurements will not be of much use for
the classification of the Rn risk of future building ground.

From the results and discussions of this paper, the following
main conclusion may be drawn:

1) The Rn-222 progeny concentration in Norwegian dwellings is on
   average higher than in most other countries.
   (United Nations, 1982.)

2) There are large regional variations, and the subsoil and bedrock
   is the most important Rn-222 source.

3) In a significant fraction of Norwegian dwellings, counter
   measures to reduce the Rn-222 progeny exposure are needed, and in
   several areas, technical measures are needed to ensure that
   future dwellings will not get enhanced levels of indoor Rn-222
   progeny.

4) Rn-220 (thoron) progeny is usually of far less importance than
   Rn-222 progeny. In normal areas, the Rn-220 progeny effective
   dose equivalent is lower than 10 % of the effective dose equivalent from Rn-222 progeny. In certain Th-232 rich area, however,
   Rn-220 progeny may give relatively high doses to the inhabitants.

5) Preliminary results suggest that Rn-222 progeny in dwellings is
   a cause of lung cancer in the general population and that there
   is a multiplicative effect between Rn progeny and cigarette
   smoking. Further studies should, however be conducted to get more
   reliable risk estimates.

6) The large regional variations in Rn-222 progeny concentrations
   in Norwegian dwellings, together with excellent data on lung
   cancer incidence and smoking habits, make it feasible to conduct
   multi-variable correlation studies of Rn-222 progeny as a cause
   of lung cancer in the Norwegian population. Such studies should
   be based on stratified random sampling based on person registers
   and data on building stock.

<u>Acknowledgment</u>

The author wishes to thank A.K.Kolstad and B.Lind for their kind
technical assistance and their friendship which makes the work
worth while.

The project was sponsored by The Royal Norwegian Council
for Scientific and Industrial Research, NTNF.

<u>References</u>

Annamäki M., 1978, Radon Measurements in Finnish Dwellings, Nordic
Society for Radiation Protection, 5th, Visby. (In Finnish.)

Archer V.E., Gillam J.D. and Wagoner J.K., 1976, Respiratory
Disease Mortality Among Uranium Miners, Am.N.Y.Acad.Sci, 271,280-293.

Cancer Registry of Norway, 1978, Incidence of Cancer in Norway
1972-1976. The Norwegian Cancer Society, Oslo.

Castrén O., Winqvist K., Mäkeläinen L and Voutilainen, A., 1984, Radon Measurements in Finnish Houses, Radiation Protection Dosimetry, 7, 33-337.

ICRP, 1984, Principles for Limiting the Exposure of the Public to Natural Sources of Radiation, ICRP Publication 39, Oxford, New York: Pergamon Press.

Jacobi, W. and Paretzke H.G.: 1985, Risk Assessment for Indoor Exposure to Radon Daughters, In Proceedings, Seminar on Exposure to Enhanced Natural RAdiation and Its Regulatory Implications, Maastricht, the Netherlands, March 25-27, Elsvier Science Publisher, Amsterdam.

Kusnetz H.L., 1956, Radon Daughters in Mine Atmospheres, a Field Method for Determing Concentrations, Ind.Hyg.Quart., 17, 85-88.

Kunz E., Sevc J. And Placek V., 1978, Lung Cancer Mortality in Uranium Miners, Health Phys., 35, 579-580.

Kunz E., Sevc J. and Placek V., 1979, Lung Cancer in Man in Relation to Different Time Distribution of Radiation Exposure, Health Phys., 36, 699-709.

Lundin R.E., Wagoner J.K. and Archer V.E., 1971, Radon Daughter Exposure and REspiratory Cancer-Quantitive and Temporal Aspects, Nat.Inst.f.Occup. Safety and Health, Nat.Inst.of Env.Health Serv. Joint Monograph No 1, US Dept of Health Education and Welfare, Washington D.C.

Müller J., Wheeler W.C., Gentleman J.F., Suranyi G. and Kusiak R.A., 1983, Study of Ontario Miners 1955-77. Pat 1, Ontario Ministry of Labor, Ontario Workers Compensation Board.

Partridge J.E., Horton T.R. and Sensitaffar E.L., 1979. A study on Radon-222 Released from Water During Household Activities, U.S. Environmental Protecttion Agency, Montgomery, AL, Technical Note ORP/EERF-79-1.

Sevc J. and Placek V., 1973, Radiation Induced Lung Cancer: Relation Between Long-Term Exposure to Radon Daughters, Proceed, 6th Conf.on Rad.Hyg.Czechoslovakia.

Sevc J., Kunz E. and Placek V., 1976, Lung Cancer in Uranium Miners and Long-Term Exposure to Radon Daughter Products, Health Phys., 30, 433-437.

Stranden E., 1976, Radioactivity of Building Materials and the Gamma Radiation in Dwellings, Phys.med Biol, 24, 921-930.

Stranden E. and Berteig L., 1980, Radon in Dwellings and Influencing Factors, Health Phys, 39, 275-284.

Stranden E., Kolstad A.K. and B.Lind., 1983, The ETB Dosemeter.
A Passive Integrating Radon Dosemeter Combining Activated Charcoal
and TLD, Radiation Protection Dosimetry, 5, 241-245.

Stranden E., 1983, Assessment of the Radiological Impact of Using
Fly Ash in Cement, Health Phys, 44, 145-153.

Stranden E., Kolstad A.K. and Lind B., 1984, Radon Exhalation:
Moisture and Temperature Dependence, Health Phys, 47, 480-484.

Stranden E., Kolstad A.K. and Lind B., 1984, The Influence of
Moisture and Temperature on Radon Exhalation, Radiation Protection
Dosimetry, 7, 55-58.

Stranden E., 1984, Thoron (Rn-220) Daughter to Radon (Rn-222)
Daughter Ratios in Thorium-Rich Areas, Health Phys, 47, 784-785.

Stranden E., Ulbak K., Edhwall H. and Jonassen N., 1985, Measure-
ment of Radon Exhalation from the Ground: A Usuable Tool for
Classification of the Radon Risk of Building Ground, Radiation
Portection Dosimetry, 12, 33-38.

Stranden E., 1985, The Radiological Impact of Mining in a Thorium-
Rich Norwegian Area, Health Phys. 48, 415-420.

Swedjemark G.A. and Mjønes L., 1984, Radon and Radon Daughter
Concentrations in Swedish Homes, Radiation Protection Dosimetry,
7, 341-345.

United Nations Scientific Committee on the Effects of Atomic
Radiation, 1982, Sources and Effects of Ionizing Radiation.
(New York, United Nations.)

Whittmore A.S. and McMillan A., 1983, Lung Cancer Mortality Among
U.S. Uranium Miners: A Reappraisal, J.Nat.Cancer Inst. 71, 489-499.

Zeiner-Henriksen T., 1976, Smoking Habits in the Norwegian Population, Tidskrift for Den norske Lægeforening, 11, 617-620,
(in Norwegian).

RECEIVED September 8, 1986

Chapter 7

Radon Levels in Swedish Homes:
A Comparison of the 1980s with the 1950s

Gun Astri Swedjemark, Agneta Burén, and Lars Mjönes

National Institute of Radiation Protection, Box 60204, S-104 01 Stockholm, Sweden

In 1980-82 a study was carried out on Swedish homes
built before 1976. The aim was to provide averages
and distributions for the radon exposure of the
Swedish population. Correlations with parameters
such as building materials and building periods
were also investigated. In 1955-56, a study of
homes built before 1946 was carried out in four
towns in central Sweden by Hultqvist with the aim
of obtaining results which were representative for
the homes in the areas that were studied. The radon
concentrations in homes from approximately the same
regions in the two studies are compared. The average was found to be four times higher for the homes
measured in 1980-82 than for those measured in
1955-56. The reason for this increase is discussed
including a thorough evaluation of the sampling and
measurement methods used in the two studies. Indications that there has been an increase in the radon
concentrations in other Swedish homes and also in
other countries in temperate regions are also
discussed.

There are many indications of an increase in radon concentration in
homes located in temperate regions during the last decades. In some
of the nordic countries the ground has a high potential for radon
exhalation and in Sweden we also have used a building material
containing a higher concentration of Ra-226 than usual building
materials. These conditions give a rather high concentration of
radon indoors and is therefore easy to measure. Increases will also
be higher resulting in more severe health effects than in regions
with low concentration of Ra-226 in the building materials or in the
ground. In Sweden measurements of the radon concentrations in homes
were made in the 1980s and was studied also in the 1950s: two
studies, which are possible to compare.

0097-6156/87/0331-0084$06.00/0
© 1987 American Chemical Society

Measuring methods

The equipment

For the country-wide study 1980-82 we used integrating instruments.
Applying the ideas from the Environmental Laboratory in New York
(Breslin and George, 1979) the radon group at our laboratory designed and built passive radon monitors named Integrating Radon
Measuring Apparatus (IRMA) based on thermoluminescence dosemeters,
TLD (Burén et al, 1982). The devices were calibrated during every
summer period in our radon chamber No I (Falk, et al, 1982;
Swedjemark, 1985) when the country-wide investigation was carried
out. The standard deviation of the measurement values between the
devices for a particular calibration period was found to be 5 - 6 %
for radon concentrations between 100 and 3 800 Bq/m^3. Humidity, 20 -
60 %, temperature, 18-26 $^\circ$C and atmospheric pressure, 970 - 1 040
mbar, were not found to influence the calibration factors. The
calibrations were made with an ion chamber with an estimated overall
uncertainty of 10 % (1 s). This uncertainty includes that of the
calibration factor of which the proportion of the systematic uncertainty is estimated to be small. The total uncertainty in a measurement value can therefore be estimated to be well below 20 %. The
sensitivity of the dosemeters and the design of the instruments
result in a greater uncertainty than that given above for radon
concentrations below 10 Bq/m^3 according to statistical calculations.
    In the studies performed by Hultqvist during the 1950s instan-
taneous sampling techniques were used. Equipment was designed by
Rolf Sievert for determining radon concentration in housing and was
described by Hultqvist (Hultqvist, 1956).
    The instrument consisted of a portable measuring box and six
ion chambers. The air was sampled by pumping the air into the ion
chamber, which was then closed. When radioactive equilibrium was
obtained between the radon and radon daughters in the chamber, a
known potential difference was applied between the outer surface of
the ion chamber and the central electrode. The charge obtained due
to the ionization was collected on the central electrode and was
measured as a change in the potential difference with an electrometer tube. This charge was directly proportional to the radon
equivalent content when the exposure time was known. The air sampled
in the ion chambers contained both Rn-222 and Rn-220 (Tn) together
with their decay products. The results, therefore, depend on the
actual combination of these two gases with decay products and was
called radon equivalent content.
    The variations in the background, the sensitivity to moisture,
the alpha activity of the chamber itself and the influence of recombination were discussed by Hultqvist. The standard deviation due to
counting statistics was estimated to be about 3 % (in a few measurements 6 %). The calibration was made by counting each alpha particle
by a proportional counter specially designed at the Department for
this purpose. The statistical uncertainty of the calibration of the
equivalent radon concentration was estimated to be 12 %.
    The activity originating from the thoron series was seldom
found to be sufficient to allow evaluation by analysis of a disintegration curve recorded with the ion chamber. The possible thoron
contribution was investigated with a corona filter and was found to
be small compared with the radon concentration.

For the two studies compared in this paper, different measuring
methods have been used. However, the systematic difference between
the results from the two types of equipment can be estimated to be
less than 20 %.

Sampling of the dwellings

The selection of the 1980-82 measurements (Swedjemark and Mjönes,
1984) was made on dwellings built before 1976 and with the aim of
determining dose distributions and the collective dose to the
Swedish population from the exposure of the short-lived radon decay
products. This was done by using the statistical selection made by
the National Institute for Building Research intended for an energy
study of the Swedish stock of houses. From a selection of 3 100
houses in 103 municipalities, 2 900 were inspected. The data was
found to be in substantial conformity with data from the land register and the population census of 1975. For the study of the radon
concentration 752 dwellings were selected at random.

The dwellings in the 1955-56 study selected for gamma measurements were built before 1946 and were located in 13 urban areas in
central Sweden. The radon measurements were made in four of these
areas, Stockholm, Norrköping, Katrineholm and Örebro. Most of the
dwellings were erected during the period 1925-1945. No distinction
was made between multi-family and detached houses. The ventilation
systems were of natural draught type in all houses as opposed to the
1980-82 investigation when many types of ventilation systems were in
use.

Sampling in the homes

At the 1980-82 study two instruments were placed in each dwelling
for two weeks, one in the living room and one in the bedroom. The
dwellings were inspected, the gamma radiation was measured and the
participants were asked to answer a questionnaire. The inhabitants
were asked to live as usual.

In the 1955-56 study the radon measurements were usually taken
in the living rooms. The gamma measurements were made in several
rooms. All samples were taken before noon. With regard to the ventilation, the houses were divided into two groups; one termed "ventilated", where the dwelling had been aired one or two hours before
the measurements. The dwellings in the group named "unventilated"
had not been aired since the day prior to the measurements.

In this comparison both ventilation groups have been used. That
seems to be correct because applying correction factors for weather-
ing, 0.9, (Swedjemark, 1978 and 1985) on the "unventilated" dwellings gave the same average for this group as for the uncorrected
average for the whole group of studied dwellings according to Table
I.

Table I. Measurements of radon concentration in air in four
        towns in Central Sweden at the beginning of the
        1950s in houses built before 1946 (Hultqvist, 1956)

| | Radon concentration,[a] Bq/m^3 | | |
	Unaired[b] houses	Aired[c] houses	Weighted average
Wood	15.2	15.5	15.2
Brick	47.4	26.0	40.0
Aerated concrete based on alum shale	133	67.0	116

[a] 1 $Bq/m^3 \approx 0.027$ pCi/liter.

[b] No thorough airing was carried out since the day before
the measurement.

[c] Dwellings were aired the morning before the sampling.

Results

Frequency distributions

Fig 1 shows the frequency distributions of the radon concentrations
weighted for the proportion of houses of different types in order to
give the whole number of dwellings in the country. The results from
the 1980-82 study are shown in Fig 1 a and from the 1955-56 study in
Fig 1 b. When comparing these distributions the higher average of
the 1980-82 study is obvious. Possible reasons and the magnitude of
the increase will be discussed in the following.

Low - intermediate - high radon levels

The material from the 1980-82 study has been divided in three intervals of the radon concentrations in order to see differences and
similarities between the levels.

(i)   The majority of dwellings are found in the group with radon
       concentrations below 150 Bq/m^3. In this group new multi-
       family houses and detached houses with mechanical ventilation
       are found. Also detached houses built during the period 1920-
       50 are found in this group.

(ii)   The intermediate group with radon concentrations of 150 -
       600 Bq/m^3 was dominated by detached houses built between 1950
       and 1975 and multi-family houses with a considerable content
       of alum-shale-based aerated concrete.

(iii) The group with radon concentrations above 600 Bq/m^3 contains
       only about ten dwellings and is very disparate. Most of them,
       however, are found on ground with exposure rates above 30 μR/h

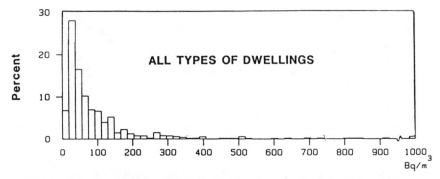

Figure 1a. Distribution of radon concentrations in the Swedish dwelling stock 1975. Indicated in the figure are also three values exceeding 1000 Bq/m^3 (1312, 1537 and 3306 Bq/m^3).

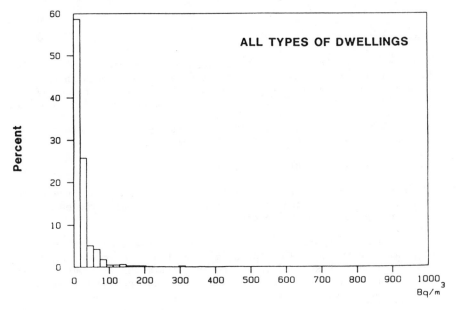

Figure 1b. Distribution of radon concentrations in the Swedish dwelling stock 1950.

according to maps made from gamma measurements carried out by
air planes (Wilson, 1984). Furthermore, a couple of these
dwellings have walls with gamma exposures above 100 µR/h measured one centimeter from the surface. About half of these
houses were built before 1900. The ground would be the most
probable radon source in these old houses because when they
were built no concrete slabs were used. The basement floors
usually consisted of gravel or laid stones, which does not
prevent the radon gas inflow from the soil. This inflow may be
very high if the air pressure indoors is made low compared to
the air pressure in the soil air and the outdoor air. This is
the case when the inflow of fresh air to the house is limited
by the weather stripping around the windows etc.

Building periods

A comparison between the radon concentrations in houses built before
1946 in the 1980-82 study and in the 1955-56 study is possible to
carry out. About 60 % of the dwellings in the multi-family houses
were built after 1950. As far as this comparison is concerned it is
important to keep in mind that the samples were different in the two
studies for this period although the houses were built during the
same period. In a large number of these houses improvements had been
made before 1980-82. In many of them the windows had been fitted
with weather stripping. The energy crisis in 1973-74 has probably
hastened a selective demolishment of older houses. It is not possible to differ between detached and multi-family houses in the
1955-56 study because it was not at that time thought to be important. Among the houses built before 1946 in the 1980-82 investigation 90 % of the detached buildings are made of wood. The
distributions of the dwellings in the 1980-82 study of all kinds of
building materials built before and from 1946 are shown in Fig 2 a
and 2 b. There is not only a higher average but also a large number
of houses with rather high radon concentrations in the houses built
before 1946.

A comparison between detached houses built before and after 1946
in the 1980-82 study shows a larger spread of the concentrations
during the later building period. There are many reasons for this

(i)   Higher amounts of more active building materials have been
       used since the 1940s. More stone-based materials have been
       used in all parts of the country, particularly in multi-family
       houses but also in detached houses. Alum-shale-based aerated
       concrete constituted about 50 % of the building materials used
       during a part of the period. Its activity concentration
       weighted for the production rate of the different factories
       has been shown to have increased (Swedjemark, 1978).

(ii)   The air change rates were reduced to save energy resulting in
       higher radon concentrations indoors (Swedjemark, 1978). The
       energy saving techniques were improved during the 1970s.

(iii) Holes around tubes for water and drainage in the bottom slab
       were required by many local building authorities during the
       1960s. The reason for this was that plastics entered the scene

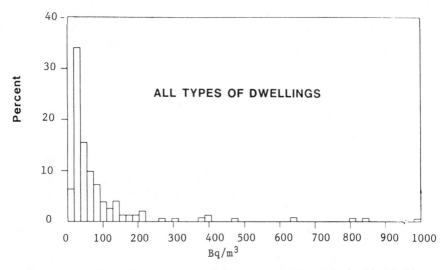

Figure 2a. Distribution of radon concentrations in houses built before 1946. Indicated in the figure is the value exceeding 1000 Bq/m^3, 1537 Bq/m^3.

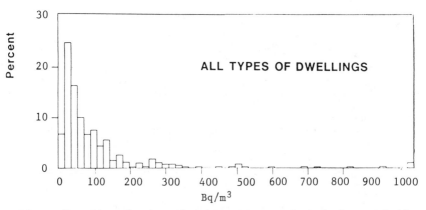

Figure 2b. Distribution of radon concentrations in houses built after 1945. Indicated in the figure are also two values exceeding 1000 Bq/m^3, 1312 and 3306 Bq/m^3.

as a building material in the water and drain installations and also the large temperature gradients which could occur between the bottom slab and the tubes during the different seasons. Installation of drainage systems are since 1970 controlled by central regulations and the type of installation mentioned above is not made today.

(iv) The low air pressure indoors compared with the pressure in the soil air and outdoors probably increased due to the decreasing of outdoor air supply in order to save energy resulting in a higher inflow of soil air.

Another difference between the 1950s and the 1980s is that people now live in other types of houses. 70 % of the population lived in wooden houses in 1950 compared with 40 % today. 3 % lived in houses built of aerated concrete based on alum-shale, in 1950. In 1980 the figure was about 10 %. These differences are also an indication on an increased collective radon concentration during the period.

The ground and the building materials

During the late 1970s there was a growing awareness of the importance of the ground as regards to radon concentrations in buildings. A number of municipalities were found to be situated in areas where problems could occur. Surveys of dwellings with high radon levels were performed by the local boards of health. Today it is assumed that approximately 10 % of the Swedish ground necessitate special arrangements for protection against penetrating radon. All towns where radon measurements took place in the 1950s have such areas. Therefore it is reasonable to compare the 1955-56 study not only to the country-wide 1980-82 study but also to those buildings in the 1980-82 study which were situated in the same areas as in the 1955-56 study.

(i) Wood has been the most common building material in the largest part of Sweden for many centuries. A study of the radon concentrations in these buildings may give a good understanding of how building techniques have influenced the concentrations in general. The frequency distribution results both for the 1955-56 study and for detached houses in the corresponding municipalities in the measurements from 1980-82 show many low but also many rather high levels with higher values for the later study. The enhanced levels are mostly caused by inflow from the ground.

(ii) For detached houses built of concrete and of brick the tendencies in the comparison are the same as for the wooden houses. Often some parts of these houses have been built of aerated concrete based on alum shale. About 10 % of the detached houses of today have parts of this material. For the multi-family dwellings the corresponding value is 35 %. In the 1980-82 study it was not possible to divide between sand-based and alum-shale-based aerated concrete at sampling. All aerated concrete for which the gamma exposure one cm from the wall was above 40 μR/h was classified as alum-shale-based aerated

concrete. The highest value measured in the 1980-82 study was
in a house of alum-shale-based aerated concrete built on
ground with high potential radon exhalation.

(iii) A good illustration of how the activity concentrations in the
building materials have increased since the 1950s is the
results of the country-wide study of gamma radiation in the
Swedish dwellings (Mjönes, 1982). The average exposure was
found to have increased by a factor of two since the 1950s
(Mjönes, 1986). The gamma radiation from the building materials reflects also other radioactive nuclides than the U-238
chain, but in Swedish building materials high concentrations
of the Th-232 chain is not common (Möre, 1985). The increase of
the gamma radiation is therefore an indication also of an
increase in radon concentration.

Comparison of averages

Table II shows that the average radon concentrations have increased
by several factors compared with the 1955-56 measurements for all
the building materials. The average, weighted for the population,
was found to have increased by a factor of four to six. Here we must
keep in mind that the ground in the four municipalities investigated
in 1955-56 has a potential for high radon exhalation compared to the
whole country. No dwellings with extremely high radon concentration
were found however. A contributing reason for this may be the diffe-
rencies of the buildings and the habits of the inhabitants now and
then. Doors and openings to cellars were by that time always closed
thereby preventing direct inlet of radon from the ground.

Table II. Arithmetic means of the radon concentration in Bq/m^3

| | Homes in houses built before 1946 four munici-palities | Building stock 1975 | |
		The whole country	The same four municipalities as in the 1955-56 investigation[a]
Wood	15.2	100	131
Concrete-brick	40	70	178
Aerated concrete based on alum shale	116	237	182
Average	24[b]	101	158[b]

[a] The borders of the municipalities have been changed since 1956
so that the regions are not quite comparable.

[b] These averages are weighted for the proportion of houses built
of different building materials in the whole country, not for the
four municipalities. There are differences in these proportions
between parts of the country.

Discussion

Radon increase in Swedish homes

The probabilities for an increase in radon concentration in
Swedish homes between the 1950s and the 1980s are based on

(i) a comparison between the studies 1980-82 for the whole country
and 1955-56 for the four municipalities in both studies.

(ii) a comparison between the radon concentration intervals in the
1980-82 study.

(iii) a comparison between the houses built before and after 1946 in
the 1980-82 study and between the houses in the 1955-56 study.

(iv) the development of the building materials and soil conditions.

All these comparisons support the hypothesis of an increase. The
arithmetic means and thereby the collective doses seem to have
increased by about a factor of four to six. If the aerated concrete
based on alum-shale had not been used, the country-wide average has
been estimated to be 30 % lower (Swedjemark 1985).
 Here we have only discussed the concentration of the radon gas.
This is because the measurements have been made of this nuclide.
However, the health effects are referred to the short-lived decay
products. The equilibrium factor depends on the ventilation rate and
the particle concentrations.
 The probability that the F-factors for Swedish dwellings
within the intervals given below are 95 % calculated from the t-
distribution based on the measurements of the day-time radon and
radon daughter levels and the air exchange rates (λ_v) in 225 dwellings (Swedjemark 1983) during the 1970s. Night-time measurements
may give a higher F-factor.

Low air exchange rate $0.28 \leq F \leq 0.74$ $\bar{F} = 0.51$
($\lambda_v < 0.30$ h^{-1}

Average air exchange rate $0.21 \leq F \leq 0.66$ $\bar{F} = 0.43$
($0.30 < \lambda_v < 0.60$ h^{-1}

High air exchange rate $0.21 \leq F \leq 0.47$ $\bar{F} = 0.33$
($\lambda_v > 0.60$ h^{-1})

All air exchange rates	$0.19 \leq F \leq 0.69$ $\bar{F} = 0.44$

The ventilation rate has decreased since the 1950s indicating a
higher equilibrium factor and thereby a higher radon daughter increase since the 1950s than the increase of the radon gas concentration. How the particle concentrations have changed is not known.

 Consequencies involving detriment to health are not discussed
in this paper.

Comparison with other countries

In no other countries measurements of radon or radon daughter concentrations have been carried out in the fifties and the sixties.
There are, however, indications that radon concentrations generally
have increased in houses in temperate regions during the last
decades. To save energy the air change rates have been decreased and
for houses in which the major radon source is the building materials
this gives an approximate inverse proportional enhancement of the
radon concentration. Tightening of windows and closing of inflow
ducts reduce the air pressure indoors in comparison with the air
pressure outdoors and in the soil air with an increased suction of
soil air into the house as a result. This increase of the radon
concentration may be very high especially for houses with a bad barrier or without a barrier between house and soil, which is especially common in old houses.
      Energy-efficient houses have been shown to have higher radon
concentrations than others. In the Netherlands, for example, rooms
with double-glazed windows were found to have radon concentrations
twice as high as those in rooms with single-pane windows (Wolfs et
al, 1984). New York energy-efficient houses with heat storage masses
had radon concentrations 1.6 times those for conventional homes
(Fleischer and Turner, 1984) and in Sweden the same enhancement in
radon concentration is found (Nyblom, 1980).

Trends

In Sweden, requirements on future building (Swedjemark, 1986) and
the limits for existing dwellings will result in lower radon
daughter concentrations in homes and to some part it already has.
For the collective dose, however, the decrease will be very slow
because of the low building rate compared with the housing stock. It
has been estimated that the average for the Swedish radon daughter
concentration, 50 Bq/m^3 EER, may be decreased to half that value
after about 100 years, if the requirements mentioned above are
fulfilled.
      In most other countries regulations or recommendations for a
decrease of the radon daughter concentrations in homes have not been
established. In USA and Canada limits have been given only for
special cases, for example building on waste from uranium and
phosphate industries (Atomic Energy Control Board, 1977; EPA, 1979;
EPA, 1980). In Finland, there are general recommendations for homes
(Finnish Radon Commission, 1982).
      A number of organizations are now working with or have submit-
ted recommendations regarding radon daughter concentrations in
homes. Examples are the International Commission on Radiological
Protection (ICRP Publication 39, 1984), the World Health Organization and the Nordic countries (The National Radiation Protection
Authorities in the Nordic Countries, 1986). In several countries
regulations are being discussed. Therefore, the radon daughter concentrations in homes may decrease in the future.

References

Atomic Energy Control Board, Criteria for radioactive clean-up in Canada, Information bulletin 77-2 (1977).

Breslin, A.J. and George, A.C., An Improved Time-Integrating Radon Monitor, Proceedings of the NEA specialist meeting, Paris Nov 1978, pp. 133-137, OECD/NEA (1979).

Burén, A., Håkansson, B., Mjönes, L., Möre, H., Nyblom, L. and Wennberg, P., An integrating radon monitor, Report a 82-06, National Institute of Radiation Protection, Stockholm (1982).

EPA 79 Federal Register, 44: 128, pp. 38665, Recommendations to State of Florida regarding Florida phosphate lands. Environmental Protection Agency, Washington DC, US (July 2, 1979).

EPA 80 Federal Register, 45: 79, pp. 87374-5. Environmental Protection Agency, Washington DC, US (April 22, 1980).

Falk, R., Hagberg, N., Håkansson, B., Nyblom, L. and Swedjemark, G.A., Calibration, Radon and Radon Daughters in Air, Report a 82-22, National Institute of Radiation Protection, Stockholm (1982) (in Swedish).

Finnish Radon Commission, Final Report 1982, Finnish Centre for Radiation and Nuclear Safety, Helsinki (in Finnish).

Fleischer, R.L. and Turner, L.G., Indoor radon measurements in the New York capital district, Health Phys, 46: 5, pp. 999-1011, (1984).

Hultqvist, B., Studies on naturally occurring ionizing radiations with special reference to radiation doses in Swedish houses of various types, Kungl. Svenska Vetenskapsakademins handlingar, Ser 4, Band 6, Nr 3, Stockholm, Sweden (1956).

ICRP Publication 39: Principles for limiting exposure of the public to natural sources of radiation. Annals of the ICRP, 14: 1, Pergamon Press, (1984).

Mjönes, L., Gamma radiation in Dwellings, Report a 81-18, National Institute of Radiation Protection, Stockholm, (1981) (in Swedish).

Mjönes, L., Gamma radiation in Swedish Dwellings (to be published in Radiation Protection Dosimetry) (1986).

Möre, H.: Radionuclides in building materials, a summary of activity concentrations in samples investigated at the National Institute of Radiation Protection, Report 85-08, National Institute of Radiation Protection, Stockholm (1985) (in Swedish).

Nyblom, L., Radon in the air in buildings, where heat accumulating is made in a stone mass, Report Ri 1980-03, Department of Radiation Physics, Karolinska Institute, Stockholm, Sweden (1980) (in Swedish).

Swedjemark, G.A., Radon in dwellings in Sweden, in Proceedings from Natural Radiation Environment III, Texas 1978 (Ed Gesell, T.F. and Lowder, W.M.), CONF-780422, pp. 1237-1259 (1980).

Swedjemark, G.A., The equilibrium factor F, Health Physics, 45: 2, pp. 453-462 (1983).

Swedjemark, G.A. and Mjönes, L.: Exposure of the Swedish population to radon daughters, in Proceedings of the 3rd International Conference on Indoor Air Quality and Climate, Stockholm, 1984, 2, pp. 37-43, Swedish Council for Building Research, Stockholm (1984).

Swedjemark, G.A., Radon and its decay products in housing - estimation of the radon daughter exposure to the Swedish population and methods for evaluation of the uncertainties in annual averages, Thesis, Department of Radiation Physics, University of Stockholm (1985).

Swedjemark, G.A., Limitation schemes to decrease the radon daughters in indoor air, Report 86-01, National Institute of Radiation Protection, Stockholm (1986).

Wilson, C., Mapping the radon risk of our environment, in Proceedings of the 3rd International Conference on Indoor Air Quality and Climate, Stockholm, 1984, 2, pp. 85-89, Swedish Council for Building Research, Stockholm (1984).

Wolfs, F., Hofstede, H., deMeijer, R.J. and Put, L.W., Measurements of radon-daughter concentrations in and around dwellings in the northern part of the Netherlands: a search for the influences of building materials costruction and ventilation. Health Phys, 47: 2, pp. 271-279 (1984).

RECEIVED August 4, 1986

Chapter 8

Indoor Radon Measurements in Finland: A Status Report

O. Castrén, I. Mäkeläinen, K. Winqvist, and A. Voutilainen

Finnish Centre for Radiation and Nuclear Safety, P.O. Box 268, SF-00101 Helsinki, Finland

Large-scale surveys indicate that the mean indoor radon concentration in Finnish dwellings is about 90 Bq/m^3. The percentages of concentrations exceeding 200, 400, 800 and 2,000 Bq/m^3 are 11, 3.9, 1.4 and 0.5 per cent, respectively. An updated version of the geographical distribution is presented. Sampling and data processing methods as well as the reasons for high concentrations are discussed.

Radon in indoor air is the main source of radiation exposure for the population of Finland. During the second half of the seventies, radon-rich tap water was thought to be the biggest source of radon in Finnish dwellings (Castrén et al.,1977, Castrén 1978). However, simultaneously with surveys in many other countries, large-scale surveys in Finland clearly demonstrated that a direct influx from the ground to the dwelling is responsible for the largest proportion of the radiation and for the highest radon levels in Finnish dwellings (Castrén et al.,1984, 1985).

The Finnish Centre for Radiation and Nuclear Safety is continuously conducting indoor radon surveys of dwellings. Already, the results from about 4500 houses are in our data register, and the number will be doubled this year. The present paper gives some new results of analyses of our data, an updated version of the previously published map of geographical distribution, and some of the principles of our monitoring strategy.

Methods

The radon measurements were performed with solid state nuclear track dosimeters. Up to 1984 we used open Kodak LR-115 dosimeters (Mäkeläinen, 1984), and thereafter electrochemically etched Makrofol polycarbonate sealed in a plastic cup (Mäkeläinen, 1986). Normally we use an integrating time of two months to reduce the variations due to changes in the weather. The necessity of a long integrating period is demonstrated by Figure 1, which shows the temporal variation in a typical high-concentration house. The dosimeters are calibrated in a controlled radon atmosphere in our laboratory and the calibration has been checked by international intercomparison. Simultaneous measurements have been made in some of the houses using

0097-6156/87/0331-0097$06.00/0

different techniques and have normally been in accordance with each
other (Figure 1).

The sometimes very large seasonal variation has also been corrected for. At least one measurement is supposed to be made in
winter (November-March). The correction is based on the knowledge
of typical winter-summer concentration ratios as shown in Figure 2.

Sampling is one of the key problems in conducting a nation-wide
survey. We think a truly random sampling procedure would increase
the cost and thus reduce the number of samples too much. The houses
to be measured were thus selected by the person distributing the
dosimeters, who tried to distribute them randomly over a given area.
In most cases the area was a municipality. The municipalities were
selected so as to cover the whole country evenly.

In constructing a distribution representative of the whole
country, we divided the country into zones so that each zone was
as homogeneous as possible with respect to the mean radon concentration of the municipalities. In the first analyses the results
for each zone were simply pooled. Population-weighting was used
when the zonal distributions were combined into a distribution for
the whole country. Later on we also used population-weighting when
combining the municipal distributions. The procedure is analogous
to a stratified random sampling in which the strata are defined on
the basis of the results. This sampling method makes it possible
to improve the accuracy by measuring more houses in zones in which
the concentrations are higher.

Results

Using the procedure described above we have constructed a distribution estimate for the concentrations in detached houses. When the
dwellings in multi-family houses are taken into account, we have
had to base our estimates on measurements in only 142 dwellings
in 15 towns. Thus, our present estimate of the mean indoor radon
concentration in Finland, 90 Bq/m^3, is still subject to revision.

According to the present data, it is estimated that the percentages of dwellings exceeding 200, 400, 800 and 2,000 Bq/m^3 are
11, 3.9, 1.4, and 0.5 per cent, respectively. There may also be
concentrations exceeding 4,000 Bq/m^3 in 0.01 - 0.1 per cent of
the dwellings. The scarcity of data for multi-family houses does
not disturb these estimates too much because most concentrations
higher than 200 Bq/m^3 are found in detached houses.

The geographical distribution is shown on a map which is updated
by adding new localities or by changing the results for previous
ones, if necessary. The latest updated map is shown in Figure 3.
A more detailed study of the geographical distribution in a high-
radon area was made in nine municipalities in the winter 1984-85
(Castrén et al.,1985). A second study, consisting of more than
3,000 houses in 32 municipalities will be completed in 1986.

Reasons for high concentrations

The main parameters determining the indoor radon concentration in
detached houses are the effective radium concentration (product of
the radium concentration and the emanation factor) and the permeability of the ground. The effects of other factors are not so easy
to ascertain from the existing data.

The geographical distribution of the external gamma background
(Figure 4) shows some correlation with the indoor radon measurements.

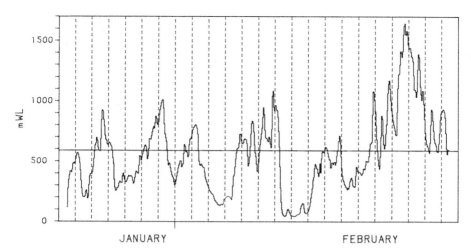

JANUARY   FEBRUARY

Figure 1. Variations in the hourly mean alpha-energy concentration during an integrating radon gas measurement of three weeks. The alpha-energy concentration calculated from the radon level (4860 Bq/m³) and the typical equilibrium factor (0.45) is also given.

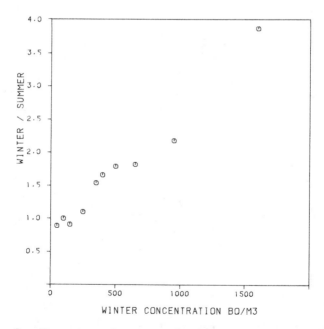

WINTER CONCENTRATION BQ/M3

Figure 2. The measured winter-summer concentration ratios in 250 detached houses (Winqvist, 1984). The houses have been divided into ten classes (at least 20 dwellings in each class) according to winter concentration. Class boundaries are 0, 90, 160, 240, 350, 400, 500, 600, 800, 1200 and 2000 Bq/m³.

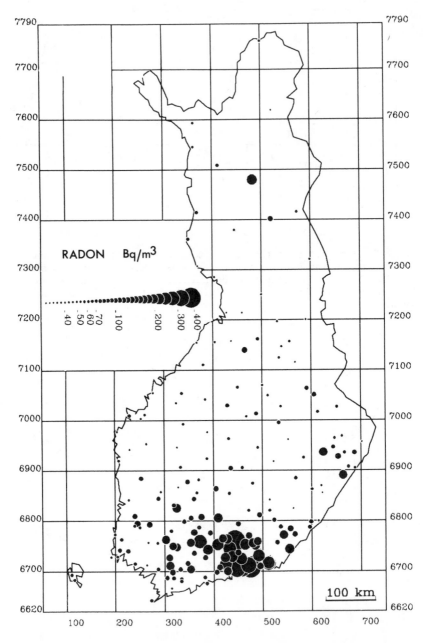

Figure 3. Radon concentration in the indoor air of detached houses. The size of the symbol indicates the magnitude of the local
geometric mean.   Measurements have been made in about 4,450
houses in 183 localities.

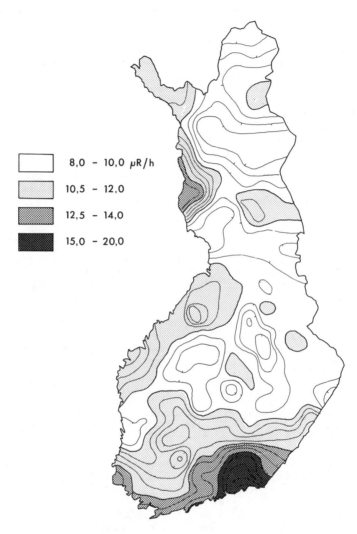

Figure 4.   The distribution of the external gamma dose rate from measurements on Finnish roads (Lemmelä, 1984).

The same is true of the geographical distribution of the
uranium concentration in glacial till, the most common soil type
in Finland (Geological Survey of Finland, 1985). The sometimes
very large effect of ground permeability is the main reason that
the details of the radon concentration distribution differ from
the distribution of other radiation parameters.
There is a very large seasonal variation in the houses with
the highest concentrations. The appreciably higher concentrations
in winter in these houses cannot be explained by lower ventilation
rates. The ventilation rate in many such houses tends to be higher
in winter. Therefore, the soil air influx caused by the stack effect
in winter must be so much larger than it is in the summer that it
overrides the effects of the higher ventilation rates. In general,
the temporal and local concentration variations in this type of
house are very large, and instantaneous values several times higher
than the annual average can be measured. Theoretical models of the
temporal variation of radon concentration in Finnish houses have
also been developed in our laboratory (Arvela and Winqvist, 1986).

Literature Cited

Arvela, H. and K. Winqvist, Influence of source type and air
exhange on variations of indoor radon concentration. Report
STUK-A51, Finnish Centre for Radiation and Nuclear Safety,
Helsinki (1986).

Castrén, O., M. Asikainen, M. Annanmäki and K. Stenstrand,
High natural radioactivity of bored wells as a radiation hy-
gienic problem in Finland, Proc. of the 4th International
Congress of International Radiation Protection Association
pp. 1033-1036, Paris (1977).

Castrén, O., The contribution of bored wells to respiratory
radon daughter exposure in Finland. Proc. of Symposium on
Natural Radiation Environment, (CONF- 780422, vol.2.)
pp. 1364-1370, Houston, Texas (1978).

Castrén, O., K. Winqvist, I. Mäkeläinen and A. Voutilainen,
Radon measurements in Finnish houses. Radiation Protection
Dosimetry 7: 333- 336 (1984).

Castrén, O., A. Voutilainen, K. Winqvist and I. Mäkeläinen,
Studies of high indoor radon areas in Finland, The Science of
the Total Environment, 45: 311-318 (1985).

Geological Survey of Finland, Geochemical Atlas of Finland,
Helsinki (1985).

Lemmelä, H., The survey of external background radiation in
Finland, Report STUK-B-VALO-32, Finnish Centre for Radiation
and Nuclear Safety, Helsinki (1984) (in Finnish).

Mäkeläinen, I., Calibration of bare LR-115 film for radon measurements in dwellings. Radiation Protection Dosimetry 7: 195-197 (1984).

Mäkeläinen, I., Experiences with track etch detectors. Proc. of the 13th International Conference on Solid State Nuclear Track Detectors, Rome (September 1985) (to be published).

Winqvist, K., Seasonal variations of indoor radon concentration in detached houses. Presented at the 7th ordinary meeting of the Nordic Society for Radiation Protection, Copenhagen (October 1984) (in Swedish).

RECEIVED August 4, 1986

Chapter 9

Concentrations in Dwellings in the United Kingdom

K. D. Cliff, A. D. Wrixon, B. M. R. Green, and J. C. H. Miles

National Radiological Protection Board, Chilton, Didcot, Oxfordshire OX11 0RQ, United Kingdom

A survey of the radon concentrations in a representative sample of more than 2000 dwellings in the UK has
been completed and provisional results are now available. On average, concentrations are 29% lower in
bedrooms than in living areas. The mean radon concentration weighted for room occupancy is 22 Bq m^{-3}.
Assuming an equilibrium factor of 0.35 and a mean
occupancy of 75%, the mean annual exposure in UK homes
is assessed as 0.08 Working Level Months (WLM) and the
mean annual effective dose equivalent as 0.43 mSv.
    Special surveys have been made in small areas
where geological conditions indicated high indoor radon
concentrations. These suggest that there is a small
number of dwellings in the UK in which the average
radon concentration may exceed 1250 Bq m^{-3}, corresponding to an annual dose of 25 mSv.

The average annual effective dose equivalent received by a member of
the UK population is currently estimated to be 2150 μSv. Of this
total, 87% arises from exposure to radiation of natural origin, the
largest single contributor being inhalation of the short-lived decay
products of radon. This exposure occurs predominantly in the home.
    A preliminary study of the concentrations of radon
decay-products in UK dwellings demonstrated the wide range of
exposures from this source (Cliff, 1978a). Cosmic ray exposures vary
by about 10% either side of the mean and terrestrial gamma dose rates
by a factor slightly higher than three about the mean. In contrast,
the range of exposures to radon decay-products in dwellings spanned
more than two orders of magnitude. This preliminary study
concentrated on dwellings in the major population centres of the UK
and no attempt was made to ensure that the sample represented the

0097-6156/87/0331-0104$06.00/0
Published 1987 American Chemical Society

entire UK housing stock. To provide more representative data, a national survey was initiated in 1982.

Design of the national survey

The national survey was designed to include about 2000 dwellings from a UK housing stock of about 20 million. Although such a sample is relatively small, it is adequate provided that selection is made properly. A representative sample of about 13,000 addresses was obtained from the postal address files held by British Telecom. This was done by selecting systematically every n^{th} address, where n is the total number of addresses divided by the required sample size. This sample was divided systematically into 26 sub-samples by taking, for the mth sub-sample, addresses numbered m + 26x, where x is an integer from 0 to 496. The sub-samples were selected representatively until the required 2000 or so houses had been obtained. Each sub-sample was exhausted before proceeding to the next (Wrixon et al., 1984).

The preliminary survey had been carried out using mechanical air samplers with gross alpha-particle counting. The national survey was conducted using passive radon gas detectors distributed and returned by post. The detectors consisted of two discs of diethylene glycol bis(allyl carbonate) polymer (CR-39), separated by a spacer and enclosed in a plastic pot, 3 cm in diameter and 3 cm high (Miles and Dew, 1982). The lid of the pot does not provide an air-tight seal, but is sufficiently close fitting to exclude radon decay-products and dust. The response of these detectors is proportional to the average radon concentration over the exposure period. Detectors were issued for exposure in the main living area and in the main bedroom of each dwelling. Exposures were for a period of six months, followed immediately by the exposure of replacement detectors for the next six month period. This permitted seasonal variations in the radon concentration to be indicated. At the completion of the survey period, the occupier of the dwelling was requested to complete a questionnaire from which information about the type of dwelling and the living habits was obtained. The national survey was completed in 1985 and a full analysis of the results is being performed. The results presented here are still therefore provisional, but it is not expected that the final outcome will differ significantly.

Local surveys

The national survey was designed to provide a realistic estimate of the exposure of the UK population to radon decay-products. Thus the estimate is weighted towards the exposure received in cities and towns. Many of these are situated in areas of sedimentary geology, where exposures to radon decay-products are generally low. Some areas of the UK were identified from geological data as likely to have high indoor radon concentrations. These areas have above average concentrations of natural radionuclides in the ground, for example, because of the presence of igneous rocks or uraniferous shale. Detailed local surveys of more than 800 dwellings were undertaken. These localities were generally small areas within larger administrative units of the UK. Thus, the results obtained from the local surveys within a county such as Cornwall are not typical of the housing stock of Cornwall as a whole.

In the local surveys the concentration of radon decay-products in room air and in outside air was determined using a Radon Decay Products Monitor. This utilises gross alpha-counting of activity collected on a filter during and after air sampling (Cliff, 1978b). The concentrations of the thoron decay-products (^{212}Pb and ^{212}Bi) were also measured using this instrument. Measurements were generally confined to one ground floor room with the door closed. One window was opened to its first fixed position to ensure that ventilation was predominantly between the room and the outside. Ventilation rates were measured using nitrous oxide as a tracer gas by monitoring the reduction in concentration over time with an infra-red gas analyser. The radon gas concentration was measured using a scintillation cell system. After surveying each dwelling by active sampling methods, passive radon gas detectors were left behind to be exposed under similar conditions to those of the national survey. Provided ventilation is solely between the room being studied and the outside, the radon source term, K, for the room can be determined (Cliff, 1978a).

$$K = j \; (1 + 0.0748j) \; (C_I - Co) \; Bq \; m^{-3} \; h^{-1}$$

where j is the ventilation rate, h^{-1}
C_I is the concentration of ^{218}Po in room air, $Bq \; m^{-3}$
Co is the concentration of ^{218}Po in outside air, $Bq \; m^{-3}$

Results

For the national survey, a total of 2519 householders agreed to participate, representing a positive response rate of just over 50% Approximately 11% of the initial participants failed to complete a full year of monitoring because of moving house or for other reasons. Results are therefore available for 2240 dwellings. Initial analysis of the distribution of dwellings by type and by age reveals a slight bias in the sample towards detached and semi-detached houses at the expense of terraced houses and dwellings in multi-occupancy buildings when compared with national statistics (OPCS, 1984). The comparative figures are given in Table 1, which also shows a similar bias with respect to age of dwelling. The bias with age is correlated with that according to type, as terraced houses are predominantly older and detached dwellings newer in the UK. These results will be reviewed and any distortions resulting from bias will be corrected for the final assessment.

An equilibrium factor of 0.35, derived from measurements made during the local surveys, has been assumed to typify conditions in UK dwellings. This value has been used to convert the average radon concentrations measured in the national survey to potential alpha-energy concentration of radon decay-products. On average, persons in the UK spend 75% of their time in their homes and 15% of their time elsewhere indoors (Brown, 1983). The occupancy factor of 0.75, together with an equilibrium factor of 0.35, results in an annual exposure of 1.3 10^{-5} J h m^{-3} (0.0037 Working Level Months, WLM) for an average indoor radon concentration of 1 Bq m^{-3}.

Results from the national survey reveal a significant difference between the average radon concentrations in the living area and that in the bedroom. The ratio of the living room concentration to that in the bedroom has a mean value for the UK of 1.4. However, the ratio varies according to area, from 1.2 in London to 1.8 in Aberdeen. On average persons in the UK spend about 34% of their time in the living room and 41% in the bedroom (Brown, 1983). Using these figures and the mean ratio of the living room concentration to the bedroom concentration given above, the average annual exposure can be calculated. Table 2 presents the average annual exposure to radon decay-products in UK dwellings, as determined by the national survey. The overall value for the UK is given, together with a selection of results for cities and counties.

Table 2 includes the capital cities of the four countries comprising the UK: Belfast (Northern Ireland), Cardiff (Wales), Edinburgh (Scotland) and London (England). Because of the small sample size in some areas, generalisations from Table 2 may not be valid.

The average exposures in dwellings in London are lower than the national mean, but those in south-west England (Devon and Cornwall) are appreciably greater than the national mean. The latter case reflects the local uranium mineralisation. Aberdeen, a city in which the majority of buildings are constructed from granite, has long been regarded as the city having the highest radiation levels in the UK. However, the national survey indicates that exposures to radon decay-products in Aberdeen are, if anything, below the national average. The reasons for the comparatively low concentrations of indoor radon in Aberdeen are not entirely clear, but there are some possible causes: low permeability of the granite; sealing of the ground below the underfloor ventilated space; good underfloor ventilation.

The parameters of the frequency distributions of radon decay-product exposure are given in Table 3. This table combines the UK data from the national survey with those from local surveys. For the local surveys, the number of dwellings shown is fewer than the number surveyed actively, because only those that completed the follow-up passive survey have been included.

Figure 1 shows the distribution of average radon concentrations in dwellings throughout the UK averaged over 100 km squares. As seen from the national survey, exposures to radon decay-products in south-west England are, on average, about four times the national mean. The local surveys, concentrating on relatively small areas within the counties of Devon and Cornwall indicate average exposures that are 15 times the national mean. Figure 2 shows, in histogram from, the distribution of average radon concentrations in the living areas of dwellings in the UK, as measured in the national survey and in the local survey of selected parts of Cornwall.

Discussion

In order to compare exposures to radon decay-products with those to other forms of ionising radiation, it is useful to assess the effective dose equivalent expressed in sieverts (Sv). A conversion coefficient of 15 Sv per J h m^{-3}, equivalent to 5.5 mSv per WLM, has been recommended (UNSCEAR, 1982). With this conversion factor, the

Table I. Representativeness of the national sample

By type of dwelling. Sample size 2199*			By age of dwelling. Sample size 1892*		
Type	National %	Sample %	Date of construction	National %	Sample %
Detached house	17	24	pre 1919	25	22
Semi-detached house	31	39	1919-1944	25	24
Terraced house	32	24	1945-1964	26	26
Purpose-built flat	14	9	post 1965	24	28
Converted flats	5	3			
Others	1	1			

* Only those dwellings for which the relevant answers to the questionnaire had been given are included

Table II. Annual average exposure to radon decay-products in UK dwellings from the national survey

Place	Number of dwellings	Annual exposure WLM
UK	2240	0.078
Aberdeen	16	0.059
Belfast	5	0.040
Birmingham	55	0.048
Cardiff	31	0.077
Edinburgh	32	0.053
Glasgow	28	0.042
London	255	0.053
Liverpool	53	0.042
Manchester	39	0.037
Newcastle	56	0.042
Sheffield	47	0.11
Devon & Cornwall	61	0.29

Table III. Annual exposures to the decay-products of radon
in UK dwellings, WLM

Parameter	UK	Cornwall*	Devon*	Uplands**	Scotland***
Dwellings surveyed	2240	250	140	140	150
Arithmetic mean, WLM	0.078	1.2	1.1	0.38	0.20
Geometric mean, WLM	0.049	0.64	0.56	0.17	0.11
Geometric SD	2.2	3.0	3.0	3.4	2.7
95% fractile, WLM	0.2	3.8	2.9	1.1	0.67
99% fractile, WLM	0.3	8.4	6.4	2.7	1.4

* Local surveys of potentially high houses.
** Parts of Yorkshire and Derbyshire.
*** Parts of Dumfries and Galloway, Grampian and Highlands.

average annual effective dose equivalent received at home by a member of the UK public from radon decay-products is 0.43 mSv.

This value of 0.43 mSv for the average annual effective dose equivalent for the UK as a whole is somewhat lower than the currently used value of 0.70 mSv (Roberts and Hughes, 1984). The difference arises partly from taking into account the lower exposure in bedrooms revealed in the national survey. Another reason is that no allowance is made in this survey (Table 2) for exposure received during the 15% of the time spent indoors at other locations nor while in the open. In assessing total population dose it seems reasonable to assume that the exposure rate during the residual 15% of time spent indoors is the same as that received at home.

The average concentration of radon in outdoor air in the UK is 2.6 Bq m^{-3}. Comprehensive data on the equilibrium factor in outdoor air in the UK is not available. Assuming equilibrium, the average exposure to radon decay products received by a member of the UK population during the 10% of time spent in the open is 0.0036 WLM, an annual effective dose equivalent of 0.02 mSv.

On the basis of a conversion coefficient of 5.5 mSv WLM^{-1}, occupants of the vast majority of dwellings in the UK receive annual effective dose equivalents less than 2 mSv. Even in the areas surveyed because of their potential for high radon exposures, the annual effective dose equivalents are unlikely to exceed a few tens of mSv. However, in certain areas of Cornwall and Devon, annual effective dose equivalents higher than 25 mSv may be received in a small percentage of dwellings. In some dwellings more than 50 mSv per year may be received.

It is noted that the ICRP has assumed a higher conversion coefficient between annual effective dose equivalent and radon concentration (ICRP, 1984) in recommending an action level for remedial measures in homes, i.e. 1 mSv y^{-1} per 10 Bq m^{-3} of equilibrium equivalent radon gas concentration (9 mSv per WLM). If this conversion coefficient were applied to our regional survey data, we would estimate, from the distribution parameters given in table 3, that about 15% of the residents of certain areas of Devon and

(Bq m-3)

≣ < 20   ▓ 40 - 60

▓ 20 - 40   ▓ > 60

Figure 1.   Radon concentrations in living areas in Great Britain.

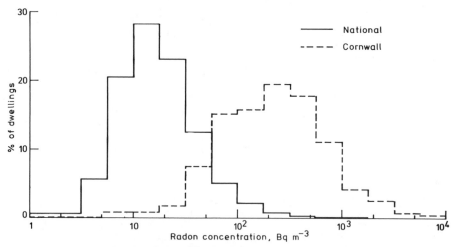

Figure 2.   Frequency distribution of radon concentration in air in
the living areas of some UK dwellings.

Cornwall receive more than 25 mSv effective dose equivalent per year.

The occurrence of high concentrations of radon decay-products in some dwellings has led members of staff of the National Radiological Protection Board to suggest a scheme to limit doses from this source of exposure (O'Riordan et al., 1983). In these recommendations a distinction is made between remedial action in existing dwellings and a design level for new buildings. It was suggested that remedial action should be undertaken in existing dwellings where the annual effective dose equivalent might exceed 25 mSv. New dwellings should be designed to ensure that the annual effective dose equivalent to the occupants should not exceed 10 mSv. These suggestions were broadly endorsed by the standing Royal Commission on Environmental Pollution in its tenth report (RCEP, 1984).

Further consideration is being given by the National Radiological Protection Board and the Department of the Environment to the need for standards in the UK and to the means of implementing them. A decision is likely to be made before the end of this year.

Acknowledgment

This work was partly funded by the Commission of the European Communities, under contract B10-F-498-82-UK.

References

Brown, L., National Radiation Surveys in the UK: Indoor Occupancy Factors, Rad. Prot. Dosim. 5: 321-323 (1983).

Cliff, K.D., Assessment of Airborne Radon Daughter Concentrations in Dwellings in Great Britain, Phys. Med. Biol. 23: 696-711 (1978a).

Cliff, K.D., The Measurement of Low Concentrations of the Daughters of Radon-222, with Emphasis on RaA Assessment, Phys. Med. Biol. 23: 55-65 (1978b).

ICRP; International Commission on Radiological Protection, Principles for Limiting Exposure of the Public to Natural Sources of Radiation, Publication 39, Ann. of ICRP 14: 1-8 (1984).

Miles, J.C.H. and J. Dew, A Passive Radon Gas Detector for use in Homes, in Proceedings 11th International Conference on Solid State Nuclear Track Detectors (P.H. Fowler and V.M. Clapham, eds.) pp 569-579, Pergamon Press (1982).

OPCS; Office of Population Censuses and Surveys, General Household Survey, HMSO, London (1984).

O'Riordan, M.C., A.C. James, S. Rae and A.D. Wrixon, Human Exposure to Radon Decay Products Inside Dwellings in the United Kingdom, NRPB-R152, HMSO, London (1983).

RCEP; Royal Commission on Environmental Pollution, Tackling Pollution – Experience and Prospects, Tenth Report, HMSO, London (1984).

Roberts, G.C. and K.S. Hughes, The Radiation Exposure of the UK Population – 1984 Review, NRPB-R173, HMSO, London (1984).

UNSCEAR; United Nations Scientific Committee on the Effects of Atomic Radiations: Sources and Biological Effects, Report to the General Assembly, Annex D, United Nations, New York (1982).

Wrixon, A.D., L. Brown, K.D. Cliff, C.M.H. Driscoll, B.M.R. Green and J.C.H. Miles, Indoor Radiation Surveys in the UK, Rad. Prot. Dosim. 7: 321–325 (1984).

RECEIVED October 31, 1986

Chapter 10

Population Doses in Ireland

J. P. McLaughlin

Physics Department, University College of Dublin, Belfield, Dublin 4, Ireland

Indoor air radon concentrations measured in a randomly
selected sample of 220 Irish houses have been found to
range from about 20 Bq/m^3 to as high as 1740 Bq/m^3 with
a median value of 61 Bq/m^3. Using current dose estimation
methods the estimated effective dose equivalents due to
radon daughter inhalation in these houses are 1.6 mSv/year
(median value) and 46 mSv/year (maximum value).
Integrating alpha track based passive detectors, which
yield both a measurement of the mean radon concentration
and of the radon daughter equilibrium (F) in each house,
are being used in this national survey.

As part of the current Radiation Protection Research Programme of
the Commission of the European Communities (C.E.C) a national
survey of indoor radiation exposure in Ireland arising from natural
radiation is being carried out. Most of the effort in this survey
is being directed towards an assessment of the doses to the general
population arising from radon and its daughters. In this paper
unless otherwise stated the term"dose" is taken to refer to effective dose equivalent. The emphasis on radon has been adopted
because a pilot survey of 278 houses, in various parts of Ireland,
carried out during 1983-85 (McAulay and McLaughlin, 1980) revealed
that indoor doses from natural radiation were dominated by those
arising from radon daughter inhalation. In the pilot survey the
estimated median value of the effective dose equivalent from radon
daughters was about 1 mSv/y while that from penetrating radiation
(gamma plus cosmic) was about 0.6 mSv/y. The doses from penetrating radiation covered a relatively short range about the median
value from approximately 0.3 mSv/y to 1 mSv/y. Those from radon
daughters were found to cover a large range having a log normal
distribution with maximum value of 31.7 mSv/y. The corresponding
indoor radon concentrations ranged from 3 Bq/m^3 to a maximum value
of about 1200 Bq/m^3. Approximately 10% of the dwellings had a
radon concentration in excess of 100 Bq/m^3 and about 4% had greater
than 200 Bq/m^3. It thus appears in Ireland that while mean values
of doses from radon daughters are likely to be low a small number

of individual households in some areas may be receiving significantly high doses. The results of the pilot survey also indicated
that indoor radon concentrations in Ireland are correlated with
the geological nature of the soil subjacent to a dwelling with
construction characteristics playing a subsidiary role. On the
basis of the findings of the pilot survey it was decided to carry
out a nationwide survey of indoor radon in a random manner.

The principal specific objectives of this study are:-

(i)   ·to obtain a representative distribution of indoor radon
       concentrations in Ireland.
(ii)   to ascertain the extent to which the radon concentrations
       correlate with various parameters such as geological charac-
       teristics, construction type, household energy usage etc.
(iii) to determine the magnitude of individual exposures and the
       likely number of households in which exposures might be so
       high as to justify the development of policies aimed at
       exposure limitation or reduction.

Survey Structure

In the 1983-85 pilot survey about half of the 278 houses investigated were chosen in selected areas of the country where uraniferous
deposits were known to exist. Within the selected areas the houses
were chosen at random. Because of the use of selected areas the
indoor radon concentration data suite obtained in the pilot survey
cannot be considered as representative of the country as a whole.
In the new national survey, in which it is planned to investigate
indoor radon concentrations in approximately 2000 households, the
households are being selected in a random manner and are thus
expected to yield data which is representative of the national
housing stock. The selection is being made in collaboration with
An Foras Forbartha (AFF), the National Institute for Physical
Planning and Construction Research, who are carrying out a C.E.C.
supported study of energy usage in Irish dwellings. Access to the
professional competence of the AFF in the housing construction
field and also to its housing energy usage study will be of
considerable advantage to the analysis phase of the indoor radon
survey.
       The selection of households is being made as follows:- A list
of 3,000 names and addresses of voters has been chosen randomly by
computer from the national electoral register. This selection is
considered to be reasonably representative of the population
distribution in Ireland both on a geographical and on a social
category basis. Each selected voter is visited by an interviewer
from the AFF. Those agreeing to participate in the AFF house
energy study are then invited to participate in the indoor radon
survey. By the end of the 1985 over 1500 households had agreed to
participate in the radon survey. The magnitude of this figure
should be considered in the context of the population size which
is presently about 3.6 million for the Republic of Ireland.
Passive radon detectors for the participating households are both
dispatched and recovered by post. Each household is sent one

detector together with instructions regarding placing it in the house. Participants are requested to place the detector in the principal bedroom or a living area of the home.

In this survey it is considered that an exposure time of about six months is the optimum for a detector to be placed in a house. It is a sufficiently long time to obtain a "representative" measure of the indoor radon concentration in a dwelling. The small number of dwellings found to have a mean radon concentration greater than about 200 Bq/m^3 are sent a second detector for a further six month period thereby obtaining a measure of the annualised mean radon level in such houses.

Each household is requested to complete a detailed questionnaire. This is designed to obtain information regarding the house and its occupants which is considered as relevant to an assessment of the radiological impact of radon and its daughters on the household. The questions range from those on the age distribution and smoking habits of the household to building construction type, ventilation and energy conservation practices used. It is made clear to each participating household that all such information obtained and the measured value of the radon concentration will be treated as strictly confidential and will be used only for statistical purposes.

Measurement Techniques

The radon detectors used in both the pilot and the main survey are of the passive time integrating alpha track plastic variety. In the pilot study each detector consisted of a single piece of CR-39 alpha track plastic mounted inside a closed small plastic container or cup. Alpha track visualisation and counting was by means of chemical etching and optical microscopy. This CR-39 based type of passive radon detector has the advantages of simplicity, reliability and excellent alpha particle registration characteristics. In common with most other similar passive detectors presently available however, it only yields information on the mean radon concentration during the exposure period. No information is obtainable from these type of detectors on the degree of equilibrium existing between radon and its short lived daughters in the indoor air of a house. An accurate value of the degree of equilibrium or F factor of the daughters is however of importance in estimating the dose to members of a household where only the mean concentration of radon is known. Here the equilibrium factor (F) is considered in respect to potential alpha energy and is defined as the ratio of the equilibrium equivalent radon concentration to the actual concentration of radon in the air. In the national surveys of indoor radon being presently carried out in the European Community countries an assumed value of F between 0.35 and 0.5 is being used for dose estimation purposes. Some recent measurements in Germany (Keller and Muth, 1985) have shown, for the houses investigated, that the F factor ranged from 0.3 to 0.35. It should however be noted that most F factor measurements given in the literature are based on short time integrating instrumental measurements in a small number of houses.

For houses in a survey found to have low radon concentrations
(say less than 100 Bq/m^3) a lack of precise knowledge of the F
factor may not be important as the values of the estimated doses
will be towards the lower end of the dose distribution for any
value of F. From a dose assessment perspective an important problem
may arise due to a lack of knowledge of the F factor for houses with
a high radon concentration (several hundred Bq/m^3 and upwards). It
is recognised that both the radon concentration and F factor values
in a house are complex functions of the radon source, building
characteristics, ventilation regimes, aerosol size distributions
etc. (Pörstendorfer, 1984). It is however reasonable to expect for
similar houses in a given area that a general tendency may exist
for both the highest radon concentrations and highest F values to
be co-associated with low ventilation rates. The use of an
assumed mean F factor, of say 0.4, for such houses where the actual
F factor may be as high as 0.6 or greater may severely underestim -
ate the doses being received by the household. It is therefore of
some importance in population dose assessment that passive time
integrating detectors should be developed capable of measuring
simultaneously both the radon concentration and the radon daughter
F factor (i.e. some form of passive "Working Level" meter is required). Investigations both experimental and theoretical, have been
made in Poland (Domanski et al., 1982) in this direction using alpha
track plastic detectors. The general conclusion from these studies
was that the particular alpha track devices used were relatively
insensitive to changes in the F factor. It has been recently
reported (Urban and Piesch, 1981; Put and Meijer, 1985) that with
modifications in the design and etching procedures of the standard
Karlsruhe passive radon dosemeter it is possible to measure simultaneously radon and radon daughter concentrations. This approach
has been used in a radon survey in the Netherlands and appears
promising but will require extensive field testing and calibration
so that its performance may be properly assessed.

Some recent developments in radon daughter dosimetry (Vanmarcke
et al., 1986; James, 1986) suggest that an approximate inverse
variation of the F factor and the unattached fraction in indoor air
leads to a relatively constant lung dose per unit concentration of
radon gas. These developments raise the question of the necessity
of making F factor measurements in addition to radon measurements
in a dwelling. To abandon F measurements, if they can be conveniently and accurately made, purely on the basis of a currently
popular lung dose model is not however advisable. Dose models apart
the value of F for a dwelling may be considered as a useful and
informative characteristic property of dwelling air and aerosol
dynamics. It is related to the interplay of the various radon
daughter production and removal mechanisms present. Where remedial
or preventative techniques are used in a dwelling the changes in F
value that occur can be used in conjunction with radon concentration
changes as a useful characteristic in assessing the effectiveness
of the techniques used.

In order to address the F factor problem in Ireland a "new"
passive detector is being used in the national radon survey. It
consists of a small closed cylindrical container or cup, approximately 7cm in height and 5.5 cm in diameter, fitted with two

pieces of LR-115 alpha track plastic. One piece of LR-115 is mounted inside the closed cup and the other is mounted on the outside of the lid. This detector is thus conventional in terms of construction but it is in its calibration and the method of alpha track analysis used that a new approach has been adopted. Following exposure to radon and its airborne daughters in a house the inner piece of LR-115 yields a track density which is proportional to the radon exposure and thus the radon concentration may be determined. The outer piece of LR-115 yields a track density which has a component arising from the radon activity in the air and also has a component arising from the alpha emitting airborne radon daughter activities in the air. A method of track density analysis has been developed, based on laboratory calibrations, by which these two components can be separated. In this way the detector yields both the mean radon concentration (using the inner LR-115 track density) and the mean F factor (using both the inner and outer LR-115 track densities). A full description of this method will be published in due course.

The plastic LR-115 was chosen as the alpha particle detecting medium, instead of the more sensitive CR-39 used in the pilot survey, to avoid obtaining a track density component on the outer piece of plastic due to plateout of radon daughters or other alpha emitters. Preliminary results from the first phase of the national survey using this new detector approach are encouraging and, as anticipated, high F values are being found to be associated with some of the highest indoor radon concentrations. It is recognised that this F factor determination method requires refinement and interference from other alpha emitters may need to be eliminated in some situations. On the other hand the current practice in Ireland as in other countries of using a single assumed F factor value in a survey is not the most satisfactory.

The mean radon concentrations determined by the passive detectors are based on calibrations using NBS standard radium-226 solutions and also from participation in the OECD (Nuclear Energy Agency)/CEC radon dosimeter intercomparisons (Commission of the European Communities, 1986) held at the U.K. National Radiological Protection Board (NRPB). Recent calibrations of the new LR-115 based detectors, in terms of response to radon concentrations and F factors have been carried out at the NRPB, which assistance is greatly appreciated. For the etching and track counting procedures used the inner LR-115 piece in the detectors has a mean sensitivity of approximately 1.6 tracks cm^{-2} kBq^{-1} m^3 hr^{-1}.

Exposure to Dose Conversion

In the radon surveys the primary quantity determined is the indoor air mean radon activity concentration. From a radiological health perspective it is the dose arising from the inhalation of radon daughters that is of interest. The conversion from radon exposure to annualised effective dose equivalent for the survey was carried out using the factors given in Table I which are similar to those being used in other European surveys. The occupancy and equilibrium factors given in this table are assumed mean values for Irish conditions and attempts are currently being made to obtain more representative values of these factors. The occupancy factor used here should be considered as an effective value, adjusted for all members of a household. Ideally it should take into account not only the fraction of time exposure to radon in a particular house but also the time fraction spent in other buildings (work place, school etc.) and outdoors exposed to mean radon concentrations appropriate to these locations.

Table I: Factors used in Radon Daughter Dose Estimations

Occupancy Factor: 0.85

Equilibrium Factor (F):   0.45

Dose Conversion Factor =   5 mSv/WLM

(1 WLM = 1 Working Level Month = 3.5×10^{-3} J.h.m^{-3})

The dose conversion factor of 5 mSv/WLM being used is that considered by a number of expert groups (OECD, 1983; UNSCEAR,1982) as appropriate for members of the public exposed to airborne radon daughters. It is however interesting to note that the International Commission on Radiological Protection (ICRP, 1984) seems to consider a dose factor as high as 10 mSv/WLM as appropriate to members of the public. Recent studies in lung dosimetry (Cohen et al., 1985) taking into account the observed enhanced deposition of inhaled radon daughters at lung airway bifurcations indicate that lung doses could be 20% greater than currently estimated. In addition doses to the upper respiratory tract (URT) from inhaled radon daughters are now beginning to receive close study. While it is difficult to predict the likely change in the dose conversion factor when URT deposition is taken into account it is clear that an increase in the conversion factor will occur.

Survey Results

The national indoor air radon survey commenced in the autumn of 1985. In this first phase of the work radon detectors were sent to 400 randomly selected households throughout Ireland. A further 400 households are scheduled to receive detectors by the end of April 1986. At the time of writing radon detectors from a total of 220 households have been recovered and processed. The principal results obtained are shown in Table II, together with the results of the earlier pilot survey and the cumulative results for both sets of data.

        The results for the first phase of the national survey are also presented in histogram form in Figures 1 and 2 together with the annualised effective dose equivalents estimated using the factors given in Table I. It is evident from the data that in the majority of households surveyed the radon concentrations and associated doses are low. In a small percentage of cases however individual households have been found with very high radon

exposure levels for which the individual risk may be higher than
what should be accepted by the individual or society.

Table II Radon concentrations (Bq/m^3) in Irish Houses

SURVEY Radon (Bq/m^3)	National (1985–) (220 Houses)	Pilot 1983–85 (278 Houses)	Combined results (498 Houses)
Median Value	61	37	53
Min Value	17	3	3
Max. Value	1740	1189	1740
95% Fractile	296	200	222

Some of the 220 detectors recently recovered have been
analysed not only for radon exposure but also to determine the
value of F (the equilibrium factor) in the houses. A preliminary
set of such F factor results, obtained by analysing the inner and
outer LR- 115 track densities of each detector, are presented in
Table III for 12 houses with mean indoor radon concentrations
greater than 200 Bq/m^3. In Table III are also presented radon
daughter doses estimated using the individually determined
equilibrium factor values F_i together with the doses estimated
on the basis of an assumed mean F factor value of 0.45.

Table III: F Factor Values for 12 Houses

Radon Conc. (Bq/m^3)	Equil Factor F_i	Dose (mSv/y) Using F_i	Dose (mSv/y) Using F = 0.45
203	0.36	4.4	5.5
217	0.36	4.7	5.9
244	0.37	5.5	6.6
266	0.42	6.8	7.2
414	0.43	10.7	11.2
522	0.29	9.2	14.1
576	0.82	28.4	15.6
607	0.27	9.8	16.5
750	0.28	12.4	20.3
815	0.39	18.9	22.0
1487	0.76	67.5	40.2
1740	0.61	63.9	47.0

It is of interest to note that the houses with the two
highest radon concentrations have determined F factor values
substantially above the assumed mean value of F = 0.45 for Irish

RADON CONCENTRATION Bq/m³

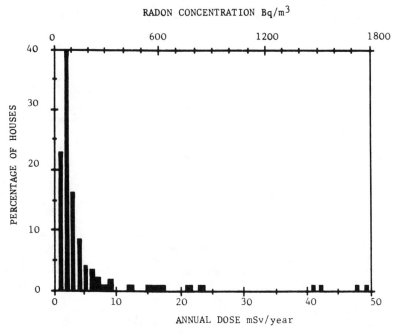

Figure 1.   Results of Phase I National Survey (Full Range)

RADON CONCENTRATION Bq/m³

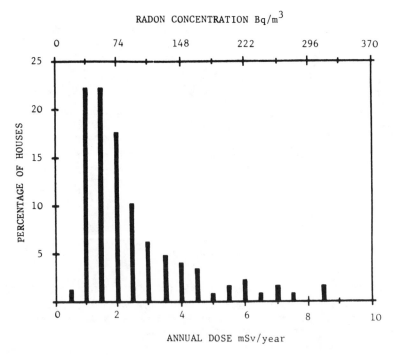

Figure 2.   Results of Phase I National Survey (0 → 10mSv/y)

houses. These high F values result in increased values of the estimated doses. The mean F factor value determined for the 12 houses equals 0.45 which is equal to the assumed values used in the survey. This is probably fortuitous. In dealing with such a small sample it is difficult to draw any general conclusions except to reiterate that where possible for a house the mean value of the F factor should be determined simultaneously with the mean radon concentration so that a dose estimate representative of each particular household may be obtained. The method now being used to determine the F factor in the Irish survey using a passive detector is being subjected to increased laboratory testing so that its reliability and accuracy may be improved.

Conclusions

A general picture of radon exposure in contemporary Irish houses is emerging from the survey date obtained to date. For a typical house the indoor air radon concentration is likely to be between 25 and 70 Bq/m^3. It is also evident that a small number of houses exist with radon levels substantially above the mean value range. Both in the pilot survey and in the ongoing present survey it appears that the principal factor governing high radon levels in Irish houses is the radon exhalation characteristics of the soil or geological structure subjacent to a house with construction characteristics playing a minor role.

It is recognised that substantial uncertainties still exist in our knowledge of the microdosimetry of inhaled radon daughters not least because significant inter – as well as intra – subject variabilities of the human lung exist (Hofmann, 1982). Notwithstanding this it is becoming increasingly obvious, from the present study and studies elsewhere, that individual members of the public are receiving continuous doses of some tens of mSv per year from radon daughters in their homes. This is a matter of substantial radiological health importance which must be addressed both from the regulatory policy stance and also from the technical remedial and preventative viewpoints. In common with other member states of the European Community (Sinnaeve et al., 1986) this problem of population radiation exposure is being seriously considered by the competent authorities in Ireland but as yet no firm policy has emerged. The findings and eventual analysis of the national radon survey described here, when it is completed, will be an essential input to the policy making process.

Acknowledgments

The radon survey of Ireland is being jointly funded by the Commission of the European Communities under Contract BI6.0117.IRL. The assistance given by the NRPB (UK), An Foras Forbartha (AFF) and M.Burke (UCD) is greatly appreciated.

Literature Cited

Cohen, B.S., Harley, N.H., Schlesinger, R.B. and Lippman, M., Nonuniform Particle Deposition on Tracheobronchial Airways: Implications for Lung Dosimetry. Presented at Second International Workshop on Lung Dosimetry, Cambridge, England, (September 1985).

Commission of the European Communities., Results of the Second CEC Intercomparison of Active and Passive Dosemeters for the Measurement of Radon and Radon Decay Products, EUR Report 10403 EN (1986).

Domanski, T., Chruscielewski, W., Swiatnicki, G., The Performance of passive differentiating track detectors containing a diffusion barrier, Rad. Prot. Dos. 2:27-32 (1982).

Hofmann, W., Cellular Lung Dosimetry for Inhaled Radon Decay Products as a Base for Radiation - Induced Lung Cancer Risk Assessment. Radiat. Environ. Biophys. 20:95-112 (1982).

International Commission of Radiological Protection, Principles of Limiting Exposure of the Public to Natural Sources of Radiation. ICRP Publication 39 (1984).

James, A.C., A Reconsideration Cells at Risk and other Key Factors in Radon Daughter Dosimetry. Presented at ACS Symposium on Radon and its Decay Products, New York (April 1986).

Keller, G. and Muth, H., Radiation Exposure in German Dwellings, some results and proposed formula for dose limitation. Sci. Total. Environ. 45:299-306 (1985).

McAulay, I.R. and McLaughlin, J.P., Indoor Natural Radiation Levels in Ireland. Sci. Total. Environ. 45:319-325 (1985).

OECD Nuclear Energy Agency., Dosimetry Aspects of Exposure to Radon and Thoron Daughter Products. Expert Report, Paris September (1983).

Pörstendorfer, J., Behavious of Radon Daughter Products in Indoor Air. Rad. Prot. Dos. 1: 107-113, (1984).

Put, L.W. and de Meijer, R.J., Measurements of Time-averaged Radon Daughter Concentrations with Passive Dosemeters. Sci. Total. Environ. 45:389-395 (1985).

Sinnaeve, J., Olast, M. and McLaughlin, J., Natural Radiation Exposure Research in the Member States of the European Community: State of the Art and Perspectives. Presented at APCA Speciality Conference on Indoor Radon Philadelphia, U.S.A. (Feb. 1986).

UNSCEAR (United Nations Scientific Committee on the Effects of Atomic Radiation)., Sources and Effects of Ionizing Radiation. Report to UN General Assembly, New York (1982).

Urban, M. and Piesch, E., Low Level Environmental Radon Dosimetry with a Passive Track Etch Detector Device, Rad. Prot. Dos. 1:97-109 (1981).

Vanmarcke, H., Janssens, A., Raes, F., Poffijn, A., Berkvens, P. and Van Dingenen, R., On the Behaviour of Radon Daughters in the Domestic Environment and its Effects on the Effective Dose Equivalent. Presented at ACS Symposium on Radon and its Decay Products, New York (April 1986).

RECEIVED October 9, 1986

Chapter 11

Long-Term Measurements of Radon Concentrations in the Living Environments in Japan

A Preliminary Report

T. Aoyama[1], H. Yonehara[1], M. Sakanoue[2], S. Kobayashi[3], T. Iwasaki[3], M. Mifune[4], E. P. Radford[5], and H. Kato[5]

[1]Department of Experimental Radiology, Shiga University of Medical Science, Otsu 520-21, Japan
[2]Low-Level Radioactivity Laboratory, Kanazawa University, Wake, Tatsunokuchi, Ishikawa 923-12, Japan
[3]National Institute for Radiological Sciences, Chiba 260, Japan
[4]Misasa Branch Hospital, Okayama University School of Medicine, Misasa, Tottori 682-02, Japan
[5]Department of Epidemiological Radiation Effects Research Foundation, Hiroshima 730, Japan

Measurements of indoor radon (Rn-222) concentrations were carried out using bare track detectors (CR-39) in Mihama, Misasa, Hiroshima and Nagasaki for one year and using Terradex SF detectors in Hokkaido for a winter. The highest median value, 31.2 Bq/m^3, was from the Mihama survey and the lowest, 10.2 Bq/m^3 was obtained from Nagasaki. The median value for Misasa (a radioactive spa area) was 24.2 Bq/m^3 and that for Hiroshima was 23.10 Bq/m^3. The values for Japanese traditional wooden houses were unexpectedly higher than these for ferro-concrete and prefarbricated houses. The difference between the first and upper floors was modest. The median value obtained from personal monitoring of 25 people living in Misasa was 23.0 Bq/m^3, very close to the value found in dwellings in this area.

Radon (Rn-222) daughter exposure at home is a potentially significant contributor to background lung cancer rates. In western countries, an increase in the number of energy-efficient homes threatens to increase the exposure of the occupants to radon and its decay products by reducing ventilation rates. This is also true in Japan, especially in winter. Japan is situated in the temperate zone, which results in a rather long hot and humid summer climate. Consequently,

0097-6156/87/0331-0124$06.00/0

in relation to their size, Japanese houses have large window areas to increase the ventilation rate. In addition, there remain many traditional wooden homes, including a clay layer in the walls and roof. Up to the present, there has been no systematic study of radon or radon daughters in Japanese homes. We have, therefore, measured the distribution of radon concentrations in houses to determine the factors that influence them in Japan. For this end, long-term measurements of radon concentrations in the home environment were undertaken as a joint-study of five reserch institutions. These preliminary observations are for 5 areas of Japan for which results are available.

Methods

The plan was to measure radon concentrations in dwellings in several cities of Honshu, including Mihama, Misasa spa, Hiroshima and in Kyushu, Nagasaki, as shown in Figure 1. A few measurements in Hokkaido have also been made. Single family houses and apartment buildings were measured. In Hiroshima and Nagasaki, about 100 dwellings were included, with particular reference to housing lived in during 1950 to 1980 by members of the Life Span Study population of the Radiation Effects Research Foundation (R.E.R.F.) in order to provide a basis for estimating the contribution of radon daughters to lung cancer risk among A-bomb survivors during the post World War II period. The survey of Hiroshima and Nagasaki (Radford et al., 1985) was carried out by R.E.R.F., Shiga Univaersity of Medical Science and Kanazawa University; the principal investigator of this survey is Prof. Edward P. Radford of R.E.R.F.. The study on the dwellings in Hokkaido was conducted by Dr. S. Kobayashi of National Institute of Radiological Sciences (N.I.R.S.). In the surveyes of Mihama and Misasa, Prof. M. Sakanoue of Kanazawa University and Dr. M. Mifune of Okayama University led the initiative, respectively.

Measurements were made using two types of passive track-etch alpha dosimeters. One of them was the bare detector of CR-39. After exposure these dosimeters were etched by 30 % NaOH at 70°C for 5 hours. The number of pits was scored under a microscope with a television camera in Shiga University of Medical Science. Methods of calibration and adjustment for deposition of radon daughters introduced by Yonehara (Yonehara et al., 1986) were adopted. The second detectors were Terradex type SF (Alter and Price, 1972). These detectors consist of a plastic cup, covered by a filter to allow entry only of gases, with a track-etch detector inside. The reading of results was carried out by Terradex Corp. in Walnut Creek, California, U.S.A.. The measurements of radon concentration were carried out by both methods in each location, except for Hokkaido where the measurements were done only by Terradex. However, the data obtained by CR-39 detectors will be mainly presented in this paper, because the two methods did not give identical results as separately reported in this proceedings by Yonehara et al. (Yonehara et al., 1986).

Detectors were usually attached to the wall of the living room of the house or apartment in an inconspicuous location about 180 cm from the floor and left for ten months in Hiroshima and Nagasaki, but collected every four months in Misasa and Mihama. Measurements were made in living rooms on the first (ground) floor, and in bedrooms on higher floors.

Figure 1. Map of Japan. The places where radon concentrations were measured are indicated on the map. The measurements were not carried out in Otsu.

Homes that were measured had three categories of construction: traditional wooden house which schematic structure is shown in Figure 2, ferro-concrete and prefabricted types. Wall construction was grouped into traditional clay, wooden or printed plywood, and concrete. Ceiling and wall materials were grouped into plasterboad and non-plasterboard. Comparisons were also made between measurements in the first (ground) floor and upper floors. Misasa spa is one of the most highly radioactive hot springs in Japan (Furuno, 1982a, 1982b, 1982c), therefore, the dwellings in Misasa were classified into houses having their own hot-spring bath, houses located close to public hot-spring baths and others. We measured also radon concentration at special places such as a bathroom of a hot spring (Misasa), a thermal grotto (Misasa), a drinking hall (Misasa), a cave (Mihama), and a cabin of a boat (Mihama). In Misasa, we also carried out personal monitoring with CR-39 detectors worn on the clothing on a continuing basis through a year.

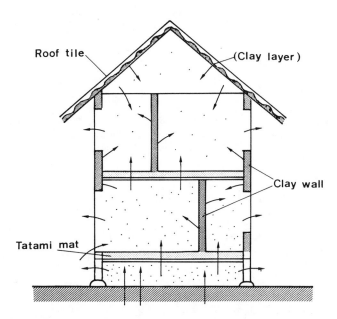

Figure 2. Schematic structure of Japanese traditional wooden house. Small dots schematically indicate radon concentrations. Arrows indicate influxes and effluxes of radon to and from the house.

Results and Discussion

Frequency distributions of indoor radon concentrations measured in
Nagasaki, Hiroshima, Misasa and Mihama are shown in Figure 3. The
data for these have approximately log-normal distributions.
Nagasaki's data show the lowest median values and Mihama's the
highest; the results from Hiroshima and Misasa surveys were
intermediate. The arithmetic means and standard deviation, geometric
means, medians and ranges of radon measurements are summarized in
Table I. The medians of the distributions for Nagasaki, Hiroshima,
Misasa and Mihama are 10.2, 23.1, 24.2 and 31.2 Bq/m^3, respectively.
These differences are possibly due to geology, construction material
and ventilation rate. The Hiroshima, Misasa and Mihama areas are
geologically of granitic rocks but the Nagasaki area is characterized
by extrusive rocks of igneous origin. This geological difference may
explain higher radon concentrations in Hiroshima, Misasa and Mihama
compared to Nagasaki. We could expect that granite would produce
more radon emanation than other ground substances. The Chyugoku
mountains are located between Hiroshima and Misasa, and the one of
the main uranium mines, Ningyotoge, in Japan is in this mountain
range, part of the divide between districts along the coasts of the
Pacific Ocean and the Japan Sea.
    The median radon value for Hokkaido is higher than those for the
other four places, probably due to the difference in the measurement

Table I. Numbers, Arithmetic Means and S.D.s, Geometric Means,
          Medians and Ranges of Rn Measurements for Dwellings in
          Different areas

	N	Mean Bq/m^3			Median Bq/m^3	Range Bq/m^3
		Arithmetic	S.D.	Geometric		
Total[*]	251	31.4 ±	40.5	18.8	17.7	0.116-289
Hiroshima	100	43.1 ±	54.0	26.2	23.1	4.59 -289
Nagasaki	93	13.6 ±	11.9	9.9	10.2	0.116- 75.6
Mihama	27	43.0 ±	32.8	32.4	31.2	8.33 -116
Misasa	31	37.2 ±	32.4	27.7	24.2	6.71 -149
Hokkaido	7	67.0 ±	71.9	40.8	35.5	11.10 -232

*: "Total" does not contain the data from Hokkaido

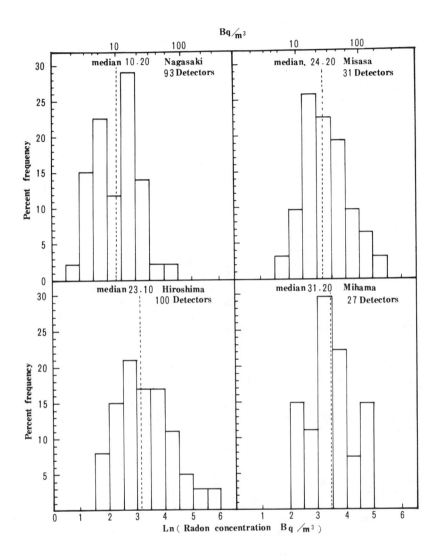

Figure 3.   Frequency distributions of measured indoor radon concentrations for Nagasaki, Hiroshima, Misasa and Mihama.

method. The survey in Hokkaido was carried out only in winter. We
also found that the average value by Terradex SF was about 2 times
higher than that by CR-39 in comparative studies in the Nagasaki,
Hiroshima, Misasa and Mihama areas. The difference in calibration
procedure for the two methods may account for these high values.
     The median of the distributions for the total of Nagasaki,
Hiroshima, Misasa and Mihama data is 17.7 Bq/m^3. This is the same as
that for central Canada, higher than found in Poland, England and
Austria, slightly lower than that for the Netherlands, and one-half
or one-third of those for West Germany, Switzerland, Finland and
Sweden (Urban et al., 1985).
     We analysed the factors which influence the radon concentration
in the dwellings. First of all, differences by house construction
are shown in Figure 4. The results were suprising because the traditional wooden house structure unexpectedly gave higher radon concentrations than the ferro-concrete and prefabricated houses.
     As shown in Figure 2, Japanese traditional wooden houses have
clay walls and a clay layer under roof tiles. If this clay contains
radium and uranium it would emanate some additional radon. Moreover
the floor of traditional Japanese houses usually is made by tatami
mats overlying loose wooden flooring, thus it is easy for radon to
come into the house from the ground. On the other hand, Japanese
houses have large windows to increase ventilation rate in summer, and
also the sub-floor ventilation rate is usually high because of open
construction around the sub-floor space. For both these reasons the
traditional houses might have reduced concentrations of radon, but
the observations indicate that the effect of construction on radon
influx may be more important. Recently these traditional wooden
house have windows with metal sash, which may have less ventilation.
It is interesting that German traditional houses also gave higher
radon concentration than houses of other types of construction (Urban
et al., 1985). In Norway, radon concentrations in wooden houses are
also higher than the values found inside brick buildings, although
those in concrete houses are highest (Stranden et al., 1979).
     When we compare the radon concentrations for the dwellings with
clay, concrete wall construction and wooden or printed plywood, the
values for clay walls are highest, as shown in Figure 5. The medians
of the distributions for clay, concrete wall structure and wooden or
printed plywood are 27.8, 18.0 and 11.2 Bq/m^3 , respectively.
     Comparison between the distribution of radon concentrations for
dwellings with plasterboard and those without plasterboard in ceiling
or wall construction is shown in Figure 6. The values of medians for
dwellings without plasterboard are higher than those with
plasterboard. The distribution for plasterboard can be divided into
two groups, but the numbers are small.
     The values of the medians of the distributions in Table II indi-
cate only a modest difference between first (ground) floors and upper
floors, with the radon levels in the latter lower by about one-third.
     Table III shows seasonal differences of mean radon concentra-
tions in the Mihama and Misasa areas. Mihama had higher mean values
in winter and summer than Misasa had, but Misasa had a higher mean
value in spring and autumn. Widespread use of air conditioners in
Mihama area may account for the high values in summer. In general
there was little seasonal variation in these two locations.

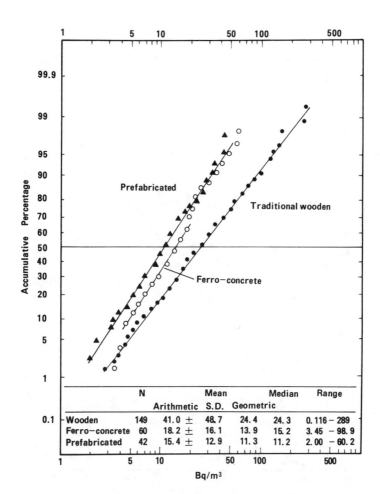

Figure 4. Cumulative frequency distributions of radon concentration for dwellings with different house constructions; traditional wooden, ferro-concrete and prefabricated. Numbers, arithmetic means and S.D.s, geometric means, medians, and ranges of radon measurements are also indicated at the bottom of the figure.

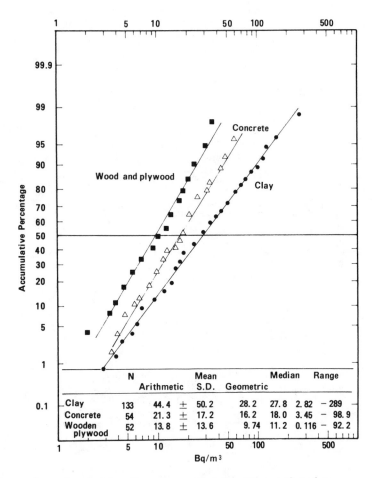

Figure 5.   Cumulative frequency distributions of radon concentration for dwellings with different wall constructions; clay, wooden and plywood, and concrete.   Numbers, arithmetic means and S.D.s, geometric means, medians, and ranges of radon measurements are also indicated at the bottom of the figure.

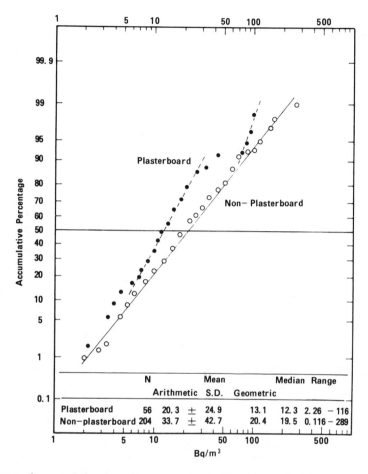

Figure 6. Cumulative frequency distributions of radon concentration for dwellings with different ceiling and wall materials; plasterboard and non-plasterboard. Numbers, arithmetic means and S.D.s, geometric means, medians and ranges of radon measurements are also indicated at the bottom of the figure.

Table II. Numbers, Arithmetic means and S.D.s, Geometric means,
Medians and Ranges of Rn Measurements for First Floor and
Second Floor or above

	N	Mean Bq/m^3			Median Bq/m^3	Range Bq/m^3
		Arithmetic	S.D.	Geometric		
First floor	205	33.6	$\pm$ 43.4	20.0	18.6	0.116-289
Second floor or above	53	18.8	$\pm$ 17.8	13.3	12.4	2.82 - 84.1

Table III. Seasonal Difference of Mean Radon Concentration
(Arithmetic) in Mihama and Misasa Area

	Winter Bq/m^3	Spring/Autumn Bq/m^3	Summer Bq/m^3
Mihama	53.4	37.9	43.1
Misasa	35.6	44.8	34.2

As mentioned before, Misasa is one of the most highly radioactive hot springs in Japan (Furuno, 1982a, 1982b, 1982c).
Nevertheless, the results from Misasa were lower than those from
Mihama and almost the same as those from Hiroshima. Of course, special places such as the thermal grotto and drinking hall in Misasa as
well as a cave in Mihama show very high radon concentrations as shown
in Table IV. It is interesting that the bathroom of a hot spring did
not show a very high radon concentration. The high ventilation rate
to reduce humidity in this room may explain this rather low radon
concentration.

Table IV. Radon concentration in special places

		N	Radon conc. Bq/m^3
Bath room of hot spring	(Misasa)	2	(Arithmetic mean) 54.3
Thermal grotto	(Misasa)	1	2080
Drinking hall	(Misasa)	1	190
Cave	(Mihama)	1	248
Cabin of a boat	(Mihama)	1	7.3

The level of radon concentration in the dwellings having their own hot-spring bath is also not high compared with conventional houses or those located close to public hot-spring bath as shown in Table V. This lack of difference may also be attributable to a high ventilation rate in the bath room.

Table V. Numbers, Arithmetic Means and S.D.s, Geometric Means, Medians, and Ranges of Rn Measurements for Dwellings with Different Relationships to Hot-spring Baths, and Persons Living in Misasa Area

	N	Mean Bq/m^3			Median Bq/m^3	Range Bq/m^3
		Arithmetic	S.D.	Geometric		
Having own hot spring	15	38.7	± 31.2	28.9	23.3	8.74 – 107
Close to public hot spring	7	41.8	± 49.1	27.8	17.3	12.5 – 149
Conventional	238	30.0	± 40.1	17.8	17.0	0.116– 289
Personal monitoring	25	31.8	± 39.5	22.4	23.0	8.30 – 205

Personal monitorings of 25 people living in the Misasa area were carried out. The median value is 23.0 Bq/m^3 which is very close to the value found for the dwellings of Misasa as shown in Table V. However, the range of the values is from 8.30 to 205 Bq/m^3 and the significant correlation between the values obtained from personal monitoring and those from dwellings measurements is not found at 5 % level. It seems, therefore, that we need more personal monitorings and to compare the results from personal with those from dwellings before making a decision whether personal monitoring is necessary or not.

Conclusions

Measurements of indoor radon concentration were carried out using bare track detectors (CR-39) in Mihama, Misasa, Hiroshima and Nagasaki for about one year and using Terradex SF in Hokkaido for a winter in Japan. The data for those have log-normal distributions. The medians of the distribution for Nagasaki, Hiroshima, Misasa and Mihama were 10.2, 23.1, 24.2 and 31.2 Bq/m^3. These differences are possibly due to geology, construction material and ventilation rate. The three Honshu areas are geologically granitic but Nagasaki is in an area of extrusive rocks. The median for Hokkaido was 35.5 Bq/m^3, but the survey was carried out only in winter and the comparative studies in the places other than Hokkaido indicated that the average value by Terradex SF was about 2 times higher than by CR-39, probably due to a different calibration procedure. The values for Japanese traditional wooden houses were unexpectedly higher than these for ferro-concrete and prefabricated houses. There was only a modest difference between first and upper floors, with the radon levels in the latter lower. Unexpectedly, the results from Misasa spa, which is one of the most highly radioactive hot springs in Japan, were lower than those from Mihama. Personal monitoring for 25 people living in the Misasa area were carried out. The median was 23.0 Bq/m³, which was very close to the median value for the dwellings. However, the significant correlation between the values from personal monitoring and those from the dwellings was not found.

Literature Cited

Alter, H.W. and P.B. Price, Radon Detection Using Track Registration Material, U.S. Pat. 3665:194 (1972).

Furuno, K., Studies on the Trace Elements in the Field of Balneology (I) Radon Contents in Spring Water, J. Jap. Assoc. Phys. Med. Balneol. Climatol. 45:37-48 (1982).

Furuno, K., Studies on the Trace Elements in the Field of Balneology (II) Determination of Radon Contents in the Air of Radioactive Spa Areas, and Excretion of Radon in Expiratory Air after using Spring Water, J. Jap. Assoc. Phys. Med. Balneol. Climatol. 45:49-67 (1982).

Furuno, K., Studies on the Trace Elements in the Field of Balneology (III) Determination of Environmental Radioactivity Spa Area in Sanin, J. Jap. Assoc. Phys. Med. Balneol. Climatol. 46:63-75 (1982).

Radford, E.P., H. Kato, K.J. Kopecky, T. Aoyama, and M. Sakanoue, Plan for Placing Radon Dosimeters in Houses in Hiroshima and Nagasaki, RERF RP 6-84:1-10 (1985).

Stranden, E., L. Berteig, and F. Ugletveit, A Study on Radon in Dwellings, Health Physics 36:413-421 (1979).

Urban, M., A. Wicke, and H. Kiefer, Bestimmung der Strahlenbelastung der Bevolkerung durch Radon und dessen kurzlebige Zerfallsprodukte in Wohnhausern und im Freien, Kernforschungszentrum Karlsruhe 3805 (1985).

Yonehara, H., H. Kimura, M. Sakanoue, E. Iwata, S. Kobayashi, K. Fujimoto, T. Aoyama, and T. Sugahara, Improvement in the Measurement of Radon Concentrations by a Track Detector, this volume (1986).

RECEIVED July 29, 1986

Chapter 12

Effects of Aerosol Polydispersity on Theoretical Calculations of Unattached Fractions of Radon Progeny

Frank Bandi, Atika Khan, and Colin R. Phillips

Department of Chemical Engineering and Applied Chemistry, University of Toronto, Toronto, Ontario M5S 1A4, Canada

Theoretical calculations of unattached fractions of radon progeny require prediction of an attachment coefficient. Average attachment coefficients for aerosols of various count median diameters, CMD, and geometric standard deviations, σ_g, are calculated using four different theories. These theories are: (1) the kinetic theory, (2) the diffusion theory, (3) the hybrid theory and (4) the kinetic-diffusion theory. Comparisons of the various calculated attachment coefficients are made and the implications of using either the kinetic or the diffusion theory to calculate unattached fractions for aerosols of various CMD and σ_g are discussed. Significant errors may arise in use of either the kinetic theory or the diffusion theory. Large and unacceptable errors arise in calculating unattached fractions of a polydisperse aerosol by characterizing the aerosol as monodisperse. Unattached fractions of RaA are calculated for two mine aerosols and a room aerosol.

Theoretical calculations of unattached fractions of radon or thoron progeny involve four important parameters, namely, 1) the count median diameter of the aerosol, 2) the geometric standard deviation of the particle size distribution, 3) the aerosol concentration, and 4) the age of the air. All of these parameters have a significant effect on the theoretical calculation of the unattached fraction and should be reported with theoretical or experimental values of the unattached fraction.

The two fundamental theories for calculating the attachment coefficient, β, are the diffusion theory for large particles and the kinetic theory for small particles. The diffusion theory predicts an attachment coefficient proportional to the diameter of the aerosol particle whereas the kinetic theory predicts an attachment coefficient proportional to the aerosol surface area. The theory

0097-6156/87/0331-0137$06.00/0
© 1987 American Chemical Society

which is usually adopted is the kinetic theory. However, when the
diameter of the aerosol particle, d, is of the order of the mean
free path, ℓ, of gas molecules (Knudsen number, $\ell/d \approx 1$), neither
the diffusion theory nor the kinetic theory is correct in its
assumptions. A hybrid theory has been used in this region and may
be applied over the whole particle-size spectrum, since it con-
verges to the kinetic theory for small particle sizes and to the
diffusion theory for large particle sizes.

    In this study, values of the average attachment coefficient,
$\bar{\beta}$, (averaged over the entire particle size spectrum) are calculated
using a combination of the above theories. Unattached fractions
are then calculated for uranium mine and indoor air particle-size
distributions and aerosol concentrations.

Classical Diffusion Theory

The original theory of diffusional coagulation of spherical aerosol
particles was developed by von Smoluchowski (1916,1917). The
underlying hypothesis in this theory is that every aerosol particle
acts as a sink for the diffusing species. The concentration of the
diffusing species at the surface of the aerosol particle is assumed
to be zero. At some distance away, the concentration is the bulk
concentration.

    Solution of this diffusion problem, assuming quasi-steady
state coagulation, leads to an attachment coefficient of the form:

$$\beta(d) = 2\pi dD \tag{1}$$

d = diameter of aerosol particle
D = diffusion coefficient of the diffusing species.

    In addition to the quasi-steady state assumption, the other
assumptions required to arrive at equation (1) are: 1. the aerosol
itself does not coagulate; 2. there is a fully developed concentration gradient around each aerosol particle; and 3. the concentration of unattached radon progeny atoms is much greater than
the concentration of aerosol particles (in order that concentration
gradients of radon progeny atoms may exist). This last assumption
is usually not valid since the radon progeny concentration is
usually much less than the aerosol concentration.

    From equation (1) it can be seen that application of the dif-
fusion theory leads to the conclusion that the rate of attachment
of radon progeny atoms to aerosol particles is directly proportional to the diameter of the aerosol particles.

Kinetic Theory

The kinetic theory of radon progeny attachment to aerosol particles
assumes that unattached atoms and aerosol particles undergo random
collisions with the gas molecules and with each other. The attachment coefficient, $\beta(d)$, is proportional to the mean relative velocities between progeny atoms and particles and to the collision
cross section (Raabe, 1968a):

$$\beta(d) = \pi(d_r/2 + d/2)^2 \sqrt{\bar{v}^2 + \bar{v}_p{}^2} \qquad (2)$$

The attachment coefficient is a function of the aerosol particle diameter, d, and mean velocity, $\bar{v}_p$, as well as the unattached progeny diameter, d_r, and its mean velocity $\bar{v}$. Since in most situations $d_r \ll d$ and $\bar{v} \gg \bar{v}_p$, equation (2) reduces to

$$\beta(d) = \pi d^2 \bar{v}/4 \qquad (3)$$

where the radon progeny mean velocity can be calculated from

$$\bar{v} = \left(\frac{8kT}{\pi m}\right)^{1/2}$$

where T is the absolute temperature, k is Boltzmann's constant and m is the mass of the unattached radon progeny.

From equation (3) is is clear that the kinetic theory predicts an attachment rate of radon daughters to aerosol particles proportional to the square of the diameter of the aerosol particle.

Hybrid Theory

When the radius of an aerosol particle, r, is of the order of the mean free path, ℓ, of gas molecules, neither the diffusion nor the kinetic theory can be considered to be strictly valid. Arendt and Kallman (1926), Lassen and Rau (1960) and Fuchs (1964) have derived attachment theories for the transition region, $r \approx \ell$, which, for very small particles, reduce to the gas kinetic theory, and, for large particles, reduce to the classical diffusion theory. The underlying assumptions of the hybrid theories are summarized by Van Pelt (1971) as follows: 1. the diffusion theory applies to the transport of unattached radon progeny across an imaginary sphere of radius $r + \ell$ centred on the aerosol particle; and 2. kinetic theory predicts the attachment of radon progeny to the particle based on a uniform concentration of radon atoms corresponding to the concentration at a radius of $r + \ell$.

The attachment coefficient, β, corresponding to the hybrid theory can be shown to be (Fuchs, 1964)

$$\beta(r_1, r_2) = 16\left(\frac{r_1 + r_2}{2}\right)\left(\frac{D_1 + D_2}{2}\right)C \qquad (4)$$

where the subscripts 1,2 represent the radon progeny and the aerosol particles respectively and

r_x = radius of species x (x = 1,2)
D_x = diffusion coefficient of species x (x = 1,2)

$$C = \cfrac{1}{\cfrac{\bar{r}}{\bar{r} + \delta r/2} + \cfrac{4\bar{D}}{G_r \bar{r}}} \quad , \text{ a correction factor, where}$$

$$\bar{r} = \frac{r1 + r2}{2} \quad , \qquad \bar{D} = \frac{D1 + D2}{2} \quad , \qquad G_r = \sqrt{G_1^2 + G_2^2} \quad ,$$

$$\delta_r = \sqrt{\delta_1^2 + \delta_2^2}$$

The thickness, δ_x, of the boundary layer corresponding to the species x, the mean velocity of x, G_x, and the diffusion coefficient, D_x, can be calculated from the equations:

$$D_x = kTB_x$$

where B_x, the mobility, is defined as:

$$B_x = (1 + \frac{A\ell}{r_x} + \frac{Q\ell}{r_x} \exp[-br_x/\ell])/6\pi\eta r_x$$

A, Q and b are the constants 1.246, 0.42 and 0.87, respectively (Fuchs, 1964)
η = viscosity of the medium
ℓ = mean free path of the gas molecules

$$G_x = \left(\frac{8kT}{\pi m_x}\right)^{1/2} \text{ where } m_x \text{ is the mass of the species x}$$

$$\delta_x = \frac{1}{6r_x \ell_x} \{(2r_x + \ell_x)^3 - (4r_x^2 + \ell_x^2)^{3/2}\} - 2r_x$$

where ℓ_x, the apparent free path of particles, is calculated from

$$\ell_x = G_x m_x B_x$$

Fuchs' attachment coefficient can be simplified by making various assumptions:

1) the velocity of radon progeny is much greater than that of the aerosol particles. ($G_1 \gg G_2$ so that $G_r \approx G_1 = \bar{v}$)

2) the radius of the aerosol particles is much greater than that of the radon progeny. ($r_2 \gg r_1$ so that $\bar{r} \approx r_2/2 = d/4$)

3) the diffusion coefficient of radon progeny is much greater than that of the aerosol particles. ($D_1 \gg D_2$ so that $\bar{D} \approx D_1/2 = D/2$)

4) δ_x, the thickness of the boundary layer is approximately equal to the mean free path, ℓ, of gas molecules in air.

Using the above assumptions, equation (4) can be written as a function of aerosol diameter:

$$\beta(d) = \frac{\pi \bar{v} d^2 D(d/2 + \ell)}{\frac{\bar{v} d^2}{4} + 4D(d/2 + \ell)} \qquad (5)$$

Kinetic-Diffusion Approximation

For small particle sizes the kinetic theory is applicable, whereas for large particle sizes the diffusion theory applies. A useful approximation is therefore to use the kinetic theory in the small particle range and the diffusion theory in the large size region. The attachment coefficient then takes the form:

$$\beta(d) = \pi d^2 \bar{v}/4 \qquad \text{for } 0 < d < d_1$$
$$\beta(d) = 2 \pi d D \qquad \text{for } d_1 < d < \infty$$

This approximation may be considered to be an alternative to the hybrid theory. The value of d_1 can be found by equating the attachment coefficients for the diffusion and kinetic theories ($d_1 = 8D/\bar{v}$).

The dependence of the attachment coefficient on diameter for the diffusion theory (eqn. (1)), kinetic theory (eqn. (3)), and hybrid theory (eqns. (4) and (5)) is shown in Figure 1 for $\sigma_g = 1$ and D, v and ℓ equal to 0.054 cm^2/s, 1.38 x 10^4 cm/s and 6.5 x 10^{-6} cm, respectively. The value of D depends on the composition of the gaseous environment since radon progeny may be encumbered by clustering of the molecules. The value of 0.054 cm^2/s is representative of the experimentally observed range of values (Table I). Similar comments may also be made about the value of $\bar{v}$.

In calculating $\beta(d)$ using Fuchs' theory, D and $\bar{v}$ for radon progeny are used as above. It is also assumed that the aerosol particles are spherical and of unit density. It may be noted that assuming a higher density for the aerosols does not change the results appreciably. Fuchs' model is calculated for

$$T = 296 \text{ K}$$
$$\eta = 18.3 \text{ x } 10^{-5} \text{ poise (viscosity of air at 296 K and 1 atm)}$$
$$k = 1.380662 \text{ x } 10^{-16} \text{ g cm}^2/\text{s}^2 \text{ K}$$

Fuchs' model the radius and mass of the radon progeny are 4.9 x 10^{-8} cm and 5.46 x 10^{-22} g, respectively, for the above mentioned values of D, v and the other parameters. These calculated values are higher than would be expected for ^{218}Po, suggesting that clustering occurs.

From Figure 1 it can be seen that Fuchs' theory and equation (5) predict similar attachment coefficients. Equation (5) is therefore used in calculating average attachment coefficients.

Although the three theories described above are not the only available theories (Table II) they are considered to be the most important.

Average Attachment Coefficients

The average attachment coefficient $\bar{\beta}$ can be calculated for the diffusion and the kinetic theory from the equation:

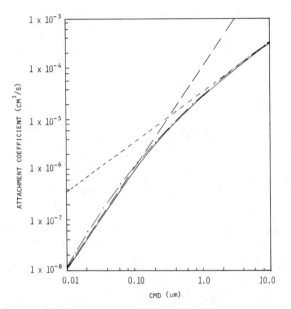

Figure 1.   Attachment coefficient vs CMD (σ_g = 1)

   ----------   diffusion theory
   — — —   kinetic theory
   ——·——·   Fuchs' theory
   —————   hybrid theory

Table I. Summary of Experimental Results for the Diffusion
Coefficient of ^{218}Po

Reference	D ($cm^2 s^{-1}$)	Gas*	Remarks**
Raabe (1968b)	0.047	air, RH=15%	D applies to neutral ^{218}Po species only.
	0.034	air, RH=35%	Radon concentration was high (5.55 x 10^3 Bq/m^3)
Thomas (1970)	0.053	dry air	The average age of ^{218}Po
	0.085	air, RH=20%	was 15-29 sec
Raghunath and			Ventilation rate:
Kotrappa	0.0648	air, RH=10%	6 air changes per hr
(1979)	0.0436	air, RH=90%	6 " " " "
	0.0811	air, RH=10%	60 " " " "
	0.0803	air, RH=90%	60 " " " "
	0.0725	air, RH=10%	6 " " " "
	0.0713	air, RH=90%	6 " " " "
	0.0955	air, RH=10%	60 " " " "
	0.0913	air, RH=90%	60 " " " "
Busigin et	0.10	air, RH=0%	D applies to neutral
al. (1982)	0.028	air, RH=8%	^{218}Po species. The
	0.043	air, RH=15%	radon concentration was
	0.046	air, RH=100%	not held constant
	0.048	air, RH=0%	
	0.035	air, RH=8%	
	0.065	air, RH=15%	
	0.076	air, RH=100%	

* Gas at room temperature (20°C) and pressure.

** Unless otherwise stated, no attempt was made to measure the
diffusion coefficient of charged and neutral species separately.

Table II. Attachment Coefficient Studies Reported in the
Literature

Attachment Coefficient	Aerosol Characteristics and Remarks
$\beta = \dfrac{\pi R^2 V(1 + \sqrt{\pi y})}{1 + \dfrac{RV}{4D}}$ where $y = q^2/2RKT$ T = temp. (K) K = Boltzmann's constant πR^2 = collision cross-section $1 + \sqrt{\pi y}$ = electrostatic enhancement factor due to image forces	Aerosols generated by electrical heating of a coil of nichrome wire Aerosol Size Range: 0.01 - 0.08 μm (radius) Aerosol Concentration: $10^2 - 10^6/cm^3$ 42.1 cm long diffusion tube preceded the filter for complete collection of unattached progeny products of ^{222}Rn Best fit to data found for 90% ^{218}Po as neutral and 10% as carrying a unit charge. Reference: McLaughlin (1972)
$\beta \propto \dfrac{R^2}{1 + hR}$ R = particle radius $h = V/4D$ V = mean gas kinetic velocity D = diffusion coeff.	Monodisperse DOP and latex spheres mixed with ^{220}Rn Aerosol Size Range: 0.04 - 0.6 μm (radius) Aerosol Concentration: $10^3 - 10^4/cm^3$ Particle sizes separated by diffusion batteries. Value of h found to = $7 \times 10^{-4}/cm$ Reference: Lassen and Rau (1960)
$\beta = \dfrac{\pi R^2 V}{1 + \dfrac{VR}{VD} \cdot \dfrac{1}{1 + \dfrac{\delta^*}{R}}}$ Theoretical mean: 1.58×10^{-6} cm^3/s Experimental mean: $(1.41 \pm 0.03) \times 10^{-6}$ cm^3/s	Natural aerosol mixed with ^{220}Rn Aerosol Size Range: 0.06 - 0.1 μm (radius) Aerosol Concentration: $3 \times 10^4/cm^3$ Attachment coefficient is for neutral daughter products. A cylindrical capacitor used as a mobility analyzer. Reference: Mohnen (1967)

Table II. Continued

Attachment Coefficient	Aerosol Characteristics and Remarks
$\beta \propto R^b$ b = 1 for 0.25–1.35 μm b = 1.34 for 0.1–0.33 μm	Polydisperse fluorescein aerosol separated into different size groups by aerosol centrifuge) Aerosol Size Range: 0.1 – 1.35 μm (radius) Aerosol Concentration: $10^5 - 10^6/cm^3$ No difference found between the attachment coefficients of charged and neutral atoms Reference: Menon et al. (1980)

* δ is a function of particle radius and mean free path of daughter products.

$$\bar{\beta} = \frac{1}{N} \int_{0}^{\infty} \beta(d)n(d)d(d)$$

where $d(d)$ = small change in diameter

$n(d) = \dfrac{dN(d)}{d(d)}$ (number distribution function)

$N(d)$ = number of particles whose diameters are less than or equal to d

Since real particle distributions tend to be log-normal, a diameter parameter u is defined as:

$$u = \ln d$$

and a distribution centred about a value $\bar{u}$ with a standard deviation σ_u is defined as:

$$n(u) = \frac{N}{\sqrt{2\pi}\,\sigma_u} \exp[-(u - \bar{u})^2/2\sigma_u^2]$$

Then, for the kinetic theory:

$$\bar{\beta} = \frac{1}{N} \int_{-\infty}^{\infty} \frac{\pi}{4}\bar{v}e^{2u} \; \frac{N \exp[-(u-\bar{u})^2/2\sigma_u^2]}{\sqrt{2\pi}\,\sigma_u} \, du \tag{6}$$

and for the diffusion theory:

$$\bar{\beta} = \frac{1}{N} \int_{-\infty}^{\infty} 2\pi D \; \frac{\exp[u] \; N \; \exp[-(u-\bar{u})^2/2]}{\sqrt{2\pi}\,\sigma_u} \, du \tag{7}$$

Noting that Count Median Diameter (CMD) = $\exp[\bar{u}]$ and Geometric Standard Deviation (σ_g) = $\exp[\sigma_u]$, equations (6) and (7) can be solved to give:

$$\bar{\beta}(\text{kinetic}) = \frac{\pi\bar{v}}{4} (\text{CMD})^2 \exp[2(\ln\sigma_g)^2] \tag{8}$$

$$\bar{\beta}(\text{diffusion}) = 2\pi D(\text{CMD})\exp[(\ln\sigma_g)^2/2] \tag{9}$$

For the diffusion-kinetic approximation, a similar approach leads to the equation:

$$\bar{\beta} = \int_{0}^{d_1} \frac{\pi}{4}\bar{v}d^2 n(d) \; d(d) + \int_{d_1}^{\infty} 2\pi dD \; n(d) \; d(d) \tag{10}$$

which can be expressed as:

$$\bar{\beta} = \frac{1}{N} \; \frac{\pi}{4} \bar{v} \; \frac{N}{\sqrt{2\pi} \; \sigma_u} \int_{-\infty}^{u_1} (\exp[u])^2 \, \exp[-(u-\bar{u})^2/2\sigma_u^2] \; du$$

$$+ \frac{2\pi DN}{\sqrt{2\pi} \; \sigma_u} \int_{u_1}^{\infty} \exp[u] \, \exp[-(u-\bar{u})^2/2\sigma_u^2] \; du \qquad (11)$$

where $u_1 = \ln(d_1)$.

Solution of equation (11) gives:

$$\bar{\beta}(\text{kin-dif}) = \frac{\bar{v}\pi}{8} (\text{CMD})^2 \exp[2(\ln\sigma_g)^2] \left\{1 + \text{erf}\left[\frac{\ln(d_1/\text{CMD}) - 2(\ln\sigma_g)^2}{\sqrt{2} \, \ln\sigma_g}\right]\right\}$$

$$+ \pi D(\text{CMD}) \exp[(\ln\sigma_g)^2/2] \left\{1 - \text{erf}\left[\frac{\ln(d_1/\text{CMD}) - (\ln\sigma_g)^2}{\sqrt{2} \, \ln\sigma_g}\right]\right\} \qquad (12)$$

For $d_1 = 0$, equation (12) reduces to the diffusion theory, and for $d_1 = \infty$, to the kinetic theory.

The average attachment coefficient for the hybrid theory is of the form:

$$\bar{\beta} = \int_{-\infty}^{\infty} \frac{\frac{\pi}{2} \bar{v}D \exp[3u]}{\frac{\bar{v}\exp[2u]}{4} + 2D\exp[u] + 4D\ell} \cdot \frac{\exp[-(u-\bar{u})^2/2\sigma_u^2]}{\sqrt{2\pi} \; \sigma_u} \; du$$

$$+ \int_{-\infty}^{\infty} \frac{\pi\bar{v}D\ell\exp[2u]}{\frac{\bar{v}\exp[2u]}{4} + 2D\exp[u] + 4D\ell} \cdot \frac{\exp[-(u-\bar{u})^2/2\sigma_u^2]}{\sqrt{2\pi} \; \sigma_u} \; du \qquad (13)$$

By letting

$$x = \frac{u - 2\sigma_u^2 - \bar{u}}{\sqrt{2}\sigma_u}$$

in the first integral, and defining z as

$$z = \frac{u - 3\sigma_u^2 - \bar{u}}{\sqrt{2}\sigma_u} \; ,$$

solution of equation (13) gives:

$$\bar{\beta}(hybrid) \ = \ \sqrt{\pi} \ \bar{v} \ D\ell(CMD)^2 \ exp[2(\ln\sigma_g)^2] \ .$$

$$\int_{-\infty}^{\infty} exp[-x^2]dx/[\ \frac{\bar{v}}{4} \ (CMD)^2 \ exp[4(\ln\sigma_g)^2] \ exp[2\sqrt{2}(\ln\sigma_g)x]$$

$$+ \ 2D(CMD)exp[2(\ln\sigma_g)^2] \ exp[\sqrt{2}(\ln\sigma_g)x] \ + \ 4D\ell]$$

$$+ \ \sqrt{\pi} \ \bar{v} \ D(CMD)^3 \ exp[9/2(\ln\sigma_g)^2] \ .$$

$$\int_{-\infty}^{\infty} exp[-z^2]dz/[\ \frac{\bar{v}}{4} \ (CMD)^2 \ exp[6(\ln\sigma_g)^2] \ exp[2\sqrt{2}(\ln\sigma_g)z]$$

$$+ \ 2D(CMD) \ exp[3(\ln\sigma_g)^2]exp[\sqrt{2}(\ln\sigma_g)z] \ + \ 4D\ell] \qquad (14)$$

These integrals can be solved by Hermite integration in which:

$$\int_{-\infty}^{\infty} e^{-x^2} \ f(x) \ dx \ \approx \ \sum_{i=1}^{n} \ w_i f(x_i)$$

where w_i = weighting factor

x_i = abscissae (zeros of Hermite polynomials)

A four point (n = 4) approximation is adequate since $\bar{\beta}$ does not change much when 5 and 6 point approximations are used. It is apparent that if σ_g = 1 (a monodisperse aerosol) equation (14) reduces to the hybrid theory with the CMD in place of the diameter.

Equations (8), (9), (12) and (14) are used to calculate the attachment coefficient for various values of σ_g and CMD.

Unattached Fractions of Radon Progeny

In order to calculate the theoretical unattached fraction of radon progeny the appropriate differential equations must be developed to describe the net formation of unattached radon progeny. The system may be visualized schematically for RaA as illustrated in Fig. 2. It is assumed that there is no flow into or out of the system or removal by electric fields. The equations which describe the system presented in Fig. 2 are:

$$\frac{dN_{2_f}}{dt} \ = \ \lambda_1 N_1 - (\lambda_s + \lambda_2 + \lambda_w^f)N_{2_f} \qquad (15)$$

$$\frac{dN_{2_a}}{dt} \ = \ \lambda_s N_{2_f} - (\lambda_2 + \lambda_w^a)N_{2_a} \qquad (16)$$

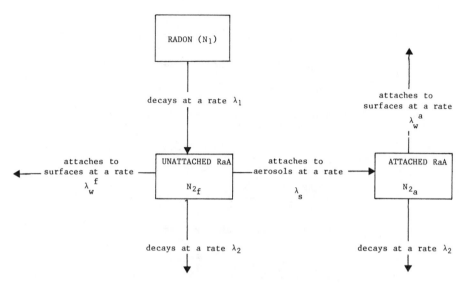

Legend

N_1 — number of radon atoms

N_{2_f} — total number of unattached RaA atoms

N_{2_a} — total number of attached RaA atoms

λ_1, λ_2 — decay constant of radon and RaA, respectively

λ_s — attachment constant of free RaA to aerosols

λ_w^a — attachment constant of attached RaA to surfaces

λ_w^f — attachment constant of free RaA to surfaces

Figure 2. Fate of RaA.

N_1 = number of radon atoms

N_{2_f} = number of unattached RaA atoms

N_{2_a} = number of attached RaA atoms

λ_1, λ_2 = decay constant of radon and RaA, respectively

λ_s = attachment constant of free RaA to aerosols

λ_w^a = attachment constant of attached RaA to surfaces

λ_w^f = attachment constant of free RaA to surfaces

Equations (15) and (16) can be solved for N_{2_f} and N_{2_a} (assuming $\lambda_1 N_1$ is approximately a constant) to give:

$$N_{2_f} = \left(N_{2_{f_o}} + \frac{\lambda_1 N_1}{K_f} \right) \exp - (K_f t) + \frac{\lambda_1 N_1}{K_f} \tag{17}$$

with $N_{2_{f_o}}$ = number of unattached RaA atoms at t = 0

$$K_f = \lambda_s + \lambda_w^f + \lambda_2$$

$$N_{2_a} = \left\{ N_{2_{a_o}} - \frac{\lambda_1 N_1 \lambda_s}{K_f K_a} - \frac{\lambda_s \left[N_{2_{f_o}} - \dfrac{\lambda_1 N_1}{K_f} \right]}{K_a - K_f} \right\} \exp(-K_a t)$$

$$+ \frac{\lambda_s \left[N_{2_{f_o}} - \dfrac{\lambda_1 N_1}{K_f} \right]}{K_a - K_f} \exp(-K_f t) + \frac{\lambda_1 N_1 \lambda_s}{K_f K_a} \tag{18}$$

with $N_{2_{ao}}$ = number of attached RaA atoms at t = 0

$$K_a = \lambda_w^a + \lambda_2$$

The unattached fraction of RaA (F_A) can be calculated as:

$$F_A = \frac{N_{2_f}}{N_{2_f} + N_{2_a}} \tag{19}$$

The unattached fraction of RaA is independent of the radon concentration since the factor $\lambda_1 N_1$ cancels out.

   If the system has attained equilibrium, the unattached fraction of RaA can be expressed as (with the use of equations (17), (18) and (19)):

$$F_A = \frac{K_a}{K_a + \lambda_s} \qquad (20)$$

Equation (20) is valid for t (the age of air) > 20 min. Equation (20) can be simplified further by assuming that $\lambda_w^a \ll \lambda_2$ ($T_{1/2}$ = 3.05 min, λ_2 = 0.00379 s^{-1}) and therefore $K_a \approx \lambda_2$. This assumption is reasonable since Pogorski and Phillips (1984) report λ_w^a for radon progeny in a chamber as 0.00008 s^{-1} and Porstendörfer (1984) estimates λ_w^a for radon progeny indoors to be between 0.000028 s^{-1} and 0.000083 s^{-1} depending on the ventilation rate.

Equation (20) can be rewritten as:

$$F_A = \frac{\lambda_2}{\lambda_2 + \lambda_s} \qquad (21)$$

The rate of attachment of radon progeny to aerosols, λ_s, is expressed as:

$$\lambda_s = \bar{\beta}N$$

where $\bar{\beta}$ = attachment coefficient (cm^3/s)

N = concentration of aerosols (particles/cm^3)

The rate of attachment can be calculated using the various attachment models presented in the previous section (for specific values of CMD, σ_g and N).

Finally, the theoretical unattached fraction for RaA can be calculated using λ_s, t (age of air) and equations (17), (18) and (19).

Results

Results are presented in Figures 3 through 7 and Table III.

Figures 3 and 4 show the variation of the attachment coefficient with count median diameter for the diffusion, kinetic, hybrid and kinetic-diffusion theory for geometric standard deviations of 2 and 3 respectively.

Figure 5 illustrates the effect of the geometric standard deviation on the attachment coefficient using the hybrid theory.

In Figs. 6 and 7 the attachment coefficient is plotted against the geometric standard deviation using the four theories, for count median diameters of 0.2 μm and 0.3 μm respectively.

Theoretical unattached fractions of RaA using average aerosol concentrations and count median diameters as found in track and trackless Canadian uranium mine are presented in Table III. The reported uranium mine aerosol properties are: N = 120,000 particles/cm^3 and CMD = 0.069 μm for a trackless mine and N = 60,000 particles/cm^3 and CMD = 0.055 μm for a track mine (Khan et al., 1986). Also presented in Table III are the theoretical unattached fractions of RaA for a typical room aerosol of CMD = 0.1 μm and aerosol concentrations of N = 10,000 particles/cm^3 and 2,000 particles/cm^3. Calculations were made assuming that the unattached RaA fraction had attained equilibrium.

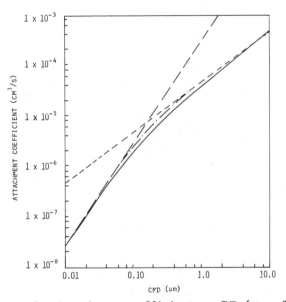

Figure 3. Attachment coefficient vs CMD (σ_g = 2)

    ------------ diffusion theory
    — — — kinetic theory
    —·——· kinetic-diffusion theory
    ———— hybrid theory

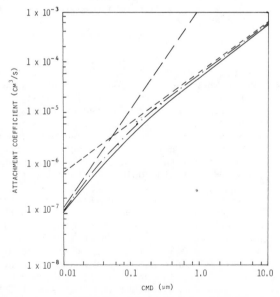

Figure 4. Attachment coefficient vs CMD (σ_g = 3)

    ------------ diffusion theory
    — — — kinetic theory
    —·——· kinetic-diffusion theory
    ———— hybrid theory

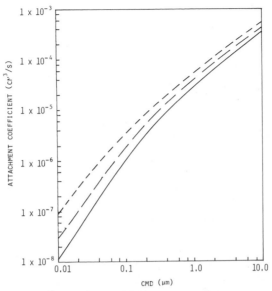

Figure 5. Attachment coefficient vs CMD using the hybrid theory for various particle distributions.

———————— hybrid theory ($\sigma_g = 1$)
— — — hybrid theory ($\sigma_g = 2$)
---------- hybrid theory ($\sigma_g = 3$)

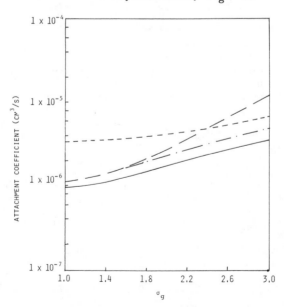

Figure 6. Attachment coefficient vs σ_g (CMD = 0.1 μm)

---------- diffusion theory
— — — kinetic theory
— · —— · kinetic-diffusion theory
———————— hybrid theory

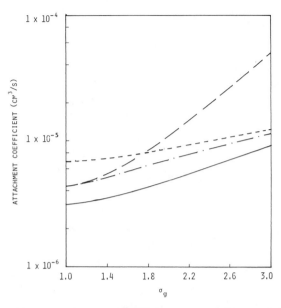

Figure 7. Attachment coefficient vs σ_g (CMD = 0.2 μm)

```
----------      diffusion theory
— — —           kinetic theory
—·——·           kinetic-diffusion theory
——————          hybrid theory
```

Table III.   Theoretical Unattached Fractions of RaA

GSD (σ_g)	Attachment theories				Conditions
	Kinetic	Diffusion	Hybrid	Kin-Diff	
1	0.161	0.033	0.168	0.161	Track mine
2	0.069	0.026	0.087	0.071	N = 60,000 particles/cm^3
3	0.017	0.018	0.040	0.030	CMD = 0.055 μm
1	0.058	0.013	0.062	0.058	Trackless mine
2	0.023	0.011	0.031	0.024	N = 120,000 particles/cm^3
3	0.005	0.007	0.015	0.011	CMD = 0.069 μm
1	0.259	0.100	0.285	0.259	Indoor air
2	0.118	0.081	0.175	0.134	N = 10,000 particles/cm^3
3	0.030	0.058	0.095	0.074	CMD = 0.1 μm
1	0.636	0.358	0.666	0.636	Indoor air
2	0.401	0.305	0.515	0.437	N = 2,000 particles/cm^3
3	0.135	0.234	0.343	0.285	CMD = 0.1 μm

Discussion

The most complete theory for aerosol coagulation is that of Fuchs (1964). Since the attachment of radon progeny to aerosols can be considered as the coagulation of radon progeny (small diameter particle) to aerosols (large diameter particle), it is reasonable to use Fuchs' theory to describe this process. The hybrid theory is an approximation to Fuchs' theory and thus can be used to describe the attachment of radon progeny to aerosols over the entire aerosol size spectrum.

If it is accepted that the hybrid theory is the most complete theory for attachment of radon progeny to aerosols, the magnitude of the error involved in using exclusively either the kinetic or the diffusion theory can be seen from Figs. 3-7 and Table III.

Figures 3 and 4 show the variation of the average attachment coefficient with CMD. It can be seen that for particles of CMD less than 0.06 μm and σ_g = 2 the kinetic theory predicts an attachment coefficient similar to the hybrid theory, whereas for CMD greater than about 1 μm the diffusion theory and the hybrid theory give approximately the same results. For a more polydisperse aerosol (σ_g = 3) the kinetic theory deviates from the hybrid theory even at a CMD = 0.01 μm. The diffusion theory is accurate for a CMD greater than about 0.6 μm. These results are easily explained since as the aerosol becomes more polydisperse, there are more large diameter particles (CMD > 0.3 μm) which attach according to the diffusion theory. In contrast, the kinetic theory becomes more inaccurate as the aerosol becomes more polydisperse.

The kinetic-diffusion approximation predicts an attachment coefficient similar to the hybrid theory for all CMDs and for both σ_g = 2 and 3 (Figs. 3 and 4). The advantage of this theory is that the average attachment coefficient can be calculated from an analytical solution; numerical techniques are not required.

Figure 5 shows the variation of the hybrid theory with CMD for various σ_g. It is obvious that assuming an aerosol to be monodisperse when it is in fact polydisperse leads to an underestimation of the attachment coefficient, leading in turn to large errors in calculation of theoretical unattached fraction.

The variation of attachment coefficient with σ_g for CMD = 0.2 μm and 0.3 μm is shown in Figures 6 and 7. Again it is apparent that the kinetic theory or diffusion theory are correct only at certain CMD and σ_g. Neither is applicable under all circumstances. It is also evident that the kinetic-diffusion theory is a good approximation to the hybrid theory under all circumstances.

Unattached fractions of RaA (at t = ∞) for two mine aerosols and for a typical room aerosol are shown in Table III. It is usually assumed that the attachment of radon progeny to aerosols of CMD < 0.1 μm follows the kinetic theory. In Table III it is apparent that the hybrid and kinetic theories predict similar unattached fractions for monodisperse aerosols. However, for more polydisperse aerosols, the kinetic theory predicts lower unattached fractions than the diffusion theory and thus the diffusion theory is the more appropriate theory to use. It is also evident that the kinetic-diffusion approximation predicts unattached fractions similar to those predicted by the hybrid theory in all cases.

Conclusions

Calculation of the attachment coefficient is required for theoretical prediction of the unattached fraction of radon progeny. The hybrid theory, which is a form of Fuchs' theory with certain justifiable assumptions, can be used to describe attachment to aerosols under all conditions of σ_g and CMD.

The kinetic theory and the diffusion theory may be used for certain aerosol distributions. However, these two theories begin to deviate from the hybrid theory as the aerosol polydispersity increases. Use of either the kinetic theory or the diffusion theory may therefore result in large errors.

Although unattached fractions predicted using the kinetic and diffusion theory for high aerosol concentrations, such as mine atmospheres, are comparable, the same cannot be said for unattached fractions predicted at low aerosol concentrations, such as indoor air. For low aerosol concentrations, neither the kinetic nor the diffusion theory predicts unattached fractions close to those predicted by the hybrid theory. Exclusive use of either of these two theories results in large errors.

Although the hybrid theory is the most correct theory to use in the prediction of unattached fractions, the error in using the kinetic-diffusion theory in place of the hybrid is small. The kinetic-diffusion theory has the advantage that the solution is in analytical form and thus is more convenient to use than the hybrid theory, which must be solved numerically.

Finally, whenever theoretical or experimental unattached fractions are reported, the aerosol characteristics σ_g, CMD and N must also be reported to allow proper interpretation of the results.

Acknowledgment

This work was supported by a grant from the Natural Sciences and Engineering Research Council of Canada.

Literature Cited

Arendt, P. and H. Kallmann, Uber den Mechanismus der Aufladung von Nebelteilchen, Zeitschrift für Physik 35:421 (1926).

Busigin, C., A. Busigin and C.R. Phillips, The Chemical Fate of ^{218}Po in Air, in Proc. Int. Conf. on Radiation Hazards in Mining (M. Gomez, ed) pp. 1043, Soc. of Mining Engineers of Amer. Inst. of Mining, Metallurgical and Petroleum Engineers, Inc., Golden, CO (1982).

Fuchs, N.A., The Mechanics of Aerosols, Mcmillan Co., New York (1964)

Khan, A., F. Bandi, C.R. Phillips and P. Duport, Underground Measurements of Aerosol and Radon and Thoron Progeny Activity Distributions, to be published in Proc. 191st American Chemical Society National Meeting, New York, April 13-18 (1986).

Lassen, L. and G. Rau, The Attachment of Radioactive Atoms to Aerosols, Zeitschrift für Physik 160:504 (1960).

McLaughlin, J.P., The Attachment of Radon Daughters to Condensation Nuclei, Proc. Royal Irish Academy 72, Sect. A., 51 (1972).

Menon, V.B., P. Kotrappa and D.P. Bhanti, A Study of the Attachment of Thoron Decay Products to Aerosols using an Aerosol Centrifuge, J. Aerosol Sci. 11:87 (1980).

Mohnen, V., Investigation of the Attachment of Neutral and Electrically Charged Emanation Decay Products to Aerosols, AERE Trans. 1106 (Doctoral Thesis) (1967).

Pogorski, S. and C.R. Phillips, The Transient Response of Radon and Thoron Chambers, Proc. Int. Conf. Occupational Radiation Safety in Mining, (H. Stocker, ed), vol. 2, 394, Can. Nucl. Assoc. (1985).

Porstendörfer, J., Behaviour of Radon Daughter Products in Indoor Air, Radiation Protection Dosimetry 7:1 (1984).

Raabe, O.G., The Adsorption of Radon Daughters to Some Polydisperse Submicron Polystyrene Aerosols, Health Phys. 14:397 (1968a).

Raabe, O.G., Measurement of the Diffusion Coefficient of RaA, Nature 217:1143 (1968b).

Raghunath, B. and P. Kotrappa, Diffusion Coefficients of Decay Products of Radon and Thoron, J. Aerosol Sci. 10:133 (1979).

Smoluchowski, M. von, Versuch einer Matematischen Theorie der Koagulationshinetik Kolloider Losungen, Zeitschrift für Physik, Chemie XCII, 129 (1916).

Smoluchowski, M. von, Drei Vortage über Diffusion, Brownsche Molekularbewegung and Koagulation von Kolloidteilchen, Physikalische Zeitschrift 17:557,585 (1917).

Thomas, J.W. and LeClaire, P.C., A Study of the Two-Filter Method for Radon-222, Health Phys. 18:113 (1970).

Van Pelt, W.R., Attachment of Rn-222 Decay Products to a Natural Aerosol as a Function of Particle Size, Ph.D. Thesis, New York University (1971).

RECEIVED September 24, 1986

MEASUREMENT METHODS

Chapter 13

Intercomparison of Different Instruments That Measure Radon Concentration in Air

Michikuni Shimo, Takao Iida, and Yukimasa Ikebe

Department of Nuclear Engineering, Faculty of Engineering, Nagoya University, Furo-cho, Chikusa-ku, Nagoya 464, Japan

An intercomparison of different instruments for measurement of radon concentration was carried out. The instruments include an ionization chamber, the charcoal-trap technique, a flow-type ionization chamber (pulse-counting technique), a two-filter method, an electrostatic collection method and a passive integrating radon monitor. All instruments except for the passive radon monitor have been calibrated independently. Measurements were performed over a concentration range from about 3.5 Bq/m^3 (in outdoor air) to 110 Bq/m^3 (in indoor air). The results obtained from these techniques, except the two-filter technique, are comparable. Radon daughter concentration measured using a filter-sampling technique was about 52 % of radon concentration.

There are various techniques for measuring radon concentration in air (Budnitz, 1974). Needless to say, a value obtained with a given instrument should be the same one with another method. The data from an instrument assured of "exchangeability" is useful. The authors have thought that an intercomparison between devices is necessary and important to ensure exchangability between instruments operating by different principle. From this viewpoint, an intercomparison between various instruments for measuring radon concentrations was planned and carried out in natural concentration levels using outdoor and indoor airs.

The measuring techniques are (see Table I): (1) an ionization chamber method -- measurement of ionization current obtained from ionization chamber which was directly filled by sample gases (DSC), (2) a charcoal-trap method -- the method for sampling radon gas with a cooled activated charcoal and for measuring activity of alpha particle from radon and its daughters with ionization chamber (ACC), (3) a flow-type ionization chamber method -- the method for directly drawn air samples into a pulse-counting flow-type ionization chamber (PFC), (4) a two-

0097-6156/87/0331-0160$06.00/0

Table I. Intercomparison of various radon instruments

Instrument/Method	Symbol	Remarks
Ionization Chamber Method	DSC	directly sampled to an evacuated chamber, current measurement
Activated Charcoal Method	ACC	charcoal: 5g, flow rate: 5 ℓ/min, sampled temp.: −73°C, sampling time: 1hr.,transfered into ionization chamber
Flow-Type Ionization Chamber	PFC	flow rate: 1~2 ℓ/min, continuously measured, current measurement and α-pulse counting
Two-Filter Method	TF	flow rate: 40 ℓ/min, sampling time: 1hr, filter: membrane filter, gross counting or α-spectrometry, Rn and Tn measured
Electrostatic Radon Monitor	ERM	flow rate: 0.5 ℓ/min, electrode: −3000 V applied, dehamidified, continuously measured, detector: ZnS(Ag)
Passive Integrating Radon Monitor	PRM	electrostatic sampling: −180 V applied, dehumidified, CN-film, exposed time: two months a measurement

filter method (TF), (5) an electrostatic radon monitor -- the method for sampling with electrostatic collection (ERM), and (6) a passive radon monitor -- the electrostatic integrating radon monitor with cellulose nitrate film (PRM). Also, measurements with air filter (namely, filter-pack method, FP) (Harley, 1953; Shimo, 1985) was simultaneously carried out and the results compared with those obtained by using the other methods.

Measurement Method

Ionization Chamber Method (DSC). The ionization chamber is one of the most common instruments for measuring radon gas. The chmaber was cylindrically shaped and had a 1.5 liter volume. Air samples were drawn through a dryer filled with $CaCl_2$ and a filter to remove water vapor, radon daughters and aerosols, so that only the parent radon gas is admitted to the chamber. The radon gas subsequently decays and reaches equilibrium with its daughters in 3.5 hours after entry into the chamber. The ionization currents are measured with a vibraing reed electrometer.

We have made the DSC the standard method for the radon measurement. The calibration procedures have been described elsewhere (Shimo et al., 1983); Briefly, the Rn-222 emanating from a standard Ra-226 hydrochloric acid solution (37 kBq) contained in a bubbling bottle entered a large stainless steel container (937 ℓ). A small part of radon gas in the container was transfered into the ionization chamber (1.5 ℓ). The radon concentration in the container was controlled by varying the storage time of radon in the bubbling bottle.

Therefore, a relationship between the radon concentration Q_{exp} (Bq/m^3) and the ionization current I (fA) was experimentally obtained;

$$Q_{exp} = 18.3 \ I \tag{1}$$

On the other hand, the radon concentration Q (Bq/m^3) can theoretically be calculated as follows:

$$Q = \frac{W}{eEVf} \ I \tag{2}$$

$$f = f_1 \times f_2 \times f_3 \tag{3}$$

where W is the energy needed for making one ion pair production by alpha particle in air, 35.5 eV; I, measured currents (fA); e, elementary unit of charge 1.602×10^{-19} coul; E, total alpha energy from Rn, RaA and RaC, 19.17 MeV; V, the effective volume of the chamber (m^3), and f, a correction factor. The factor f is $f_1 \times f_2 \times f_3$; f_1: correction for decay during the measurement, f_2: correction for the wall loss (loss of ionization due to the wall of chamber) (Tachino et al. 1974) and f_3: correction from the columnar recombination (Boag, 1966). For the DSC using a 1.5 ℓ volume chamber, the value of f is 0.46 (f_1=0.987, f_2=0.47, and f_3=0.985). Therefore, the calculated radon concentration Q_{cal} (Bq/m^3) is

$$Q_{cal} = 16.8 \text{ I} \tag{4}$$

From equations (1) and (4), it is clear that the results of calibration agreed with the theoretical calculation within 8.4 percent. Finally, we have used the experimental results for the relationship between radon concentration and ionization current. The sensitivity of the DSC is about 40 Bq/m^3.

Charcoal-Trap Method (ACC). The Charcoal-Trap Method has been used for sampling radon gas in the atmospheric air and for measuring radon emanation rate from the soil surface by some researchers (e. g. Kawano and Nakatani, 1964; Megumi and Mamuro, 1972). The trap contained about 5 grams of activated granular charcoal in a brass tube. The coolant system consisted of a mixture of dry-ice and ethanol to keep the temperature at -73°C during air sampling at 5 ℓ/min. The absorbed radon gas was purged using argon gas at 800°C and then transfered into an ionization chamber. The radon activity in the chamber was measured at about 4 hours after transference in the same manner as the DSC (Shimo et al., 1983). The collection coefficient of the ACC was evaluated to be 0.88±0.05.

Flow-Type Ionization Chamber (PFC). The Flow-Type Ionization Chamber Method (PFC) has been developed for continuously measuring radon gas in the atmospheric air. The detail of the device has been described elsewhere (Shimo, 1985). Briefly, measurements are continuously carried out by drawing air through the detector at 1.0 ~ 2.0 ℓ/min. The ionization currentdue to alpha particles from radon and its daughters is detected with a vibrating reed electrometer (VRE) in the same manner as the DSC. The sensitivity of the current method of this device was about 10 Bq/m^3. Thus to detect a lower concentration of radon, a pulse-counting technique was used; The output signal of the VRE was detected into a differential amplifier, and a signal from an alpha particle is distinguished from background currents due to gamma-radiation and cosmic rays and counted. The alpha-counts in an hour, C is obtained by the following formula:

$$C = QTVf \tag{5}$$

where Q is the radon concentration (Bq/m^3); T, the counting time (sec); V, the volume of chamber (m^3) (12.72 x 10^{-3}m^3) and f, a correction factor.

The current measurement was calibrated using the same way as the DSC, and good agreement between the expreiment and theoretical calculation was obtained. The PFC was simultaneously able to detect over the range of 10 ~ 100 Bq/cm^3 of radon concentration using the current and pulse-counting methods. Therfore, the calibration of the pulse-counting technique was performed by using the result of

the current measurement carried out simultaneously with convienient
radon concentration; This gave the relationship between the radon
concentration Q (Bg/cm^3) and counts C (counts/hr):

$$Q = 0.0418 \ C \qquad\qquad (6)$$

The sensitivity of the pulse-counting method is about 0.2
Bq/m^3.

Two-Filter Method (TF). The Two Filter Method was first introduced
by Thomas and LeClare (Thomas and LeClare, 1970). A modified 66.3
liter decay chamber was used with a filter on each end. Air was
sampled at 40 ℓ/min. The inlet filter removed all of the radon
daughters and aerosol particles but allowed radon gas to pass.
During the transport in the chamber, radon atoms decay to form RaA
atoms. The RaA atoms except those diffused to the wall of the
chamber are collected on the exit filter. The sample collected on
the exit filter is removed and counted. The present method is able
to simultaneously measure radon and thoron concentrations by
alpha spectroscopic technique (Ikebe et al., 1979).

The radon concentration in air, Q (Bq/m^3), is obtained from the
equation

$$Q = \frac{C}{\varepsilon\zeta\eta qF} \ \frac{1}{k} \qquad\qquad (7)$$

where C is the total alpha counts in the measured time; ε, the
counting efficiency; ζ, the emerging efficiency of the alpha
particles from the filter; η, the collection efficiency; q, the flow
rate; F, a constant decided from the decay constants of radon or
thoron and progeny, the flow rate and sampling and counting times,
and k, a calibration factor.

The calibration of this method was performed with Rn-222 and
Rn-220 emanating from standard Ra-226 and Th-228 solutions,
respectively. It is defined that the calibration factor is the
ratio of experimental counts to one expected from used radon and/or
thoron gases by using Equation (7); The calibration factors of
radon and thoron were evaluated to be 0.81 and 0.51, respectively.

Electrostatic Radon Monitor (ERM). The Flow-Type Electrostatic
Sampling Instrument (the electrostatic radon monitor) was first
developed by Dalu and Dalu, 1971 (Dalu and Dalu, 1971). In the
present study a modified version developed by Iida (1985) was used.

A 15 liter, semispherical sampling vessel with a flow rate of
0.5 ℓ/min was used. The electrode (38 mm diameter) in front of the
ZnS(Ag) scintillator was placed in the center of the bottom and was
set at -3,000 V relative to the vessel wall. Since the ERM is
sensitive to water vapor (Porstendörfer et al, 1980; Dalu et al.,
1983), the air sample was passed through a dehumidefier to maintain
the relative humidity in the chamber less than 2.9 %.

The radon concentration in air Q (Bq/m^3) is estimated from the
following equation:

$$Q = \frac{C(I) - \sum_{i=1}^{5} C'(I-i)}{VF(I)} \cdot \frac{1}{k} \qquad (8)$$

where C(I) is the total alpha counts in the measured time interval I; C'(I-i), the expected alpha counts from previous sampling period i; V, the volume of sampling vessel (m); F((I), a constant determined from the decay constants of radon progeny, the flow rate and sampling and counting times, and k, the calibration factor.

The instrument was calibrated using a 1.5 liter flow-type ionization chamber: The calibration factor was evaluated to be 0.318.

Passive Integrating Radon Monitor(PRM). Many types of passive integrating radon monitors have been developed (Urban and Piesch, 1981; Alter and Fleischer, 1981; Iida, 1985). The authors developed a sensitive device, PRM, for measuring natural radon concentration levels by electrically collecting Rn-222 progeny on an electrode behind which a cellulose nitrate (CN) films was set (Ikebe et al., 1985). A window for ventilating air was provided in the bottom wall. The PRM detects alpha particles only from Po-218 and does not detect those from Po-214 because of an aluminized mylar sheet in front of the electrode, the geometrical arrangement between the aluminized mylar sheet and CN films, and eching time. An electrode at the center of the cylindrical vessel (140 mm diameter, 100 mm height) is set at -180 V relative to the vessel wall. The PRM is also sensitive to water vapor for the same reasonin as the case of the ERM (Negro and Watnick, 1978; Kotrappa et al., 1982; Annanmaki et al., 1983). Therefore, approximate 150 g of drying agent (P_2O_5 powder) was put on the bottom of the vessel for remeving water vapor from the sampled air.

The radon concentration in air $Q(Bq/m^3)$ is obtained from the equation

$$Q = \frac{D-B}{KT} \qquad (9)$$

where D (cm^{-2}) is etched pit density; B (cm^{-2}), background density; K, the conversion factor, and T (h), exposure time. The conversion factor was estimated to be $(4.75 \pm 0.41) \times 10^{-2}$ ($cm^{-2}/(Bg\ mm^{-3}\ hr)$) from comparing with the value by using the DSC.

Detection Limits of Instruments. The measuring range of those six instruments are shown in Figure 1. All of the instruments except for the DSC are able to detect the minimum radon concentration at 1 Bq/m^3. Since the radon concentration in outdoor air is several Bq/m^3 (e.g. Shimo and Ikebe, 1979), five out of six techniques (excluding the DSC) are comparable of detecting radon in the outdoor air. The DSC can only be used when the radon concentration is higher than 40 Bg/m^3.

Experimental Procedure

Several measurements were simultaneously carried out with four
instruments, excluding the ERM and the PRM. The experimental
arrangement is shown in Figure 2. The pathways 1 and 2 were used
for sampling outdoor and indoor air, respectively. Air sampled from
a 66.3 m^3 room was exhaused back to indoors to keep indoor radon
concentration relatively constant. The indoor radon concentration
ranged from 5 Bq/m^3 to 110 Bq/m^3 (pathway 2 in Figure 2).
    The sampling time of the DSC was very short because the air was
sampled into an evacuated chamber (1.5 liter volume); However, the
ACC and the TF were sampled during 1 hour at the flow rate of
5 ℓ/min and 40 ℓ/min, respectively. The current measurements of the
DSC and the ACC were made beginning at 3 hours after end of
sampling; The measurement of the TF was performed during the 4096
seconds immediately after the end of sampling. The PFC was operated
continuously in this work and the radon concentration has been
estimated from the counts per hour. The FP was operated with a 15
min sampling-time and a 40 min counting-time, so that a value was
obtained once an hour.
    The simultaneous measurements were performed in 5 rounds with
more than 8 measurements for the four instruments except the DSC
(one measurement) made in each round. The indoor measurement with
the ERM was continuously made in the laboratory together with the
PFC for several months afer making simultaneous measurements using
the five devices. These results were compared with each other.
    The measurements with the PRM was performed over the two years
following the period mentioned above. The PRM was exposed for two
months for each measurement. This approach gave an average
concentration in two months. The results obtained in outdoor and
indoor air were compared with the corresponding data from the PFC
and the ERM.

Results and Discussion

Figure 3 is an example of simultaneous measurements of radon
concentration in indoor air. The measurements shown in this figure
were made under changing condition; the indoor radon concentration
increased after the room had been closed tightly.
    Table 2 is a summary of the experimental results from all
methods; This table gives the ratio of each concentration obtained
from the DSC, the PFC, the TF, and the FP to that from the ACC, and
the ratio of each value from the ERM and the PRM to that from the
PFC. The results in Table 2 show that good agreement was obtained
between the DSC and the ACC. However, the ratio of the results from
the PFC to that of the ACC was 1.15 ± 0.33. Furthermore, the value
obtained from the TF was usually smaller than that from the ACC, and
the mean value was 0.68 ± 0.18. The value obtained from the FP was
smaller than that obtained from the ACC. This results is because
radon and its daughters are usually at non-equilibrium state in the
atmosphere.
    On the other hand, a good agreement between the ERM and the PFC
was obtained. Also, the value from the PRM agreed with that from
the PFC, whereas the simultaneous measurements between the ERM and

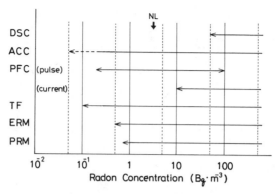

Figure 1. The detection range of various instruments, NL: Natural radon concentration level.

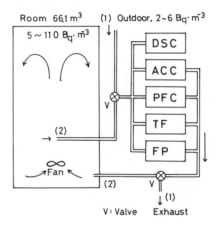

Figure 2. Schematic diagram of the simultaneous sampling system.

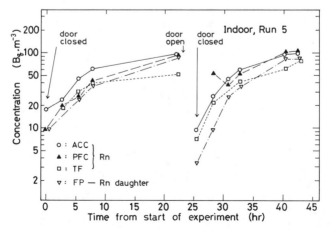

Figure 3. Radon and daughter concentrations obtained from various instruments.

Table II. The ratio of radon concentration obtained from various instruments

Round	DSC/ACC	PFC/ACC	TF/ACC	FP/ACC	ERM/PFC	PRM/PFC	No. of meas.	Remarks
1	—	—	0.72±0.08	—	—	—	8	Indoors
2	—	1.30±0.27	—	0.52±0.09	—	—	10	Indoors
3	—	—	0.54±0.44	0.43±0.09	—	—	12	Outdoors
4	—	1.19±0.20	0.78±0.21	—	—	—	12	Indoors
5	1.03	0.82±0.19	0.70±0.09	0.62±0.20	—	—	10	Indoors
6	—	—	—	—	1.04±0.21	—	45	Indoors
7	—	—	—	—	—	1.15±0.36	8	Outdoors
mean ± SD	1.03	1.15±0.33	0.68±0.18	0.52±0.15	1.04±0.21	1.15±0.36		
RC	—	0.97	0.91	0.97	0.92	0.30		

SD: standard deviation, RC: Relative coefficient

the PFC and between the PRM and the PFC were independently performed from other devices. One can indirectly compare values obtained from these two methods to those from other methods through the PFC. Finally, it was concluded that the values obtained by the DSC, the ACC, the PFC, the ERM and the PRM except fro the TF agreed with each other.

Conclusion

An intercomparison between different instruments for measurement of radon concentration was carried out using indoor and outdoor air as sampling gas. Consequently, the values obtained by using the PFC, the ACC, the ERM and the PRM based on the DSC agreed with each other. However, the value obtained with the TF was smaller than those obtained with other methods.

The availability and suitability of those devices are briefly descrived as follows; In high radon concentration, greater than 10 times outdoor levels, the DSC is suitable because air can simply be sampled in a short time. On the other hand, the ACC is useful for very low concentration but the operation is troublesome. For continuously measurement of the radon concentration over long period, the PFC and the ERM are available. The TF can be used for obtaing the radon and thoron concentrations. The PRM is useful for simultaneously obtaining many radon values at various locations. The FP may be used to estimate a rough radon concentration.

References

Alter, H. W. and R. L. Fleischer, Passive integrating radon monitor for environmental monitoring, Health Physics, 40: 693–702 (1981).

Annanmaki, M., H. Koskela, M. Koponen, and O. Parviainen, RADOK: An Integrating, Passive Radon Monitor, Health Physics, 44: 413–416 (1983).

Boag, J. W., Ionization Chamber, in Radiation Dosimetry (F. H. Attix, W. C. Roesch, and E. Tochilin ed) vol. II, pp. 30–36, Academic Press, New York (1966).

Budnitz, R. J., Radon-222 and its daughters --- A Review of Instrumentation for Occupational and Environmental Monitoring, Health Physics, 26: 145–163 (1974).

Dalu, G. and G. A. Dalu, An Automatic Counter for Direct Measurements of Radon Concentration, Aerosol Science, 2: 247–255 (1971).

Dua, S. K., P. Kotrappa, and P. C. Gupta, Infuluence of Relative Humidity on the Charged Fraction of Decay Products of Radon and Thoron, Health Physics, 45: 152–157 (1983).

Herley, J. H., Sampling and Measurement of Airborne Daughter Products of Radon, Nucleonics, 11: 12–15 (July) (1953).

Iida, T., An Electrostatic Radon Monotor for the Continuous
Measurement of Environmental Radon, in Atmospheric Radon Families
and Environmental Radioactivity (S. Okabe, ed) pp. 65-73, Atomic
Energy Society of Japan, Tokyo (1985) (in Japanese).

Iida, T., Passive Integrating Radon Monitor, Hoken-butsuri, 20:
407-415 (1985).

Ikebe. Y., M. Shimo, and T. Iida, A study on the Behavior of
Atmospheric Radioactive Aerosols in Facilities, Department of
Nuclear Engineering, Faculty of Engineering, Nagoya University,
Nagoya(1979) (in Japanese).

Ikebe, Y., T. Iida, M. Shimo, H. Ogawa, J. Maeda, T. Hattori, S.
Minato, and S. Abe, Evaluation by alpha track Detectors of Rn
Concentrations and f Values in the Natural Environment, Health
Physics, 49: 992-995 (1985).

Kawano, M. and S. Nakatani, Some properties of Natural Radioactivity
in the Atnmosphere, in The natural Radiation Evviroment, (J. A. S.
Adams and W. M. Lowder, ed) pp. 291-312, Univ. Chicago Press,
Chicago (1964).

Kotrappa, P., S. K. Dua, N. S. Pimpal, P. C. Gupta,K. S. V. Nambi,
A. M. Bhagwat, and S. D. Soman, Passive Measurement of Radon and
Thoron Using TLD of SSNTD on Electrets, Health Physics, 43: 399-404
(1982).

Megumi, K. and T. Mamuro, A Method for Measuring Radon and Thoron
Exhalation from the Ground, J. Geophysical Redsearch, 77: 3051-3056
(1972).

Negro, V. C. and S. Watnick, "FUNGI" A Radon Measuring Instrument
with Fast Response, IEEE Trans. Nuclear Science, 25: 757-761 (1975).

Porstendörfer, J. , A. Wicke, and A. Schraub, Method for a
Continuous Registration of Radon, Thoron, and Their Decay Products
Indoors and Outdoors, in The Natural Radiation Environment III, (T.
F. Gesell and W. M. Lowder, ed) pp. 1293-1307, Technical Information
Center/U.S. DOE, Springfield (1980).

Shimo, M. and Y. Ikebe, Radon-222, Radon-220 and Their Short-Lived
Daughter Concentrations in the Atmosphere, Hoken-Butsuri, 14:
251-259 (1979) (in Japanese).

Shimo, M., Y. Ikebe, J. Maeda, R. Kamimura, K. Hayashi, and A.
Ishiguro, Experimental Study of Charcoal Adsorptive Technique for
Measurement of Radon in Air, J. Atomic Energy of Japan, 25: 562-570
(1983) (in Japanese).

Shimo, M., A Continuous Measuring Apparatus using Filter-Sampling
Technique for Environmental Radon Daughters, Research Letters on
Atmospheric Electricity, 4: 63-70 (9184).

Shimo, M., A Flow-Type Ionization Chamber for Measuring Radon Concentration in the Atmospheric Air, in <u>Atmospheric Radon Familiers and Environmental Radioactivity</u> (S. Okada, ed) pp. 37-42, Atomic Energy Society of Japan, Tokyo (1985) (in Japanese).

Tachino, T., Y. Ikebe and M. Shimo, Wall Effect of Ionization Chambers in Measurement of Ionization Current due to Several Gaseous Alpha Radioactive Sources --- Monte Carlo Calculation, <u>J. Atomic Energy Society of Japan</u>, 16: 626-631 (1974) (in Japanese).

Thomas, J. W. and P. C. LeClare, A Study of the Two-Filter Method for Radon-222, <u>Health Physics</u>, 18: 113-122 (1970).

Urban, M. and E. Piesch, Low Level Environmental Radon Dosimetry with a Passive Trach Etch Detector Device, <u>Radiat. Prot. Dosim.</u>, 1: 97-109 (1981).

RECEIVED August 20, 1986

Chapter 14

Improving Bare-Track-Detector Measurements of Radon Concentrations

H. Yonehara[1], H. Kimura[1], M. Sakanoue[2], E. Iwata[2], S. Kobayashi[3], K. Fujimoto[3], T. Aoyama[1], and T. Sugahara[4]

[1]Department of Experimental Radiology, Shiga University of Medical Science, Otsu 520-21, Japan
[2]Low-Level Radioactivity Laboratory, Kanazawa University, Wake, Tatsunokuchi, Ishikawa 923-12, Japan
[3]National Institute for Radiological Sciences, Chiba 260, Japan
[4]Health Research Foundation, Kyoto 602, Japan

The accuracy of the measurement of radon concentrations with bare track detectors was found to be unsatisfactory due mainly to the changes of the deposition rate of radon progeny onto the detector as a result of air turbulence. In this work, therefore, a method was developed which can correct the contributions of the deposition to the track densities by classifying the etched tracks according to their appearance, i.e. round or wedge shaped. Using this method, about 30% improvement in the error of measurements was achieved. The calibration coefficient obtained by experiment was 0.00424 tracks/cm^2/h/(Bq/m^3), which agreed well with the calculated value. Comparison was also made of the present method with other passive methods, charcoal and Terradex, as to their performance under the same atmosphere.

Various types of dosimeters have been developed for the estimation of exposures to radon (Rn-222) and its progeny in dwellings. Most of the dosimeters are the passive integrating monitor type using α-track etch detectors. They are classified into bare detectors, detectors with a diffusion chamber, and those using electrostatic devices. The detectors with a diffusion chamber (Alter and Fleischer, 1981; Urban and Piesch, 1981) are covered with a filter or a membrane in order to prevent the entrance of radon progeny and thoron (Rn-220), and they are considered to be accurate. However, they are somewhat expensive and their sensitivity is not so high. While those using electrostatic devices (Kotrappa et al., 1982; Ikebe et al., 1984) are useful for precise measurements of low level radon, they are too expensive to make long-term measurements at numerous locations. The bare track detectors had been studied before the two other types of detectors were developed (Rock et al., 1969; Alter and Fleischer, 1981). The devices are suitable for measurements in a large number of dwellings, because the detectors are very inexpensive and can be sent and set up easily. The materials used for alpha particle detection are an allyl diglycol carbonate, called CR-39, or cellulose nitrate (CN). CR-39 is the most sensitive material at present and the energy range of α-rays which can produce tracks is wide. The bare detectors are sensitive not only to airborne activities but also to activities deposited on the surface of detectors. As the deposition rate of radon progeny onto the surface of detectors varies largely with the turbulence of air and the unattached fraction of radon daughters, the calibration coefficient is not constant in various locations. In this work a method was developed which can correct the errors due to the variation of the deposition rate of daughters and the equilibrium factor F in the measurements by the bare CR-39 detector. Measurements of radon concentrations using this method were performed at 251 dwellings in Japan for about one year (Aoyama et al., 1986).

Outline of the Method

The sources of α-rays which produce the tracks on the bare CR-39 detectors are divided into airborne activity and activity deposited on the surface of the detectors. The relationship between time-averaged radon concentration (C_0) and the track density (T) on the bare track detector is represented by

$$T = K \, C_0 \, t = (K_{df} + K_{da} + K_a) \, C_0 \, t$$

and K_{df}, K_{da}, K_a are given by

$$K_{df} = \{ G_1 + (1 - P) \, G_3 \} \, V_g^f \, f_1 \, \frac{C_1}{\lambda_1 C_0} + G_3 \, V_g^f \, f_2 \, \frac{C_2}{\lambda_2 C_0}$$

$$+ G_3 \, V_g^f \, f_3 \, \frac{C_3}{\lambda_3 C_0} \tag{1}$$

$$K_{da} = \{ G_1 + (1 - P) \, G_3 \} \, V_g^a \, (1 - f_1) \, \frac{C_1}{\lambda_1 C_0}$$

$$+ G_3 \, V_g^a \, (1 - f_2) \, \frac{C_2}{\lambda_2 C_0} + G_3 \, V_g^a \, (1 - f_3) \, \frac{C_3}{\lambda_3 C_0} \tag{2}$$

$$K_a = E_0 + E_1 \, \frac{C_1}{C_0} + E_3 \, \frac{C_3}{C_0} \tag{3}$$

where $E_1 = E_1^f \, f_1 + E_1^a \, (1 - f_1)$,

$E_3 = E_3^f \, f_3 + E_3^a \, (1 - f_3)$,

K = total calibration coefficient,

K_{df} = calibration coefficient for activity of unattached daughters deposited on the detector,

K_{da} = calibration coefficient for activity of attached
        daughters deposited on the detector,

K_a = calibration coefficient for activity of airborne
      nuclides,

C_i = time-averaged activity concentration of nuclide i,

λ_i = decay rate constant of nuclide i,

G_i = geometrical detection factor for a deposited nuclide i,

E_i = geometrical detection factor for an airborne nuclide i
      depending on a variation of the activity concentration
      with distance from the detector,

$E_i^{\ f}$ = E_i for unattached daughters,

$E_i^{\ a}$ = E_i for attached daughters,

f_i = unattached fraction of a daughter i,

$V_g^{\ f}$ = deposition velocity for unattached daughters,

$V_g^{\ a}$ = deposition velocity for attached daughters,

P = permanent recoil loss probability from the detector for
    Pb-214 atom,

t = exposure time,

the subscripts i = 0, 1, 2, and 3 denote Rn-222, Po-218, Pb-214
and Bi-214 (or Po-214), respectively.

According to the UNSCEAR 1982 report, it can be assumed that
the equilibrium factor F ranges from 0.3 to 0.7 and f_1 ranges from
0.01 to 0.1. It has been reported that $V_g^{\ f}$ ranges from 0.1 to 0.5
cm/sec (Scott, 1983). Taking the variations of F, f_1 and $V_g^{\ f}$ into
consideration, the variation of K is too large to be neglected.
Thus the bare detector can be applied only in the dwellings where
the variations of those factors are small.

Figure 1 shows a microphotograph of tracks produced by radon
and its progeny. It is evident from Figure 1 that the tracks can be
classified into round (R) and wedge-shaped (W) tracks. The characteristic for the formation of the two types of tracks was studied by
experiment and the results are shown in Figure 2. This figure shows
the cross section of the solid in which the α-rays produced the two
types of tracks. In this figure, the distance corresponds to the
residual range of α-ray at the incidence point P in the unit of cm
of air equivalent thickness (r). If an α-ray which enters the
detector through the point P stops in region R, it produces a round

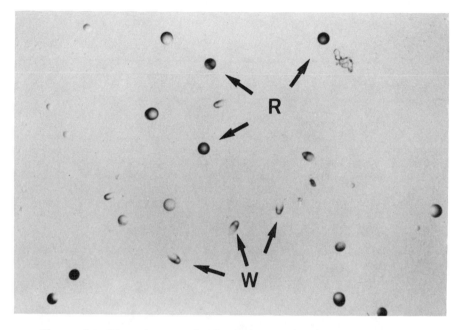

Figure 1. Microphotograph of tracks on the detector produced by radon and its progeny. R shows round tracks and W shows wedge-shaped tracks.

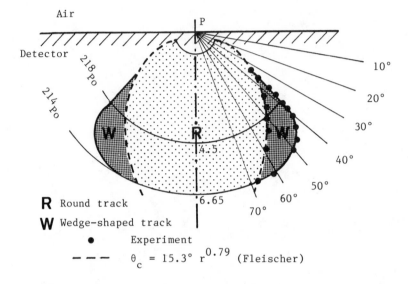

Figure 2. Characteristic for the formation of the two types of track. The solid circle shows the result obtained by the experiment, and the dashed line is an approximation presented by Fleischer (Fleischer, 1984). (θ_c :critical etching angle; r:residual range of alpha particle at the incidence point P in cm of air).

track. If it stops in region W, it produces a wedge-shaped track.
It can be seen from this figure that whether a round track or a
wedge-shaped track is produced is related to the incidence angle and
the residual range of the α -ray at the incidence point on the surface of detector. It can be shown that for the tracks produced by
the deposited Po-218 and Po-214, the ratios of the round track density and the wedge-shaped track density to the total track density
are 0.60 and 0.40, respectively. If the respective ratios for airborne activities are denoted as a and b, the round track density
(T_r) and the wedge-shaped track density (T_w) are represented by

$$T_r = 0.60 \ T_d + a \ T_a$$

$$T_w = 0.40 \ T_d + b \ T_a$$

where T_d and T_a denote total track densities produced by deposited
activities and airborne activities, respectively. From the above
simultaneous equations, T_a is given by

$$T_a = \frac{1}{a - 1.5 \ b} (T_r - 1.5 T_w) \tag{4}$$

While T_a is also represented by

$$T_a = (E'_0 + E'_1 \frac{C_1}{C_0} + E'_3 \frac{C_3}{C_0}) \ C_0 \ t \tag{5}$$

where $E_1' = E_1'^f \ f_1 + E_1'^a \ (1 - f_1)$
     $E_3' = E_3'^f \ f_3 + E_3'^a \ (1 - f_3)$

and E'_i denotes detection factor for the total track density caused
by airborne nuclide i, including that for the density of wedge-
shaped track; $E_i'^f$ and $E_i'^a$ denote E_i' for unattached and attached
daughters, respectively. We define here the corrected track density
(T_c) as

$$T_c = T_r - 1.5 \ T_w \tag{6}$$

From Equations 4, 5, and 6, the relationship between T_c and C_0 can
be represented by

$$T_c = K_c \ C_0 \ t$$

$$K_c = (a - 1.5b) \ (E'_0 + E'_1 \ \frac{C_1}{C_0} + E'_3 \ \frac{C_3}{C_0} \ )$$

The calibration coefficients K and K_c were calculated on the following assumptions:

(1) The concentration of radon gas in air is assumed to be constant regardless of the distance from the surface of detector.

(2) The variation of concentration of a daughter i with distance (x) from the surface of a detector is assumed to be given by

$$C_i'(x) = C_i \ (1 - e^{-x \sqrt{\lambda_i / D}}) \tag{7}$$

where $C_i'(x)$ = concentration of a daughter i at the point which is a distance x away from the surface of the detector,

C_i = concentration of a nuclide i in free air,

D   = diffusion coefficient.

Equation 7 is an approximation used by Fleischer (Fleischer, 1984).

(3) The value of D is assumed to be 0.05 cm^2/s for unattached daughters, and assumed to be 1 x 10^{-5} cm^2/s for attached daughters.

(4) The concentrations of unattached Pb-214 and Bi-214 (Po-214) are assumed to be negligible ($f_2 = 0$, $f_3 = 0$).

(5) The deposition velocity of attached daughters (V_g^a) is assumed to be constantly 1% of V_g^f (Jacobi, 1972).

(6) The permanent recoil loss probability from the detector for Pb-214 atom is assumed to be 0.5 (P = 0.5) (Jonassen, 1976).

The values of the detection factors obtained by calculation using these assumptions and the data of Figure 2 are shown in Table I.

Table I. The Detection Factors Obtained by Calculation

E_0	E_1^f	E_1^a	E_3^a	E_0'	$E_1'^f$
0.00176	0.000614	0.00195	0.00249	0.00182	0.000625

$E_1'^a$	$E_3'^a$	G_1	G_3	G_1'	G_3'
0.00207	0.00308	0.116	0.0367	0.0763	0.0260

In Table I, G_i denotes the geometrical detection factor for the round track produced by a deposited daughter i and G_i' denotes that for the wedge-shaped track. The variation of K_c and K calculated using the values in Table I are shown in Figure 3. In the calculation of K, it was assumed that the wedge-shaped tracks are not counted in the conventional method of track counting. Although K varies largely with the equilibrium factor (F), the unattached fraction of Po-218 (f_1), and the deposition velocity of unattached daughters (V_g^f), K_c varies slightly with these factors. According to the previous studies, it can be assumed that the mean value and standard deviation (S.D.) for F are 0.5 and 0.2, those for f_1 are 0.06 and 0.05, and those for V_g^f are 0.3 and 0.2 cm/sec, respectively. Under these assumptions, the mean value of K is 0.00737 tracks/cm^2/h/(Bq/m^3) and the relative errors (1 S.D.) due to the variations of F, f_1 and V_g^f (V_g^a) are 17%, 22%, and 28%, respectively, and the combined error due to the variations of these factors is estimated to be 39%. On the other hand the mean value of K_c is 0.00375 and the errors due to F, f_1, and V_g^f (V_g^a) are 13%, 1%, and 0%, respectively, and the combined error is 13%.

Detector and Procedure after Exposure

The detectors were 5 x 2.5 cm pieces of CR-39 (Solar Optical Japan, Osaka, Japan) which were fixed on cardboard. They were set up on the wall or in the other places in the dwellings to be investigated. Following exposure, the CR-39 pieces were etched in 30% solution of NaOH at 70°C for 5 hours. The tracks were scored at total magnification of x1210 using T.V. assisted optical microscopy. The counting area was 3 cm^2 at most, corresponding to 750 microscope fields.

Experimental Calibration

The six experiments for calibration were made in two chambers. One was a laboratory with a volume of about 23 m^3 and the other was a cellar with a volume of about 15 m^3. Both chambers were closed tightly in order to reduce changes in the radon concentration. The radon concentrations were measured by a grab sampling method using activated charcoal in a column set in dry ice and ethanol (Shimo et al., 1983). The activities of Pb-214 were measured with a solid state detector(ORTEC Ge-LEPS). The detector was calibrated with reference sources and the total system of measurement was cross-calibrated with a grab sampling method using an ion chamber. The concentrations of the radon daughters were also measured with a filter method using ZnS scintillation counter. The counter was

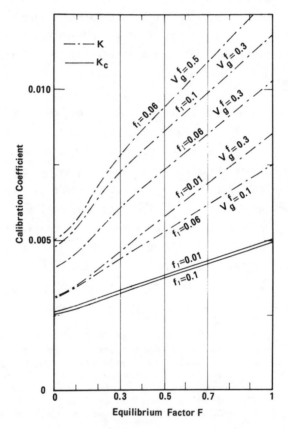

Figure 3. Variation of calibration coefficients, Kc and K, derived from the calculation.

calibrated with a standard uranium source. These measurements were
made every day or at least every third day. In the two experiments,
a fan was used in order to observe the effect of the increase of the
deposition velocity of the daughters. The velocity of wind of the
fan was 171 cm/sec. The conditions and results of the experiments
are shown in Table II. The relationships between the time-averaged
radon concentrations and the track densities per unit time are shown
in Figure 4. It can be seen from this figure that a linear
relationship exists between the average radon concentration and the
corrected track density (T_c) obtained by the present method.
However, as regards the conventional method, a linearity exists only
in the case when the fan was off, and the track densities T in the
turbulent atmosphere were higher than those in the still atmosphere.
The calibration coefficients K_c and K were calculated with a multi-
step method of linear regression analysis (Skinner and Nyberg,
1983). According to the experimental data, the calibration coefficients K_c and K were 0.00424 and 0.0172 tracks/cm^2/h/(Bq/m^3),
respectively. The mean value of K_c determined from the experimental
data is higher by 13% than that previously derived purely on the
basis of calculation.

Error Estimation

The total error for the measurements with the present method is made

up of various errors as follows:

(1)The statistical error of the track counting,

(2)The error due to ambiguities in classification and discrimination

of tracks,

(3)The error due to the parameters such as F, f_1, and V_g^f (V_g^a),

(4)The error due to the variations of the background, the condition

of chemical etching, and the batch variation of the materials,

(5)The error due to the ratio of thoron concentration to radon.

As there are few data on the ratio of thoron concentration to radon,

we have assumed that the value is constant and is equal to 0.14,

Error(4) and Error(5) are assumed to be negligible for the error

calculation. The error due to ambiguities in classification and

discrimination of tracks according to the shape or the size was

assessed by experiment and the relative error (1 S.D.) was less than

10%. The relative error due to the variations of F, f_1, and V_g^f

(V_g^a) is less than 13% as described in the previous section. The

combined error due to these factors is estimated to be 16%. The to-

tal error can be obtained by combining this error and Error(1). If

we use the definition that the lowest detectable limit is the radon

concentration at 50 % relative S.D., the lowest detectable

Table II. Conditions and Results of Calibration Experiment

Volume of Chamber (m^3)	Duration (Days)	Average Radon Conc. $\pm$ S.E. (Bq/m^3)	F	Fan	Tc $\pm$ S.E. (tracks/cm^2/h)	T $\pm$ S.E. (tracks/cm^2/h)
23	20	98.0 $\pm$ 9.2	0.44	off	0.386 $\pm$ 0.016	1.12 $\pm$ 0.05
23	25	70.7 $\pm$ 7.8	0.54	off	0.304 $\pm$ 0.018	0.733 $\pm$ 0.010
23	14	60.7 $\pm$ 4.6	0.61	off	0.271 $\pm$ 0.021	0.672 $\pm$ 0.025
23	15	40.3 $\pm$ 2.4	0.37	on	0.164 $\pm$ 0.014	1.25 $\pm$ 0.06
15	5	688 $\pm$ 73	0.62	off	3.03 $\pm$ 0.28	7.54 $\pm$ 0.624
15	8	540 $\pm$ 32	0.37	on	2.17 $\pm$ 0.18	13.9 $\pm$ 0.411

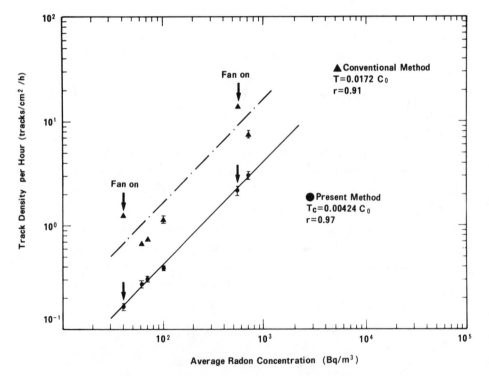

Figure 4. Relationships between the time-averaged radon
concentration and the track densities per unit time. Error
bars represent standard error of track counting.

limit after exposure over 3 months is estimated to be 1.2 Bq/m^3. In this calculation it has been assumed that the area of counting fields for the sample is 3 cm^2, that for the background is 11 cm^2, the round track density due to background is 22 tracks/cm^2, and the wedge-shaped track density due to background is 0 tracks/cm^2, which are the practical or the average values for application in our experiments.

Comparison with Other Methods

In order to assess the accuracy of the present method, we compared it with two other methods. One was the Track Etch detector manufactured by the Terradex Corp. (type SF). Simultaneous measurements with our detectors and the Terradex detectors in 207 locations were made over 10 months. The correlation coefficient between radon concentrations derived from these methods was 0.875, but the mean value by the Terradex method was about twice that by our detectors. The other method used was the passive integrated detector using activated charcoal which is in a canister (Iwata, 1986). After 24 hour exposure, the amount of radon absorbed in the charcoal was measured with NaI (Tl) scintillation counter. The method was calibrated with the grab sampling method using activated charcoal in the coolant and cross-calibrated with other methods. Measurements for comparison with the bare track detector were made in 57 indoor locations. The correlation coefficient between the results by the two methods was 0.323. In the case of comparisons in five locations where frequent measurements with the charcoal method were made or where the radon concentration was approximately constant, the correlation coefficient was 0.996 and mean value by the charcoal method was higher by only 12% than that by the present method.

Discussion

Calibration Coefficients. Two mean values of the calibration coefficient K$_c$ are presented. The mean value of K$_c$ obtained by calculation was 0.00375 tracks/cm^2/h/(Bq/m^3), and that obtained by experiment was 0.00424. Taking account of the ambiguities in the assumptions for the calculation and the experimental errors, there is no significant difference between the two values of K$_c$. If the ambiguities and the experimental errors are negligible, the difference is due to the variation of the ratio of thoron concentration to radon. When the ratio is assumed equal to 0.14, the mean value of K$_c$ obtained by calculation agrees with the experimental result. There are very few data for this ratio in dwellings. From a small study, Schery (1985) reported that the mean value and S.D. for the ratio were 0.23 and 0.37, respectively. If the values could be applied to a dwelling which has a general distribution, the mean value of K$_c$ is calculated to be 0.00458 and the combined error due to the variations of the ratio, F, and f$_1$ is calculated to be 29%.

Comparison with Other Methods. We compared the present method with other methods. The reason for poor correlation between the charcoal

method and the present method seems to be due to the large variation
in radon concentration with change in period and position of the
measurement and due to change in ratio of thoron concentration to
radon. Because these two methods agree with each other at various
locations where the measurement with charcoal method were frequently
made or the radon concentration was approximately constant, the effect of the variation in ratio of thoron concentration to radon
seemed not to be so significant. However, we need to further investigate the reason for the difference in these methods.

On other hand, we found a good correlation between the results
by the present method and those by the Terradex detector. However,
the mean value obtained by the Terradex detector were about twice
those by the present method. The reasons for this significant difference are unknown and may be due to errors in the calibration experiments and in the conditions during the measurements. In the
calibration experiments the effect of existence of thoron in the
chambers could be one of the reasons. As regards the condition in
the measurements, methods for subtraction of background tracks and
deposition of dust onto the bare detector could be candidates.
However, we do not have enough data to determine the reasons for
some of this difference.

Conclusion

Using the present method, the bare CR-39 detector becomes useful for
the long-term measurements of radon concentrations in various
dwellings. The present method is useful for laboratories which wish
to measure radon concentrations at low cost. The practical calibration coefficient K_c equals to 0.00424 tracks/cm^2/h/(Bq/m^3).

Acknowledgments

We would like to thank Dr. Ikebe and Dr. Iida of Nagoya University
who provided helpful discussions and made comparison measurements.
We are also grateful to Adjunct Professor E. P. Radford of University of Pittsburgh for his critical review of the manuscript.

Literature Cited

Alter, H.W. and R.L. Fleischer, Passive Integrating Radon Monitor
for Environmental Monitoring, Health Physics 40:693-702 (1981).

Aoyama, T., H. Yonehara, M. Sakanoue, S. Kobayashi, T. Iwasaki, M.
Mifune, E.P. Radford and H. Kato, Long-Term Measurements of Radon
Concentration in the Living Environments in Japan: A Preliminary
Report, this volume (1986).

Fleischer, R.L., Theory of Passive Measurement of Radon Daughter and
Working Levels by the Nuclear Track Technique, Health Physics
47:263-270 (1984).

Ikebe, Y., T. Iida, M. Shimo, H. Ogawa, J. Maeda, T. Hattori, S.
Minato, and S. Abe, Evaluation by α-track Detectors of Rn Concentrations and f Values in the Natural Environment, Health Physics
49:992-995 (1985).

Iwata, E., Measurement of Gaseous Radioactive Nuclides in the Atmosphere, Faculty of Science, Kanazawa University, Master Thesis (1986).

Jacobi, W., Activity and Potential α-Energy of ^{222}Radon- and ^{220}Radon-daughters in Different Air Atmospheres, Health Physics 22:441-450 (1972).

Jonassen, N. and J.P. McLaughlin, On the Recoil of RaB from Membrane Filters, J. Aerosol Sci. 7:141-149 (1976).

Kotrappa, P., S.K. Dua, N.S. Pimpale, P.C. Gupta, K.S.V. Nambi, A.M. Bhagwat, and S.D. Soman, Passive Measurement of Radon and Thoron Using TLD or SSNTD on Electrets, Health Physics 43:399-404 (1982).

Rock, R.L., D.B. Lovett, and S.C. Nelson, Radon-daughter Exposure Measurement with Track Etch Films, Health Physics 16:617-621 (1969).

Schery, S.D., Measurements of Airborne ^{212}Pb and ^{220}Rn at Varied Indoor Locations Within the United States, Health Physics 49:1061-1067 (1985).

Scott, A.G., Radon Daughter Deposition Velocities Estimated from Field Measurements, Health Physics 45:481-485 (1983).

Shimo, M., Y. Ikebe, J. Maeda, R. Kamimura, K. Hayashi, and A. Ishiguro, Experimental Study of Charcoal Adsorptive Technique for Measurement of Radon in Air, Journal of Atomic Energy Society of Japan 25:562-570 (1983).

Skinner, S.W. and P.C. Nyberg, Method for the Calculation of Radon Response Characteristics of Integrating Detectors, Health Physics 45:544-550 (1983).

United Nations Scientific Committee on the Effect of Atomic Radiation, Ionizing Radiation: Sources and Biological Effects, 1982 Report to the General Assembly (1982).

Urban, M. and E. Piesch, Low Level Environmental Radon Dosimetry with Passive Track Etch Detector Device, Radiation Protection Dosimetry 1:97-109 (1981).

RECEIVED August 4, 1986

Chapter 15

Determination of Radon-220 and Radon-222 Concentrations in Fumarolic Gases

Kyouko Yamashita, Hideki Yoshikawa, Makoto Yanaga, Kazutoyo Endo, and Hiromichi Nakahara

Department of Chemistry, Tokyo Metropolitan University, Setagaya-ku, Tokyo 158, Japan

A radiochemical method for the determination of Rn-220 in fumarolic gas is studied. Both condensed water and non-condensing gas are collected together and Pb-212 is precipitated as PbS. After dissolving the precipitate in conc.HCl, it is mixed with an emulsion scintillator solution for activity measurements. As Pb-214 is simultaneously measured, the observed ratio of Pb-212 /Pb-214 gives Rn-220/Rn-222. This method is superior to the method of directly measuring Rn-220 for the samples in which Rn-220/Rn-222 ratios are less than unity. This method and the previously proposed direct method were applied in the field, and new data obtained. An attempt was also made to understand the formation and transport of radon underground.

The Rn-222 concentrations in the soil gas, fumarolic gas, atmosphere, and in the underground water have been measured extensively for the studies of seismology, uranium mining, environmental science and geochemistry. It has been known that its concentrations are often very high in fumarolic gases and in the underground water, the reason for which is, however, not clarified yet.

Rn-220 is another isotope of radon and belongs to the thorium decay series. Due to its short half life of 55.6 s, reports on its concentrations in those gases and in natural water are still scant. They are also important for a better estimate of our exposure to natural radioactivity and also for the geochemical study of the formation of those radon isotopes and their underground movement.

In the previous paper(Yoshikawa et al., 1986), we proposed a

0097-6156/87/0331-0186$06.00/0

method for the direct determination of Rn-220 and Rn-222 in fumarolic gases by use of the toluene extraction of radon followed by radioactivity counting with a portable liquid scintillation counter that could be carried to the field. However, because of the finite sampling time and its short half life, an assumption had to be made in order to obtain an absolute concentration at the time of ejection to the ground surface as discussed in detail in ref.(Yoshikawa et al., 1986) and described briefly in the following.

In the present work, an indirect method for the determination of Rn-220 in fumarolic gases has been studied that measures essentially the activity of Pb-212, the progeny of Rn-220, together with a further investigation on Rn-220 and Rn-222 concentrations at various sites. An attempt is also made for understanding of the Rn-220 and Rn-222 data, namely, the formation and transport of the radon gas underground.

Direct determination of Rn-220 and Rn-222

The direct method reported in ref.(Yoshikawa et al., 1986) will be explained briefly so that the reader may understand the reason why an indirect method is proposed in the present work.

A fumarolic gas sample quickly collected into a syringe is shaken with a toluene-PPO scintillator solution, and the radioactivity of the extracted radon is measured by a portable liquid scintillation counter that has three single channel analysers with an automatic print-out and recycle counting system. The decay and growth of the radioactivity is followed for about 20 min during which the activities due to Rn-220, Po-216, and Rn-222, Po-218, Pb-214, Bi-214, Po-214 are mainly observed. The absolute value of the sum of these activities can be obtained by the integral counting method even when the sample contains some quenchers(Homma and Murakami, 1977; Murakami and Horiuchi, 1979). The observed decay and growth curve is then analyzed by the following equation.

$$A(t) = A_0^{Tn} f(t) + A_0^{Rn} g(t) \tag{1}$$

where $A(t)$ is the activity observed at time t; A_0^{Tn} and A_0^{Rn} are the activities of Rn-220 and Rn-222 at the end of the solvent extraction, and $f(t)$, $g(t)$ are the functions that take care of the decay and growth of Rn-220, Rn-222 and their progenies.

As the process of the gas sampling and the solvent extraction has to be carried out quickly (within 3 min at the longest) for the direct measurement of the Rn-220 decay, accurate measurements of the gas volume and the temperature are not possible. Therefore, the coexisting Rn-222 is used as a tracer for the determination of the gas volume and the solvent extraction efficiency; that is, another gas sampling is carried out slowly with a cooling device and the gas volume of the non-condensing gas is accurately measured. Then, the solvent extraction is repeated twice to obtain the extraction efficiency of radon into the toluene. As the Rn-222 concentration should be the same for the rapidly and slowly collected samples, the activity of Rn-220 per unit volume of the non-condensing gas A_{total}^{Tn} can be evaluated as follows :

$$A_{total}^{Tn} = \frac{A_o^{Tn}}{A_o^{Rn}} \times A_{total}^{Rn} \tag{2}$$

where A_{total}^{Tn} and A_{total}^{Rn} are further corrected for the decay to give the Rn-220 and Rn-222 concentrations at the end of the gas sampling. In order to obtain the Rn-220 concentration in the fumarolic gas at the time of the ejection into the atmosphere, a reasonable assumption has to be made on the amount of Rn-220 decayed during the gas sampling. In ref.(Yoshikawa et al., 1986) it is assumed that the ejection rate of the non-condensing gas is constant with the concentrations of Rn-220 and Rn-222 unvaried. Then, the correction factor is

$$\lambda_{Tn} \tau / (1 - \exp(-\lambda_{Tn} \tau)) \tag{3}$$

where τ is the sampling time. As this factor becomes as large as 2.5 for $\tau = 180$ s, it may introduce the largest ambiguity in the direct determination of Rn-220. Some results of field work performed in this work are shown in Table I., and the decay and growth of the total activity is shown in Figure 1 as an example.

Indirect determination of Rn-220

As discussed above, the correction for the decay of Rn-220 during the gas sampling introduces the largest ambiguity in the direct determination of Rn-220. Therefore, another indirect method of measuring the radioactivities of Pb-212 and its progeny has been investigated in the present work. An advantage of this method is that the gas sampling does not have to be carried out fast. Thus the gas volume can be made large so long as the condensed water as well as the non-condensing gas is collected into a sampling bottle quickly. Moreover, the radioactivity measurement does not have to be performed at the sampling site. However, the disadvantage is the decreased sensitivity in the activity measurement due to the longer half life of Pb-212, i.e., the half life of 10.64 hr is about 690 times larger than that of Rn-220. The measurement of Pb-212 actually has a meaning somewhat different from the direct measurement of Rn-220, and it is expected to give some in-

Table I   Determination of Rn-220 and Rn-222 concentrations in volcanic gases

Location	Data (1985)	Temp. (°C)	Collection time (s/100ml)	R-gas (%)	Rn-220 (kBq/l)	Rn-222 (kBq/l)	Rn-220/Rn-222
KUSATSU HOT SPRING (GUNMA PREFEC.)							
Sasshogawara	6.13	95.0	23	1.5	0.04±0.11	0.0274±0.003	1.3
	6.13		10		0.07±0.18		2.1
Yubatake	6.14	82.5	180	31.4	N.D.	0.0103±0.0003	-
USU VOLCANO (HOKKAIDO PREFEC.)							
fumarole I	6.27	648	213	-	N.D.	0.0148±0.003	-
fumarole II	6.27	704	95	-	N.D.	0.03±0.01	-
KIRISHIMA (KAGOSHIMA PREFEC.)							
Tearai hot spring	8.04	98.0	85	-	N.D.	2.814±0.003	-
Kurino hot spring							
Hachiman-Jigoku I	8.05	86.0	75	-	0.9±0.3	0.407±0.003	2.0
Hachiman-Jigoku II	8.05	74.5	105	11	5 ± 1	0.315±0.003	18
	8.05		160		12 ± 3		39
Maeno-Jigoku	8.06	43.0	215	-	4.1±0.3	0.444±0.003	8.8
Ramune-Yu	8.06	49.8	138	-	N.D.	0.55±0.003	-

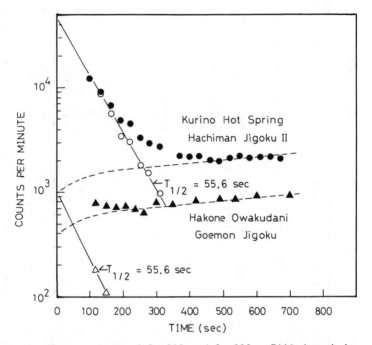

Figure 1.   Decay curves of Rn-222 and Rn-220.   Filled symbols and
open one correspond to the observed activities, and those of Rn-
220 obtained by subtraction of the broken line from the filled
symbols, respectively.   Broken lines and straight lines represent
calculated activities of Rn-222 and its daughters, and decay
curves of Rn-220, respectively.

formation on the amount of Rn-220 initially drifted from the source
into the up-stream of the gas if the decay products of Rn-220 stay in
the gas stream and come up to the surface.(See below later section,
" Formation and transport of radon") In the present work, lead
isotopes were chemically separated from the sample gas as lead sulfide
since the formation of lead sulfide was inevitable under the presence
of H_2S in the fumarolic gas. The lead sulfide was then dissolved in a
small amount of concentrated HCl and mixed with the Insta Gel(emulsion
scintillator solution, Insta Gel, Packard Inc.) for the liquid scintillation counting. The chemical yield and the volume of the collected non-condensing gas were obtained from the measurement of the
activities of Pb-214 and its progeny which were in radioequilibrium
with their precursor Rn-222 whose concentration was determined
separately by the direct method.

Gas collection and sample preparation. The fumarolic gas was collected either by a separatory funnel half-immersed into the water or
by a doubly sealed stainless pipe driven into the vent of the volcanic
gas. It was led through a short polyethylene tube into a vacuum
bottle in which was added beforehand 25 ml of the 0.2 N HCl solution
containing 20 mg Pb^{2+} as a carrier. After the collection of the non-
condensing gas and the condensed water, the bottle was tightly sealed
and left for more than 4 hrs until the radioequlibrium was established
within the bottle between the Pb-212 and its decay products. During
this standing time, the Pb-214 and its decay products initially
present at the time of the sampling would decay out and they would attain a new radioequilibrium with their precursor Rn-222. The bottle
was then vigorously shaken and, when necessary, a few milliliters of
the 0.2 N HCl solution saturated with H_2S were added to ensure the
lead sulfide precipitation. The solution was filtered with a filter
paper with suction and an aliquot amount of the 0.2 N HCl solution
saturated with H_2S was further used to wash the bottle and the
precipitate quickly. The time of the separation of Pb-214 from Rn-222
was recorded at this point. The lead sulfide precipitate was then
dissolved in a small amount of hot conc.HCl (less than 2 ml in
volume) on the filter paper and the filtrate was collected in a 25 ml
low-potassium counting vial. After adjusting the volume to 10 ml with
water, 12 ml of Insta Gel was added and the vial was shaken vigorously
until a single layer of transparent emulsion was obtained. In the
above chemical procedure, Pb and Bi are expected to behave similarly.

Radioactivity measurement The radioactivities of lead isotopes and
their decay products were measured with TRICARB # 3380 liquid scintillation counter (Packard Inc.). The radioisotopes concerned, and their
decay charateristics are shown in Figure 2. In the case of the direct
method, the absolute radioactivity can be obtained by the integral

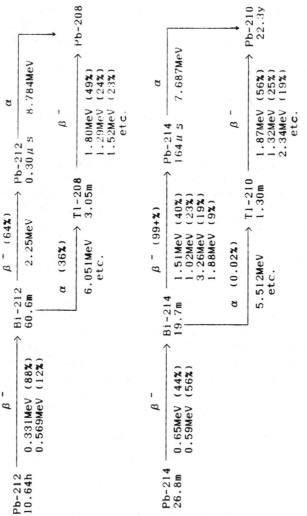

Figure 2.   Decay characteristics of Pb-212 and Pb-214 and their progenies.

counting method(Homma and Murakami, 1977; Murakami and Horiuchi, 1979) in which the activities of Rn-220(α) - Po-216(α) and Rn-222(α) - Po-218(α) - Pb-214(β^-) - Bi-214(β^-) - Po-214(α) are measured. However, in the present case it is doubtful if the integral counting method can be applied to the absolute measurement of Pb-212 and its decay products because the maximum β^--ray energy of Pb-212 is only 0.33 MeV. According to ref.(Horrocks, 1966), this method does not give an absolute activity for β^--ray energies less than 0.25 MeV. Furthermore, the half life of Po-212 is short compared to the gate width of the sum-coincidence circuit, and its α-ray pulse is suspected to be summed up with the β-ray pulse of Bi-212.

In order to know the quenching effect and the number of decays actually observable per decay of Pb-212 for the Insta Gel sample which contains Pb-212 and its decay products in radioequilibrium, the following experiments were performed.

(i) The Rn-220 gas produced from a Th-228 source (confirmed to be free from the nuclides belonging to the uranium series) was extracted into a toluene-PPO solution(Yoshikawa et al., 1986) and the activities of Rn-220-Po-216 and Pb-212-its decay products in radioequilibrium were measured by the liquid scintillation counter using the integral counting method. The result was

$$A_{Pb} \cdot \lambda_{Pb} / A_{Tn} \cdot \lambda_{Tn} = 1.47 \pm 0.07 \qquad (4)$$

where A_{Tn} is the activity of Rn-220 ; A_{Pb} is the sum of the activities of Pb-212 and its decay products, and λ_{Tn} and λ_{Pb} are the decay constants of Rn-220 and Pb-212, respectively. As expected, only 1.47 decays per decay of Pb-212 was found counted in the case of the toluene-PPO solution. It is to be noted that Pb-212 as well as Rn-220 are all extracted if the HCl -Insta Gel solution is used.

(ii) The Pb-212 was extracted into a 1.5 M HDEHP(Di-2-ethylhexyl phosphoric acid)-toluene solution from the 0.01 M HCl solution containing Ra-224. In order to know the effect of the composition of the scintillator solution on the counting efficiency, 100 μl each of the Pb-212-HDEHP toluene solution was added to the toluene-PPO vials and HCl-Insta Gel vials for counting by the integral counting method. The average ratio of the count observed for the HCl-Insta Gel (cocktail of 2 ml conc.HCl + 8 ml H_2O + 12 ml Insta Gel) to that for the toluene -PPO solution was found to be 1.03 ± 0.047.

(iii) In order to see the quenching effect, the amount of HCl in the HCl-Insta Gel cocktail was varied from 1.0 ml to 2.5 ml, and the activities of Pb-212 and its decay products were counted by the integral counting method. It was found that the quenching effect could be corrected by the integral counting method within the above variation of the HCl content.

From these experiments it was concluded that when Pb-212 and its

decay products were in radioequilibrium within the HCl-Insta Gel
vial,(1.45 ± 0.10) decays per decay of Pb-212 were actually counted
by the integral counting method. In Fig.3 is shown an example of the
decay-growth curve of Pb-214, Pb-212 and their progenies. Pb-212 and
its progeny decay with the half life of Pb-212 while Pb-214 and its
progeny decay with the half lives of Pb-214 and Bi-214, namely, 26.8 m
and 19.7 m. The latter decay can be expressed by

$$A(t) = A_1^o \exp(-\lambda_1 t) + 2A_2^o \exp(-\lambda_2 t)$$

$$+ \frac{2\lambda_2}{\lambda_2 - \lambda_1} A^o \{\exp(-\lambda_1 t) - \exp(-\lambda_2 t)\} \tag{5}$$

where A_1^o and A_2^o are the activities of Pb-214 and Bi-214 at the time of
the separation from Rn-222, and λ_1 and λ_2 are the decay constants of
Pb-214 and Bi-214, respectively. In the present chemical procedure,
A_1^o and A_2^o can be considered to be equal. The first part of the decay
curve in Figure 3. was analyzed by Equation 5. to obtain A_1^o. The long
component decayed with the half life of Pb-212 and gave the absolute
activity of Pb-212 when divided by a factor of 1.45 as discussed
above. From the activity ratio of Pb-212 / Pb-214 multiplied by 690
(λ_{Pb} / λ_{Tn} in Equation 4.), the activity ratio of Rn-220 / Rn-222 at
the time of sampling could be evaluated by correcting for the decay of
Rn-222 during the time between the gas sampling and the chemical
separation of Pb-214. Finally, the Rn-220 concentration in the gas
was obtained by multiplying this ratio with the Rn-222 concentration
observed by the direct method.

Results and discussion Some results of the indirect method obtained
in the field work are shown in Table II., and they are compared with
those of the direct method. Results of the two methods are barely in
agreement within one standard deviation. The correction factor of
Equation 3. used in the direct method was found not in a large error
when the sampling time was less than about two half lives of Rn-220.
It was also demonstrated that the indirect method was superior for the
determination of Rn-220 in a sample in which the ratio Rn-220 / Rn-222
was small. In the direct method, the decay analysis of Equation 1.
where the build-up of Po-218 ($T_{1/2}$ = 3.05 m), Pb-214 ($T_{1/2}$ = 26.8 m),
and Bi-214 ($T_{1/2}$ = 19.7 m) is taken into account brings about a large
uncertainty in the evaluation of the Rn-220 concentration even when
the activity itself is not small whereas the decay analysis of Equation 5 gives an unambiguous result since the half lives of Pb-214 and
Bi-214 are quite different from that of Pb-212. The lower limit of
the Rn-220 concentration that can be determined by the indirect method
is calculated to be 0.4 nCi/l for the condition that one-liter of gas
is sampled and that the radioactivity measurement of Pb-212 is con-

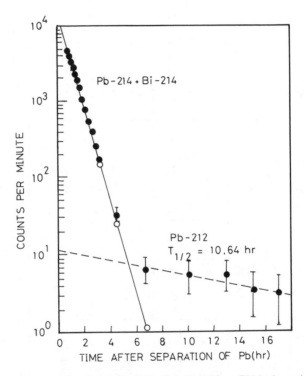

Figure 3.   Decay curves of Pb-212 and Pb-214. Filled symbols and open ones correspond to the observed activities, and those of Pb-214 and Bi-214 obtained by subtraction of the broken line from the filled symbols, respectively. Straight line and broken line represent the decay of Pb-214 and Bi-214, and the decay of Pb-212, respectively.

Table II. Comparison of the direct and indirect method

Location		TAMAGAWA HOT SPRING		HAKONE HOT SPRING
		Ohbuki	Higashimori	Goemon-Jigoku
Temperature (°C)		96	97	84
Gas collection method		orifice of hot spring	gas vent	orifice of hot spring
	Sampling time(s)	79	135~140	4~8
Direct Method	Rn-222 (kBq/l)	0.048 ± 0.003	0.085 ± 0.003	0.066 ± 0.003
	Rn-220 (kBq/l)	5.6 ± 0.4	1.4 ± 0.4 1.1 ± 0.2	0.03 ± 0.07 0.06 ± 0.04 0.09 ± 0.03 0.05 ± 0.07
	Rn-220/Rn-222	115 ± 12	16 ± 4 13 ± 3	0.5 ± 1.1 0.9 ± 0.6 1.3 ± 0.5 0.8 ± 1.1
Indirect Method	Rn-220/Rn-222	136 ± 13	11 ± 1 9 ± 1 11 ± 1	1.5 ± 0.3 2.0 ± 0.3
	Rn-220 (kBq/l)	6.5 ± 0.7	0.93 ± 0.07 0.78 ± 0.07 0.93 ± 0.07	0.10 ± 0.02 0.13 ± 0.02

tinued for 50 min and the net count should be more than two standard
deviations of the background (45 cpm in this work) of the liquid scintillation counter.

Formation and transport of radon

There have been some speculations on the origin of radon in spring gas
and volcanic gas (Iwasaki, 1969). As shown in Table I, concentrations
of Rn-222 and Rn-220 are found as high as 76 nCi/l and 1300 nCi/l,
respectively, and the Rn-220/Rn-222 ratio becomes more than 100 at
some places. It has been known that the elemental content of Th-232
in igneous rocks is about 4 times as large as that of U-238 on the
average and in terms of radioactivity, their concentrations are both
about 1.3 pCi/g(Evans and Raitt, 1935). Besides, the effective mean
free path of Rn-220 in water and in air is much smaller than that of
Rn-222(Fleischer and Mogro-Campero,1977). The question is, then, how
to explain the high concentrations of Rn-220, Rn-222 and the large
  Rn-220/Rn-222 ratio of more than unity in fumarolic gas. In the following an attempt is made to explain the formation and transport of
radon in the fumarolic gas from the data of this work and those in
literature observed at Tamagawa Hot Spring located in Akita
Prefecture, Japan. This hot spring has been well known for the
radioactive mineral, Hokutolite, which is a mixed crystal of $BaSO_4$ and
$PbSO_4$(Minami, 1964). The fumarolic gas is ejected from the orifices
of the hot spring at the bed of a stream at several locations along
the flow. The temperature and the pH of the stream water is 97°C and
1.2, respectively, and Hokutolite grows on the surface of andesite
rocks and on the bank of the stream. An enormous amount of sulfurous
sinter is precipitated in a wide area at the bottom of the stream.
Many investigations have been carried out on chemical compositions of
the water and the gas(Iwasaki, 1959) and also on radioactivities. The
major cations in the water are Al^{3+}(116-230 mg/l), Ca^{2+}(127-227 mg/l),
Fe^{3+}(73-113 mg/l), Mg^{2+}(49-101 mg/l), Na^+(64-114 mg/l), K^+(15-78 mg/l)
whereas anions are Cl^-(2670-3410 mg/l) and SO_4^{2-}(1142-1440 mg/l). The
chemical composition of the noncondensing gas is CO_2(60-85 %), H_2S(4-
37 %), SO_2(less than 0.2 %), and non-acidic gas(1-36 %, hereafter
called R-gas). Radioactivities of radium isotopes in the water and in
Hokutolite were measured by Saito et al.(Saito et al., 1963) and their
results are cited in Table III. The activities contained in the sulfurous sinter were measured in this work by γ-ray spectrometry and
the results are shown in Table IV. From the results in Tables III and
IV, the activity ratios of Ra-224 to Ra-226 and Ra-228 to Ra-226 are
found to be not much different in the water, and in the sulfurous
sinter. The smaller ratio in Hokutolite is known to be due to the
slow sedimentation rate of the mineral which results in the survival
of Ra-228 only at the surface of the mineral(Saito et al., 1963). The

Table III. Activities of radium isotopes at Tamagawa hot spring

	Ra-226	Ra-224/Ra-226	Ra-228/Ra-226
Spring water	0.51~0.73 dps/l	8.0~9.4	14
Hokutolite (5yr)	58 dps/g	6.4	-

Table IV. Activities in sulfurous sinter

Th-232 series Nuclide	dps/g	U-238 series Nuclide	dps/g
Ac-228 (Ra-228)	18 ± 2	Ra-226	1.1 ± 0.2
Pb-212	18 ± 1	Pb-214	1.4 ± 0.1
Bi-212	18 ± 0.7	Bi-214	1.26 ± 0.09
Tl-208	15 ± 2		
Ave.	18 ± 0.5	Ave.	1.29 ± 0.09

constant ratios of the radium isotopes in water and in the deposit
reveal that the ordinary chemical distribution of radium between two
phases takes place. It is highly probable that the sediment at the
bottom of the stream or underground is the source of radon isotopes in
the fumarolic gas. As the half life of Rn-220 is very short, it is
difficult to explain the observed high concentration of Rn-220 in the
gas unless it comes out to the gas phase quickly without the process
of diffusion within solids or water. The fraction of radon atoms that
came out to the air or to the water from fine particles was observed
in the present work as shown in Table V for monazite sand and the sulfurous sinter. The fact that the fraction found in the air is the
same for both Rn-220 and Rn-222 clearly indicates that radon atoms
come out of the solid by the recoil process. If it is assumed that
the sediment which contains a large amount of Ra-228 and Ra-226 with
the activity ratio of roughly 14 to 1 is the source of radon found in
the gas, the Rn-220/Rn-222 ratio is expected to be at most about 10
when radioequilibrium is assumed within the respective decay series.
The decay products of Ra-228, namely, Ac-228, Th-228 and Ra-224 are
all expected to co-precipitate with $PbSO_4$ and $BaSO_4$ under the chemical
condition described above(Kahn, 1951). Then, the large ratio of Rn-
220/Rn-222 has to be explained by the build-up time of radon in the
gas. Let us assume that radon and thoron atoms recoil out into the
gas phase with the build-up time τ_b from the sediment which contains
their parents Ra-226 and Ra-228. The true meaning of τ_b would be the
contact time of the gas with the source sediment. Small gas bubbles
will grow larger as they go through the porous sediment and finally
loose contact with the source, joining into the main stream of the gas
that comes up from the bottom. When the time required for the
transport of the radon-containing gas to the ground surface τ_t after
loosing the contact with the source is taken into account, the following equation is obtained for the activity ratio of Rn-220(T) and Rn-
222(R).

$$\frac{A_T}{A_R} = \frac{n_T}{n_R} \cdot \frac{1 - \exp(-\lambda_T \tau_b)}{1 - \exp(-\lambda_R \tau_b)} \cdot \frac{\exp(-\lambda_T \tau_t)}{\exp(-\lambda_R \tau_t)} \tag{6}$$

where n_T and n_R are the average number of radon atoms transfered into
a unit volume of gas bubbles per unit time. The ratio n_T/n_R may be
taken to be the activity ratio of Ra-228/Ra-226 in the source if a
direct recoiling or a fast transfer of radon atoms into the gas phase
is assumed. The small gas bubbles can be either the ones formed
within the sediment or those strayed from the main up-stream of the

gas into a huge body of the porous sediment. After spending the time τ_b within the sediment, the bubbles will drift toward the main stream of the fumarolic gas and are diluted with the bulk of the gas. Equation 6. can be further simplified if $\exp(-\lambda_R \tau_t) \doteqdot 1$, and $\{ 1 - \exp(-\lambda_T \tau_b) \} \doteqdot 1$. ($\tau_b$ is long enough for the complete build-up of Rn-220 and τ_t is much shorter than the half life of Rn-222.)

$$\frac{A_T \exp(\lambda_T \tau_t)}{A_R} = \frac{n_T}{n_R} \cdot \frac{1}{1 - \exp(-\lambda_R \tau_b)} \qquad (7)$$

If the activity ratio Rn-220/Rn-222 of 136 observed by the indirect method at Ohbuki (see Table II) is taken to be the value of the term on the left hand side of the equation, and the Ra-228/Ra-226 ratio of 14 in Tables III and IV is assumed for n_T/n_R, the build-up time τ_b is evaluated to be 14 hours. If n_T/n_R is as small as one, τ_b becomes one hour. For explanation of the absolute concentration of Rn-220 and Rn-222, a knowledge of n_T, n_R is required. Let us assume, using the result of Table V, that one out of two hundred decays of Ra-224 and Ra-226 within the sediment brings about one atom of Rn-220 and Rn-222 in the gas bubble that has strayed from the main gas stream into the sediment. Activities of Rn-220 and Rn-222 per cubic centimeter of the non-condensing gas observed at Ohbuki are 6.5 dps and 0.048 dps, respectively (see Table II), and those of Ra-228 and Ra-226 per gram of the sulfurous sinter are 18 dps and 1.3 dps, respectively (Table IV). Then, as n_T = 6.5 atoms/cm^3·sec and n_R = 0.47 atoms/cm^3·sec (if τ_b = 14 hrs), the amount of the sediment required for one cubic centimeter of the non-condensing gas to make contact within a unit time in order to explain the observed concentrations is only 72 g for Rn-220 and Rn-222. If the buildup time is 1 hr (n_T/n_R = 1), n_R becomes 6.5 atoms/cm^3·sec, and, therefore, the amount of the sediment required is 1000 g.

As the time required for the gas bubbles to be in contact with the sediment, and the amount of the sediment from which radon atoms have to be gathered into the gas bubble per unit time are not unrealistic, the concentrations of Rn-220 and Rn-222 observed at Ohbuki, Tamagawa hot spring may be explained by the model discussed above, namely, by the model of the presence of the source enriched in Ra-228 and Ra-226 near the surface supplying Rn-220 and Rn-222 to the gas coming up from somewhere deep underground although the explanation of the observed Ra-228/Ra-226 ratio in the sediment has to be sought for as a next problem.

Table V. Fraction of radon atoms escaped from fine particles‡

	Concentration of Rn parent		Fraction of Rn atoms found in air‡‡		Fraction of Rn atoms found in water
	Th-232(dps/g)	U-238(dps/g)	Rn-222	Rn-220	Rn-222
Monazite Sand 1	135 ± 2	16.2 ± 0.8	1/160	1/170	1/90
(Malaysian) 2	133 ± 2	20.3 ± 0.9	1/170	1/110	1/90
3	125 ± 2	14.3 ± 0.7	1/160	1/150	1/170
Sulfurous sinter (Tamagawa)	4.1 ± 0.1	0.30 ± 0.02	1/200	-	-

‡ Fraction is defined as the number of Rn atoms found divided the number of Rn atoms expected from that of the parent

‡‡ Samples were dried before the experiment

Literature Cited

Evans, R. D. and R. W. Raitt, The Radioactivity of the Earth's Crust
and Its Influence on Cosmic-Ray Electroscope Observations Made Near
Ground Level, Phys. Rev. 48(3):171-176 (1935).

Fleischer, R. L. and A. Mogro-Campero, Mapping of Integrated Radon
Emanation for Detection of Long-Distance Migration of Gases Within the
Earth: Techniques and Principles, J. Geophys. Res. 83(B7):3539-3549
(1977).

Homma, Y. and Y. Murakami, Study on the Applicability of the Integral
Counting Method for the Determination of ^{226}Ra in Various Sample Forms
Using a Liquid Scintillation Counter, J. Radioanal. Chem. 36:173-184
(1977).

Horrocks, D. L., Low-Level Alpha Disintegration Rate Determinations
with a One-Multiplier Phototube Liquid Scintillation Spectrometer,
Int. J. Appl. Radiat. Isotopes 17:441-446 (1966).

Iwasaki, I., Geochemical Studies of the Tamagawa Hot Springs, Onsen
Kagaku (Journal of the Balneological Society of Japan) 14(2):27-37 (in
Japanese) (1959).

Iwasaki, I., Onsen Kogaku Kaishi (Journal of the Society of Engineers
for Mineral Springs, Japan) 7:109-114 (in Japanese) (1969).

Kahn, M., Coprecipitation, Deposition, and Radiocolloid Formation of
Carrier-Free Tracers, Radioactivity applied to Chemistry(A. C. Wahl,
and N. A. Bonner, ed) pp. 403-433, John Wiley & Sons, Inc., New York
(1951).

Minami, E., Hokutolite from the Hot Spring Tamagawa, Kobutsugaku Zas-
shi (J. Mineralogical Soc. Japan) 2:1-23(in Japanese) (1964).

Murakami, Y. and K. Horiuchi, Simultaneous Determination Method of
Radon-222 and Radon-220 by a Toluene Extraction-Liquid Scintillation
Counter, J. Radioanal. Chem. 52(2): 275-283, (1979).

Saito, N., Y. Sasaki, and H. Sakai, Radiochemical Interpretation on
the Formation of Hokutolite, Geochemistry of the Tamagawa Hot Springs
(E.Minami, ed), pp. 182-198, Japan (1963).

Yoshikawa, H., M. Yanaga, K. Endo, and H. Nakahara, A Method for
Determinig ^{220}Rn Concentrations in the Field, Health Phys.(in press
1986).

RECEIVED November 10, 1986

Chapter 16

A Critical Assessment of Radon-222 Exhalation Measurements Using the Closed-Can Method

Christer Samuelsson

Department of Radiation Physics, University Hospital, Lund University, 221 85 Lund, Sweden

The immediate response to closing an exhalation can is a depressed radon exhalation rate of the sample enclosed. Ignorance of this phenomenon, which is predicted by the time-dependent diffusion theory, may cause serious underestimation of the free and undisturbed exhalation rate of the sample. The rate of change in exhalation rate is initially maximal. For intermediate sample thicknesses ($0.2L < d < 0.5L$, L= diffusion length) the closed-can method does not give the free exhalation rate if the radon growth is measured after a few hours because of the rapid change from free to steady state bound exhalation. If only a decay correction is applied to a leaking can, the free exhalation rate may be underestimated by more than the factor $\{1+(pd/h)\}$, where p and h are the sample porosity and outer volume height of the can, respectively. The temporal evolution of the exhalation after closing a leaking can is complicated and the use of completely radon-tight cans is highly recommended. Systematic experimental investigations of the temporal behavior of the radon exhalation from porous materials in closed cans are badly lacking.

In a large survey of radon transport phenomena (Collé et al., 1981) several exhalation measurement methodologies were reviewed. One of the laboratory techniques in that review, perhaps the most common one for exhalation measurements of small porous samples, will be scrutinized in this contribution. I am referring to the closed-can accumulation method, which means that the sample to be investigated is enclosed in a can and the exhalation of radon is determined from the radon growth in the air inside the can.

At the time of the NBS review (Collé et al., 1981) in 1981 no one had applied the results of time-dependent diffusion theory to the accumulation closed-can method. Therefore neither the review nor the earlier contemporary state-of-the-art papers (e.g. Jonassen, 1983) could describe properly or quantify the influence from radon

0097-6156/87/0331-0203$06.00/0
© 1987 American Chemical Society

leakage or 'back diffusion'. A phenomenological description of the temporal behaviour of the radon exhalation from a sample inside an accumulating can, is the main issue of this paper.

The time-dependent diffusion theory, with boundary conditions relevant to the accumulating closed-can method, was already at hand for radon-tight cans in 1971 (Krisiuk et al., 1971), but the practical consequences were not drawn until 1983 (Samuelsson and Pettersson, 1984). Lamm and collaborators (Lamm et al.,1983) verified the theoretical results of Krisiuk and extended the theory to leaking vessels. For the purpose of diffusion-length determination the time-dependent diffusion theory model has been utilized by Zapalac (1983). Dallimore and Holub (1982) have fitted their radon exhalation results from closed-can samples allowing radon equilibrium buildup, to a time-dependent diffusion-flow model. The flow was measured and calculated for a sudden opening of the can into a larger volume system. In their paper the pressure difference over the sample was always zero.

The objectives in the following talk are to discuss qualitatively the time behaviour of exhalation in closed cans and then compare these qualitative reasonings with the exact results of the time-dependent diffusion theory. The theoretical results are, to a large extent, presented in diagram form only. Readers interested in the exact equations behind these diagrams are referred to the paper by Lamm (Lamm et al., 1983).

The Sample/Can Geometry and Conditions before Closure

In order to simplify the situation, we assume that our porous sample under investigation covers the bottom of an open straight-walled can and fills it to a height d (Figure 1). Such a sample will exhibit the same areal exhalation rate as a free semi-infinite sample of thickness 2d, as long as the walls and the bottom of the can are impermeable and non-absorbant for radon. A one-dimensional analysis of the diffusion of radon from the sample is perfectly adequate under these conditions. To idealize the conditions a bit further we assume that diffusion is the only transport mechanism of radon out from the sample, and that this diffusive transport is governed by Fick's first law. Fick's law applied to a porous medium says that the areal exhalation rate is proportional to the (radon) concentration gradient in the pores at the sample-air interface

$$E(t)= -D_e \frac{\partial C(z,t)}{\partial z} \tag{1}$$

where D_e is the effective diffusion coefficient
           z is the height within the sample (cf. Figure 1)
           $C(z,t)$ is the concentration of radon in the pore
               volume and
           t is the time.

If we assume that the air surrounding the sample is perfectly mixed, the concentration of radon in the outermost pores (z=d) is always the same as in the air outside the sample. Let us leave the sample in the open can long enough to stabilize the initial conditions. The

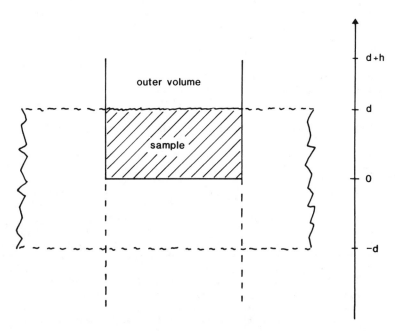

Figure 1. A semi-infinite slab of thickness 2d in open air has the same areal exhalation rate as the open surface of the sub-sample of thickness d in an open can. The height coordinate is denoted by z. When the can is closed the outer volume height is h.

sample is then in a steady state exhaling into free air and the radon concentration at z=d and outside the sample is zero. Our task is now to determine this so-called free exhalation rate, an exhalation undisturbed by the can. We will try to execute our task by closing the can at time zero (t=0), with the intention that, from the initial increase in radon inside the can, we shall be able to calculate the free and undisturbed exhalation rate. How we collect the radon from the can without disturbing the ongoing exhalation is not our main concern for the moment. We simply postulate that this can be done, without introducing pressure changes, dilution etc., and turn back to our main issue: does the time-dependent diffusion theory permit us to measure the free exhalation rate by the technique outlined? This question will be addressed in the following sections.

The Initial Phase after Closure in a Non-Leaking Can

At time zero, when the can is closed we have an depth concentration gradient, the 'driving source' so to speak, in the outer layer of the sample corresponding to free exhalation rate. Let us now imagine what happens during the first few seconds after closing the can. Assume, just for simplicity, that the free exhalation rate corresponds to one thousand radon atoms per second. In a first approximation we have then, after one second, 10^3 radon atoms in the outer volume of the can. The decay probability per atom is only $2.1 \cdot 10^{-6}$ s^{-1}. This means that, as a very good approximation, we can neglect the decay correction during the first few, say 5 minutes. After these 5 minutes we have accumulated $3 \cdot 10^5$ radon atoms in the outer volume of the can. The corresponding concentration is $3 \cdot 10^7$ atoms (or about 100 Bq) per cubic metre, if the outer volume (dead space volume) is say 10 litres. Will this radon concentration significantly affect the free exhalation rate? Well, we cannot settle that question until we know the radon concentration distribution in space and time in the pore volume, and to obtain that we must solve the time-dependent diffusion equation. One obvious conclusion however, can be drawn from the qualitative discussion above: one cannot exclude the possibility that the free exhalation rate is significantly lowered during the first few minutes after closing the can.

Consulting theoretical calculations we see that this is exactly what the diffusion theory predicts (cf. Figure 2). In the example in Figure 2 the free exhalation rate has decreased by about 20 % within 4 minutes after closure. The corresponding radon concentration-depth profile is shown in Figure 3. A significant change in the concentration gradient at the surface is seen. The deeper situated pores are not affected at all in the short time perspective shown.

From Figure 3 we have that the radon concentration after 4 minutes is about 6.5 Bq m^{-3} in the outer volume of the can. Let us assume that the can has a horizontal cross section area of 0.2 m^2. The outer volume is then 20 litres and the concentration 6.5 Bq m^{-3} corresponds to a mean exhalation rate of 2.7 mBq m^{-2} s^{-1}, which is lower than the free exhalation rate by a factor 3.15/2.7= 1.17.

If our mission is an experimental determination of the free exhalation rate we are in difficulties. Still with the sample in Figure 2 as an illustration, we see that even if we take a grab

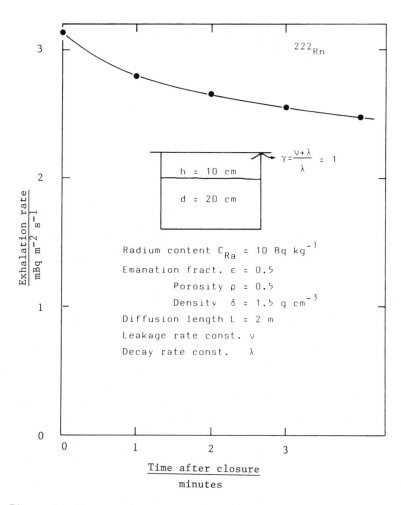

Figure 2.   The areal exhalation rate as a function of time after closing the can. Sample geometry and data as shown (theory).

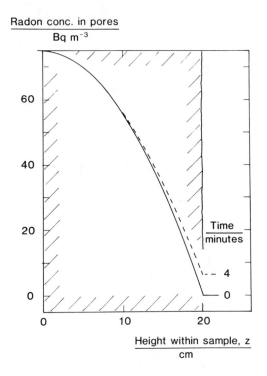

Figure 3. The internal radon concentration of the sample in Figure 2 at time of closure (corresponding to free exhalation rate) and four minutes after closure of the can (theory).

sample from the can volume as early as 4 minutes after closure we underestimate the free exhalation rate systematically. We may in addition run into sensitivity problems. In total we have only 0.12 Bq of radon in 20 litres of air. Which is a measurable amount as a whole, for instance in a large ionization chamber or concentrated into a small ZnS bottle, but poor counting statistics would be expected unless we have detectors of extremely low background and long counting times at our disposal. The only 'light in the dark' is the fact that the radium content of the fictive sample in Figure 2 is rather low. 'Problematic' building materials may contain a factor of ten, or even a hundred, higher concentrations of radium.

The conclusion from the experiment illustrated in Figures 2 and 3 is thus that the exhalation rate will change so rapidly after closure that only a mean exhalation rate, significantly lower than the free exhalation rate, can be measured. This problem can be solved if we choose our sample geometry differently or make another approach to the grab sampling from the can. These things will be dealt with in the next section.

The Intermediate and Final States in a Non-Leaking Can

One, or actually two questions must be raised in connection with the exhalation behaviour exemplified in Figure 2. For how long will the exhalation decrease and what is the final steady-state exhalation value? The latter question is the easier one, as the quotient free exhalation rate, E_0, to final steady-state exhalation with the can closed, E_h, is (cf. Figure 2 for the notation):

$$E_0/E_h = 1 + \{ pL \tanh(d/L) \}/h \qquad (2)$$

For a thin sample (d<0.5L), like that in Figure 2, the quotient is equal to $(1+\alpha)/\alpha$ where α equals $h/(pd)$, the outer to inner volume ratio. Applying Equation 2 to our case in Figure 2 the steady-state exhalation rate is simply half of the free exhalation rate.

In order to answer the former question, how long does it take to change from the free to final steady-state exhalation, we must again consult detailed diffusion theory. It can be shown (Samuelsson, 1984) that the reshaping of the radon depth-concentration gradient is characterized by an exponential sum of the form

$$\sum_i \exp\{-(y_iL/d)^2\lambda t\} \exp\{-\lambda t\} \qquad (3)$$

where the y_i:s are the non-trivial solutions to the transcendental equation $-\alpha y_i = \tan y_i$. (For all non-leakage cases the roots y_i increase monotonically with a difference approximately equal to π, starting somewhere in the interval $\pi/2 \dashrightarrow \pi$). For thin samples, $L/d \gg 1$, all terms in the sum die out very rapidly. This is equivalent to a very rapid change from the free to final steady-state exhalation rate in the closed can. As the time-dependent exhalation rate in the can asymptotically approaches the final steady-state exhalation, a convenient measure is the 95 or 99% (gradient) reshaping time, which is the time it takes for the exhalation to reach within 5 and 1%, respectively, of its final value. The sample illustrated in Figure 2 has a 99% reshaping time of 80 minutes. Theoretically calculated 95% reshaping times as well as the ratio E_0/E_h

(cf. Equation 2) are exemplified in Table I. (Samuelsson and Pettersson, 1984). For samples comparable in thickness, or thicker than
the diffusion length, the reshaping times can be of the order of
days.

Table I. The 95% gradient reshaping times (see text) for different
combinations of sample thickness, d, outer volume height, h, and
diffusion length, L. The quotient free exhalation rate, E_0, and the
final steady-state exhalation rate, E_h, is also given.

Sample Thickness	Outer volume height	Can height	Diffus. length	Free to final steady state exhalation	Reshaping time
d	h	h+d	L	E_0/E_h	t(95%)
cm	cm	cm	cm		minutes
2	48	50	100	1.02	5
10	40	50	100	1.13	80
40	10	50	100	2.90	570
2	48	50	20	1.02	70
10	40	50	20	1.12	1840
40	10	50	20	1.96	6260
2	13	15	100	1.08	4
10	5	15	100	2.00	50
2	13	15	20	1.08	90
10	5	15	20	1.92	1180

(Reproduced with permission from Ref. Samuelsson and Pettersson,
1984. Copyright 1984, Nuclear Technology Publishing).

If we continue to explore what happens to our thin (thin relative to the diffusion length that is) sample in Figure 2, there is
one feature of the time-dependent diffusion theory that is very
important, and perhaps a bit unexpected. The outer volume radon
concentration continues to grow even if the final steady state
exhalation has been reached. If we, for a moment, forget that we had
an exhalation up to twice the final one during the initial 80 minutes, we can use the rule of thumb for a constant supply of atoms
into a closed space: it takes about three half-lives until we reach
saturation activity. For radon it thus takes nearly two weeks for
the maximum concentration in the outer volume to be reached. This
maximum is close to 7500 Bq m^{-3} for the model sample illustrated in
Figure 2. One can say that after the reshaping of the depth-concentration gradient in the pores, the system is in a sort of equilibrium with a parallell increase in concentration in the pores and
the outer volume. During this period the radon concentration in all
pores, irrespective of depth, and in the outer volume as well is
increased by the same absolute amount during an arbitrary time
interval. In other words, the radon concentration-depth gradient

(the concentration derivative with respect to z) is kept constant, but the whole radon concentration distribution in the pores increases uniformly for all depths z until the maximum saturated value is reached. We shall have this time-behaviour of the pore concentration distribution in mind, when we in the next section come to explain the exhalation behaviour in the case of a leaking-can system.

The reshaping time for the sample in Figure 2 is of the order of one hour. The mean exhalation rate during time periods about one magnitude larger than the reshaping time will deviate insignificantly from the final steady-state exhalation rate, E_h. This fact opens up the possibility of an indirect determination of the free exhalation rate, E_0. In our case in Figure 2 we for instance take a grab air sample from the can 10 hours after closure, calculate the mean exhalation rate, which will be very close to E_h, and then Equation 2 will give the free exhalation rate, E_0. The knowledge necessary to use this method is the outer to inner volume ratio, α, and the thickness d of the sample in relation to the diffusion length, L.

If we choose α much larger than 1 (thin samples d<0.5L) or h>>pL (thick samples d>>L), the final steady-state exhalation deviates very little from the free exhalation rate and we do not need to know the reshaping time or use Equation 2 for corrections. An air grab sample taken at any time (and corrected for radioactive decay if necessary) after closure, will yield the free exhalation rate to a good approximation, provided that the can is perfectly radon-tight.

The 'thin sample' case is illustrated in Figure 4, in which the time behavior of the exhalation rate is given with the outer volume height as parameter.

The Exhalation in Leaking Cans

For a leaking can (leak rate constant, ν s^{-1}) the time behaviour of the exhalation is complicated. Equation 2 has to be revised to

$$E_0/E_h = 1 + \{pL \tanh(d/L)\}/h\gamma \qquad (4)$$

where γ is the leakage factor (defined in Figure 2). As in the non-leaking case we assume that the air in the outer volume is perfectly mixed. It follows from this that the elimination of atoms in the outer volume is a first order process, with an elimination rate constant equal to $\nu + \lambda$.

From Equation 4 we see that the final steady-state exhalation, E_h, is dependent on the leakage factor, γ, in a simple fashion. An increased outer height, h, is simply exchangeable with the same factorial increase in leakage factor: a fact we will use later on. The effect of leakage on the reshaping process is more complicated, and it would take too much time and space here to give the detailed analytical results. I will instead present some plausible guesses of how the exhalation rate ought to behave after closing a leaking can, and then check these guesses with the predicted results from time-dependent diffusion theory.

Let us again start from the 20 cm thick sample depicted in Figure 2. The best thing to do is to stay with this sample just as it is, i.e. as a non-leakage case, and then compare it with another

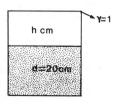

EXHALATION (mBqm^{-2} s^{-1})

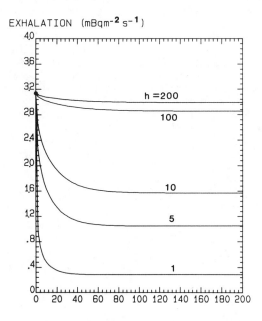

TIME (min)

Figure 4. The areal exhalation rate as a function of time after closing the can, for various value of outer volume height, h. The sample is the same as in Figure 2 (theory).

sample having the same free exhalation and the same final steady-state exhalation rate, E_h, but in a leaking can. Say that we choose a leakage factor, γ, equal to 2. According to Equation 4 we must reduce the free air height above the sample to 5 cm, if the two E_h:s are to be the same. The two cases just described are denoted by 'a' and 'b' in Figure 5.

The initial behaviour of the exhalation during the few minutes in case 'a' have been dealt with above (cf. Figure 2). In the leaking case 'b', the initial depression of the exhalation rate must be even faster and more pronounced due to the smaller outer volume. In fact, during the first few minutes, when the probability of both decay and leakage can be neglected, our leaking can 'b' will behave similarly to the corresponding non-leaking can, in Figure 5 denoted by 'c'.

In the time interval following the fast initial reduction in the exhalation rate, the leaking can 'b' will not be in equilibrium with the outer volume radon concentration increase, as was the case for the radon-tight can 'a'. The reason for this is that the increase in radon concentration in the outer volume, compared with the non-leakage case, is not fast enough due to radon atoms leaking out. The net result of this is that the radon concentration-depth gradient in the sample pores will start to slowly increase again, i.e. the exhalation increases. This process continues until the final steady-state exhalation is reached.

Theoretically obtained exhalation curves versus time for the three cases 'a', 'b', and 'c' just discussed, are shown in Figure 5. The logarithmic scale should be noted, extending the actually very fast initial decrease in exhalation disproportionately. The complicated behaviour of exhalation in leaking cans, together with the difficulties connected with the construction of cans with a reproduceable leakage factor, lead me to the conclusion that in all serious closed-can exhalation measurements, leaking cans must be avoided. To be considered radon-tight a can must have a leakage rate constant that is negligable compared with the decay rate constant, λ, for radon-222 ($\lambda = 2.1 \ 10^{-6} \ s^{-1}$).

Figure 6 displays exhalation rate curves versus time for the sample in Figure 2, with the leakage factor γ as the variable parameter. Large leaks make the final steady-state exhalation rate deviate less from the free exhalation rate, in accordance with Equation 4. It must be remembered, however, that the radon activity accumulating in the outer volume is dependent on γ, the exhalation rate is only the strength of the radon 'source' feeding this volume.

Exhalation Experiments

What kind of experiments can be done to elucidate the special features of the time-dependent diffusion theory as applied above? One of several possibilities is to choose a sample/can geometry in such a way that the initial change in slope of the outer volume concentration as a function of time is pronounced. This is equivalent to saying that the final exhalation rate should be significantly lower than the free exhalation rate and that the gradient reshaping time should be fairly short, of the order of hours. Results from such an experiment are displayed in Figure 7. Notice the characteristic constant slope (about 11 kBq m^{-3} h^{-1}) of

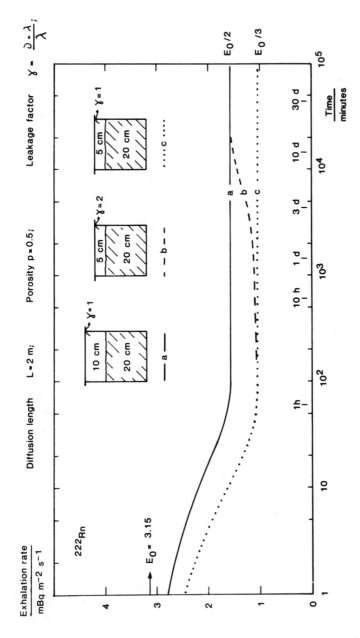

Figure 5. The areal exhalation rate from the porous sample in Figure 2, enclosed in three different exhalation cans. Two of them ('a' and 'c') are completely radon-tight and the third ('b') has a radon leak rate constant ν, numerically equal to the radon decay rate constant ($\nu = \lambda = 2.1 \cdot 10^{-6}$ s^{-1}). The cans are closed at time zero. The radon exhalation evolution as a function of time is discussed in the text (theory).

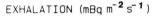

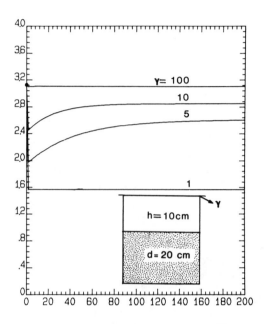

Figure 6. The areal exhalation rate during the first 200 hours after closure, from the porous sample in Figure 2, with the leakage factor γ as the variable parameter (γ is defined in Figure 2) (theory).

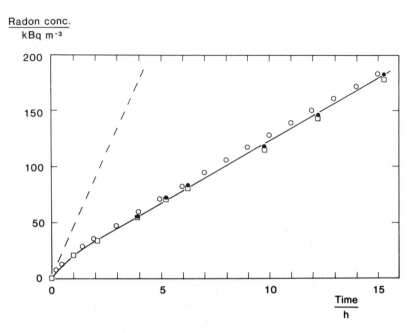

Figure 7. Radon concentration growth in the outer volume during the first fifteen hours after closure. The exhalation can is radon-tight ($\gamma = 1$). The exhalation material is dry sand mixed with 11 % ground uranium ore by weight. The diffusion length, L, is 1.4 m, the sample thickness, d, is 26 cm and the outer volume height, h, is 4.0 cm. Other parameters of the sample are as follows: porosity= 0.47, radium concentration= 1180 Bq kg^{-1}, emanation fraction= 0.33, bulk density= 1710 kg m^{-3} (experiment + theory).

□   radon concentration in grab samples taken from the can (expt.)

——●——   grab samples corrected for dilution effects in the can

O   outer volume radon concentration as predicted by time-dependent diffusion theory

— — —   outer volume radon growth corresponding to free exhalation rate

the growth curve, after the initial gradient reshaping period (about one hour in the figure). This slope corresponds to the final steady-state exhalation rate, E_h. Using Equation 2 we can conclude that the free exhalation rate indicates a slope of about 46 kBq m^{-3} h^{-1}. As expected from the time-dependent diffusion theory discussed above, this value cannot be obtained from the type of single experiments exemplified in Figure 7. The initial depression of the free exhalation rate is far too rapid for that.

In order to experimentally check the validity of the time-dependent diffusion theory as applied to the closed-can method, extensive systematic investigations of radon exhalation, with care-fully controlled parameters, are needed. To the author's knowledge no such investigations have yet been performed.

Conclusions

Increased insight into the temporal behaviour of the radon exhalation from porous materials, after closing an exhalation vessel, can be gained from mere qualitative arguments, based on Ficks first law. Assuming instantaneous outer volume mixing, the exact solution of the time-dependent diffusion equation fully supports the predictions arrived at qualitatively and provides the exact time scale for the initial reduction in the free exhalation rate. For certain common sample/can geometries the initial change in radon exhalation after closing the exhalation can is far too rapid to permit a direct experimental determination of the free exhalation rate from the initial radon growth in the outer volume of the can. This problem can be avoided, sacrificing sensitivity, if the outer volume is made so large that the final steady-state exhalation rate approximates the free exhalation rate. The free exhalation rate can also be determined indirectly by applying a correction factor to a measured final steady state exhalation rate. This latter aproach is not necessarily 'slow', since for certain exhalation can geometries, the final steady state exhalation rate is achieved within one to a few hours after closing the can. The temporal evolution of the exhalation after closing a leaking can is complicated and the use of completely radon-tight cans is highly recommended. Systematic experimental investigations of the temporal behaviour of the radon exhalation from porous materials in closed cans are badly lacking.

Acknowledgments

Dr Inger-Lena Lamm and Mr Håkan Pettersson are recognized for vol-untary contribution of their invaluable expertise, discussing details of this manuscript. I am indebted to Mrs Margaretha Wolff for technical assistance and to the Swedish Natural Science Research Council for their financial support.

References

Dallimore P.J. and R.F. Holub, General Time-Dependent Solutions for Radon Diffusion from Samples Containing Radium, Report of Investigation 8765, Bureau of Mines, United States Department of the Interior, Denver (1982).

Collè R., R.J. Rubin, L.I. Knab and J.M.R. Hutchinson, Radon Transport through and Exhalation from Building Materials: A Review and Assessement, NBS Technical Note 1139, National Bureau of Standards, U.S. Department of Energy, Washington D.C. (1981).

Jonassen N., The Determination of Radon Exhalation Rates, Health Physics 45:369-376 (1983).

Krisiuk E.M., S.I. Tarasov, V.L. Shamov, N.I. Shalak, E.P. Lisachenko, and L.G. Gomelsky, A Study on Radioactivity of Building Materials, Leningrad Research Institute for Radiation Hygiene, Leningrad (1971).

Lamm I-L, C. Samuelsson, and H. Pettersson, Solution of the One-Dimensional Time-Dependent Diffusion Equation with Boundary Conditions Applicable to Radon Exhalation from Porous Materials, Coden Report LUNFD6(NFRA3042), Lund University, Lund (1983).

Samuelsson C. and H. Pettersson, Exhalation of ^{222}Rn from Porous Materials, CONF831049, Rad. Prot. Dosim. 7:95-100 (1984).

Samuelsson C., ^{222}Rn and Decay Products in Outdoor and Indoor Environments; Assessment Techniques Applied to Exhalation, Air-Cleaning and Arctic Air, Dissertation, Lund University, Lund, Sweden (1984).

Zapalac G.H., A Time-Dependent Method for Characterizing the Diffusion of ^{222}Rn in Concrete, Health Physics 45:377-383 (1983).

RECEIVED August 4, 1986

Chapter 17

Underground Measurements of Aerosol and Activity Distributions of Radon and Thoron Progeny

Atika Khan[1], Frank Bandi[1], Richard W. Gleasure[1], Colin R. Phillips[1], and Philippe Duport[2]

[1]Department of Chemical Engineering and Applied Chemistry, University of Toronto, Toronto, Ontario M5S 1A4, Canada
[2]Atomic Energy Control Board, P.O. Box 1046, Ottawa, Ontario K1P 5S9, Canada

Aerosol and activity distributions of ^{218}Po, ^{214}Pb, ^{214}Bi and ^{212}Pb were determined in two different underground mining environments by means of an optimized time delay counting scheme and diffusion batteries. In one environment, diesel equipment was operating, and in the other electrically powered equipment. The two environments differed significantly in total aerosol concentration. In the diesel environment, in particular, aerosol concentrations were unsteady and fluctuated with vehicular traffic and mining activities. As measured by radon progeny disequilibrium, the age of the air ranged from about 25 min to 60 min. Thoron Working Levels were of the same order as radon Working Levels. Data are reported on the aerosol and activity size distributions in both the diesel and the electric mine.

Most previous measurements of aerosol and activity size distributions in uranium mines have focussed on radon progeny (Chapuis et al., 1970; Cooper et al., 1973; George et al., 1977). Thoron progeny measurements have been made in Canadian uranium mines, but for trackless mine conditions only (Busigin et al., 1981). Since the characteristics of aerosols and aerosol concentrations in trackless mines are generally different from those in track mines because of the presence of diesel exhaust in the former, separate measurements of aerosol size and activity distributions are needed in the two types of mines. The measurements presented here were undertaken in order to evaluate the aerosol concentrations, radon and thoron progeny Working Levels, aerosol size distributions and activity size distributions of ^{218}Po, ^{214}Pb, ^{214}Bi and ^{212}Pb in two operational hard rock uranium mines, one a track type mine and the other a trackless type. The age of the air was obtained approximately from the radon progeny concentrations.

0097-6156/87/0331-0219$06.00/0

Experimental Technique

The aerosol size distribution for mine aerosols can be represented adequately by the function (Davies, 1974):

$$n(d) = \frac{1}{d\sqrt{2\pi} \ln \sigma_g} \exp\left(-\frac{1}{2}\left(\frac{\ln(d/\bar{d})}{\ln \sigma_g}\right)^2\right) \tag{1}$$

An aerosol size distribution can, therefore, be described in terms of the count median diameter, d, and the geometric standard deviation, σ_g. These parameters were obtained from experimental data using a diffusion battery method (Busigin et al., 1980). A diffusion battery is an assembly of a number of cylindrical or rectangular channels. The relative penetration of aerosols through different sizes of diffusion batteries at specified flow rates allows the aerosol size distribution to be calculated.

Three diffusion batteries (Busigin et al., 1981) were used in conjunction with a portable (Environment One) condensation nuclei counter (minimum detectable particle size of 0.0025 μm) and a flow-regulated Dupont pump (Model P-4000). The number of diffusion batteries used was a compromise between the ideal and the practical. Although the ideal number of diffusion batteries is greater than three, the practical limit was three owing to the available temporary arrangements for making measurements in operating uranium mines, and the number of available personnel, pumps and counters for simultaneous operation of all batteries. For the measurements, air was first drawn through the sample port of the condensation nuclei counter by the pump to record the total aerosol concentration. The sample port was then connected to each of the three diffusion batteries in turn and the number of aerosol particles penetrating each battery obtained. The penetrations g_3, g_4 and g_5 through the diffusion batteries (nos. 3, 4 and 5) were then calculated from the ratio of the number of aerosols measured with and without a battery. An optimal fit of the Kth approximation of the size distribution $n^{(K)}(d)$ to the actual distribution $n(d)$ was then obtained using diffusion battery data. An approximation $n^K(d)$ to $n(d)$ is obtained as

$$n_q^K(d) = \sum_{\alpha=1}^{q} \frac{h_\alpha^K}{\sqrt{2\pi}\, d \ln \sigma_\alpha^K} \exp\left(-\frac{1}{2}\left(\frac{\ln\left(d/d_\alpha^K\right)}{\ln \sigma_\alpha^K}\right)^2\right) \tag{2}$$

where $\sum_{\alpha=1}^{q} h_\alpha^{(K)} = 1$, q is the number of lognormal distributions summed and $r_\alpha^{(K)}$, $\sigma_\alpha^{(K)}$ and $h_\alpha^{(K)}$ are the parameters associated with the lognormal size distributions of the Kth approximation.

The penetration probability of a particle passing through the diffusion battery is given by (Busigin et al., 1980)

$$P = \sum_{i=1}^{\infty} a_i \, e^{-b_i \mu} \tag{3}$$

where μ = 2 WLD/FH for a rectangular plate diffusion battery
  W = width of each rectangular plate (cm)
  L = length of each rectangular plate (cm)
  D = diffusion coefficient of the particle (cm^2/s)
  F = volume flow rate through a single channel in the
      diffusion battery (cm^3/s)
  2H = spacing between two plates (cm) (2H $\ll$ W)

The penetration constants a_i and b_i for parallel plate diffusion batteries are:

a_1 = 0.91035, a_2 = 0.5314, a_3 = 0.01529, a_4 = 0.0068

b_1 = 1.88517, b_2 = 21.4317, b_3 = 62.3166, b_4 = 124.5

The experimental diffusion battery penetration, g_j, is related to the theoretically calculated penetration probability, $P_j(d)$ by

$$g_j = \int_0^{\infty} P_j(d) \, n(d) \, d(d) \tag{4}$$

where   n(d) is the required aerosol size distribution given by eqn. (1).

The optimization calculation is aimed at minimizing the difference between g_j as given by equation (4) and the approximation $g_j^{(K)}$ where

$$g_j^{(K)} = \int_0^{\infty} P_j(d) \, n^{(K)}(d) \, d(d) \tag{5}$$

The dimensions of the diffusion batteries are given in Table I. At a sampling rate of 3 L/min, diffusion battery (D.B.) no. 3 would remove 50% of the 0.03 μm particles, D.B. no. 4 would remove 50% of the 0.08 μm particles and D.B. no. 5 would remove 50% of the 0.2 μm particles.
The count median diameter, d, and the geometric standard deviation, σ_g, were calculated by using the experimental diffusion battery penetration values, g_j, as input for a computer program based on the optimization procedure described earlier (Busigin et al., 1980).
Three different experimental protocols were used throughout the work. Two protocols were used to determine the activity size distribution, the first protocol, based on individual progeny, for the bulk of the measurement days (Jan 29 to Feb 6) and the second, based on Working Level, for the last two days (Feb 7 and 8). The third experimental protocol was used for general determination of Working Levels from Jan 29 to Feb 6.

Table I. Dimensions of Diffusion Batteries

D.B. no.	No. of channels	Width (cm)	Length (cm)	Spacing between plates (cm)
3	25	10	19	0.08
4	50	10	30	0.04
5	140	10	43	0.04

Activity Size Distribution − First Experimental Protocol. For
determination of the activity size distribution of the individual
progeny in the first experimental protocol, the concentration of
the radioactive species of interest is used to find the penetration
fraction through the diffusion batteries. The experimental method
consists of drawing air through 0.8 μm pore size Millipore Type AA
filters placed in magnetic filter holders at the air exit ends of
the diffusion batteries. An open face filter holder with a similar
filter is used as a reference. Air is drawn simultaneously through
the reference filter and the three diffusion batteries for ten
minutes. The filters are then counted in Trimet alpha counters
according to a counting schedule based on optimization with respect
to minimum uncertainty in measurement of each of ^{218}Po, ^{214}Pb,
^{214}Bi and ^{212}Pb. The fluctuations in flow rate and concentrations
during sampling were taken as 5% each. Two samples per day could
be obtained using this schedule of 10 minutes sampling and subsequent counting at 2−4, 9−12, 85−95 and 240−250 minutes after
sampling. Equations (5a−d) were used to calculate the concentrations C(1), C(2), C(3) and C(4) for ^{218}Po, ^{214}Pb, ^{214}Bi and
^{212}Pb, respectively.

$\eta VC(1) =$
    0.146577 I(1) − 0.135175 I(2) + 0.039193 I(3) − 0.02596 I(4)
$\eta VC(2) =$
    −0.0167797 I(1) + 0.009946 I(2) + 0.022673 I(3) − 0.017961 I(4)
$\eta VC(3) =$
    −0.0072511 I(1) + 0.035622 I(2) − 0.017533 I(3) + 0.012187 I(4)
$\eta VC(4) =$
    −0.0000119 4I(1) + 0.0000433I(2) − 0.0001868 I(3) + 0.0059551(4)

$$(5a-d)$$

    In equations (5a−d), the I(j) are the counts obtained in the
jth counting interval, and η and V are the counting efficiency
(cpm/dpm) and the sampling flow rate (L/min) respectively.

Activity Size Distributions − Second Experimental Protocol. For
the last two days of measurements (Feb 7 and 8) the second experimental protocol was used to determine the activity size distribution in terms of Working Levels. In this, the sampling and
counting schedules were adjusted in order to increase the number of
samples obtained per day from two each to four each, for both radon
and thoron progeny. The scheme consisted of 5 minute sampling
followed by 2−7 minute counting for radon progeny and 15 minute
sampling followed by 270−285 or 330−345 minute counting for thoron
progeny.
    The radon progeny Working Levels for calculation of activity
distributions were calculated using the equation:

$$RnWL = \frac{I\ (2-7)}{5 \times 5 \times 235\ \eta V} \qquad (6)$$

where 235 is the weighting factor dependent on the decay equations
and sampling and counting periods were 5 min each.

The thoron progeny Working Levels for calculation of activity distributions were calculated as:

$$TnWL = \frac{I\ (330-345)}{15 \times 15 \times 13\ \eta V} \tag{7}$$

where 13 is the weighting factor dependent on the decay equations and sampling and counting periods were 15 min each.
The first sample of the day was counted at 270–285 minutes after sampling in order to allow counting to be completed in the underground working period. Equation (7) was used to calculate the thoron progeny Working Level in this case, with $I(330-345)$ replaced by $I(270-285)$.
The concentration of ^{212}Pb was calculated from:

$$C_{212_{Pb}} = \frac{I\ (330-345)}{2.22 \times V \times 15 \times 15 \times \exp(-\lambda_{212_{Pb}} \times 337.5) \times \eta}$$

$$= \frac{I(330-345)}{345.78\ V\eta} \tag{8}$$

where 1/2.22 is the conversion factor from dpm to pCi/L, 337.5 min is the time from the end of sampling to the mid-point of the counting interval, and sampling and counting periods are 15 minutes each.
        For the first sample of the day, this concentration was given by:

$$C_{212_{Pb}} = \frac{I(270-285)}{369.14\ V\eta} \tag{9}$$

where the time elapsed from the end of sampling to mid-counting interval is 277.5 min and other parameters are the same as for eqn. (8).
        The activity size distributions were determined from the calculated penetration values in the diffusion batteries using the method outlined for aerosol size measurement (equation (6) for RnWL and equations (8) and (9) for ^{222}Pb concentration).

Working Level Measurement – Third Experimental Protocol. Working Levels of radon and thoron progeny were determined separately from those used for activity size distributions four times per day during the entire measurement period except the last two days. The method consisted of sampling air for ten minutes through an open face filter holder containing a 0.8 μm pore size filter. The filter was then counted at 2–7 and 340–350 minutes after sampling in accordance with an optimized two-count scheme for determining radon and thoron progeny Working Level (Khan and Phillips, 1986). The following equations were used for the calculations:

$$RnWL = \frac{I\ (2-7)}{10 \times 5 \times 235 \times V\eta} = \frac{I(2-7)}{11750\ V\eta} \tag{10}$$

where 235 is the weighting factor dependent on the decay equations, the sampling period is 10 min and the counting period is 5 min.

$$TnWL = \frac{I\ (340-350)}{10 \times 10 \times 13 \times V\eta} = \frac{I(340-350)}{1300\ V\eta} \tag{11}$$

where 13 is the weighting factor dependent on the decay equations and sampling period and counting period are 10 min each.

The I's represent the counts obtained in the counting interval (minutes) given in brackets, and V and η are the pump flowrate and the counter efficiency, respectively.

On the last two days, Working Levels determined by equations (6) and (7) for activity size distribution were used.

Results

The Working Level data for both track and trackless mines are listed in Tables II(a) and II(b). The aerosol concentrations recorded frequently (every ten to twenty minutes) during each measurement day are plotted in Figs. 1 and 2 for the track mine and Figs. 3 and 4 for the trackless mine. The concentrations of ^{218}Po, ^{214}Pb, ^{214}Bi and ^{212}Pb obtained from the reference filter sample for activity size distribution measurements are listed in Table III. The approximate age of the air based on the radon progeny concentration ratios (Evans, 1969) is also given in Table III. Because air streams from different airways are likely to mix in the mine, the age of the air calculated in this way reflects only the progeny ratios, not the physical age of the air stream or streams. The activity size distribution and aerosol size distributions found in the two mines are shown in Table IV.

The mean values of the activity and aerosol median diameters together with the best estimate of the standard deviation, σ_{n-1}, based on the total number of measurements made for each parameter, are listed in Table V. Figures 5-8 show representative size distributions.

Discussion

It is only possible to compare results from the track mine with results from the trackless mine on the understanding that neither results can be considered to be typical of the relevant mine since the variation within each mine, as a result of vehicular traffic, non-similar sampling locations, etc., is extremely great. For purposes of identification, the data are however referred to as pertaining to either the track or the trackless mine.

The thoron progeny Working Level was found to be about the same as the radon progeny Working Level in the track mine but lower than the RnWL found in the trackless mine (Table II). The aerosol concentrations were much lower (40,000-80,000 aerosols/cm^3) in the track mine than those found in the trackless mine (80,000 to >300,000/cm^3). Further, the aerosol concentrations were fairly stable in the track mine (Figs. 1 and 2) whereas in the trackless mine (Figs. 3 and 4) sharp transients occurred due to heavy vehicular traffic. Measurements in the trackless mine are therefore

Table II(a). Working Levels of Radon and Thoron Progeny
(based on eqns. (10) and (11))

Mine	Date	Time	RnWL	TnWL
Track	Jan 29	11.25	0.09	0.12
		13.40	0.08	0.09
	Jan 30	10.10	0.11	0.15
		12.25	0.09	0.12
	Jan 31	11.20	0.14	0.17
		13.35	0.20	0.20
	Feb 1	11.50	0.55	0.49
		14.05	0.48	0.35
Trackless	Feb 4*	11.05	0.36	--
	Feb 5	10.00	0.24	0.21
		12.15	0.29	0.22
	Feb 6	11.25	0.31	0.20
		13.40	0.31	0.19

* Filter blown away with wind before the second count for TnWL
could be obtained.

Table II(b). Working Levels of Radon and Thoron Progeny (based on eqns. (6) and (7))

Date	Time	RnWL	TnWL
	8.40	--	0.22*
	8.55	--	0.22*
	9.10	0.37	--
	9.20	0.36	--
	9.40	0.33	--
	9.50	0.30	--
	10.10	0.33	--
Feb 7	10.20	0.32	--
	10.40	0.28	--
	10.55	0.34	--
	12.05	--	0.20
	12.20	--	0.20
	12.40	--	0.20
	13.15	--	0.20
	13.40	--	0.19
	13.55	--	0.20
	8.30	--	0.20*
	8.45	0.27	--
	8.55	0.30	--
	9.15	0.28	--
Feb 8	9.35	0.28	--
	9.55	0.34	--
	12.00	--	0.20
	12.20	--	0.19
	13.30	--	0.19

* Slightly overestimated due to modified counting interval in order to enable counting within the allowed working period.

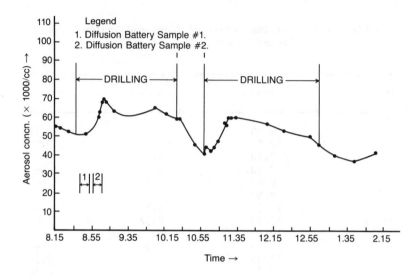

Figure 1. Aerosol concentrations for Jan. 30 (track mine)

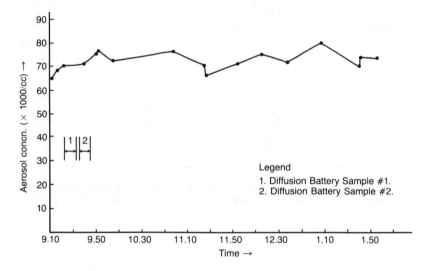

Figure 2. Aerosol concentrations for Feb. 1 (track mine).

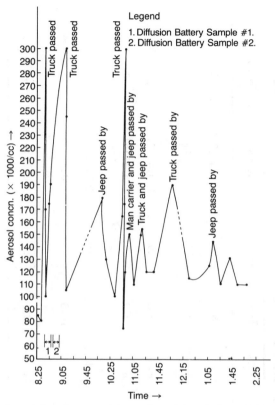

Figure 3. Aerosol concentrations for Feb. 4 (trackless mine).

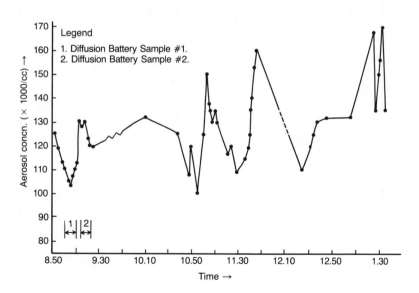

Figure 4. Aerosol concentrations for Feb. 6 (trackless mine).

Table III. Concentrations of Radon/Thoron Progeny

Mine	Date	Time	Concentrations (pCi/L)				Radon progeny concn. ratios	Age of air (min)
			^{218}Po	^{214}Pb	^{214}Bi	^{212}Pb		
Track	Jan 29	9.00	25.8	11.6	5.9	1.25	100:45:22	30
		9.15	22.6	8.4	8.3	1.42	100:37:37	25
	Jan 30	8.45	21.2	8.5	4.5	1.09	100:40:21	25
		9:00	13	8	7.9	1.13	100:61:61	40
	Jan 31	8.55	24.4	12.9	7.8	1.21	100:53:32	45
		9.10	16.5	11	10.3	1.25	100:67:62	45
	Feb 1	9.25	58	44	35	3.25	100:76:60	60
Trackless	Feb 4	8.40	72.2	36.3	24.4	2.41	100:50:34	35
	Feb 5	9:00	53.4	30.5	20	1.72	100:57:37	40

Table IV. Activity and Aerosol Size Distributions, Count Median Diameter (CMD) and the Geometric Standard Deviation (indicated in brackets)

Mine	Date	Activity size distributions, AMD (μm)					Aerosol size distribution	
		Time	^{218}Po	^{214}Pb	^{214}Bi	^{212}Pb	Time	CMD (μm)
Track	Jan 30	8.45	0.053 (2.21)	0.047 (1.59)	0.043 (2.43)	0.057 (2.30)	9.40	0.051 (2.34)
		9.00	0.063 (2.62)	0.056 (1.51)	0.059 (1.34)	0.072 (2.47)	11.40	0.053 (1.52)
							12.50	0.050 (2.68)
							13.50	0.051 (2.14)
	Jan 31*	8.55	—	0.054 (2.16)	0.057 (2.40)	0.082 (2.55)	9.35	0.062 (2.42)
		9.10	—	0.044	0.060	0.10	10.00	0.065 (1.40)
				(2.47)	(2.19)	(1.75)	11.50	0.058 (2.19)
							13.00	0.061 (1.72)
							14.05	0.072 (2.56)
	Feb 1	9.40	0.047 (2.29)	0.047 (1.54)	0.053 (1.45)	0.058 (1.89)	10.25	0.046 (1.60)
							12.20	0.050 (1.76)
							13.25	0.048 (1.56)
Track-less	Feb 4*	8.40	0.084 (2.06)	0.091 (1.48)	0.096 (2.09)	—	9.30	0.088 (1.84)
							12.40	0.094 (1.43)
	Feb 5*	8.40	—	0.053 (2.76)	0.078 (2.50)	0.092 (1.29)	9.20	0.070 (1.15)
							11.50	0.084 (1.48)
							12.50	0.032 (1.41)

Continued on the next page

Table IV. Continued

Mine	Date	Activity size distributions, AMD (μm)					Aerosol size distribution	
		Time	^{218}Po	^{214}Pb	^{214}Bi	^{212}Pb	Time	CMD (μm)
	Feb 6	9.00	0.066 (2.40)	0.073 (1.81)	0.077 (1.75)	0.09 (2.23)	9.45	0.062 (1.55)
							12.00	0.047 (1.98)
							13.05	0.076 (1.41)

		Time	RnWL	Time	Pb-212
	Feb 7 [†]	9.10	0.097 (1.84)	8.40	0.125 (1.68)
		9.40	0.085 (1.86)	13.15	0.107 (1.65)
		10.10	0.089 (2.55)	13.40	0.106 (1.62)
		10.40	0.096 (1.81)		
	Feb 8**	8.45	0.079 (1.65)	--	--
		9.15	0.089 (1.58)	--	--

* Blanks indicate that the penetration fractions obtained were physically impossible, for example, the concentration after passage through D.B.#3 calculated as greater than concentration calculated for the reference filter.

** Pb-212 activity size distribution data had to be rejected due to a pump failure during measurements.

† Aerosol size distribution measurements were not made.

Table V. Median Diameters for all Measured Size Distributions in the Present Work

Mine	Mean value of median diameter (μm) $\pm \sigma_{n-1}$					
	^{218}Po	^{214}Pb	^{214}Bi	^{212}Pb	RnWL	Aerosol
Track	0.054 ± 0.008	0.05 ± 0.005	0.054 ± 0.007	0.07 ± 0.015	—	0.055 ± 0.008
Trackless	0.073 ± 0.01	0.07 ± 0.02	0.085 ± 0.009	0.1 ± 0.013	0.089 ± 0.007	0.069 ± 0.019

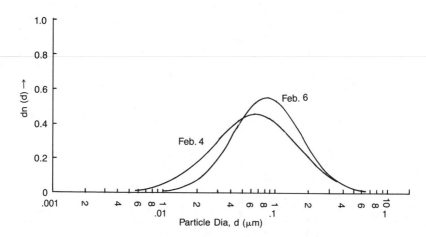

Figure 5. Activity size distributions for ^{218}Po (trackless mine).

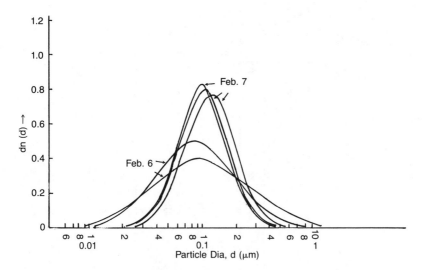

Figure 6. Activity size distributions for ^{212}Pb (trackless mine).

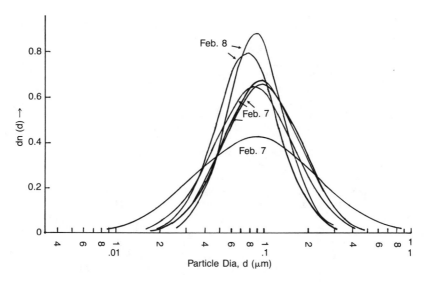

Figure 7. Activity size distributions for RnWL (trackless mine).

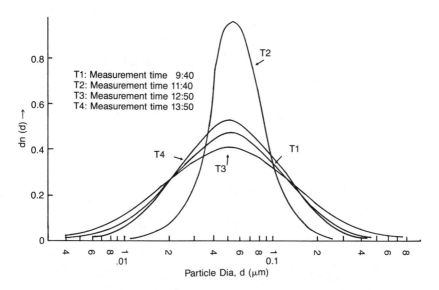

Figure 8. Aerosol size distributions for Jan. 30 (track mine).

subject to greater uncertainties due to the rapid changes in the sampling environment over short periods of time (<10 minutes). The greater uncertainty is reflected in the measurement data listed in Tables II-IV where fewer data are reported for the trackless mines than for the track mine. Because the duration of some transients were similar to the measurement period, some data in the trackless mine had to be rejected.

The aerosol and activity size distribution data shown in Table V indicate that the activity median diameters for aerosols to which ^{218}Po and ^{214}Pb are attached is about the same as the count median diameter of the general aerosol itself in both track and trackless mines. The activity median diameters of ^{212}Pb and radon progeny (in terms of RnWL) are higher by about 30-40% than the aerosol diameter. The activity median diameter of the aerosols carrying ^{214}Bi is the same as the overall aerosol distribution count median diameter for the track mine but higher than the aerosol diameter for the trackless mine. The RnWL activity size distribution was not determined for the track mine. The results in Table V indicate, however, that the RnWL distribution would probably be the same as the aerosol size distribution in the track mine but somewhat higher than the aerosol size distribution in the trackless mine because of the larger ^{214}Bi attached aerosol diameter (Table V).

The aerosol distributions are calculated in terms of a single mode, without attempting to resolve them into a major large mode and a minor very small (unattached) mode. The unattached mode is very much smaller in diameter (of molecular cluster dimensions) than the major mode of the aerosol and in underground mines its peak height is very small. To resolve such a mode would require more than the three diffusion batteries used for the measurements.

The larger size of ^{212}Pb in both the track and trackless mine may be due to the longer half-life of 212Pb, which allows ^{212}Pb atoms to spend more time in the vicinity of aerosols than any of the other measured radon decay products. A longer contact time would lead to increased coagulation of aerosols and larger particle sizes.

Conclusions

The conclusions may be summarized as follows:

a) Aerosol concentrations measured in the track mine were found to be consistently lower (40,000-80,000/cm^3) and more stable than those measured in the trackless mine (80,000 to >300,000/cm^3).

b) The thoron progeny Working Levels found in the track mine (0.09-0.49) were in general about the same as the radon progeny Working Levels (0.08-0.55). In the trackless mine, however, the thoron progeny Working Levels (0.19-0.22) were lower than the measured radon progeny Working Levels (0.24-0.36).

c) The average age of air determined from progeny ratios was 25–60 minutes in the track mine and 35–40 minutes in the trackless mine.

d) The radon progeny activity size distributions in the track mine were in general similar to the aerosol size distribution, with average activity median diameters of 0.054 μm for ^{218}Po and ^{214}Bi aerosols and 0.05 μm for ^{214}Pb aerosols. The count median diameter for the aerosol was 0.055 μm. In the trackless mine, the average activity median diameters of 0.073 μm and 0.07 μm for ^{218}Po and ^{214}Pb aerosols were about the same as the average count median diameter of the aerosol (0.069 μm). However, the average activity median diameter of ^{214}Pb (0.085 μm) aerosols and the average activity median diameter of aerosols to which radon progeny were attached (0.089 μm) were both greater than the aerosol diameter.

e) The ^{212}Pb activity median diameter was larger than the aerosol diameter for both the track mine (0.07 μm compared to the aerosol diameter of 0.055 μm) and trackless mine (0.1 μm compared to the aerosol diameter of 0.069 μm).

f) All aerosol and activity size distribution activity median diameters were in the range 0.05 to 0.1 μm for both track and trackless mines.

Acknowledgments

This work was supported by a contract from the Atomic Energy Control Board of Canada. The co-operation of mine management and personnel is gratefully acknowledged.

Literature Cited

Busigin, A., A.W. van der Vooren and C.R. Phillips, A Technique for Calculation of Aerosol Particle Size Distributions from Indirect Measurements, J. Aeros. Sci. 11:359–366 (1980).

Busigin, A., A.W. van der Vooren and C.R. Phillips, Measurement of the Total and Radioactive Aerosol Size Distributions in a Canadian Uranium Mine, Amer. Ind. Hyg. Assoc. J. 42:310–314 (1981).

Chapuis, A., A. Lopez, J. Fontan, F. Billard, and G.J. Madelaine, Spectre Granulometrique des Aerosols Radioactifs dans une Mine d'Uranium, Aerosol Sci. 1:243–253 (1970.

Cooper, J.A., P.O. Jackson, J.C. Langford, M.R. Peterson and B.O. Stuart, Characteristics of Attached Rn-222 Daughters under both Laboratory and Field Conditions with Particular Emphasis upon Underground Uranium Mine Environments, Batelle Pacific NorthWest Laboratories, 1973.

Davies, C.N., Size Distribution of Atmospheric Particles, J. Aeros. Sci. 5:293–300 (1974).

Evans, R.D. Engineers Guide to the Elementary Behaviour of Radon Daughters, Health Phys. 17: 229-252 (1969).

George, A.C., L. Hinchliffe and R. Sladowski, Size Distribution of Radon Daughter Particles in Uranium Mine Atmospheres, HASL-326, 1977.

Khan, A. and C.R. Phillips, A Simple Two-Count Method for Routine Monitoring of Radon and Thoron Progeny Working Levels in Uranium Mines, Health Phys. 50:381-388 (1986).

RECEIVED October 16, 1986

PROPERTIES

Chapter 18

Chemical Properties of Radon

Lawrence Stein

Chemistry Division, Argonne National Laboratory, Argonne, IL 60439

Radon is frequently regarded as a totally inert element. It is, however, a "metalloid" -- an element which lies on the diagonal of the Periodic Table between the true metals and nonmetals and which exhibits some of the characteristics of both. It reacts with fluorine, halogen fluorides, dioxygenyl salts, fluoronitrogen salts, and halogen fluoride-metal fluoride complexes to form ionic compounds. Several of the solid reagents can be used to collect radon from air but must be protected from moisture, since they hydrolyze readily. Recently, solutions of nonvolatile, cationic radon have been produced in nonaqueous solvents. Ion-exchange studies have shown that the radon can be quantitatively collected on columns packed with either Nafion resins or complex salts. In its ionic state, radon is able to displace H^+, Na^+, K^+, Cs^+, Ca^{2+}, and Ba^{2+} ions from a number of solid materials.

Since the discovery of the first noble gas compound, $Xe^+PtF_6^-$ (Bartlett, 1962), a number of compounds of krypton, xenon, and radon have been prepared. Xenon has been shown to have a very rich chemistry, encompassing simple fluorides, XeF_2, XeF_4, and XeF_6; oxides, XeO_3 and XeO_4; oxyfluorides, $XeOF_2$, $XeOF_4$, and XeO_2F_2; perxenates; perchlorates; fluorosulfates; and many adducts with Lewis acids and bases (Bartlett and Sladky, 1973). Krypton compounds are less stable than xenon compounds, hence only about a dozen have been prepared: KrF_2 and derivatives of KrF_2, such as $KrF^+SbF_6^-$, $KrF^+VF_6^-$, and $KrF^+Ta_2F_{11}^-$. The chemistry of radon has been studied by radioactive tracer methods, since there are no stable isotopes of this element, and it has been deduced that radon also forms a difluoride and several complex salts. In this paper, some of the methods of preparation and properties of radon compounds are described. For further information concerning the chemistry, the reader is referred to a recent review (Stein, 1983).

0097-6156/87/0331-0240$06.00/0

Clathrate Compounds

Radon forms a series of clathrate compounds (inclusion compounds) similar to those of argon, krypton, and xenon. These can be prepared by mixing trace amounts of radon with macro amounts of host substances and allowing the mixtures to crystallize. No chemical bonds are formed; the radon is merely trapped in the lattice of surrounding atoms; it therefore escapes when the host crystal melts or dissolves. Compounds prepared in this manner include radon hydrate, $Rn\ 6H_2O$ (Nikitin, 1936); radon-phenol clathrate, $Rn\ 3C_6H_5OH$ (Nikitin and Kovalskaya, 1952); radon-p-chlorophenol clathrate, $Rn\ 3p\text{-}ClC_6H_4OH$ (Nikitin and Ioffe, 1952); and radon-p-cresol clathrate, $Rn\ 6p\text{-}CH_3C_6H_4OH$ (Trofimov and Kazankin, 1966). Radon has also been reported to co-crystallize with sulfur dioxide, carbon dioxide, hydrogen chloride, and hydrogen sulfide (Nikitin, 1939).

Radon Difluoride

Most chemical experiments with radon have been carried out with isotope ^{222}Rn (half-life 3.82 days), which decays by α-emission as shown in Figure 1. The $\beta-$ and γ-emitting daughters ^{214}Pb and ^{214}Bi, as well as the α-emitting daughters ^{218}Po and ^{214}Po, grow into radioactive equilibrium with the parent within four hours. We have used the 1.8 MeV γ-ray of ^{214}Bi, which can be measured conveniently through the walls of glass or metal apparatus, to determine the position of radon in tracer experiments, since the bismuth follows the radon as it moves from one location to another.

Figure 2 shows a Monel vacuum line, constructed with Autoclave Engineers Type 30 VM6071 valves and a 0-1000 torr Helicoid pressure gauge, which is currently used for the preparation of radon compounds. The radon is obtained from a Pyrex bulb containing an aqueous solution of radium chloride (approximately 30 mCi of $^{226}RaCl_2$ in dilute HCl). The bulb is suspended in a plastic cup surrounded by lead bricks and is attached to the line by a Kovar-Pyrex seal. In a typical experiment, approximately 0.1 to 1.0 mCi of ^{222}Rn is pumped from the bulb through a bed of Drierite, condensed in a cold trap at $-195°C$, distilled in vacuum from the trap at $23°C$ to the reaction vessel at $-195°C$, mixed with a fluorinating agent, and allowed to react at either room temperature or an elevated temperature. The reaction vessel may be a closed-end Monel tube, as shown, or a Kel-F plastic test tube.

When radon is heated to $400°C$ with fluorine, a nonvolatile fluoride is formed (Fields et al., 1962, 1963). It has been deduced from the chemical behavior that the product is radon difluoride, RnF_2. (Products of the tracer experiments have not been analyzed because of their small mass and intense radioactivity.)

$$Rn + F_2 \xrightarrow[400°C]{} RnF_2 \tag{1}$$

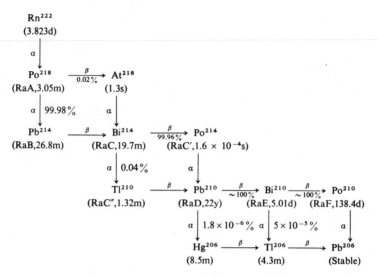

Figure 1.   The decay scheme of radon–222.

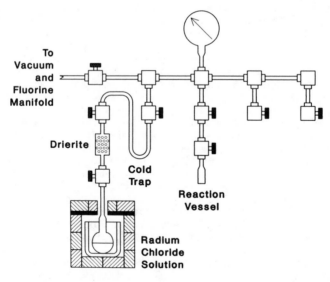

Figure 2.   Monel vacuum line for the preparation of radon compounds.

The compound can be reduced with hydrogen at 500°C to quantitatively recover elemental radon.

$$RnF_2 + H_2 \xrightarrow[500°C]{} Rn + 2HF \tag{2}$$

If a large amount of radon, such as 10 mCi, is mixed with fluorine in a small vessel (e.g., a 30 ml Kel-F test tube), the fluorination occurs spontaneously in the gas phase at room temperature or in liquid fluorine at -195°C. The activation is provided by the intense α radiation, which produces large numbers of ions and excited atoms.

Complex Fluorides

Radon reacts spontaneously at room temperature with many solid compounds that contain oxidizing cations, such as BrF_2^+, IF_6^+, O_2^+, and N_2F^+ (Stein, 1972, 1973, 1974; Stein and Hohorst, 1982). Xenon also reacts with a few compounds of this type which have very high oxidation potentials (Stein, 1973, 1974). The xenon products have been analyzed by Raman and mass spectrometric methods and shown to consist of xenon (II) complex fluorides.

$$Xe + 2O_2^+SbF_6^- \longrightarrow XeF^+Sb_2F_{11}^- + 2O_2 \tag{3}$$

$$Xe + N_2F^+SbF_6^- \longrightarrow XeF^+SbF_6^- + N_2 \tag{4}$$

$$Xe + BrF_6^+AsF_6^- \longrightarrow XeF^+AsF_6^- + BrF_5 \tag{5}$$

By analogy, radon is believed to form the following products.

$$Rn + 2O_2^+SbF_6^- \longrightarrow RnF^+Sb_2F_{11}^- + 2O_2 \tag{6}$$

$$Rn + N_2F^+SbF_6^- \longrightarrow RnF^+SbF_6^- + N_2 \tag{7}$$

$$Rn + N_2F_3^+Sb_2F_{11}^- \longrightarrow RnF^+Sb_2F_{11}^- + N_2F_2 \tag{8}$$

$$Rn + ClF_2^+SbF_6^- \longrightarrow RnF^+SbF_6^- + ClF \tag{9}$$

$$Rn + BrF_2^+SbF_6^- \longrightarrow RnF^+SbF_6^- + BrF \tag{10}$$

$$Rn + BrF_2^+TaF_6^- \longrightarrow RnF^+TaF_6^- + BrF \tag{11}$$

$$Rn + BrF_2^+BiF_6^- \longrightarrow RnF^+BiF_6^- + BrF \tag{12}$$

$$Rn + BrF_4^+Sb_2F_{11}^- \longrightarrow RnF^+SbF_6^- + BrF_2^+SbF_6^- \tag{13}$$

$$Rn + IF_6^+SbF_6^- \longrightarrow RnF^+SbF_6^- + IF_5 \tag{14}$$

Hydrolytic Reactions of Radon Compounds

Radon difluoride is quantitatively reduced to elemental radon by water in a reaction which is analogous to the reactions of krypton difluoride and xenon difluoride with water. Complex salts of radon also hydrolyze in this fashion.

$$KrF_2 + H_2O \longrightarrow Kr + \tfrac{1}{2}O_2 + 2HF \tag{15}$$

$$XeF_2 + H_2O \longrightarrow Xe + \tfrac{1}{2}O_2 + 2HF \tag{16}$$

$$RnF_2 + H_2O \longrightarrow Rn + \tfrac{1}{2}O_2 + 2HF \tag{17}$$

$$RnF^+SbF_6^- + H_2O \longrightarrow Rn + \tfrac{1}{2}O_2 + HF + HSbF_6 \tag{18}$$

This behavior provides evidence that in each of the compounds, radon is in the +2 oxidation state. When higher-valent xenon compounds, such as XeF_4 and XeF_6, are hydrolyzed, water-soluble xenon species (XeO_3 and XeO_6^{4-}) are produced (Malm and Appelman, 1969). We have observed no radon species corresponding to these xenon species in hydrolysis experiments.

Russian scientists (Avrorin et al., 1981, 1985) have reported that reactions of complex mixtures of radon, xenon, metal fluorides, bromine pentafluoride, and fluorine yield a higher fluoride of radon which hydrolyzes to form RnO_3. However, efforts to confirm these findings have been unsuccessful. In similar experiments which have been carried out at Argonne National Laboratory (Stein, 1984), it has been found that radon in the hydrolysate is merely trapped in undissolved solids; centrifugation removes the radon from the liquid phase completely. This is in marked contrast to the behavior of a solution of XeO_3, which can be filtered or centrifuged without loss of the xenon compound. Hence there is no reliable evidence at present for the existence of a higher oxidation state of radon or for radon compounds or ions in aqueous solutions. Earlier reports of the preparation of oxidized radon species in aqueous solutions (Haseltine and Moser, 1967; Haseltine, 1967) have also been shown to be erroneous (Flohr and Appelman, 1968; Gusev and Kirin, 1971).

Solutions of Ionic Radon

Stable solutions of radon difluoride can be prepared in nonaqueous solvents, such as halogen fluorides and hydrogen fluoride (Stein, 1969, 1970). Radon reacts spontaneously at 25°C or at lower temperatures with each of the halogen fluorides except IF_5. It also reacts with mixed solvent-oxidant pairs, such as $HF-BrF_3$, $HF-BrF_5$, and IF_5-BrF_3, and solutions of K_2NiF_6 in HF.

$$Rn + 2ClF \longrightarrow RnF_2 + Cl_2 \tag{19}$$

$$Rn + BrF_3 \longrightarrow RnF_2 + BrF \tag{20}$$

$$Rn + BrF_5 \longrightarrow RnF_2 + BrF_3 \tag{21}$$

$$Rn + IF_7 \longrightarrow RnF_2 + IF_5 \tag{22}$$

$$Rn + NiF_6^{2-} \longrightarrow RnF_2 + NiF_2 + 2F^- \tag{23}$$

Electromigration studies, which have been carried out with a Kel-F plastic cell (Figure 3), have shown that radon is present in many of these solutions as a cation (Stein, 1970, 1974). When the cell

is filled with inactive solvent and a solution of radon difluoride
is added to the center leg, the application of a D.C. voltage to
the nickel electrodes causes the radon to move to the cathode. In
a pure conducting solvent, such as bromine trifluoride, the following ionization occurs:

$$RnF_2 \rightleftharpoons RnF^+ + F^- \tag{24}$$

$$RnF^+ \rightleftharpoons Rn^{2+} + F^- \tag{25}$$

In a solution containing added electrolyte, however, dissociation
may be suppressed, and the radon may form anionic complexes.

Recently, a search has been conducted for solvents which are
less hazardous to handle than halogen fluorides and hydrogen fluoride and which can be used to prepare solutions of the cationic
species. Two have been found that are highly oxidation-resistant
and suitable for this purpose: 1,1,2-trichlorotrifluoroethane and
sulfuryl chloride. Solutions of cationic radon prepared by oxidizing ^{222}Rn with halogen fluorides in these solvents have been shown
to be stable when stored in capped FEP Teflon bottles at room temperature for several weeks (Stein, 1985). Because of their ease of
preparation and relative safety, solutions of this type can be used
most readily to study reactions of radon.

Ion-Exchange Reactions of Cationic Radon

Figure 4 shows the behavior that is observed when a solution of
cationic radon in 1,1,2-trichlorotrifluoroethane is passed through
a column packed with KPF_6. The radon displaces potassium ion and
adheres in a narrow band at the top of the column. It can be
washed repeatedly with dilute BrF_3 in the halocarbon solvent, then
eluted rapidly with 1.0 M BrF_3 in sulfuryl chloride. The radon
daughters remain on the column during elution and decay in situ;
new daughters are generated in the radon-containing eluant fractions.

We have found that similar behavior occurs when the column is
packed with other salts of Group I elements, such as $NaSbF_6$,
Na_3AlF_6, or K_2NiF_6, or with Nafion ion-exchange resins (H^+ or K^+
forms). In batch equilibration experiments, using 1 g amounts of
solids stirred with 5-15 ml volumes of solutions, we have found
that the radon ions can also be collected on the compounds $CsBrF_4$,
$Ca(BrF_4)_2$, and $Ba(BrF_4)_2$. Thus it is apparent that, in its oxidized state, radon can displace H^+, Na^+, K^+, Cs^+, Ca^{2+}, and Ba^{2+}
ions from a number of solid materials.

By measuring the distribution coefficient, K_d, of cationic
radon on Nafion resin (H^+ form) in BrF_3-trichlorotrifluoroethane
solutions as a function of the concentration of BrF_3, we have been
able to show that the charge on the radon cation is +2 and that the
parent molecule is RnF_2. This physico-chemical method makes use of
the fact that BrF_3 produces the univalent cation BrF_2^+, which com-
petes with Rn^{2+} for sites on the resin. The following equilibria
occur in this system (R^- represents the anion of the resin):

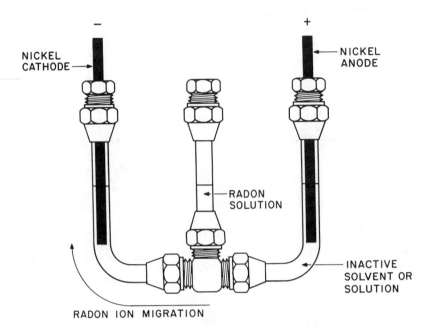

Figure 3. Kel-F plastic electrolysis cell. Reproduced with permission from Stein, L., Ionic Radon Solutions. Copyright 1970
the American Association for the Advancement of Science.

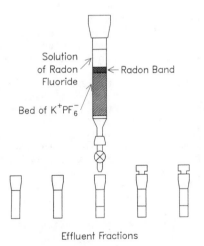

Figure 4. The ion-exchange behavior of cationic radon on a
column packed with $K^+PF_6^-$.

$$Rn^{2+} + 2H^+R^- \rightleftharpoons Rn^{2+}(R^-)_2 + 2H^+ \tag{26}$$

$$BrF_2^+ + H^+R^- \rightleftharpoons BrF_2^+R^- + H^+ \tag{27}$$

$$2BrF_2^+ + Rn^{2+}(R^-)_2 \rightleftharpoons 2BrF_2^+R^- + Rn^{2+} \tag{28}$$

Many finely divided solids will partly remove trace amounts of radon from solution by physical adsorption, but we have found that removal by ion-exchange is much more efficient. The distribution coefficient, K_d, of cationic radon ranges from about 90 to 4000 ml/g on ion-exchange materials that have been tested thus far in dilute BrF_3-trichlorotrifluoroethane solutions. In contrast, the coefficient is less than 10 ml/g on materials which do not undergo ion-exchange with the radon species, such as LiF, MgF_2, PbF_2 Al_2O_3, and CeO_2.

These studies show that radon can be classified as a metalloid element, together with boron, silicon, germanium, arsenic, antimony, tellurium, polonium, and astatine. Like these elements, radon lies on the diagonal of the Periodic Table between the true metals and nonmetals (Figure 5) and exhibits some of the characteristics of both (Stein, 1985).

Possible Applications of Radon Chemistry

Tests in the laboratory with radon-air mixtures have shown that two reagents, dioxygenyl hexafluoroantimonate ($O_2^+SbF_6^-$) and hexafluoro-iodine hexafluoroantimonate ($IF_6^+SbF_6^-$) are particularly well suited for trapping radon (Stein et at., 1977; Stein and Hohorst, 1982). In the form of powders or pellets, both compounds remove more than 99% of the radon from air at low or moderate flow rates. The air must first be dried by passage through a bed of desiccant, however, since the compounds are decomposed by moisture. Figure 6 shows one type of metal and plastic cartridge containing 1.0-1.5 g of $O_2^+SbF_6^-$ powder that has been tested as a device for analyzing ^{222}Rn. To perform an analysis with this system, a measured volume of air is drawn by a battery-operated pump through a drying tube and the cartridge, which captures the radon as a nonvolatile compound. After radioactive equilibrium has been established between radon and its short-lived daughters (approximately 4 hours), the γ-emission of the cartridge is measured with a well-type scintillation counter. The amount of radon is then calculated from the γ-emission rate. In tests conducted with samples of air containing 3.5 to 11,700 pCi/l of ^{222}Rn (simulating the wide range of radon concentrations that can occur in a uranium mine), this type has been found to have a count rate of 2.74 counts/min/pCi of ^{222}Rn (all γ-ray energies).

An advantage of this method over the collection of radon with charcoal is that no refrigerant is required; both $O_2^+SbF_6^-$ and $IF_6^+SbF_6^-$ have been shown to operate efficiently at temperatures ranging from 10° to 40°C and probably can also be used at higher or lower temperatures. An advantage over the grab sampling technique is that only the radon is collected; a large volume of air (10-50 liters, for example) can be sampled, all of the radon retained, and the air discarded. This reduces the volume of sample that must be transported to a laboratory for analysis. A disadvantage of the

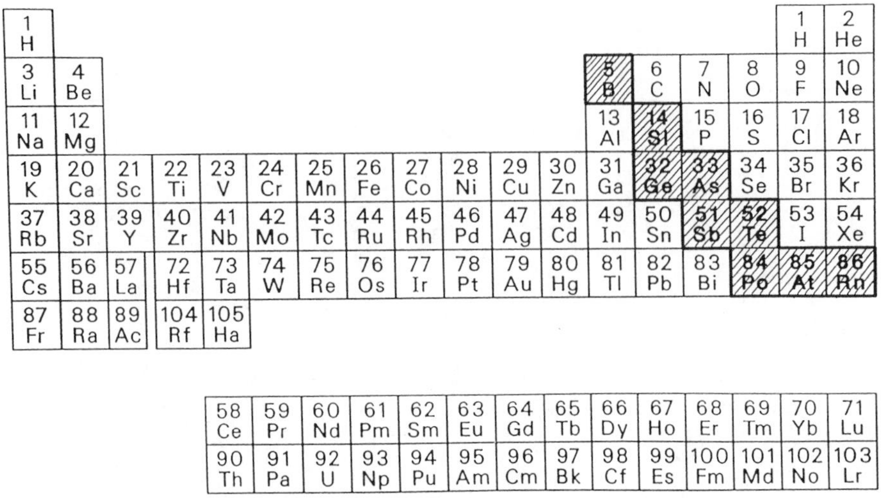

1 H
1 H
2 He
3 Li
4 Be
5 B
6 C
7 N
8 O
9 F
10 Ne
11 Na
12 Mg
13 Al
14 Si
15 P
16 S
17 Cl
18 Ar
19 K
20 Ca
21 Sc
22 Ti
23 V
24 Cr
25 Mn
26 Fe
27 Co
28 Ni
29 Cu
30 Zn
31 Ga
32 Ge
33 As
34 Se
35 Br
36 Kr
37 Rb
38 Sr
39 Y
40 Zr
41 Nb
42 Mo
43 Tc
44 Ru
45 Rh
46 Pd
47 Ag
48 Cd
49 In
50 Sn
51 Sb
52 Te
53 I
54 Xe
55 Cs
56 Ba
57 La
72 Hf
73 Ta
74 W
75 Re
76 Os
77 Ir
78 Pt
79 Au
80 Hg
81 Tl
82 Pb
83 Bi
84 Po
85 At
86 Rn
87 Fr
88 Ra
89 Ac
104 Rf
105 Ha
58 Ce
59 Pr
60 Nd
61 Pm
62 Sm
63 Eu
64 Gd
65 Tb
66 Dy
67 Ho
68 Er
69 Tm
70 Yb
71 Lu
90 Th
91 Pa
92 U
93 Np
94 Pu
95 Am
96 Cm
97 Bk
98 Cf
99 Es
100 Fm
101 Md
102 No
103 Lr

Figure 5. The arrangement of the metalloid elements (dark shading) in the Periodic Table.

KEL-F INLET TUBE
COPPER-NICKEL OUTLET TUBE
BRASS PLUG
BRASS CAP
"O" RING A
"O" RING B
KEL-F FILTER ASSEMBLY
REAGENT
CYLINDRICAL KEL-F BODY
0
1
2
3
CENTIMETERS

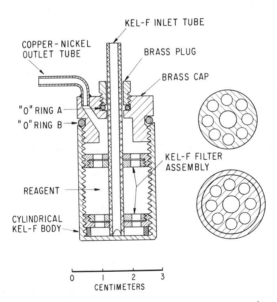

Figure 6. Kel-F plastic cartridge containing $O_2^+SbF_6^-$ reagent for the analysis of radon gas (from Stein et al., 1977).

method is that the background count rate of a γ-scintillation counter is usually higher than that of a Lucas flask-photomultiplier combination, a very sensitive α counter. However, the background count rate can be reduced by using lead shielding, several scintillators, and anti-coincidence circuitry.

Chemical methods have been proposed for purifying radon-laden air in uranium mines (Stein, 1975, 1983). The cost of operating a full-scale air purification system using $O_2^+SbF_6^-$ reagent in a mine has been estimated at 0.245 U.S. dollars per 1000 standard cubic ft of treated air (1975 dollars) (Lindsay et al., 1975). This includes the cost of drying the air beforehand with a refrigerant-desiccant system. Comparable costs for removing the radon by physical methods have been estimated as follows: cryogenic condensation, $0.117; adsorption on charcoal at $-80°C$, $0.170; adsorption on Molecular Sieve at $-80°C$, $0.376; and membrane permeation, $0.402 per 1000 standard cubic ft of treated air. All of these methods are considered to be too expensive, in comparison to ventilation, to be used at present.

Some mixtures of noble gases can be separated by chemical methods. A mixture of radon and xenon, for example, can be separated by selectively oxidizing the radon with $ClF_2^+SbF_6^-$, $BrF_2^+SbF_6^-$, or $IF_6^+SbF_6^-$. Xenon can be trapped with stronger oxidants, such as $O_2^+SbF_6^-$, $O_2^+PtF_6^-$, or $N_2F^+SbF_6^-$, and thus separated from krypton and lighter noble gases. Ternary mixtures of krypton, xenon, and radon can be separated by successive oxidation of the radon and xenon.

Acknowledgments

Work performed under the auspices of the Office of Basic Energy Sciences, Division of Chemical Sciences, U. S. Department of Energy, under Contract W-31-109-Eng-38.

Literature Cited

Avrorin, V. V., Krasikova, R. N., Nefedov, V. D., and Toropova, M. A., Production of Higher Fluorides and Oxides of Radon, Radiokhim. 23:879-883 (1981).

Avrorin, V. V., Krasikova, R. N., Nefedov, V. D., and Toropova, M. A., Coprecipitation of Radon Oxide with Cesium Fluoroxenate, Radiokhim. 27:511-514 (1985).

Bartlett, N., Xenon Hexafluoroplatinate (V), $Xe^+PtF_6^-$, Proc. Chem. Soc. 218 (1962).

Bartlett, N. and Sladky, F. O., The Chemistry of Krypton, Xenon, and Radon, in Comprehensive Inorganic Chemistry (A. F. Trotman-Dickenson, ed), Vol. 1, pp. 213-330, Pergamon Press, Oxford, England (1973).

Fields, P. R., Stein, L., and Zirin, M. H., Radon Fluoride, J. Am. Chem. Soc. 84:4164 (1962).

Fields, P. R., Stein, L., and Zirin, M. H., Radon Fluoride: Further Tracer Experiments with Radon, in Noble-Gas Compounds (H. H. Hyman, ed), pp. 113-119, University of Chicago Press, Chicago, IL (1963).

Flohr, K. and Appelman, E. H., The Resistance of Radon to Oxidation in Aqueous Solution, J. Am. Chem. Soc. 90:3584 (1968).

Gusev, Yu. K. and Kirin, I. S., The Chemical State of ^{222}Rn in the α-Decay of ^{226}Ra in Certain Solutions and Crystals, Radiokhim. 13:916-918 (1971).

Haseltine, M. W. and Moser, H. C., Oxidation of Radon in Aqueous Solutions, J. Am. Chem. Soc. 89:2497-2498 (1967).

Haseltine, M. W., Some Solution Chemistry of Radon, Dissert. Abstr. B28, No. 7, 2755 (1968).

Lindsay, D. B., Whittier, A. J., Watson, W. I., Evans, R. D., Schroeder, G. L., Maletskos, C. J., Dale, S. E., Carlisle, A. J., Field, E. L., Igive, B. U., Lawter, J. R., Interess, E., Neill, J. K. O., Ketteringham, J. M., Nelson, L. L., McMahon, H. O., Santhanam, C. J., and Johnston, R. H., Advanced Techniques for Radon Gas Removal, Report USBM-H0230022, prepared for the U. S. Bureau of Mines by Arthur D. Little, Inc., Cambridge, Mass., May 1975, 170 pp.

Malm, J. G. and Appelman, E. H., Chemical Compounds of Xenon and Other Noble Gases, Atomic Energy Rev. 7, No. 3, 3-48 (1969).

Nikitin, B. A., Radon Hydrate, Z. anorg. allgem. chem. 227:81-93 (1936).

Nikitin, B. A., Chemistry of the Inert Gases. IV. Solid-Solution Formation Between Inert Gases and Other Substances, Doklad. Akad. Nauk SSSR 24:562-564 (1939).

Nikitin, B. A. and Ioffe, E. M., A Compound of Radon with p-Chlorophenol, Doklad. Akad. Nauk. SSSR 85: 809-10 (1952).

Nikitin, B. A. and Kovalskaya, M. P., Compounds of Inert Gases and Their Analogs with Phenol, Izvest. Akad. Nauk SSSR, Otdel Khim. Nauk 24-30 (1952).

Stein, L., Oxidized Radon in Halogen Fluoride Solutions, J. Am. Chem. Soc. 91:5396 (1969).

Stein, L., Ionic Radon Solutions, Science 168:362-364 (April 17, 1970).

Stein, L., Chemical Methods for Removing Radon and Radon Daughters from Air, Science 175:1463-1465 (March 31, 1972).

Stein, L., Removal of Xenon and Radon from Contaminated Atmospheres with Dioxygenyl Hexafluoroantimonate, O_2SbF_6, Nature 243, No. 5401, 30-32 (May 4, 1973).

Stein, L., Noble Gas Compounds: New Methods for Collecting Radon and Xenon, Chemistry 47, No. 9, 15-20 (1974).

Stein, L., Shearer, J. A., Hohorst, F. A., and Markun, F., Development of a Radiochemical Method for Analyzing Radon Gas in Uranium Mines, Report USBM-HO252019, prepared for the U. S. Bureau of Mines by Argonne National Laboratory, Argonne, IL, January 1977, 78 pp.

Stein, L. and Hohorst, F. A., Collection of Radon with Solid Oxidizing Reagents, Environ. Sci. Technol. 16:419-422 (1982).

Stein, L., The Chemistry of Radon, Radiochim. Acta 32:163-171 (1983).

Stein, L., Hydrolytic Reactions of Radon Fluorides, Inorg. Chem. 23:3670 (1984).

Stein, L., New Evidence that Radon Is a Metalloid Element: Ion-Exchange Reactions of Cationic Radon, J. Chem. Soc., Chem. Comm. 1631-1632 (1985).

Trofimov, A. M. and Kazankin, Yu. N., Clathrate Compounds of p-Cresol with Noble Gases. II. Compound of p-Cresol with Krypton and Radon, Radiokhim. 8:720-723 (1966).

RECEIVED August 5, 1986

Chapter 19

Effect of Radon on Some Electrical Properties of Indoor Air

Marvin Wilkening

Department of Physics and Geophysical Research Center, New Mexico Institute of Mining and Technology, Socorro, NM 87801

The radon isotopes and their decay products are important to the study of the indoor environment not only for their contribution to the internal dose, but also for their effects upon the electrical nature of indoor air. Ions of special interest are those formed after the decay of a radon or daughter atom. They have a mass and electrical mobility comparable to ordinary atmospheric small ions. An ionization rate of 25×10^6 ion pairs m^{-3} s^{-3} can be expected for typical indoor radon concentrations. The resulting positive ions in indoor air outnumber the radon daughter positive ions by a factor of 10^5. Measurements in a room on campus where the radon concentration averaged 23 Bq m^{-3} yielded a Potential Alpha Energy Concentration (PAEC) of 0.003 WL. The PAEC due to the positively charged daughter ions was only about 3% of that due to attached daughters. Aerosol and Rn-222 concentrations, and PAEC (WL) levels peaked in the morning hours in the indoor environment while the mean aerosol diameter remained essentially constant at about 0.04 μm.

Charge States of the Daughter Products of Radon in Air

Perturbation theory was applied to the ionization of atoms accompanying alpha and beta decay soon after the advent of quantum theory (Migdal, 1941). Migdal concluded that the probability for ionization in the outer shells of naturally radioactive atoms at the instant of decay is of the order of unity. This was confirmed by Knipp and Teller (1941) and experimental verification of this result was provided later by Wexler (1965) and others.

When Rn-222 decays with the emission of a 5.49 Mev alpha particle the recoiling energy of the newly formed Po-218 atom is 0.10 Mev which is sufficient for a range of about 50 micrometers in air. However, the velocity of the recoiling Po-218 atom is still small compared with that of the orbital electrons.

0097-6156/87/0331-0252$06.00/0
© 1987 American Chemical Society

What is of even greater importance is the charge of the heavy
ion at or near the end of the recoil range when thermal energies are
approached. Of particular interest in this connection is the work of
Fite and Irving (1972) on collision and chemi-ionization processes
involving uranium atoms combined with oxygen atoms and molecules.
Certain heavy metal molecular ions, e.g., UO_2^+, are found to be
practically immune to neutralization by electrons in the low energy
range.   The first ionization potentials of Po (8.4 ev) and the other
alpha emitting daughters are less than those of N_2 (15.6 ev) and O_2
(12.1 ev) and other trace gases in the air.   Hence, on a simple
collision basis, less energy must be expended to ionize a neutral
heavy metal ion than one of the surrounding molecules in the air.
     The details of the manner in which the daughter ions remain in
their positive charge states, combine with other atoms in associative
or chemi-ionization processes, and provide "clustering" with polar
molecules such as water and sulfur dioxide are only beginning to be
understood (Wilkening, 1973; Mohnen, 1977; Rudnick et al, 1983).
Regardless of the mechanisms involved, the fact that radon daughter
ions begin their existence in the positive ion state makes them an
interesting component of the atmospheric ion population.

Characteristics of the Daughter Ions

The general characteristics of these ions have been reviewed by
Bricard and Pradel (1966).   The attachment of radon and thoron
daughter ions to aerosols including condensation nuclei has been
studied by a number of investigators (Raabe, 1968; McLaughlin, 1972;
Porstendorfer and Mercer, 1979; and Busigin, et al., 1981), and will
not be considered further in this paper.   We turn now to some experimental results that bear directly on the characteristics of the
radon daughter ions.

Mobility. The mobility of an ion in air is an important property
both for characterizing ions and for removal considerations.
Measurement of the mobility of the daughter ions in our laboratory
(Wilkening, et al., 1966) showed that on the average 75% had mobilities in the range 0.25 to 1.50 x 10^{-4} m^2 v^{-1} s^{-1}. Previous
measurements (Jonassen and Wilkening, 1965) of the mobilities of
ordinary atmospheric small ions gave 1.6 x 10^{-4} m^2 v^{-1} s^{-1} for
positive ions and 2.0 x 10^{-4} m^2 v^{-1} s^{-1} for negative ions in the same
locale.   Hoppel (1970) found a mean of 1.35 x 10^{-4} m^2 v^{-1} s^{-1} for
positive ordinary ions and a range of 1.7 to 2.1 x 10^{-4} m^2 v^{-1} s^{-1}
for negative ions dependent upon the presence of water vapor.   These
results clearly show that the electrical mobilities and hence, the
masses of the radon daughter positive ions fall in the same general
range as those for ordinary atmospheric small ions.

Mean Life, Attachment, Recombination and Plateout. The approximate mean life (τ) of Po-218 in ion form can be represented as ($\lambda + \alpha n + \beta N + po$)$^{-1}$ where λ is the decay constant for Po-218, α is the recombination coefficient for Po-218 and ordinary ions of negative charge in the atmosphere, n is the concentration of small negative ions, β is the attachment coefficient, N is the number of condensation nuclei per unit volume, and po is plateout on solid surfaces. For $\lambda = 3.79 \times 10^{-3}$ s^{-1}, $\alpha = 1.4 \times 10^{-12}$ m^3 s^{-1}, n = 550 $\times 10^6$ ions m^{-3}, $\beta = 2 \times 10^{-12}$ m^3 s^{-1}, and N = 40 $\times 10^9$ nuclei m^{-3}, the mean life is 12 seconds in a typical outdoor situation (no plateout). If N is decreased to 10^9 nuclei m^{-3}, the mean life increases to 40 seconds illustrating the sensitivity to condensation nuclei concentrations and the relatively long life of these ions. From the ratio $\alpha n / \beta N = 10^{-2}$, it is seen that attachment processes dominate over recombination effects in a typical situation (Roffman, 1972). Plateout has received recent attention by George, et al., (1983), Knutson, et al., (1983) and others.

Ionization due to Rn-222, Rn-220 and their Decay Chains

During the radioactive decay of the radon isotopes and their daughters, alpha and beta particles and gamma rays are given off. These radiations give up their energy by ionization and excitation of atoms and molecules which they encounter. A single alpha particle from the decay of Rn-222 will produce about 150,000 ion pairs with an average expenditure of 34 ev per pair (Israel, 1973). The complicated processes that are involved in the transformation of these ion pairs, or those created by other ionizing agents, into atmospheric small ions have been described by Mohnen (1977). The radon decay chains in the atmosphere are the chief ionizing agents in the lower atmosphere, exceeding terrestrial radiation and ionization from cosmic rays. Since indoor levels of Rn-222 and Rn-220 and their daughters exceed those of the outdoors by factors of up to 10, it is clear that there is a significant source of ionization in dwellings due primarily to the presence of the radon isotopes and their decay products.

Ionization rates from radon can vary remarkably ranging from 10 x $_6 10^6$ ion pairs m$_3^{-3}$ s$_1^{-1}$ in an outdoor mountain environment to 2300 x 10^6 ion pairs m^{-3} s^{-1} in the Carlsbad Caverns in New Mexico where radon levels are quite high (Wilkening, 1985). Recent measurements of Rn-222 levels in houses in the southwestern U.S. gave mean Rn-222 concentrations of 63 Bq m^{-3} and a Potential Alpha Energy Concentration (PAEC) of 0.007 Working Level (Wilkening and Wicke, 1986). An ionization rate of 26 x 10^6 ion pairs m^{-3} s^{-1} can be expected from such concentrations. Allowance for typical Working Levels for Rn-220 (thoron) and its daughters in indoor environments (Schery, 1985) would increase this figure by perhaps one-third or more. Definitions and comparison values for PAEC and WL can be found elsewhere (for example, Lawrence Berkeley Laboratory, 1983; NCRP, 1984). The difference between the radioactive daughter ions which may exist in ion form and the ordinary atmospheric ions resulting from air molecules being ionized by the decay of the daughter nuclides in the radon/thoron decay series should be re-emphasized at this point.

Atmospheric Electrical Parameters

The ionization of nitrogen, oxygen and trace gas molecules in the air due to the presence of the natural radioelements in the soil and air and the cosmic radiation has a direct effect upon the electrical characteristics of the atmosphere.

Ions in the Atmosphere. A key feature of any atmospheric ion is its electrical mobility which is a measure of its change in drift velocity when placed in an electric field. The mobility depends upon the charge of the ion and inversely upon the mass. The diffusion coefficient, another important characteristic, for an ion is the product of the mobility and its thermal energy. It is mobility that forms the basis for classification of ions (Chalmers, 1967; and Pruppacher and Klett, 1978). The atmospheric "small ions" have a mobility in the range of 1 to 2 x 10^{-4} meters/second per volt/meter. They generally consist of a singly ionized atom or molecule with water or other polar molecules clustered about it similar to the radon daughter positive ions described earlier. Atmospheric "large ions" are essentially one or more ions attached to small particles or condensation nuclei in the air. These particles, whether in charged or uncharged states, can be removed by filters. They constitute a major source of the internal dose from radon and thoron daughters in air.

Experiments have shown (Wilkening, et al, 1966) that radon daughter ions of positive sign were found to account for only about 30 ppm of the total positive small-ion concentration in clean outdoor air (about 10 ppm for indoor air in this study). Simultaneous measurements of the concentrations of the Rn-222 daughter ions and the total atmospheric small ions of the same sign yield a correlation coefficient of 0.8 or better for fair weather conditions. This correlation was confirmed by experiments that have shown that the electrical conductivity of the atmosphere (product of the small-ion density, the electronic charge and the mean mobility of the ions) is directly proportional to the concentration of the total small-ion concentration at both mountain and valley sites in New Mexico (Jonassen and Wilkening, 1965). Hence, the validity of using radon-daughter ions as tracers in atmospheric electrical studies was confirmed.

The concentration of small ions in the atmosphere is determined by 1) the rate of ion-pair production by the cosmic rays and radio-active decay due to natural radioactive substances, 2) recombination with negative ions, 3) attachment to condensation nuclei, 4) precipitation scavenging, and 5) transport processes including convection, advection, eddy diffusion, sedimentation, and ion migration under the influence of electric fields. A detailed differential equation for the concentration of short-lived Rn-222 daughter ions including these terms as well as those pertaining to the rate of formation of the radioactive positive ions by decay of the parent nuclide and the decay of the ions themselves has been given by Roffman (1972) together with appropriate numerical solutions.

Methods of Measurement. The positive and negative small-ion concentrations, the electrical conductivity, and ion mobilities were measured with a Gerdien-type instrument. From analyses of current-versus-voltage curves obtained with the apparatus operated at constant volume flow, values for the ion density, the conductivity, and the mean mobility of the ions can be determined (Wilkening and Romero, 1981).

Radon concentrations were measured by use of calibrated Lucas scintillation flasks, while radon and thoron daughters and the resulting potential alpha energy concentration (PAEC) were determined using filter samples (Thomas, 1972) and a continuous electrostatic precipitator (Andrews et al., 1984). The radon daughter positive ions were collected in a Gerdien-type apparatus similar to the conductivity meter except that the inner collector electrode is replaced by a lucite cylinder covered by a removable aluminum foil. Positive small ions with preset lower mobility limits are collected on the foil for a 10-minute period after which the daughter ion concentrations are determined by the Thomas method.

Aerosol concentrations were determined by a condensation nucleus counter (CN) (General Electric). For the most recent work, an electrical aerosol analyzer (EAA) (TSI 3800) was used. The CN counter is sensitive to aerosols in the 0.001 to 0.1 µm range, while the aerosol analyzer has a range of from 0.0032 to 1.0 µm. The latter instrument yields both particle density and mean diameters over preset size intervals. Data taken simultaneously indoors with both instruments yielded a high correlation coefficient. A meteorological station (Weather Measure-M800) was operated outdoors near the building being studied with continuous recording of temperature, pressure, humidity, precipitation, and wind speed and direction.

Experimental Results

Measurements reported herein were made in a first-floor laboratory room in a two-story classroom building having concrete floors and walls. Windows were not open and no artificial ventilation was used during the period of study.

Diurnal Patterns. Radon and its daughter products are known to follow typical diurnal patterns in the outdoors with high levels in the early morning and lower concentrations in the afternoons (Gesell, 1983; UNSCEAR, 1982). Seasonal variation is subject to a variety of geographic and meteorological influence and does not fit a clear-cut pattern.

Much can be learned about radon and its daughters and their interactions with aerosols and other atmospheric components under conditions of consistent time variation as seen in normal day-night regimes. Typical examples are shown in Figure 1. This study includes data for seven 24-hour periods during September of 1985 taken in the first-floor classroom building. No windows were open, no artificial ventilation was used and access was limited to occasional entry from laboratory personnel during the days when

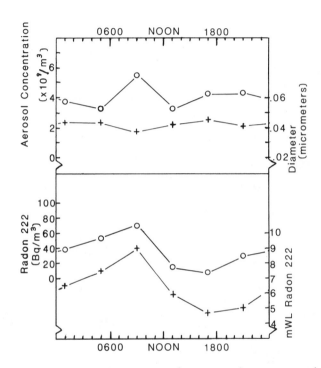

Figure 1.   Mean diurnal patterns for aerosol concentrations
(—o—) and size (—+—) top, and Rn-222 concentrations (—o—)
with PAEC (Working Level) (—+—) bottom, for seven days in
September 1985.

equipment was running. Mean values for 4-hour intervals are plotted although the EAA aerosol spectrometer provides samples at two minute intervals.

Aerosol concentrations, radon, and PAEC (in Working Levels) clearly peak in early hours after sunrise and at about the time of morning human use of the building which is coincident with the outdoor peak in radon and fine aerosols due to their overnight accumulation near ground level under local temperature inversions.

The daily average aerosol concentration is 4060 x 10^6 particles per m^3 with a fluctuation of $\pm$ 50 percent based on the standard deviation of the mean. The mean particle size likewise shows large variation with a mean of 0.04 $\pm$ 0.01 μm. There was some evidence in the aerosol analyzer data for major and minor modes in the size distribution as found by George et al. (1980, 1984). The mean diameter of the particles remained constant as seen in the graph. Both Rn-222 and PAEC (WL) values followed similar diurnal patterns. It is reasonable to expect higher WL levels during the morning due both to increase in radon and aerosol concentrations. The mean diurnal PAEC is 0.006 WL for this seven-day period in September.

Radon and Atmospheric Parameters. Aerosol concentrations, total small-ion densities, electrical conductivity, and Rn-222 with its short-lived daughters are given in Table I. The data were taken on 5 days during February of 1985. Days with rain or high winds are not included. As for the data in Figure 1 the windows were closed with minimum entry through the door. Heating was from a hot water exchange unit that did not permit air exchange with the outside, but it did have a fan which was on sufficiently to cause some increase in aerosol concentration over that measured in the September period. Errors indicated are again the standard deviation from the mean over the days of measurement.

Aerosol concentrations were high for this period averaging 104 x 10^9 m^{-3}. It should be noted that these are condensation nuclei counts, not data from the aerosol spectrometer. The CN instrument does respond to a smaller particle size which may explain in part the relatively high levels. Radon-222 levels in the room averaged 23 Bq m^{-3} which was a factor of about 7 higher than outdoor levels, about as expected for a room of relatively low ventilation rate. The radon daughters Po-218, Pb-214 and Bi-214 show equilibrium factors of 0.9, 0.4 and 0.3 which combined yield a PAEC of 0.0027 WL while the Rn-222 daughter ions have a WL of 7 x 10^{-5} or only about 3% of that due to attached daughters.

It is of interest to compare the numbers of radon daughter positive ions with the total population of positive and negative small ions in the atmosphere. The activity due to the daughters in Bq m^{-3} is converted to numbers of the radioactive daughters by dividing the Bq m^{-3} by λ, the appropriate decay constant for the isotope involved. The data in Table I show that there are 800 positive ions per m^3 while there are 10^8 total positive small ions present i.e., 8 radon daughter positive ions per million total small ions. Since the negative small ions have an abundance roughly equal to the positively charged ones, one can see that the positive ions from radon do not affect the overall character of the electrical nature of the atmosphere. Any physiological effects of ions inhaled by humans would not be affected by presence of the daughter positive ions insofar as their charge is concerned.

Table I: Comparison of Radon and Daughter Products with Aerosols and
Atmospheric Electrical Characteristics in a First-floor Room.   Errors
Shown are the Standard Deviation Over a Five Day Period
(February 1985).

Measurement	Mean Value
Aerosol particles $\times 10^9$ (m^{-3})	104 ± 70
Ion density, $\times 10^9$	
+ions (m^{-3})	0.10 ± 0.03
−ions (m^{-3})	0.14 ± 0.07
Conductivity, $\times 10^{-14}$	
+(ohm m)$^{-1}$	3.2 ± 1.1
−(ohm m)$^{-1}$	2.4 ± 0.7
Radon 222 $(Bq\ m^{-3})$	23 ± 7
Radon 222 daughters $(Bq\ m^{-3})$	
Po 218	22 ± 15 (0.9)*
Pb 214	$10 + 4$ (0.4)
Bi 214	7 ± 4 (0.3)
Radon 222 daughter + ions, $(Bq\ m^{-3})$	
Po 218	1.5 ± 1.3 (0.06)*
Pb 214	0.15 ± 0.02 (0.006)
Bi 214	0.03 ± 0.07 (0.001)
Radon 222 daughter + ions (m^{-3})	
Po 218	400 ± 340
Pb 214	350 ± 50
Bi 214	50 ± 120
Total Rn + ions (m^{-3})	800

*Equilibrium factor relative to Rn-222

The atmosphere is a good insulator but it does have a conductivity which like the ion density is not affected by the radon
daughter ions. It is of interest to note that the electrical conductivity is balanced with respect to positive and negative polarity
within the accuracies indicated.

The most significant information given in Table I is that the
radon daughter ions do contribute to the total internal dose if only
at about 3% in this study. The effect of free ions in contrast with
daughters attached to aerosols is substantial when the mechanics of
deposition in the bronchial tree and lungs are taken into account.
The dose from "unattached" Po-218, which includes the Po-218 positive
ions as a major component, can be from 3 to 40 times that of the
attached Po-218 (NCRP, 1984). Current dosimetry models allow for the
important role played by Po-218 in small-ion form. Hence, their
effect is significantly greater than the 3% contribution to the PAEC
(WL) shown in Table I.

Summary

Radon daughter ions and the ionization caused by the decay chains of
radon and thoron in indoor air play important roles both from the
contribution made by the daughter product positive ions to internal
dose and from the effects of ion-pair production on the indoor
atmospheric electrical parameters.

Aerosol and Rn-222 concentrations and the PAEC (WL) levels peak
in the indoor environment in the morning hours. Radon 222
measurements averaged 23 Bq m^{-3}, about 7 times the ambient outdoor
level. Activities from the individual short-lived daughters lead to
an overall PAEC of 0.003 Working Level. About 3% of this was due to
the contributions from the radon daughter positive ions which are
known to be more effective than the radon daughters attached to
aerosols in terms of internal dose because of their high attachment
rate in the upper bronchii.

Indoor small ion concentrations for both positive and negative
ions averaged about 0.1×10^9 ions m^{-3}, which was measurably less
than outdoor concentrations given in the literature (0.5×10^9 m^{-3}).
The relatively high aerosol concentrations were a contributing
factor. The atmospheric conductivity values for both positive and
negative charge carriers were the same within limits of measurement.
It is interesting to note that the total radon daughter positive ions
make up only about 7 ppm of the total ions of the same sign in the
indoor environment. Again their contribution to radiation dose is
much greater than their proportion by numbers.

Further studies of the relation between radon levels and
atmospheric electrical parameters in the indoor environment are
underway.

Acknowledgment

The author is indebted to Edward McNamee who collected much of the
data as part of his program leading to the M.S. degree in Physics, to
Dr. Luis Quindos of the Department of Medical Physics of the
University of Santander, Spain; and to Dr. S.D. Schery of the New
Mexico Institute of Mining and Technology.

The work was supported by contract DE AS04 84ER60216 of the Office of Health and Environmental Radiation of the U.S. Department of Energy and by the Geophysical Research Center of the New Mexico Institute of Mining and Technology.

Literature Cited

Andrews, L.L., S.D. Schery, and M.H. Wilkening, An Electrostatic Precipitator for the Study of Airborne Radioactivity, Health Phys. 46:801-808 (1984).

Bricard, J. and J. Pradel, Electric Charge and Radioactivity of Naturally Occurring Aerosols, Aerosol Science:91-104 (1966).

Busigin, A., A.W. vanderVooren, J.C. Babcock, and C.R. Phillips, The Nature of Unattached RaA (^{218}Po) Particles, Health Phys. 40:335-343 (1981).

Chalmers, J.A., Atmospheric Electricity, Pergamon Press, New York, NY (1967).

Fite, W.L. and P. Irving, Chemi-ionization in Collisions of Uranium Atoms with Oxygen Atoms and Molecules, J. of Chem. Phys. 56:4227-4428 (1972).

George, A.C. and A.J. Breslin, The Distribution of Ambient Radon and Radon Daughters in Residential Buildings in the New Jersey-New York Area, in Natural Radiation Environment III, Vol. 2 (T. Gesell, W. Lowder, eds.) pp. 1272-1292, Tech. Information Center, U.S. Dept. of Energy, Springfield, Va. (1980).

George, A.C., E.O. Knutson, and K.W. Tu, Radon Daughter Plateout-I Measurements, Health Phys. 45:439-444 (1983).

George, A.C., M.H. Wilkening, E.O. Knutson, D. Sinclair, and L. Andrews, Measurements of Radon and Radon Daughter Aerosols in Socorro, New Mexico, Aerosol Sci. and Tech. 3:277-281 (1984).

Gesell, T.F., Background Atmospheric ^{222}Rn Concentrations Outdoors and Indoors: A Review, Health Phys. 45:2 (1983).

Hoppel, W.A., Measurement of the Mobility Distribution of Tropospheric Ions, Pure and App. Geophys. 81:230-245 (1970).

Israel, H., Atmospheric Electricity, Vol. II, National Science Foundation, Program for Scientific Translation, U.S. Dept. of Commerce, Springfield, Va. (1973).

Jonassen, N. and M.H. Wilkening, Airborne Measurements of Radon 222 Daughter Ions in the Atmosphere, J. Geophys. Res. 75:9 (1970).

Jonassen, N. and M.H. Wilkening, Conductivity and Concentration of Small Ions in the Lower Atmosphere, J. Geophys. Res. 70:4 (1965).

Knipp, J. and E. Teller, On the Energy Loss of Heavy Ions, Phys. Rev. 59:659-669 (1941).

Knutson, E.O., A.C. George, J.J. Frey, and B.R. Koh, Radon Daughter Plateout-II, Prediction Model, Health Phys. 45:445-452 (1983).

Lawrence Berkeley Laboratory, Instrumentation for Environmental Monitoring, Vol. 1, John Wiley and Sons, New York, NY (1983).

McLaughlin, J.P., The Attachment of Radon Daughter Products to Condensation Nuclei, Proc. of Royal Irish Academy, 72:Sec. A, 51-70 (1972).

Mohnen, V.A., Formation, Nature, and Mobility of Ions of Atmospheric Importance, in Electrical Processes in Atmospheres (H. Dolezalek and R. Reiter, eds.) pp. 1-17, Steinkopff, Darmstadt, Federal Republic of Germany (1977).

Migdal, A., Ionization of Atoms Accompanying Alpha and Beta Decay, J. of Phys. IV:449-453 (1941).

NCRP, Report 78, Evaluation of Occupational and Environmental Exposures to Radon and Radon Daughters in the United States, National Council on Radiation Protection and Measurements, Bethesda, Md. (1984).

Porstendorfer, J. and T.T. Mercer, Influence of Electric Charge and Humidity upon the Diffusion Coefficient of Radon Decay Products, Health Phys. 37:191-199 (1979).

Pruppacher, H.R. and J.D. Klett, Microphysics of Clouds and Precipitation, D. Reidell Publishing Company, Boston, Mass. (1978).

Raabe, O.G., Concerning the Interactions that Occur Between Radon Decay Products and Aerosols, Health Phys. 17:177 (1969).

Roffman, A. Short-lived Daughter Ions of Radon 222 in Relation to Some Atmospheric Processes, J. Geophys. Res. 77:30 (1972).

Rudnick, S.N., W.C. Hinds, E.F. Maher, and M.W. First, Effect of Plateout, Air Motion and Dust Removal on Radon Decay Product Concentration in a Simulated Residence, Health Phys. 45:463-470 (1983).

Schery, S.D., Measurements of Airborne ^{212}Pb and ^{220}Rn at Varied Indoor Locations within the United States, Health Phys. 49:1061-1067 (1985).

Thomas, J.W., Measurements of Radon Daughters in Air, Health Phys. 23:783-789 (1972).

UNSCEAR, Ionizing Radiation: Sources and Biological Effects, United Nations, New York, NY (1982).

Wexler, S., Primary Physical and Chemical Effects Associated with Emission of Radiation in Nuclear Processes, Acta Chem. Biol. Radiat. 3:148 (1965).

Wilkening, M.H., Daughter Ions of ^{222}Rn in the Atmosphere, 54th Annual Meeting Bulletin, American Geophysical Union, Washington, D.C., April 1973.

Wilkening, M.H., M. Kawano, and C. Lane, Radon Daughter Ions and Their Relation to Some Electrical Properties of the Atmosphere, Tellus 18:679-684 (1966).

Wilkening, M.H. and V. Romero, ^{222}Rn and Atmospheric Electrical Parameters in the Carlsbad Caverns, J. Geophys. Res. 86:9111-9916 (1981).

Wilkening, M.H. Characteristics of Atmospheric Ions in Contrasting Environments, J. Geophys. Res. 90:5933-5935 (1985).

Wilkening, M.H. and A.W. Wicke, Seasonal Variation of Indoor Radon at a Location in the Southwestern United States, accepted for publication, Health Phys. (1986).

RECEIVED August 4, 1986

Chapter 20

The Effect of Filtration and Exposure to Electric Fields on Airborne Radon Progeny

Niels Jonassen

Laboratory of Applied Physics I, Technical University of Denmark, 2800 Lyngby, Denmark

The level of airborne short-lived decay products of ^{222}Rn in indoor air is for a given radon feed determined by a series of removal processes like ventilation, filtration and plateout, both passive and electric field-induced. The paper describes a series of investigations, where various types of filters, mechanical and electrical, have been operated in rooms with radon levels ranging from about 100 $Bq \cdot m^{-3}$ to about 5000 $Bq \cdot m^{-3}$ and with aerosol concentrations up to 1 to $2 \cdot 10^{11}$ m^{-3}. Measurements of individual radon progeny concentrations and their unattached fractions show that by using filtration rates of up to 3 to 4 h^{-1} it is possible to lower the potential alpha energy concentration by a factor of 5 to 6, while the radiological doses to a certain part of the respiratory tract, according to the normally accepted dose models, for the same range of filtration rates will on the average only be reduced to about 50 % of the value applicable to the air in the untreated state. The electrical properties, such as polarity, charged fraction and mobility of both attached and unattached radon progeny has also been studied, and some results of the effect on the radon progeny of exposing the air to an electric field are also reported.

Since the emergence a couple of decades ago of radon as a radiological problem in many ordinary living and working environments several procedures have been suggested for reducing the exposure due to radon and more specifically its airborne short-lived progeny, the so-called radon daughters.

Attempts of keeping the radon level in indoor air low can either be directed at keeping the radon feed low by reducing the entry of radon (from the soil) or to remove radon (and its progeny) from the indoor air by ventilating with outdoor radon-free or radon-poor air.

Ventilation

The radon concentration C_r at a given ventilation rate r is theoretically given by

$$C_r = C_o \cdot \lambda_o / (\lambda_o + r) \tag{1}$$

where C_o is the radon concentration at a ventilation rate of zero and $\lambda_o = 7.5 \cdot 10^{-3} \ h^{-1}$ is the decay constant of radon.

In a normal dwelling with ventilation rates above, say, $0.3 \ h^{-1}$ equation (1) may be written

$$C_r \simeq C_o \cdot \lambda_o / r \tag{2}$$

This latter equation shows even for a low ventilation rate of say $0.3 \ h^{-1}$ that $C_{o.3} \simeq 2.5 \ \%$ of C_o. At first sight this seems to indicate a very powerful effect of ventilation at reducing indoor radon concentrations (and as a consequence the potential alpha energy concentration, PAEC, from its progeny). In practice, however, a house, which has a radon problem, already will have its own normal ventilation rate r, which is greater than r = 0. If increased ventilation is to be considered as part of the remedial strategy then its effect must be considered within a reasonable range of ventilation rates rather than with respect to the somewhat academic situation of r = 0. An increase of the ventilation rate from for instance 0.3 to $2 \ h^{-1}$ for remedial purposes reduces, according to equation (2), the radon concentration to approximately 15 % of the initial value. If a dwelling has a severe radon problem as found in the high radon level houses in Eastern Pennsylvania, Finland or Sweden such a reduction may be inadequate. In such high radon level houses the ventilation rates required to adequately reduce PAEC may exceed the ventilation rates compatible with household comfort and energy budgeting.

In the case of ventilation consideration of energy conservation must be taken into account. Some work in recent years, however, indicate that the conflicting requirements of energy conservation by means of making dwellings more airtight and increased ventilation for radon reduction may be accomodated by the use of heat exchangers.

Removal of airborne radon progeny

In practice radon may only be removed from a given atmosphere by radioactive decay and by substitution of the radon-laden air with (outdoor) air with low radon content (ventilation).

A series of other removal processes, such as filtration, diffusional plateout and electrostatic deposition, may, however, be effective in the case of the airborne radon progeny, and since the major part of the radiological exposure in a given environment connected with radon is due to the radon progeny rather than to the radon gas itself, it may be of interest to look into these processes.

Filtration

When air is passed through a filter, be it electro- or mechanical, airborne, inactive as well radioactive, particles are removed with

an efficency depending upon the type of filter and the size distribution of the particles. In many ways filtration and ventilation
with radon free air are thus equivalent, but it should be kept in
mind, that filtration on one hand does not affect the radon concentration but on the other hand does also not involve the energy-con-
suming side effect of introducing cold air and may further cause a
change in the partitioning of the radon progeny to species with
higher diffusion ability (unattached progeny) and thus increase the
total removal rate of radioactive material.

The radiological efficiency of a given removal process may be
evaluated in terms of reduction of exposure or of recieved dose.

<u>Exposure</u>

Let us consider an atmosphere where the (activity) concentrations of
radon, ^{222}Rn, and its three short-lived daughters, ^{218}Po, ^{214}Pb and
^{214}Bi, are C_0, C_1, C_2 and C_3, the decay constants and unattached
fractions of the daughter products, λ_1, λ_2, λ_3, f_1, f_2 and f_3 respectively, and E' and E" the energies of the alpha particles from
^{218}Po and ^{214}Po (the fourth short(est)-lived radon daughter).

The exposure rate characteristic for this atmosphere is then
given by the potential alpha energy concentration (PAEC) E_p

$$E_p = (E'+E'')\cdot C_1/\lambda_1 + E''\cdot(C_2/\lambda_2+C_3/\lambda_3) \qquad (3)$$

The maximum PAEC E_{po} which can be supported by the radon concentration C_0 is

$$E_{po} = C_0\cdot((E'+E'')/\lambda_1 + E''\cdot(1/\lambda_2+1/\lambda_3)) \qquad (4)$$

The degree of equilibrium of the airborne daughter products with the
radon in the air is characterized by the equilibrium factor, F

$$F = E_p/E_{po} = 0.105\cdot c_1 + 0.516\cdot c_2 + 0.380\cdot c_3 \qquad (5)$$

where $c_1 = C_1/C_0$, $c_2 = C_2/C_0$ and $c_3 = C_3/C_0$ are the relative daughter concentrations.

The effect of a given process such as plateout, filtration or
field deposition on the PAEC is, however, better expressed by the
ratio of the PAEC at a process level, r, to the PAEC at a process
level of r = 0. This quantity, which like the equilibrium factor is
independent of the radon concentration, can adequately be named the
ERF (energy- (or exposure) reduction factor)

$$ERF(r) = E_{p,r=r>}/E_{p,r=o>} \qquad (6)$$

<u>Dose</u>

The radiological impact of being exposed to a given atmosphere is,
however, only partly determined by the PAEC, but also by the likelihood of deposition of the inhaled activity at specific sites in the
respiratory tract and thus for a given value of PAEC upon the fraction of the progeny in both the attached and the unattached state.
Several models for the radiological dose received over a given time

by a given part of the respiratory tract have been proposed. The most general accepted models at present are the Jacobi-Eisfeld (Jacobi and Eisfeld, 1981), the James-Birchall (James et al., 1980) and the Harley-Pasternack (Harley and Pasternack, 1972, 1982). Generally the dose D can be written

$$D = R \cdot t \cdot (A' \cdot f_1 \cdot C_1 + B' \cdot f_2 \cdot C_2 + C' \cdot f_3 \cdot C_3$$
$$+ A \cdot (1-f_1) \cdot C_1 + B \cdot (1-f_2) \cdot C_2 + C \cdot (1-f_3) \cdot C_3) \qquad (7)$$

where R is the breathing rate characteristic of the group of individuals considered, t the exposure time, A, B, C, A', B' and C' are constants characteristic for the model, group of individuals exposed and aerosol distribution, and the rest of the quantities have the same meaning as in equations (3) and (4).

The doses predicted by the different models vary considerably both from model to model and grom group to group. The variation from one model to another derives primarily from the much higher (but different) doseweights ascribed to the unattached fractions of the radon daughters in the James-Birchall and the Harley-Pasternack-models than in the Jacobi-Eisfeld-model, and in order to find a common convenient measure for the effect of a process on the dose we define a dose reduction factor, DRF, in analogy with the ERF, equation (6), as the ratio between the dose (for a given group of individuals) at a given process level, r, to the same quantity at a process level r = 0, averaged over all groups and models considered

$$DRF(R) = (D_{r=r} / D_{r=o})_{ave} \qquad (8)$$

The question of the effect of filtration will be illustrated in the following by some results taken from a series of papers on this topic (Jonassen and McLaughlin, 1982, 1984, 1985, Jonassen 1984a, 1986). In a certain set of experiments the air in a virtually unventilated room with a volume of about 300 m^3 was passed through an electro-filter with effective filtration rates up to about 3 h^{-1}. The concentration of radon and its short-lived daughters were measured as well as the unattached fractions of the daughter products and the aerosol concentration.

In Figure 1 are plotted the ERF and DRF as a function of the filtration rate from measurements in an atmosphere with an aerosol concentration in the range 50.000-100.000 cm^{-3}, with an average aerosol diameter of 0.10 micrometer. The doses are calculated for the groups: infants 1 year, children 6 and 10 years, and adults, breathing rates 0.45, 0.75 and 1.20 $m^3 \cdot h^{-1}$. It appears that it is possible to lower the PAEC, as expressed by the ERF, to about 15 % of the unfiltrated value, while at the same time the average dose, represented by the DRF, is reduced to about 50 % of the unfiltrated value.

The reduction in PAEC is considerably higher than can be accounted for by filtration directly (Jonassen, 1984a, Jonassen and McLaughlin, 1984), because the filtration as suggested, not only removes radioactive material, but also through removal of aerosol particles increases the unattached fractions and thus enhances the plateout-removal of airborne radioactivity. On the other hand the filtration-induced increase in unattached fractions will increase the average radiological dose per unit of PAEC and this is responsible

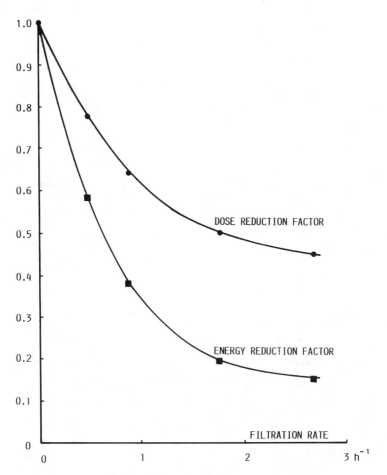

Figure 1. Effect of filtration on radiological dose and exposure rate in an atmosphere with intermediate aerosol concentration.

for the fact that the reduction in DRF is less than can be explained from the filtration caused removal of radioactive material.

Electric fields and ions

Already in the early days of radon research around the turn of the century, it was realized that collection of radon progeny on metal plates or wires could be made more efficient if the collectors were kept at a negative potential with respect to the surroundings (Briggs, 1921). It has later been shown that the activity this way collected primarily consists of positively charged unattached progeny, primarily ^{218}Po. But also part of the attached progeny exist in charged states. The fraction of each progeny product which is in a charged state is only known with a great deal of uncertainty. Values of 10-20 % of positively charged ^{218}Po and somewhat higher for ^{214}Pb have been reported (Jonassen, 1985) together with values of 60-70 % of the attached progeny more or less evenly distributed between positive and negative charges. The unattached and attached ions also differ in terms of mobility, as unattached progeny ions have mobilities around 10^{-4} $m^2V^{-1}s^{-1}$ like ordinary small ions and the attached ions have mobilities in the range $10^{-8} - 10^{-7}$ $m^2V^{-1}s^{-1}$ like ordinary large ions (charged aerosol particles).

Those airborne radon progeny products that are charged (or can be polarized) may be extracted from the air by being exposed to an electric field. This has been shown experimentally by the collection of progeny activity on discs placed at a negative potential in a radon-rich atmosphere (Khan and Philips, 1984, Jonassen, 1984a, 1984b).

The possible use of electrostatic deposition for lowering the level of radon progeny activity in a room has been investigated in a couple of preliminary studies (Jonassen, 1983, 1984b). A thin metal wire was placed approximately 1 m from the walls of a 150 m^3 room and kept at various negative voltages. At a voltage of -5 kV and at low aerosol concentrations it was found that the PAEC was lowered to about (ERF =) 0.52 and the average dose to about (DRF =) 0.66. At high aerosol concentrations the corresponding figures were ERF = 0.72 and DRF = 0.70.

This arrangement was also tried in combination with various household electro-filters giving a total ERF $\simeq$ 0.1-0.2 and a total DRF $\simeq$ 0.36 -0.44 at low aerosol concentrations and ERF $\simeq$ 0.3-0.4 and DRF $\simeq$ 0.45 at high aerosol concentrations.

In some of the experiments various types of ion generators were also in use. Bigu (1983, 1984) has demonstrated that it is possible to lower the PAEC of both radon and thoron progeny in a relatively small box ($\sim$ 3 m^3) by operating a negative ion generator in the box. No explanation of the mechanism involved, however, is given, and since neither individual daughter concentrations nor their unattached fractions were reported, it is only possible to suggest an explanation, but it seems resonable to assume that the decrease in potential alpha energy concentration is primarily caused by a field induced reduction in the aerosol concentration. This will obviously in it self reduce the PAEC but also increase the unattached fractions and consequently the plateout (both field induced and diffusional) of the unattached species.

In the work reported here (in the 150 m^3 room) it was found that if
an ion generator was operated without the assistance of a filtration
device an effect could be detected only if the generator was of the
open field type, i.e. if the generator not only produces ions but also exposes the surrounding air to a static field. Typical values for
a 150 m^3 room were for both ERF and DRF $\simeq$ 0.8-0.9. If the ion generators were operated simultaneously with a filtering device the performance of the filters were always increased typically by 5-10 %.

It should be stressed that the field-ions related measurements
are still of a very preliminary nature. Work is in progress both on
determining the fundamental electrical properties of the radon progeny and based on this on the influence on progeny removal of electrode configuration etc.

Conclusion

It has been demonstrated that it is possible to lower the level of
airborne radon progeny by filtering and/or expose the air to an electric field. If the radiological risk is measured by the potential
alpha energy concentration (PAEC) or exposure rate, for instance expressed in J·m^{-3} or WL, the level may be lowered to about 10-20 % of
the value in untreated air, while the reduced level may only be about
40-50 % of the untreated one if the average dose to a certain part
of the respiratory tract, for instance expressed in Gy·year^{-1}, is
considered.

The question of whether exposure rates or doses should be used
in evaluating all types of remedial procedures has so far received
very little attention. For reasons of convenience the PAEC is normally used, sometimes even estimated from the radon concentration
assuming a rather arbitrarily chosen value of the equilibrium factor.
It seems reasonable to assume that this in certain cases may give a
rather misleading description of the radiological conditions.

References

Bigu, J., On the Effect of a Negative Ion-Generator and a Mixing
Fan on the Plate-Out of Radon Decay Products in a Radon Box,
Health Phys., 44, no.3, 259-266, (1983).
Bigu, J., On the Effect of a Negative Ion-Generator and a Mixing
Fan on the Attachment of Thoron-Decay Products in a Thoron Box,
Health Phys., 46, no.4, 933-939, (1984).
Briggs, G.H., Distribution of the Active Deposits of Radium,
Thorium and Actinium in Electric Fields, Phil. Mag. 41, 357,
1921.
Harley, N.H., and P.S. Pasternack, Experimental Absorption
Applied to Lung Dose from Radon Daughters, Health Phys.,23,
771-782, (1972).
Harley, N.H., and P.S. Pasternack, Environmental Radon Daughter
Alpha Dose Factors in a Five-lobed Human Lung, Health Phys., 42,
789-799, (1982).
Jacobi, W., and K. Eisfeld, Internal Dosimetry of Radon-222,
Radon-220 and their short-lived daughters, in Proc. Nat. Rad.
Env., Bombay, (1981), 131-143, (Vohra, Mishra, Pillai, and
Sadasivan (eds)), Wiley East, Ltd, (1982).

James, A.C., Greenhalgh, J.R., and A. Birchall, A Dosimetric
Model for Tissues of the Human Respiratory Tract at Risk from
Inhaled Radon and Radon Daughters – A Systematic Approach to
Safety, Proc. 5th Cong. IRPA, Jerusalem 1980, Perg. Press,
Oxford, vol 2, 1045–1048, (1980).
Jonassen, N., The Effect of Electric Fields on ^{222}Rn Daughter
Products in Indoor Air, Health Phys., 45, (2), pp. 487–491,
(1983).
Jonassen, N., Removal of Radon Daughters by Filtration and
Electric Fields, Rad. Prot. Dos., 7, nos. 1–4, pp. 407–411,
(1984).
Jonassen, N., Electrical Properties of Radon Daughters in
Proc. Int. Conf. Occupational Radiation Safety in Mining,
(H. Stocker, ed), 561–564. Canadian Nuclear Association,
Toronto, (1984).
Jonassen, N., Radon Daughter Levels in Indoor Air, Effects of
Filtration and Circulation, Progress Report VIII, Laboratory
of Applied Physics I, Technical University of Denmark, (1985).
Jonassen, N., Removal of Radon Daughters from Indoor Air,
ASHRAE-transactions, 91, part 2, (1986).
Jonassen, N., and J.P. McLaughlin, Air Filtration and Radon
Daughter Levels, Env. Int., 8, 71–75, (1982).
Jonassen, N., and J.P. McLaughlin, Airborne Radon Daughters,
Behavior and Removal, in Proc. 3rd Int. Conf. Indoor Air
Quality and Climate, vol 2, 21–27, Stockholm (1984).
Jonassen, N., and J.P. McLaughlin, The Reduction of Indoor Air
Concentrations of Radon Daughters without the Use of Ventila-
tion, Sci. Total Envir., 45, 485–492, (1985).
Khan, A., and C.R. Philips, Electrets for Passive Radon Daughter
Dosimetry, Health Phys., 46, 141–149, (1984).

RECEIVED August 4, 1986

Chapter 21

Plate-Out of Radon and Thoron Progeny on Large Surfaces

J. Bigu

Elliot Lake Laboratory, Canada Centre for Mineral and Energy Technology, Energy, Mines, and Resources Canada, P.O. Box 100, Elliot Lake, Ontario P5A 2J6, Canada

The radon and thoron progeny plate-out characteristics of several materials have been investigated in a large (26 m^3) radon/thoron test facility (RTTF). Discs (about 0.5 mm thickness, and 25 mm diameter) were made up of different materials such as metals, plastics, filter materials, fabrics, and powders. The discs were placed on the wall, and/or horizontal trays, of the RTTF for 24 hours. After exposure the materials were removed from the RTTF and their surface α-activity was measured (differential α-count, gross α-count and α-spectrometry). Experiments were also conducted using specially designed Ra-226 and Th-228 reference sources. Plate-out studies were conducted preferentially at low relative humidity and low aerosol concentrations. Significant differences in the measured surface α-activity were found between different materials. This work is relevant to the determination of radon and thoron progeny deposition velocities on, and attachment rates to, large surfaces.

The short-lived decay products of radon and thoron are initially formed in a positively charged atomic state of great diffusivity. These decay products readily attach themselves to small surfaces such as aerosols, and to large surfaces such as walls. Attachment of particles to large surfaces is commonly referred to as plate-out. The degree of plate-out depends on environmental conditions including aerosol concentration, air moisture content, the presence of tracer gases, and most probably, on the properties of the surface of the material. Although theoretical and experimental data are available on the attachment of the radon and thoron progeny to aerosols, less is known on the attachment to walls with regards to the chemical nature and physical characteristics of the wall surface.

Data on the rate of attachment or deposition, i.e., plate-out of radioactive particles on walls can be used to calculate the particle deposition velocity. Deposition rates can be determined experimentally by measuring the surface activity on some samples

0097-6156/87/0331-0272$06.00/0

designed for the purpose. These samples usually assume the form of small 'discs' which are attached to the surface where plate-out is to be determined. The samples are used because plate-out measurements carried out directly on the surface of interest may not be practical, and in most cases quite difficult. Furthermore, because of practical considerations, a limited number of materials are used for this purpose under the assumption that the mechanism, or extent, of deposition is independent of the type of material used. As there is insufficient evidence to support or deny this view, work in this area is relevant.

This paper deals with the plate-out characteristics of a variety of materials such as metals, plastics, fabrics and powders to the decay products of radon and thoron under laboratory-controlled conditions. In a previous paper, the author reported on measurements on the attachment rate and deposition velocity of radon and thoron decay products (Bigu, 1985). In these experiments, stainless steel discs and filter paper were used. At the time, the assumption was made that the surface α-activity measured was independent of the chemical and physical nature, and conditions, of the surface on which the products were deposited. The present work was partly aimed at verifying this assumption.

Experimental Procedure

Experiments were conducted in a large ($\sim$26 m^3) radon/thoron test facility (RTTF) designed for calibration purposes and simulation studies (Bigu, 1984). A number of different materials were exposed in the RTTF to a radon/radon progeny or thoron/thoron progeny atmosphere. Exposure of the materials was carried out under laboratory-controlled conditions of radiation level, aerosol concentration, air moisture content and temperature. The materials used were in the form of circular 'discs' of the same thickness ($\sim$0.5 mm) and diameter ($\sim$25 mm), and they were placed at different locations on the walls of the RTTF at about 1.6 m above the floor. Other samples were placed on horizontal trays. Samples (discs) of different materials were arranged in sets of 3 to 4; they were placed very close to one another to ensure exposure under identical conditions. Exposure time was at least 24 hours to ensure surface activity equilibrium, or near equilibrium, conditions.

After a given exposure time, the discs were removed from the RTTF and their surface α-activity was measured as a function of time, up to 2 hours, using standard radiation instrumentation and methods. Counting of samples began 2 min after their removal from the RTTF. For the thoron progeny, samples were counted for 15 min followed by 5 min counts with 1 min interval between counts. For the radon progeny two different counting routines were used, namely: 2-min counts with 1-min interval, and 5-min counts with 1-min interval. The former counting routine was intended for more precise Po-218 measurements. In addition, gross α-count (30 min counts) and α-spectrometry measurements were also done.

The surface α-activity on the samples for each exposed set of samples was compared. Each set of samples was exposed, and measured, a minimum of about five times to improve overall statistics. Experiments were conducted in the temperature range 19-22°C and at low relative humidity (10 to 40%). The aerosol concentration, mainly natural atmospheric aerosols, was kept low, i.e., 5×10^2 to 9×10^3 cm^{-3}. No attempt was made to maintain the aerosol concentration at any particular level at these low aerosol concentrations. The values attained were those due to controlled make-up unfiltered air, purposely leaking into the test facility used for the experiments. The aerosol concentration varied with barometric pressure.

The range of radiation conditions in the RTTF was as follows. Radon gas concentration, [Rn-222]: 40-200 pCi/L (1.48×10^3 - 7.4×10^3 Bq/m^3; radon progeny Working Level, WL(Rn): 20-400 mWL; thoron gas concentration [Rn-220]: 350-1700 pCi/L (1.295×10^4 - 6.29×10^4 Bq/m^3); thoron progeny Working Level, WL(Tn): 0.15-14 WL. (The square brackets are used here to denote activity concentration.)

The materials investigated included several metals (copper, Cu; aluminum, Al; stainless steel, SS; and galvanized steel, GS); filter materials (Glass Fiber, Millipore and Varsopor); plastic materials (Plexiglas); 'powders' (activated carbon, talcum); and other materials such as sandpaper, emery cloth and several fabrics. Activated carbon and talcum powder samples were prepared by depositing a thin layer (~0.5 mm) of the powder on filter paper on which some binding material (glue) had been added to its surface to retain the powder.

Further plate-out studies were conducted using radon progeny and thoron progeny reference sources, models Rn-190 and Th-190, respectively, manufactured by Pylon Electronic Development (Ottawa), hereafter referred to as Pylon sources, for simplicity. These are small cylindrical containers (<40 cm^3 volume) provided with a Ra-226 source or Th-228 source. The containers can be opened at their base and some suitable material can be placed in it for exposure purposes (Vandrish et al., 1984). The Ra-226 and Th-228 sources decay, respectively, to Rn-222 and Rn-220 which in turn, decay into their progeny. In this respect, the above sources can be considered miniature RTTFs quite suitable for plate-out studies, in which air flow pattern effects are minimized.

Radioactivity measurements in the above case were similar to the measurements indicated for the test facility, except that only one sample per source could be investigated at a time. In order to facilitate the study, four sources were used, two Ra-226 sources and two Th-228 sources. The strengths of the two Ra-226 sources were different. The same applies to the two Th-228 sources.

Because the reproducibility of α-particle counting with the Pylon sources was quite good, this study was conducted as follows. Four identical samples, i.e., same material, were placed in the decay product standard sources for a period of 24 h at a time. The four samples were then removed and measured. Gross α-particle count and α-spectrometry analysis of the samples were done.

The bulk of the data presented here was obtained in a follow-up to a plate-out study reported earlier elsewhere (Bigu and Frattini, 1985). It should be noted that some qualitative and quantitative differences between earlier studies and the most recent follow-up may be observed depending upon experimental conditions. Although the conditions of the surface of the material, environmental conditions and air flow patterns are believed to play a significant role in plate-out phenomena, their effect on the latter, and their relationships, are far from being understood. Furthermore, the chemical nature of the material, and of the radon and thoron progeny, could play a significant role in plate-out phenomena.

Results and Discussion

The data obtained in this study are presented in Figures 1 to 5, and Tables 1 to 3. For simplicity, the data will be presented according to the location where they were obtained, namely the RTTF, and the Pylon reference sources.

Large Radon/Thoron Test Facility (RTTF). Figures 1 to 5 show the surface α-activity measured on several materials exposed to a radon progeny atmosphere (Figures 1 and 2), and to a thoron progeny atmosphere (Figures 3 to 5).

Figure 1 shows no significant difference between the three metals, namely: aluminum (Al), copper (Cu), and galvanized steel (GS), exposed to a radon progeny atmosphere. Figure 2 shows a significant plated-out activity difference between Fiberglas filters, and coarse sandpaper and emery cloth exposed to a radon progeny atmosphere. No appreciable difference was found between sandpaper and emery cloth.

Figure 3 shows a significant difference in thoron progeny activity plated-out on laboratory coat cotton cloth samples and Millipore filters (0.8 μm), and emery cloth. Differences between the two latter sample materials are not clear in these measurements.

Figures 4 and 5 show significant differences in thoron progeny activity plated-out on cotton cloth and emery paper (Figure 4), and cotton cloth and Fiberglas filters (Figure 5). Differences between the other pair of materials were less pronounced or difficult to ascertain (<7%).

As expected, Figures 1 and 2 show that the surface α-activity on the materials examined decreased with increasing time after exposure. The approximate half-life corresponding to the surface α-activity was in the range of 35 to 50 min. These values roughly agree with the 'combined' half-life of the short-lived decay products of radon. The differences in the decay curves for different materials and for a given material under different environmental conditions, i.e., radiation level, temperature, relative humidity and aerosol concentration are mainly due to the different radioisotope make-up about the surface of the materials investigated. The radioisotope make-up very much depends on the radon progeny disequilibrium ratios, i.e., [Pb-214]/[Po-218] and [Bi-214]/[Po-218]. The radon progeny disequilibrium ratios are highly dependent on the air flow characteristics, relative humidity and aerosol concentration in the RTTF.

Figures 1 and 2 also show a rapid decrease in α-activity during the first few minutes after removal of the samples from the RTTF followed by a less pronounced decrease in the activity. This behaviour is due to the rapid decay of Po-218 with a half-life of about 3 min.

Figures 3 to 5 show that the surface thoron progeny α-activity remained fairly constant during the counting period, i.e., about 100-min. This result indicates that the α-activity measured on the surface of the materials was mainly due to Bi-212, and Po-212 in equilibrium with Bi-212, which was in equilibrium with the relatively long-lived, 10.6 hour half-life, Pb-212. Hence, Bi-212 decayed with the half-life of Pb-212.

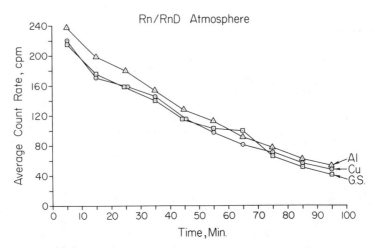

Figure 1   Alpha activity versus time from radon progeny plated-out on several materials.

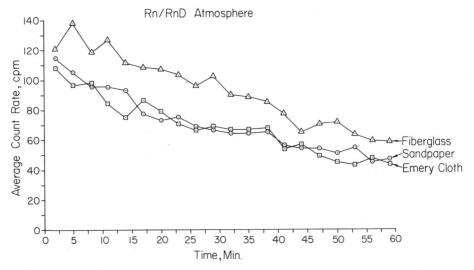

Figure 2   Alpha activity versus time from radon progeny plated-out on several materials.

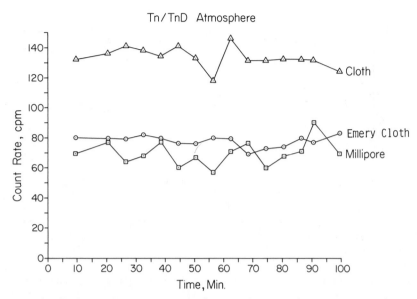

Figure 3   Alpha activity versus time from thoron progeny plated-
out on several materials.

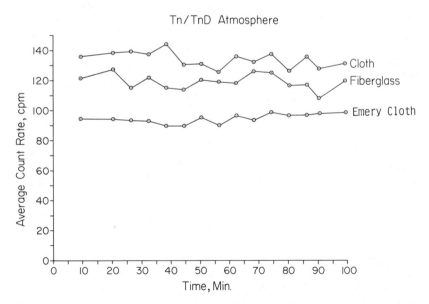

Figure 4   Alpha activity versus time from thoron progeny plated-
out on several materials.

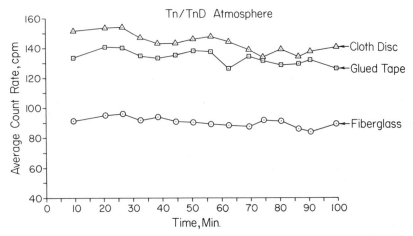

Figure 5   Alpha activity versus time from thoron progeny plated-out on several materials.

Table I   Thoron progeny activity plated-out on different materials

Material	Gross α-particle Count	R*	Qualitative Agreement with Previous Data	Remarks
Copper	9521±1216	0.947		
Galvanized steel	10051±1972	0.999	} Yes	N/A
Aluminum	10052±2093	1.000		
Copper	9469±3393	0.976		
Plexiglas	9508±3580	0.980	} Partial	N/A
Aluminum	9701±2822	1.000		
Plexiglas	8914±2330	0.898		
Aluminum	9854±3286	0.993	} Marginal	N/A
Galvanized steel	9926±3297	1.000		
Millipore	5199±1101	0.730		See
Emery cloth	5208±721	0.731	} Yes	Fig. 3
Cotton cloth	7122±903	1.000		
Fiberglas	5820±447	0.818		See
Tape	7041±748	0.989	} Yes	Fig. 5
Cotton cloth	7115±750	1.000		
Emery cloth	5414±445	0.667		See
Fiberglas	6973±714	0.860	} Yes	Fig. 4
Cotton cloth	8110±230	1.000		
Carbon powder	5436±519	0.786		
Cardboard	5823±628	0.842	} Yes	N/A
Talcum powder	6918±77	1.000		

* R represents the ratio of the activity of the material to the maximum
   activity measured within each set of three materials.

Table II - Radon progeny plate-out on different materials using Pylon source Standards model Rn-190

Material	N_1	N_2	$N_{1,2}^+$	N_3	N_4	$N_{3,4}^+$	$N_{1,2,3,4}^+$	S.R.*
Carbon	49198	24735	73933	115678	60133	175801	249734	—
Copper	20127	11056	31183	42966	23537	66503	97686	4.39
Cotton cloth	20763	11025	31788	40024	21789	61813	93601	2.94
Nuclepore	21102	10954	32056	40901	21687	62588	94644	4.02
Fiberglas	19248	10201	29449	39238	21348	60586	90035	3.70
Emery cloth	18486	9872	28358	35357	19479	54836	83194	3.37
Sandpaper	16539	8834	25373	32701	17405	50106	75479	3.30

* S.R. stands for spectral ratio, i.e., the α-particle count ratio between $^{214}Bi(\equiv {}^{214}Po)$ and ^{218}Po.

+ $N_{1,2} = N_1 + N_2$; $N_{3,4} = N_3 + N_4$; and $N_{1,2,3,4} = N_{1,2} + N_{3,4}$

Note: the symbol N denotes gross α-count. S.R. refers to N_1 and N_3 only.

Table III— Thoron progeny plate-out on different materials using Pylon source standards model Th-190

Material	N_1	N_2	$N_{1,2}^+$	N_3	N_4	$N_{3,4}^+$	$N_{1,2,3,4}^+$	S.R.*
Carbon	16620	16024	32644	431120	422542	853662	886306	—
Fiberglas	11586	11495	23081	262294	261282	523576	546657	1.673
Copper	11258	10993	22251	259513	259678	519191	541442	1.769
Nuclepore	11543	11485	23028	258653	257761	516414	539442	1.768
Cotton cloth	11056	11025	22081	243801	242296	486097	508178	1.704
Emery cloth	10123	9982	20105	224460	224178	448638	468743	1.697
Sandpaper	9807	9781	19588	215702	214943	430645	450233	1.678

* S.R. stands for spectral ratio, i.e., the α-particle count ratio between ^{212}Po and ^{212}Bi.

+ $N_{1,2} = N_1 + N_2$; $N_{3,4} = N_3 + N_4$; and $N_{1,2,3,4} = N_{1,2} + N_{3,4}$

Note: the symbol N denotes gross α-count.

In order to verify the results reported above, an independent series of measurements in the large RTTF was carried out. Samples were exposed to a thoron progeny atmosphere, and surface gross α-particle activity was measured this time for periods of 30 min. The environmental conditions during this experimental phase were in the following range. Temperature: 24-27°C, relative humidity: 40-55%, aerosol concentration: $1.2 \times 10^3 - 3.4 \times 10^3$ cm^{-3}. Some of the data obtained are reproduced in Table I.

Table I shows that, in most cases, qualitative agreement with previously reported data is attained (Bigu and Frattini, 1985). Absolute values, of course, change according to experimental conditions. The large standard deviations quoted in Table I arise from experiments carried out under significantly different radon (or thoron) progeny concentrations. The values quoted in the Table are average values calculated from several measurements conducted on different days under different experimental conditions. Data for each set of materials have been arranged according to increasing surface α-particle activity measured on the material. The variable R represents the ratio of minimum to maximum surface activity measured on each set of three different materials. Gross α-particle 30-min counts are given as absolute α-counts by dividing the instrument reading by its α-particle counting efficiency. Each set of materials was exposed and measured several times on different days, and an average value (shown in the Table) was calculated, as indicated above.

The effect of surface electrostatic charge on a material on the attachment of the decay products of radon has been known since pioneering work on atomic structure by Rutherford. Extensive research into this area for the radon and thoron progeny has been conducted in this laboratory for environmental monitoring purposes. Several authors have reported on the effect of electrostatic charge on the collecting characteristics of copper for the radon progeny for exploration purposes (Card and Bell, 1979).

In a series of recent thoron progeny experiments in a small and large RTTF, dramatic differences in the α-activity collected on copper discs were observed when a positive charge, a negative charge, or a zero charge, i.e., discs at ground potential, was applied. Discs positively charged ($\sim +270$ V) collected 1.4 to 2.8 (ave.:2.0) times more α-activity than discs at zero potential. Discs negatively charged (~ -270 V) collected 11 to 21 (ave.:17) times more α-activity than discs at zero potential. The α-activity ratios between discs charged at -270 V and $+270$ V were in the range 4.5 to 15.5 (ave.: 9.5). The above figures, and figures for the radon progeny obtained in the past, varied quite substantially with environmental conditions in a rather complex and little understood fashion.

The data on the electrostatic effect discussed above is relevant in the context of this paper because certain materials can esily acquire electrostatic charge which could, therefore, alter or mask their plate-out characteristics to the decay products of radon and thoron.

Pylon Radon (Thoron) Progeny Reference Sources. Plate-out studies using the Pylon reference sources are shown in Tables II and III. In this case the materials (discs) were placed in the reference sources where they were exposed for 24-h, after which they were analyzed by gross α-particle count and α-spectrometry. Two counts, and spectra, were taken (1000 s counting time) on each sample exposed to each reference source. Measurements were carried out 1-min and 40-min after exposure. The first count and spectrum using the radon progeny reference source were intended to study the deposition of the short-lived radioisotope Po-218. Ambient temperature and relative humidity at the time the samples were placed in the reference sources were carefully noted. The range of values was as follows: 20.4-23.1°C (temperature), and 10.7-33.7% (relative humidity).

Tables II (radon progeny) and III (thoron progeny) present some of the data obtained. In these Tables N_1 and N_2, obtained with the weak reference sources, stand for the gross α-particle count measured 1-min and 40-min after exposure, respectively. The symbols N_3 and N_4 represent the same as N_1 and N_2, respectively, but for the strong references sources. Furthermore, in the above Tables: $N_{1,2} = N_1 + N_2$, $N_{3,4} = N_3 + N_4$, and $N_{1,2,3,4} = N_{1,2} + N_{3,4}$.

Tables II and III show that the α-activity plated-out on the materials investigated can roughly be divided into three groups, namely, low activity (emery cloth and sandpaper), medium activity (Fiberglas filter, Nuclepore filter, cotton cloth and copper), and high activity (activated carbon deposited on suitable material).

The only data from Tables II and III that can be compared directly with Figures 1 to 5, and Table I are that corresponding to the materials emery cloth, Fiberglas filter and cotton cloth. It is seen that the data corresponding to the radon progeny in Table II agree, qualitatively speaking, with Figure 4, for the thoron progeny. There is agreement between the thoron progeny data of Figure 4, Table I and Table III for emery cloth as having the lowest plated-out activity. However, there is disagreement between cotton cloth and Fiberglas. This difference shows with both the weak and strong reference sources (see Table III). This experimental observation seems to rule out spurious or systematic errors.

The spectral ratios (S.R.) given in Tables II and III show that Nuclepore filters and copper discs have the highest S.R., indicating that less degradation of the spectra, and hence, highest energy resolution, occurs. It should be noted, that except for cotton cloth (Table II), there seems to be a correlation between the S.R. and the activity plated-out on the material.

The α-particle activity measured on activated carbon samples is much higher than that corresponding to any other material (see Tables II and III). This result was expected because of the well known adsorptive properties of carbon for radon and thoron. However, the activated carbon data on Tables II and III is in sharp disagreement with previous, and numerous, data obtained in the large RTTF (see Table I). This topic is under investigation. It should be noted that there are substantial quantitative differences between the behaviour of the radon and thoron progeny relative to activated carbon.

The data presented here suggest some qualitative and/or quantitative differences in behaviour between the radon progeny and thoron progeny relative to some materials, and for some materials relative to either the radon progeny or the thoron progeny.

Other data, not presented here, also suggest that the plate-out characteristics of some materials depend on air moisture content, aerosol and dust concentration and size distribution and other variables in a rather complex, not clearly understood, fashion. Hence, substantial differences may arise when the same set of materials are investigated under different environmental conditions.

At present, only some speculation can be offered regarding the cause for the observed plate-out characteristics of different materials, as estimated from surface α-activity measurements. Some factors that might account for the observed data have been indicated elsewhere (Bigu and Frattini, 1985).

Not previously considered in the above reference is α-particle self-absorption phenomena. However, attempts are at present being made to estimate radon and thoron progeny deposited on the surface of materials by γ-spectrometry and gross γ-count in order to eliminate self-absorption effects.

Conclusions

Significant differences in the radon and thoron progeny plate-out characteristics of different materials have been found. As radon and thoron progeny deposition velocities are normally determined from surface α-activity measurements using some material as a collector of activity, the choice of collecting surface is important in view of our results. The arbitrary choice of a material as an 'activity' collector may lead to substantial errors in the determination of deposition velocities, and other variables, if the data obtained for this particular material are to be used as representative of the deposition velocity on a large surface made of a different material. It should be noted that there are some practical difficulties in determining radon and thoron progeny activity directly on the surface of interest.

Some of the data reported here, and unpublished data, seem to indicate that there might be some qualitative difference between the behaviour of the radon progeny and thoron progeny towards a given material.

The underlying physical and/or chemical mechanisms responsible for the differences observed between the radon progeny and the thoron progeny as related to different materials are not clearly understood. Finally, it should be pointed out that the main thrust in this paper was to determine differences in surface α-activity measured on different materials with the same geometrical characteristics exposed to identical radioactive atmospheres. The calculation of deposition velocities and attachment rates, although it follows from surface α-activity measurements, was not the intent of this paper. This topic is dealt with elsewhere (Bigu, 1985).

The nature of the data discussed here is in part consistent with recent work by other authors who suggested a sticking coefficient of less than unity, under certain conditions, for the attachment of Po-218 to monodisperse aerosols (Ho et al., 1982), and to surfaces (Holub, 1984).

Acknowledgment

The author would like to thank A. Frattini, E. Edwardson and D. Irish (Waterloo University, Co-op student) for conducting some of the measurements in this report.

References

Bigu, J., A Walk-In Radon/Thoron Test Facility, Am. Ind. Hyg. Assoc. J. 45:525-532 (1984).

Bigu, J., Radon Daughter and Thoron Daughter Deposition Velocity and Unattached Fraction Under Laboratory-Controlled Conditions and in Underground Uranium Mines, J. Aerosol Sci., 16:157-165 (1985).

Bigu, J. and A. Frattini, Radon Progeny and Thoron Progeny Plate-Out on a Variety of Materials, Division Report MRP/MRL 85-72(TR), CANMET, Energy, Mines and Resources Canada (1985).

Card, J.W, and K. Bell, Radon Decay Products and their Application to Uranium Exploration, CIM Bull., 72:81-87 (1979).

Ho, W.,P. Hopke, and J.J. Stukel, The Attachment of RaA (^{218}Po) to Monodisperse Aerosol, Atmospheric Environment 16:825-836 (1982).

Holub, R.F., Turbulent Plate-Out of Radon Daughters, Rad. Prot. Dosimetry 7:155-158 (1984).

Vandrish, G., K. Theriault and F. Ryan, The Pylon 190 Standard: a Novel Filter Calibration Standard for Alpha-Spectrometry, in Proceedings International Conference on Occupational Radiation Safety in Mining (H. Stocker, ed) vol. 2, pp. 390-393, Canadian Nuclear Association, Toronto (1984).

RECEIVED September 30, 1986

Chapter 22

Free Fractions, Attachment Rates, and Plate-Out Rates of Radon Daughters in Houses

J. Porstendörfer, A. Reineking, and K. H. Becker

Isotopenlaboratorium für Biologische und Medizinische, Forschung der Georg-August-Universität, Burckhardtweg 2, D-3400 Göttingen, Federal Republic of Germany

The paper summarizes the experimental data on the equilibrium factor, F, the free fraction, f_p, the attachment rate to the room air aerosol, $\bar{X}$, the recoil factor, r_1, and the plateout rates of the free, q^f, and the attached, q^a, radon daughters, determined in eight rooms of different houses. In each room several measurements were carried out at different times, with different aerosol sources (cigarette smoke, stove heating etc.) and under low ($v < 0.3$ h^{-1}) and moderate ($0.3 < v < 1$ h^{-1}) ventilations.

The mean value of the equilibrium factor F measured in houses without aerosol sources was 0.3 ± 0.1 and increased up to 0.5 by additional aerosol particles in the room air. The fraction of the free radon daughters had values between $f_p = 0.06$–0.15 with a mean value near 0.1. Only additional aerosol sources led to a decrease of f_p - values below 0.05.

Besides the attachment rate the plateout rates had the greatest influence on F and f_p. The plateout rates of the free, q^f, and the attached, q^a, radon daughters obtained in rooms with low ventilation varied between 20 and 100 h^{-1} and 0.1 and 0.4 h^{-1} with average values of about 40 h^{-1} and 0.2 h^{-1}, respectively. The recoil factor r_1 describing the desorption probability of Pb-214 atoms had a value of 0.50 ± 0.15, calculated from the measured data.

The shortlived radon daughters Po-218 (RaA), Pb-214 (RaB), Bi-214 (RaC) and Po-214 (RaC') are the main cause of natural radiation exposure. Meanwhile a number of monitoring programs for radon (c_0^i) and radon daughter products (c_j^i) in indoor air have been conducted in several countries (BMI, 1985; Brunner et al., 1982; Hogeweg et al., 1984; Mc Gregor et at., 1980; O'Riordon et al., 1983; Swedjemark and Mjönes, 1984), which are on an average 2-10 times higher than in the free atmosphere (c_0^a, c_j^a). The measured medium

0097-6156/87/0331-0285$06.00/0
© 1987 American Chemical Society

values of the equilibrium equivalent radon concentration c_{eq}^i are between 10-40 Bqm^{-3} with a large variation of more than two orders of magnitude in different houses of the same country.

For estimating the lung dose after inhalation by means of models three parameters of air activity are necessary:

1. The activity concentration of the radon daughters

   c_j (j=1: Po-218; j=2: Pb-214; j=3: Bi-214)
   or the equilibrium equivalent radon concentration:
   c_{eq} =0.105c_1 + 0.516c_2 + 0.379c_3
   The equilibrium factor F = c_{eq}/c_0(c_0 : radon concentration) describes the state of equilibrium between radon and its daughters.

2. Unattached fraction for total potential α -energy of the daughter mixture: f_p = c_{eq}^f/c_{eq}

3. The activity size distribution of the radon daughter aerosols.

The concentration of radon decay products and therefore the factors F and f_p are influenced by the basic processes of the attachment, recoil and deposition (plateout) and by room specific parameters of radon emanation and ventilation (Fig. 1).

The attachment rate to the atmospheric aerosol $\overline{X}=\overline{\beta}*Z$, is a linear function of the particle concentration Z. Values of $5*10^{-3}$ cm^{3}h^{-1} for the average attachment coefficient $\overline{\beta}$ measured in laboratory rooms were reported by Mohnen (1969) and Porstendörfer and Mercer (1978).

The recoil factors r_1 define the probability of whether an attached radioactive atom desorbs from the particle surface in consequence of an alpha decay or not. Mercer and Strowe (1971) found a recoil factor r_1 = 0.81 in their chamber studies in contradiction to the value of $r_1 \sim$ 0.4 measured by Kolerski et al. (1973). No other results about the recoil factor are available in the literature.

Surface deposition is the most important parameter in reduction of the free and aerosol attached radon decay products in room air. If V is the volume of a room and S is the surface area available for deposition (walls, furniture etc), the rate of removal (plateout rate) q is v_g*S/V, always assuming well mixed room air. v_g is the deposition velocity.

Taking into account the results of wind tunnel experiments the average deposition velocities for the free (v_g^f = 2 m h^{-1}) and the attached (v_g^a = 0.02 m h^{-1}) radon decay products can be derived (Porstendörfer, 1984). Knutson et al. (1983) measured similar results in their chamber investigation. The results show that the values of the deposition velocity of the free radon daughters are about 100 times those of the aerosol radon progeny. But there are no information about the effective deposition surface S of a furnished room for the calculation of the plateout rates q^f and q^a by means of v_g^f and v_g^a . For this reason the direct measurements of the plateout rates in rooms are necessary. Only Israeli (1985) determined the plateout rates in houses with values between q^f = 3-12 h^{-1} and q^a = 0.4-2.0 h^{-1}, which give only a low value of the ratio v_g^f /v_g^a = 6-8.

This paper will summerize our experimental data on the equilibrium factor (F), the free fraction (f_p), the attachment rate to the room air aerosol ($\overline{X}$), the recoil factor r_1 and the plateout

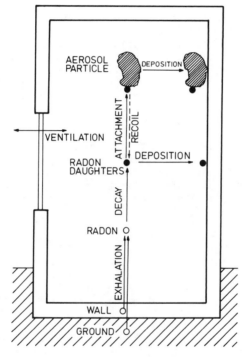

Figure 1. The basic processes influencing the activity balance of radon and radon daughters in houses.

rates of the free (q^f) and attached (q^a) radon daughters, obtained in houses of an area with a higher radon concentration level (North-East of Bavaria). The aim of these investigations was to find out the size of the equilibrium factor F, the unattached fraction f_p and what is the influence of the attachment on the room aerosol and the plateout processes on the concentration of the free and attached radon daughters.

Measurement technique

The activity concentration of radon and of the free and aerosol attached radon daughters were measured simultaneously. During these measurements the aerosol particle concentration was registered by means of a condensation nuclei counter (General Electric).

The radon in the air was measured continuously by electroprecipitation of the positively charged Po-218 ions in an electric field (10 kV) on a surface barrier detector (Porstendörfer, et al., 1980). For this purpose the air was dried, filtered and sucked into an aluminium sphere ($\sim$2 l) with a flowrate of 0.5 $lmin^{-1}$. The counts due to Po-218 and Po-214 were proportional to the radon activity concentration. Their disintegrations were directly detected by alpha spectroscopy with an energy resolution of about 80 keV. The monitor could detect down to 5 Bq m^{-3} with a two hour counting time and 30 % statistical accuracy.

For the determination of the free fraction the attached and the total radon daughter activity concentration were measured. The separation of the free radon daughters from the total daughter concentration was carried out by means of a high-volume diffusion screen battery (Reineking and Porstendörfer, 1986) with a 50 % penetration efficiency for 4 nm particles and for a flowrate of 2.4 $m^3 h^{-1}$.

Each serie of measurements consisted of two parallel samples with counting during and after sampling, one with the screen diffusion battery and the second as the reference sample, so that the fractional free radon daughters could be calculated. The radon daughters are collected on a membrane filter (filter diameter 25 mm, pore diameter 1.2 μm) and the decays of Po-218 and Po-214 are counted by means of alpha spectrometry with a surface barrier detector (area 300 mm^2).

For the determination of the radon daughter concentrations down to 1 Bq $* m^{-3}$ a counting and sampling time of 60 minutes and a counting time of 20-40 minutes after the end of sampling were used. From the counts of Po-218 and Po-214 of the two samples measured in different intervals the concentrations of the free and attached radon daughters were calculated (Wicke and Porstendörfer, 1983).

From the measured activity concentrations the ratios of the free activity to the total activity $f_j = c_j^f/c_j$ (j = 1,2), the corresponding values of f_p and the equilibrium equivalent radon concentration c_{eq} and F, respectively, were calculated.

Evaluation of the measured data

Room model calculations (Porstendörfer et al., 1978; Porstendörfer, 1984) showed the extent of the influence of the different processes and room specific parameters on the radon and the radon daughters indoors. By means of the measured data of radon and free and attached radon daughters and under consideration of this room model the following parameter were determined:

1. Radon emanation e ($Bqm^{-3}h^{-1}$)
2. Ventilation rate v (h^{-1})
3. Attachment rate to the room aerosol $\overline{X}$ (h^{-1})
4. Plateout rate of the free radon daughters on walls and other room surfaces q^f (h^{-1})
5. Plateout rate of radon daughter aerosols on walls and other room surfaces q^a (h^{-1})
6. Recoil factor r_1

The radon emanation and the ventilation rate of a room can be derived from the increase of the radon concentration by the radon exhalation and from the steady state condition between exhalation and air exchange with the free atmosphere. In Fig. 2 the variation of the radon concentration as function of time is shown measured in two houses with different radon emanations and ventilation rates.

The different air exchange rates in a room were obtained by closed, and partly opened windows.

The increase of radon concentration just after thoroughly ventilating the room and closing doors and windows can be described with equation (Porstendörfer et al., 1980):

$$c_0^i(t) = \frac{e + vc_0^a}{v + \lambda_0}\left[1 - e^{-(v+\lambda_0)t}\right] + c_0^a * e^{-(\lambda_0+v)t} \tag{1}$$

which can be written in a simpler form

$$c_0^i(t) = \frac{e}{v}(1 - e^{-vt}) + c_0^a \tag{2}$$

with $\lambda_0 \ll v$. c_0^i and c_0^a are the radon activity concentrations indoors and outdoors, respectively.

In the case of $t < 1/v$ the radon emanation can be determined:

$$e = \Delta c_0^i(t) / \Delta t \tag{3}$$

In a certain period of time there is a steady state condition in a room and a constant radon activity concentration $c_0^i(\infty)$ could be measured. By means of this value and equations (2) (with $t = \infty$) and (3) the actual ventilation rate of a room can be calculated:

$$v = (e - \lambda_0 c_0^i(\infty)) / (c_0^i(\infty) - c_0^a) \tag{4}$$

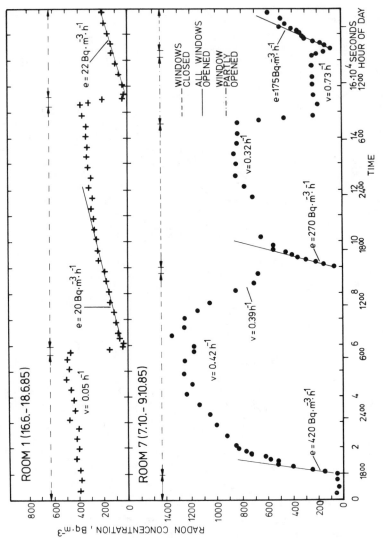

Figure 2. The radon activity concentration in two rooms as function of the time under different ventilation rates.

All our measurements were carried out in rooms with $v < 1\ h^{-1}$ and the radon concentration outdoors ($c_0^a \sim 5\ Bq\ m^{-3}$) was always much smaller than indoors ($c_0^i > 100\ Bq\ m^{-3}$). Assuming a constant radon emanation, a homogeneous activity distribution and 100 % prefiltering of the free fractions of the incoming air, the radon, the free and attached radon daughter activity concentrations indoors (c_0^i; c_j^{if}; c_j^{ia}) and outdoors (c_0^a; c_j^{af}; c_j^{aa}) under steady conditions are:

$$c_0^i = (e + vc_0^a)/(\lambda_0 + v) \tag{5}$$

$$c_j^{if} = (\lambda_j c_{j-1}^{if} + r_{j-1}\lambda_j c_j^{ia})/(v + \lambda_j + q^f + \bar{X}) \tag{6}$$

$$c_j^{ia} = (vc_j^{aa} + (1 - r_{j-1})\lambda_j c_j^{ia} + \bar{X}c_j^{if})/(v + \lambda_j + q^a) \tag{7}$$

with $\quad c_0^{if} = c_0^i$, $c_0^{ia} = 0$, $c_j^{af} = 0$

Under the conditions $v < 1\ h^{-1}$ and $c_j^i \gg c_j^a$ the attachment rate $\bar{X}$ could be calculated by means of the measured c_1^{ia} and c_1^{if} values:

$$\bar{X} = \frac{c_1^{ia}}{c_1^{if}}(\lambda_1 + v + q^a) \approx \frac{c_1^{ia}}{c_1^{if}}(\lambda_1 + v) \tag{8}$$

with $\quad (\lambda_1 + v) \gg q^a$

Then the plateout rate q^f of the free radon daughters can be determined from the activity concentration of the free Po-218 and of the radon:

$$q^f = \lambda_1 \frac{c_0^i}{c_1^{if}} - v - \lambda_1 - \bar{X} \tag{9}$$

In addition an equation of q^a can be evaluated:

$$q^a = \lambda_3 \frac{c_2^{ia}}{c_3^{ia}} - \lambda_3 \frac{c_2^{if}}{c_3^{ia}} \frac{\bar{X}}{v + \lambda_3 + q^f + \bar{X}} - \lambda_3 - v \tag{10}$$

which permits to calculate the plateout of the radon daughter aerosol by means of the measured values of c_2^{ia} , c_1^{ia} and c_1^{if}.

Finally with the measured values of c_2^{ia}, c_1^{ia} and c_1^{if} the recoil factor r_1 can be calculated:

$$r_1 = \frac{1 - \dfrac{c_2^{ia}}{c_1^{ia}} \dfrac{v+\lambda_2+q^a}{\lambda_2} + \dfrac{c_1^{if}}{c_1^{ia}} \dfrac{\bar{X}}{v+\lambda_2+q^f+\bar{X}}}{1 - \dfrac{\bar{X}}{v+\lambda_2+q^f+\bar{X}}} \qquad (11)$$

This equation was derived using c_2^{ia} and c_2^{if} of equations (6) and (5).

The concentrations of radon (c_o^i) and the free (c_1^{if}, c_2^{if}) and on aerosol attached (c_1^{ia}, c_2^{ia}, c_3^{ia}) radon daughters were measured and with these data the equilibrium factor F and the free fraction of the radon daughters f_p were calculated. The room parameters (e, v) and the parameters of radon daughter transport processes ($\bar{X}$, q^f, q^a, r_1) were evaluated by means of equations (3), (4), (8), (9), (10) and (11) using the measured data.

Results

The studies were performed in eight rooms of different houses with higher radon activity concentrations of 300-1600 Bq m^{-3}. All the rooms were furnished. Several measurements were carried out in each room at different times and under different room conditions. If the doors and the windows were closed most of the rooms were lowly ventilated with v $\leq$ 0.3 h^{-1}. Measurements were carried out in a room under different ventilation values. The ventilation of a room was changed by opening a window. In some houses additional aerosol particles were generated by cigarette smoke, stove heating, electrical motor and candle light.

The rooms without aerosol sources and low ventilation rate (v $\leq$ 0.3 h^{-1}) had low aerosol concentrations (2*10^3 - 10^4 cm^{-3}) due to the small influence of the higher aerosol concentrations outdoors (aerosols by traffic and combustions) (Table Ia). In this case the aerosol in the room air was aged by coagulation and plateout and had less condensation nuclei of smaller sizes (d<100 nm). Rooms with a moderate ventilation show higher particle concentrations ((1-5)*10^4 cm^{-3}) (Table IIa). With aerosol sources in a room (Table III) the aerosol concentrations can increase to 5*10^5 particles/cm^3. The relative error of the measured particle concentration is in the order of 15% primary determined by the uncertainties of the absolute calibrations of the condensation nuclei counter.

The equilibrium factor F in low ventilated rooms without aerosol sources varied between 0.2 and 0.4 (Table Ia) with an average value near 0.30; a similar value as reported by Keller and Folkert, 1983, and by Wicke and Porstendörfer, 1982. In rooms with additional aerosol sources an average F-value between 0.4 and 0.5 was obtained (Table III). An error of about 20 % can be estimated for the equilibrium factor.

The free fraction of the radon daughters f_p measured in rooms with low ventilation and no aerosol sources shows values between 0.06 - 0.15 (Table Ib) with a mean value near 0.10. In this case the values of the attachment rates $\bar{X}$ range between 20 h^{-1} and 40 h^{-1}. The f_p-values < 0.05 were obtained in rooms with aerosol sources, which always had values of the attachment rate > 100 h^{-1} (Table III).

In Fig. 3 all measured data of $f_j = c_j^{if}/c_j^i$ and f_p are presented as function of the attachment rate. The absolute errors of f_j and f_p are in the order of 0.02 and 0.01, respectively. These errors were calculated from the errors due to the alpha counting and due to the mathematical procedure of the filter activity evaluation.

The calculated parameters $\bar{X}$, q^f, q^a and r_1 are presented in Tables Ib, IIb and III. In addition the attachment equivalent diameter $\bar{d}$ was determined from the attachment coefficient $\bar{\beta} = \bar{X}/Z$ by means of the attachment theory (Porstendörfer et al., 1979).

In poorly ventilated rooms the average value of the attachment coefficient with $7.4*10^{-3}$ $cm^3 h^{-1}$ (Table Ib) was significantly higher than the value in rooms with moderate ventilation ($2.4*10^{-3}$ $cm^3 h^{-1}$) (Table IIb) corresponding to diameters $\bar{d}$ = 117 nm and $\bar{d}$ = 6.5 nm, respectively. This aged aerosols in poorly ventilated rooms did not only show lower particle concentrations but also greater particle diameters.

The plateout rates of the free and attached radon daughters obtained in rooms with low ventilation varied between 20 and 100 h^{-1} (q^f) and 0.1 and 0.4 h^{-1} (q^a) with average values of about 40 h^{-1} and 0.2 h^{-1}, respectively (Table Ib). Spectacular is the increase of the plateout rates of the free radon daughters in rooms with cigarette smoke and candle light (Table III). In this case the organic vapors in the air could have an increase of the diffusion by neutralisation (charge transfer) of the positively charged radon daughters (Frey et al., 1981). This can also be the reason for the equilibrium factor F being relatively low with cigarette smoke (0.4-0.5), in spite of the great attachment rate to smoke particles. The errors for the determination of q^f and q^a can be estimated with 15 % and 30 %, respectively. The ratio of the deposition velocities $v_g^f/v_g^a = q^f/q^a$ (Table Ib) had average values near 200 in contradiction to the value 6-8 reported by Israeli (1985). Knutson et al. (1983) determined a value of v_g^f/v_g^a = 100.

The calculated recoil factor r_1 had an average value of about 0.50 ±0.15, including the results with particles of cigarette smoke, stove heating etc. In spite of an error for the determination of about 30 %, this value is significantly lower than determined by Mercer (Mercer and Stowe, 1971; Mercer, 1976).

Conclusion

The results of these measurements show that the fraction of the free radon daughters in rooms with low and moderate ventilation and without any aerosol sources are higher (f_p = 0.06 - 0.15) than proposed in literature (Jacobi and Eisfeld, 1980; ICRP 32, 1981; NEA, 1983). A mean value of 10 % (f_p = 0.1) was determined. Only additional aerosol sources in a room such as cigarette smoke, cooking, candle light or stove heating led to a decrease of the f_p-value below 0.05.

Table Ia. The room parameters (v, e, S), the aerosol concentration (Z), the activity concentrations of radon (c_o^i) and the free (c_1^{if}, c_2^{if}) and attached $(c_1^{ia}, c_2^{ia}, c_3^{ia})$ radon daughters, measured in closed rooms (low ventilation).

room	v (h^{-1})	e (Bqm^{-3}h^{-1})	S (m^2)+ / S/V(m^{-1})	Z(cm^{-3}) *10^3	c_o^i (Bqm^{-3})	c_1^{ia}(Bqm^{-3})	c_1^{if}(Bqm^{-3})	c_2^{ia}(Bqm^{-3})	c_2^{if}(Bqm^{-3})	c_3^{ia}(Bqm^{-3})
1	0.03	19	182 / 1.64	2.3 (1.3-4.1)	411 (395 - 429)	144 (112-169)	60 (45 - 88)	134 (113-152)	7 (0-13)	127 (106-149)
2	0.08	139	53 / 1.70	1.9 (1.6-2.1)	1559 (1359-1726)	367 (310-413)	141 (114-178)	288 (272-311)	5 (0-16)	272 (250-293)
3	0.10	41	119 / 1.53	13 (10 - 16)	500	283 (143-380)	124 (94 -160)	229 (120-311)	14 (2-40)	203 (102-287)
4	0.13	43	53 / 2.06	4.0 (3.1-5.0)	300 (266 - 327)	72 (62 - 81)	47 (42 - 51)	49 (35 - 64)	7 (0-14)	45 (34 - 61)
5	0.14	50	133 / 1.94	4.3 (2.0-6.1)	331 (293 - 398)	139 (115-160)	87 (61 -114)	118 (105-128)	8 (1-14)	113 (86 -115)
6	0.16	129	180 / 1.64	9.3 (1.8 -18)	764 (651 - 850)	312 (183-409)	157 (115-221)	223 (158-277)	21 (0-49)	179 (129-228)
7	0.31	216	61 / 2.00	4.4 (3.6-6.3)	946 (675 -1297)	308 (187-449)	172 (122-223)	211 (116-370)	13 (0-31)	168 (93 -312)
8	0.44	350	54 / 2.07	6.6 (4.3-9.7)	785 (695 - 929)	196 (158-239)	93 (57 -119)	127 (109-160)	11 (9-14)	101 (85 -141)

+ S = room surface (walls, ceiling, floor), V = room volume

Table Ib. The equilibrium factor (F), the free fraction (f_p), the attachment parameters ($\bar{X}, \bar{\beta}, \bar{d}$), the plateout rates ($q^f$, q^a) and the recoil factor (r_1), calculated from the measured data of Table Ia (low ventilation).

room	f_p (Bq m^{-3})	F	$\bar{X}$ (h^{-1})	$\bar{\beta}$ (cm^3 h^{-1}) $*10^{-3}$	$\bar{d}$ (nm)	q^f (h^{-1})	q^a (h^{-1})	q^f/q^a	r_1
1	0.068 (0.040-0.126)	0.35 (0.29-0.41)	33 (19-47)	14 (10 - 22)	173	46 (29- 63)	0.09	511	0.28
2	0.061 (0.046-0.070)	0.20 (0.19-0.22)	36 (24-49)	19 (12 - 27)	228	101 (66-139)	0.06	1700	0.34
3	0.083 (0.051-0.139)	0.49	32 (21-42)	2.4 (1.4-4.2)	64	-	0.42	-	-
4	0.144 (0.074-0.199)	0.19 (0.17-0.21)	21 (18-23)	5.3 (4.0-7.4)	100	52 (46- 58)	0.10	520	0.56
5	0.099 (0.062-0.146)	0.40 (0.33-0.42)	22 (14-31)	5.2 (3.5-7.0)	99	16 (11- 23)	0.15	110	0.73
6	0.116 (0.046-0.169)	0.32 (0.27-0.37)	28 (12-41)	3.0 (2.3-6.7)	72	25 (15- 32)	0.43	58	0.54
7	0.116 (0.077-0.147)	0.24 (0.20-0.30)	25 (16-32)	5.7 (4.4-7.8)	105	36 (27- 45)	0.28	130	0.47
8	0.112 (0.094-0.138)	0.18 (0.15-0.19)	30 (19-53)	4.5 (4.4-5.5)	91	71 (61- 99)	0.12	590	0.36
mean	0.100	0.30	28	7.4	117	43	0.21	205	0.47

Table IIa. The aerosol concentration (Z), the activity concentrations of radon (c_0^i) and the free (c_1^f, c_2^f) and attached (c_1^{ia}, c_2^{ia}, c_3^{ia}) radon daughters, measured in rooms with moderate ventilation (partly opened windows).

room	ν (h⁻¹)	Z(cm⁻³) *10³	c_0^i(Bqm⁻³)	c_1^{ia}(Bqm⁻³)	c_1^f(Bqm⁻³)	c_2^{ia}(Bqm⁻³)	c_2^f(Bqm⁻³)	c_3^{ia}(Bqm⁻³)
1	0.40	37 (31 – 43)	52 (44 – 67)	22 (18 – 29)	4 (3 – 5)	13 (13 – 15)	0 (0 – 2)	13 (12 – 16)
2	0.53	32 (8.2 – 47)	260 (174 – 320)	79 (53 – 100)	13 (12 – 13)	55 (39 – 70)	2 (0 – 4)	52 (36 – 67)
7	0.89	12 (10 – 14)	247 (216 – 327)	92 (72 – 115)	55 (43 – 84)	45 (29 – 62)	0 (0 – 5)	31 (16 – 53)

Table IIb. The equilibrium factor (F), the free fraction (f_p), the attachment parameters ($\bar{X}, \bar{\beta}, \bar{d}$), the plateout rates ($q^f$, q^a) and the recoil factor (r_1), calculated from the measured data of Table IIa (moderate ventilation).

room	f_p (Bqm⁻³)	F	$\bar{X}$ (h⁻¹)	$\bar{\beta}$ (cm³h⁻¹) *10³	$\bar{d}$ (nm)	q^f (h⁻¹)	q^a (h⁻¹)	q^f/q^a	r_1
1	0.025 (0.023-0.080)	0.29 (0.27-0.30)	88 (78- 96)	2.5 (1.8-3.1)	66	98 (89 -110)	0.05	–	0.74
2	0.038 (0.019-0.076)	0.23 (0.21-0.24)	90 (65-110)	2.8 (2.3-7.9)	70	180 (120-214)	0.05	–	0.35
7	0.128 (0.085-0.157)	0.21 (0.15-0.26)	24 (20- 32)	2.0 (1.6-2.7)	58	22 (13 – 31)	0.17	130	1.06
mean	0.064	0.24	67	2.4	65	100	–	–	0.72

Table III. The aerosol particle concentration (Z), the equilibrium factor (F), the free fraction (f_p), the attachment parameters ($\bar{X}, \bar{\beta}, \bar{d}$), the plateout rates ($q^f$, q^a) and the recoil factor (r_l), obtained in lowly ventilated rooms with aerosol sources.

room	aerosol source	$Z(cm^{-3})$ $*10^3$	f_p (Bqm^{-3})	F	$\bar{X}$ (h^{-1})	$\bar{\beta}(cm^3 h^{-1})$ $*10^{-3}$	$\bar{d}(nm)$	q^f (h^{-1})	q^a (h^{-1})	r_l
1	candle light	260 (250-260)	0.014 (0.014-0.029)	0.34 (0.31-0.36)	121 (108-138)	0.5 (0.4-0.5)	28	165 (149-181)	0.05	0.30
	candle light sooty particle	34 (16-68)	0.005	0.43 (0.41-0.43)	710 (530-1400)	21 (8.5-74)	243	700 (520-1500)	0.05	0.29
	during smoking cigarettes	240 (130-260)	0.006	0.46 (0.37-0.50)	1000	4.2	90	900	–	0.41
5	electrical motor	20 (7.5-45)	0.097 (0.090-0.103)	0.40 (0.39-0.41)	28 (26-29)	1.4 (0.6-4)	48	15 (15-15)	0.23	1.03
	tiled stove	30 (11-54)	0.010 (0.009-0.016)	0.47 (0.43-0.53)	210 (143-450)	7.0 (4.9-16)	118	85 (48-120)	0.37	0.39
	after smoking cigarettes	200	0.006	0.52	380	18	220	26	0.47	1.00
7	during smoking cigarettes	280 (130-500)	0.03	–	470 (230-1000)	2.3 (0.5-4.8)	62	–	–	–
	after smoking cigarettes	44 (9-110)	0.02	–	560 (130-1700)	12 (6.0-15)	163	–	–	–

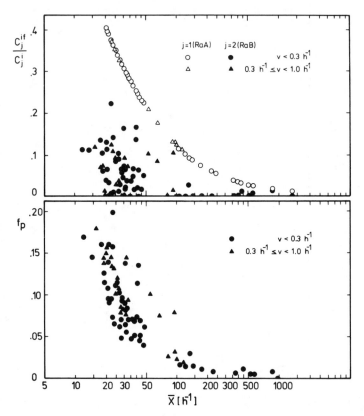

Figure 3. Ratios of the free to the total activity concentrations of RaA and RaB and f_p as function of the attachment rate $\bar{X}$ measured in eight rooms of different houses.

The mean value of the equilibrium factor F in houses was 0.3 ± 0.1 without aerosol sources and can increase up to 0.5 with cigarette smoke in the room air.

The plateout rates of the free radon decay products on room surfaces are about 200 times higher than the values of the aerosol radon daughters and have a great influence on the radon daughter activity concentrations indoors.

In addition this investigation showed that the desorption probability by recoiling of Pb-214 atoms from particle surfaces of atmospheric aerosol is 0.50 ± 0.15.

Acknowledgments

This work is supported by the Commission of the European Communities, contract no. BI-6-0130-D(B)

References

Brunner, H., W. Burkart, E. Nagel and C. Wernli, Radon in Wohnräumen in der Schweiz - Ergebnisse der Vorstudie 1981/1982, Eidg. Institut für Reaktorforschung, Würenlingen, Report RM-81-82-11.

Bundesminister des Innern, Radon in Wohnungen und im Freien, Erhebungsmessungen in der Bundesrepublik Deutschland, BMI-Report (1985).

Frey, G., P.K. Hopke and J. Stukel, Effects of Trace Gases and Water Vapor on the Diffusion Coefficient of Polonium-218, Science 211:480-481 (1981).

Hogeweg, B., B.F. Bosnjakovic and W.M. Willart, Radiation Aspects of Indoor Environments and Related Radioecological Problems: a Study of the Situation in the Netherlands, Rad. Prot. Dos. 7:327-331 (1984).

Israeli, M., Deposition Rates of Rn Progeny in Houses, Health Physics 49:1069-1083 (1985).

Keller, G. and K.H. Folkerts, Ra-222 Concentrations and Decay Products Equilibrium in Dwellings and in the Open Air, Health Physics 47:385-398 (1984).

Knutson, E.O., A.C. George, J.J. Frey and E. Koh, Radon Daughter Plateout, Part II. Prediction Model, Health Physics 45:445-452 (1983).

Kolerskii, S.V., Yu.V. Kusnetsov, N.M. Polev and L.S. Ruzer, Effect of Recoil Nuclei being Knocked of Aerosol Particles onto Free-Atom Concentrations of Daughter Emanation Products, Izmeyitel naya Tekhnika 10:57-58 (1973).

Mc Gregor, R.G., P. Vasudev, E.G. Letourneau, R.S. Mc Cullough, F.A. Prantl and H. Taniguchi, Background concentrations of radon and radon daughters in Canadian homes, Health Physics 39:285-289 (1980).

Mercer, T.T., The Effects of Particle Size on the Escape of Recoiling the RaB Atoms from Particulate Surfaces, Health Physics 31:173-175 (1976).

Mercer, T.T. and W.A. Stowe, Radioactive Aerosols Produced by Radon in Room Air: Inhaled Particles III (Edited by Walton, W.H.) pp. 839-850, Old Waking: Unwin Bros. Ltd., (1971).

Mohnen, V.A., Die radioaktive Markierung von Aerosolen, Physik 229:109-122 (1969).

O'Riordon, A.C. James, G. Rae and A.D. Wrixon, Human Exposure to Radon Decay Products Inside Dwellings in the United Kingdom, National Radiological Protection Board, Chilton, Report NRPB-R152 (1983).

Porstendörfer, J., A. Wicke, A. Schraub, The Influence of Exhalation, Ventilation and Deposition Processes upon the Concentration of Radon (Rn-222), Thoron (Rn-220) and their Decay Products in Room Air, Health Physics 34:465-473 (1978).

Porstendörfer, J. and T.T. Mercer, Influence of Nuclei Concentration and Humidity upon the Attachment Rate of Atoms in the Atmosphere, Atmospheric Environment 12:2223-2228 (1978).

Porstendörfer, J., G. Röhrig and A. Ahmed, Experimental Determination of the Attachment Coefficients of Atoms and Ions on Monodisperse Aerosols, J. of Aerosol Science 10:21-28 (1979).

Porstendörfer, J., A. Wicke ad A. Schraub, Methods for a Continuous Registration of Radon, Thoron, and their Decay Products Indoors and Outdoors, in: Natural Radiation Environment III, (Eds.: T.F. Gesell and W.M. Lowder) published by Technical Information Center/US Department of Energy, CONF-780422 (Vol. 2) pp. 1293-1307 (1980).

Porstendörfer, J., Behaviour of Radon Daughter Products in Indoor Air, Rad. Prot. Dos. 7:107-113 (1984).

Reineking, A. and J. Porstendörfer, High-volume Screen Diffusion Batteries and the Alpha Spectroscopy for Measurements of the Radon Daughter Activity Size Distributions in the Environment, J. of Aerosol Science 17 (1986) (accepted for publication).

Swedjemark, G.A. and L. Mjönes, Radon and Radon Daughter Concentrations in Swedish Homes, Rad. Prot. Dos. 7:341-345 (1984).

Wicke, A. and J. Porstendörfer, Radon Daughter Euqilibrium in Dwellings, in: Radiation Environment (K.G. Vohra et al., ed.) pp. 481-488, Wiley Eastern Limited, New Delhi (1982).

Wicke, A. and J. Porstendörfer, Application of Surface Barrier Detectors for the Measurement of Environmental Radon and Thoron Daughters in Air, in: Proceedings of the International Meeting on Radon-Radon Progeny Measurements, EPA 520/5-83/021 (1983).

RECEIVED August 4, 1986

Chapter 23

The Behavior of Radon Daughters in the Domestic Environment
Effect on the Effective Dose Equivalent

H. Vanmarcke[1], A. Janssens[1], F. Raes[1], A. Poffijn[2], P. Berkvens[1], and R. Van Dingenen[1]

[1]Nuclear Physics Laboratory, State University of Gent, Proeftuinstraat 86, B-9000 Gent, Belgium
[2]Physics Laboratory (II), State University of Gent, Proeftuinstraat 86, B-9000 Gent, Belgium

Simultaneous measurements of the radon daughter concentrations, the ventilation rate and the size distribution of the inactive aerosol have been performed in two bedrooms, a living room and a cellar. The measured radon daughter concentrations were fitted by the room model to optimize the deposition rate of the unattached daughters. The mean value was 18/h in the rooms and 8/h in the cellar. Then the unattached fraction was calculated in each measurement and was found to be between .05 and .15 without aerosol sources in the room and below .05 in the presence of aerosol sources. The effective dose equivalent was computed with the Jacobi-Eisfeld model and with the James-Birchall model and was more related to the radon concentration than to the equilibrium equivalent radon concentration. On the basis of our analysis a constant conversion factor per unit radon concentration of 5.6 $(nSv/h)/(Bq/m^3)$ or 50 $(\mu Sv/y)/(Bq/m^3)$ was estimated.

The radiation problem due to the presence of radon (Rn-222), thoron (Rn-220) and their respective daughter products in underground mines has been recognised many years ago and has been extensively studied. There has been a growing consensus that in a mining environment the quantity, exposure to radon daughter potential alpha energy (thoron daughter), can be transformed into adequate radiation protection. On historical grounds, this way of assessing the dose is known as "the working level concept".

Some years ago it was realized that the indoor inhalation of the short-lived radon daughters constitutes the most important contribution to the radiation exposure of the general population (Unscear, 1982). The working level concept has been introduced in the domestic environment due to the success of the concept in the occupational environment and due to a lack of experimental data on the relative and absolute magnitudes of the transformation and

0097-6156/87/0331-0301$06.75/0

removal processes determining the fate of the radon daughters in the
domestic air. These processes are : radioactive decay, ventilation,
attachment on and detachment from the ambient aerosol and deposition
on walls and furniture. The rate of deposition is strongly dependent
on the physical characteristics of the daughters. Although the
description of the physico-chemical transformation of the daughters
is very complex (Raes et al.,1985a; Raes,1985b) we still stick to
the simple classification into a fraction that is attached to the
ambient aerosol and a fraction that contains small clusters which
for convenience is called the unattached fraction. The deposition of
the latter and the attachment to the room aerosol are fast
competitive processes. They determine to a large extend the internal
radon daughter equilibrium in houses. The other processes :
ventilation, deposition of the attached fraction and detachment of
Pb-214 from the room aerosol due to the recoil energy, are less
critical to the equilibrium. According to Mercer (1976) a recoil
factor .83 is used. In turbulent mixed air, the concentrations of
the radon daughters can be expressed as a series of differential
equations. The steady-state equilibrium version is known as the room
model. The model was first applied to underground mines
(Raabe,1969;Jacobi,1972) and later on adapted to the domestic
environment. The Porstendörfer version (1984) includes prefiltering
of the unattached concentrations in the incoming air. In the present
paper his version has been changed slightly, assuming the ratio's of
the "outroom" to the "inroom" daughter concentrations to be equal.
The error coming from this assumption is small due to the limited
impact of ventilation on the daughter equilibrium. The term
"outroom" is preferred to outdoor because part of the incoming air
comes from adjacent rooms.

       Our research project has the ambition to contribute to the
knowledge of the parameters of the room model and to calculate the
activity median diameter of the attached daughters (AMD) which is an
important parameter in the lung dosimetry models. However the main
reason for application of the room model is to investigate the
unattached fraction, which yields a much higher dose to the
bronchial epithelium than the attached fraction. A direct
measurement of the unattached fraction in dwellings is only possible
with a high volume sampler, since the concentrations of radon
daughters in dwellings are in general rather low. This could disturb
the steady-state conditions in the room. Besides there are
difficulties in achieving an adequate separation of the attached and
unattached fractions by collection on a gauze.
       The working level concept evaluates the unattached fraction
and the activity median diameter in an indirect way, through the
dose conversion factor. This paper will show that in the domestic
environment this is mostly inaccurate to estimate the dose.

Experimental apparatus

The apparatus for measuring the low radon daughter concentrations,
occuring in normal indoor conditions, involves sucking air at a
constant rate through a filter and counting the activity by means of
alpha spectroscopy during sampling, and during a decay time
interval.The detector configuration has been calibrated through

simultaneous α and γ counting (Vanmarcke,1984) and the analysis of the spectrum was standardised. The timing has been optimized by minimizing the mean of the minimum measurable radon daughter concentrations (MMC), which is defined by Nazaroff (1984), as the concentrations at which the relative standard deviation in the measurement due to counting statistics in 20%. The flow rate and the detector efficiency are taken to be 28 l/min and 0.127 respectively, our standard conditions. It was found that a long sampling time and a delay time, longer than the minimum time needed to transfer the spectrum to a storage medium, are favourable. As an example Figure 1 shows the optimized MMC's as a function of the sampling and delay time for a measuring time of 60 min.

The same kind of optimization has been performed for the thoron daughters. In the calculations the sampling period was set at 30 min and the first decay time interval is started after the decay of the radon daughters (270 min). For a total measurement time of 16 hours the optimized MMC of Pb-212 and Bi-212 are respectively 0.02 Bq/m^3 and 60 Bq/m^3 (270-370 min, 540-960 min). Better results for Bi-212 are obtained with only one decay time interval and an estimation of the ratio of Pb-212 to Bi-212 out of the removal processes (ventilation and deposition of the attached thoron daughters). The influence of the removal rate on the potential alpha energy concentration is small. For the decay interval (270-960 min) the MMC of Pb-212 is 0.014 Bq/m^3, assuming the sum of the removal rates to be 0.6 ± 0.5/h.

Measuring the Po-218 and Po-214 alpha decay during sampling improves significantly the precision (Cliff,1978), in particular for Po-218 and thus for the unattached fraction. However, at the same time a part of the unattached activity is lost in the complex sampling-head geometry. This loss has been determined in our 1 m^3 radon chamber as a function of the flow rate during sampling (Figure 2). The daughters were kept in the unattached state by a forced ventilation of 0.46/h over an absolute filter. The turbulence, induced by the ventilation, reduces the mean residence time of the daughters in a reproducible way. The daughters were alternately sampled with and without the complex sampling-head and counted during decay. In Figure 2 it can be seen that, at 28 l/min the loss of the unattached daughters equals 22 ± 2%. An independent way of assessing the order of magnitude of this loss is through an analysis of the results to the last quality assurance excercise at NRPB (Miles et al.,1984). Comparing our measurements with the measurements of participants using a bare filter gives a loss of the unattached activity of about 30%.

The deposition velocities of the unattached daughters were calculated from the measurements in the radon chamber with the bare filter (Figure 2) and found to equal $.095 \pm .007$ cm/s for Po-218, $.085 \pm .012$ cm/s for Pb-214 and $.045 \pm .015$ cm/s for Bi-214. This decrease in deposition velocity is one of the most important sources of error in the room model.

Together with the radon daughter measurements, nearly continuous measurements of the ventilation rate are performed by means of the release of N_2O tracer gas and observation of its decay with an infrared spectrometer (Miran 101). Furthermore the aerosol concentration and size distribution are monitored every 20 to 30 min with an automated aerosol spectrometer (Raes et al.,1984).

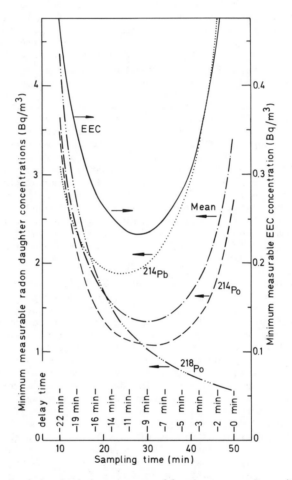

Figure 1. Optimized minimum measurable concentrations (MMC's) of radon progeny as a function of sampling time. The mean of the MMC's is used as optimisation parameter and the total measurement time is fixed at 60 min.

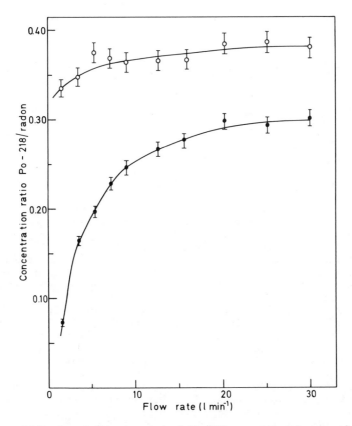

Figure 2. Ratio of the unattached Po-218 concentration to the radon concentration for different flow rates during sampling. The daughters are alternately sampled with (●) and without (O) the detection head.

Measurements

The measurements were performed at four different locations with
radon concentrations ranging from 7 to 111 Bq/m^3 and with low to
moderate ventilation rates. The main characteristics of the
locations are listed in Table I.

Table I. Main characteristics of the four locations studied.

Number	:	1	2	3	4
Type	:	det. house	lab.	det. house	semi-det.
Location	:	quiet area	busy road	quiet area	railway
Floor level	:	gr. floor	basement	1st floor	1st floor
Type	:	bedroom	cellar	bedroom	liv. room
No days	:	7	4	4	3
No meas.	:	32	12	15	13
Vent. (1/h)	:	.28	.74	.17	.46
range	:	.13-.67	.35-2.5	.11-.20	.38-1.26
Rn (Bq/m^3)	:	58	75	30	9
range	:	31-103	44-111	20-41	7-21

The first house and the room chosen for the measurement are the same
as referred to in earlier studies (Raes et al.,1984;Vanmarcke et
al.,1985). Eightythree radon daughter measurements were carried out
on 18 different days. Eleven measurements were rejected from the
optimization excercise because they were performed within 2 hours
after the generation of a high aerosol concentration, so that no
steady-state was reached.
        As an example, Figure 3 shows the data determined during one
day in each of the four locations. The indicated ventilation rates
are best fits to the N_2O decay curves and are considered as
representative for the indicated time periods. There is some
arbitrariness in the fittings leading to discontinuities in the
results. The accuracy is believed to be of the order of 20%. High
aerosol concentrations with different size distributions were
produced by burning a joss-stick, a bit of paper or by smoking or
cooking. The calculated active size distributions corresponding with
these disturbancies are plotted in Figure 4. The attachment rate is
calculated out of the aerosol size distribution as explained in
(Raes et al.,1984) and has a systematic uncertainty of about 50%.
The lower part of Figure 3 show the measured radon daughter
concentrations.

Analysis

The deposition rate of the attached fraction, plotted in Figure 3,
is calculated from the aerosol size distribution assuming diffusion
and electrophoresis to be the most important deposition mechanisms
(Raes et al.,1985a). The accuracy of the absolute values was checked
by forming the aerosol mass balance after the generation of a high
aerosol concentration.In Table II is compared the decay of the

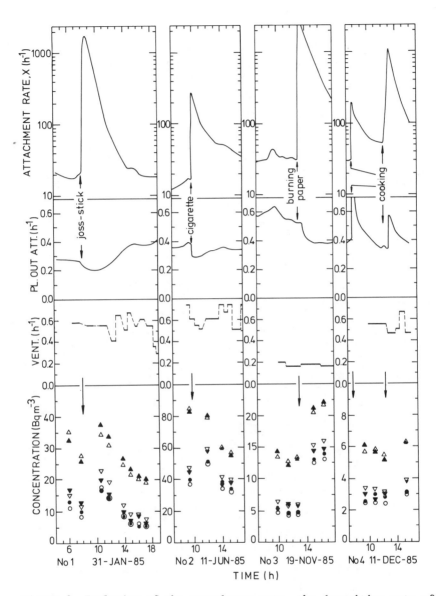

Figure 3. Evolution of the attachment rate, the deposition rate of the attached daughters, the ventilation rate and the radon daughter concentrations (▲ Po-218 ▼ Pb-214 ● Bi-214 measured, △ ▽ ○ fitted) during one day in each of the four locations.

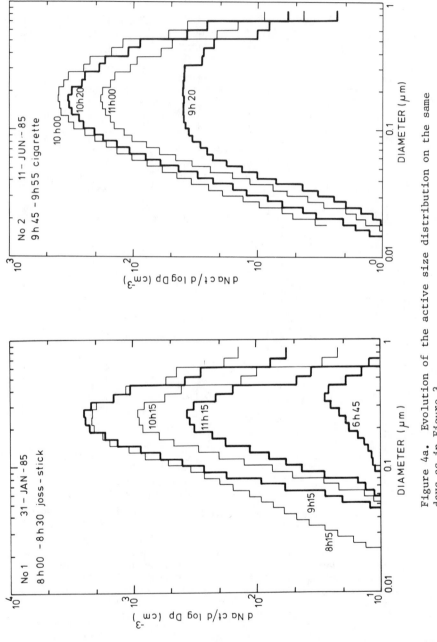

Figure 4a. Evolution of the active size distribution on the same days as in Figure 3.

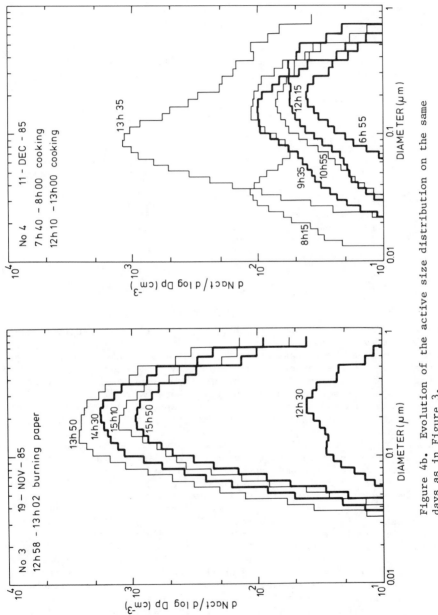

Figure 4b. Evolution of the active size distribution on the same days as in Figure 3.

aerosol volume (λ_{vol}) to the sum (λ_{sum}) of the measured ventilation (λ_{vent}) and the calculated aerosol volume deposition (λ_{dep}). Although the corresponding deposition rate of the attached daughters (λ_a) tend to be somewhat larger than reported in the literature(Table V), the analysis would indicate even higher values.

Table II. The measured decay of the aerosol volume (λ_{vol}) is fitted
          by an exponential curve to the data and compared to the
          sum of the calculated aerosol volume deposition (λ_{dep}) and
          the measured ventilation rate (λ_{vent}).

Loc. : No :	date	generation	deposition				decay
			λ_a	λ_{dep}	λ_{vent}	λ_{sum}	λ_{vol}
			1/h	1/h	1/h	1/h	1/h
1 :	31/01	joss-stick	.22	-	.55	<.77	1.00
1 :	01/02	joss-stick	.19	-	.35	<.54	.74
2 :	11/06	smoking	.30	.18	.55	.73	.81
2 :	13/06	burn.paper	.24	.18	.55	.73	.73
3 :	19/11	burn.paper	.38	.26	.18	.44	.53
3 :	20/11	burn.paper	.62	.35	.12	.47	.50
4 :	09/12	cooking	.37	.27	.38	.65	.78
4 :	11/12	cooking	.45	.24	.50	.74	.61

    The measured radon daughter concentrations can be fitted by
the room model with parameters assumed to be known (measurement,
literature) and with parameters for which values are optimized. The
evaluation of the latter is carried out by a very general programme
for the optimization of any function. At present this function is
defined as the squared sum of the difference between the measured
and calculated radon daughter concentrations, weighted as a function
of their counting statistics. With the present input data it was
found that only one parameter could be fitted with reasonable
accuracy. However, still two parameters have to be determined : the
deposition rate of the unattached daughters and the ratio of the
"outroom" to the "inroom" daughter concentrations.
    The uncertainty of the fitted values of these two parameters
has been estimated objectively by means of a Monte-Carlo simulation
model. The data points on each curve in Figure 5 are the mean of 100
calculated points and each point is the "best-fit" of the parameter
to a simulated measurement in a simulated indoor environment in
which allowance is made for fluctuations of the parameters.
    The complete description of the model is beyond the scope of
the paper. In Figure 5 it is shown that at low attachment rates the
deposition rate of the unattached daughters (λ_{un}) can be evaluated
and at high attachment rates the ventilation rate.
    On this basis the ventilation rate was fitted to the data at
high attachment rates and compared to the measured ventilation rate
so that the ratio of the "outroom" to the "inroom" daughter
concentration (P) could be calculated (Table III). Afterwards the
deposition rate of the unattached daughters was calculated from the

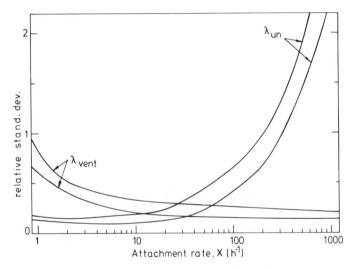

Figure 5. Relative standard deviation on the fitting of the deposition rate of the unattached daughters (λ_{un}) and on the fitting of the ventilation rate (λ_{vent}), calculated by means of a Monte-Carlo simulation model. The lower curve is obtained with counting statistics alone. The upper curve includes one hour time fluctuations on the input parameters, with 10% rel. stand. dev. on X, λ_{un} (15/h), λ_a(.35/h), λ_{vent}(.45/h) and radon conc. (50 bq/m^3) and 2% on recoil factor (.83), penetration unattached (.78) and flow rate (28 1/min).

data with low attachment rates. The mean values together with their standard deviations of the mean are indicated in Table III. The overall error however is dominated by the systematic uncertainty on the calculated attachment rates and is estimated at 40 %.

Table III. Optimization of the ratio of the "outroom" to the "in-room" daughter concentration (P) and of the deposition rate of the unattached daughters (λ_{un}).

Loc. : No :	P	Att.rate interval 1/h	Number of meas.	λ_{un} 1/h	s.d.m. 1/h	S/V 1/m	v_{un} cm/s
1 :	0.20	17-31	17	17.0	1.4	2.7	.17
2 :	0.90	9-30	5	8.3	1.3	2.8	.08
3 :	0.35	32-37	6	24.2	1.3	3.5	.19
4 :	0.60	54-89	4	17.8	7.0	2.3	.21

The corresponding deposition velocities are about .19 cm/s for the rooms and .08 cm/s for the cellar.

The surface-to-volume ratio's indicated in Table III include the furniture. Their standard deviation is estimated at 20 % and comes mainly from the surface determination of the furniture.These ratio's are rather high compared with values reported in the literature (see Table V), although our rooms had normal dimensions.

All the parameters of the room model are now fixed which means that one radon concentration corresponds to only one set of daughter concentrations. Then the radon concentration for which the corresponding daughter concentrations are closest to the measured concentrations is determined in each measurement. The agreement between calculated and measured daughter concentrations is rather good, as can be seen in the lower part of Figure 3. The unattached fraction is determined from the measured daughter concentrations using the ratio's of the calculated attached daughters to the calculated unattached daughters. Finally the equilibrium factor has been assessed from the calculated radon concentration and the measured daughter concentrations.

Evaluation

The fraction of unattached daughters (fp), the equilibrium factor (F) and the activity median diameter (AMD) are plotted in Figure 6 for all the measurements. The AMD is derived from the aerosol measurements. These three parameters are important in the dosimetric models. At the top of Figure 6 the effective dose equivalent is plotted, computed with two models called the J-E (Jacobi-Eisfeld) and J-B (James-Birchall) models in the NEA-report (1983, table 2.9, linear interpolation between AMD=0.1 and 0.2 μm). The figure also shows the effective dose equivalent calculated from the equilibrium equivalent radon concentrations with the NEA dose conversion factor (NEA,1983, table 2.11).

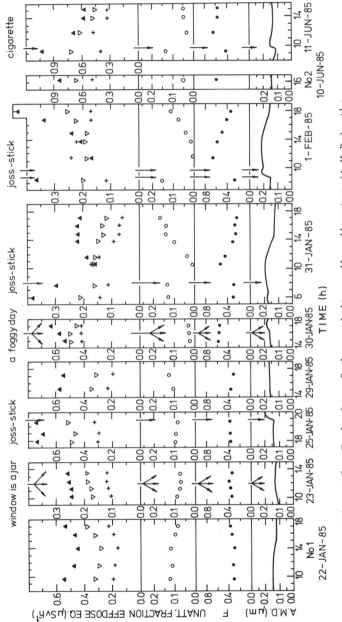

Figure 6a. Evolution of the activity median diameter (A.M.D.), the equilibrium factor (F), the unattached fraction and the effective dose equivalent (▲ J-B, ▽ J-E, + NEA) during the case studies.

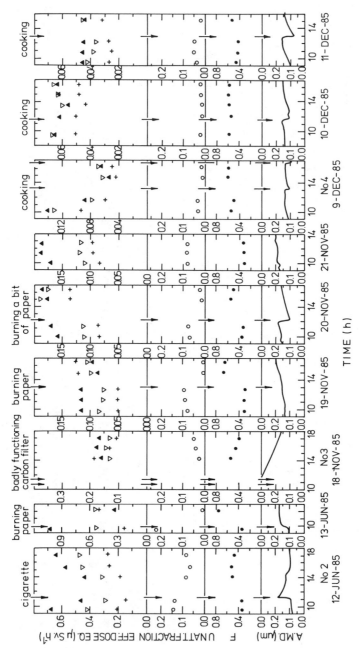

Figure 6b. Figure 6a continued.

The doses computed with the J-B model are up to a factor two larger than the ones computed with the J-E model, due to the higher dose conversion factor for the unattached fraction in the first model. Both models yield higher doses than given by the NEA conversion factor, because the unattached fraction (see Figure 6) is much higher than estimated (about 2%, NEA,1983). The difference is even more striking in Figure 7 where the doses per unit radon concentration are plotted as a function of the equilibrium factor. The full lines are calculated for fixed parameters and changing attachment rates. As fixed parameters the mean values of the measurements and two deposition rates of the unattached daughters are taken. The conversion factors of the NEA and the ICRP (NEA,1983;ICRP 39,1984) are proportional to the equilibrium factor and show a very different relationship compared to the J-E model and the J-B model. The main reason for this is that a low equilibrium factor is connected with a high unattached fraction. With the J-B model the dose per unit radon concentration even decreases with increasing equilibrium factor. Therefore it would be more adequate to introduce a constant conversion factor to the radon concentration.

In Figure 8 the doses per unit radon concentration are plotted as a function of the measured ventilation rate. The NEA conversion factor for low and moderate ventilation (NEA,1983, table 2.10) is multiplied by the appropriate equilibrium factor. In the figure no influence of the ventilation rate on the doses is found.

In Figure 9 it is shown that the attachment rate is the dominating factor for the unattached fraction. The equilibrium factor however, is also strongly influenced by the deposition rate of the unattached daughters. The curves are calculated as in Figure 7. The limited fluctuations of the actual data illustrate the importance of the attachment rate.

Thoron

Almost every day a thoron daughter measurement was performed, after the decay of the radon daughters. The results are listed in Table IV. The daughters were calculated from one decay time interval and an estimation of the ratio Pb-212 to Bi-212 out of the removal processes (ventilation and deposition of the attached daughters (λ_a)). The ratio of the potential α-energy of the thoron daughters to the radon daughters is .43, which is about the same as reported by Stranden (1980) for Norwegian dwellings .48 and somewhat less than found by Scherry (1985) in the U.S.A. .62 . The average dose equivalent due to the thoron daughters is about 17% of that due to the radon daughters, using the NEA dose conversion factors (NEA, 1983).

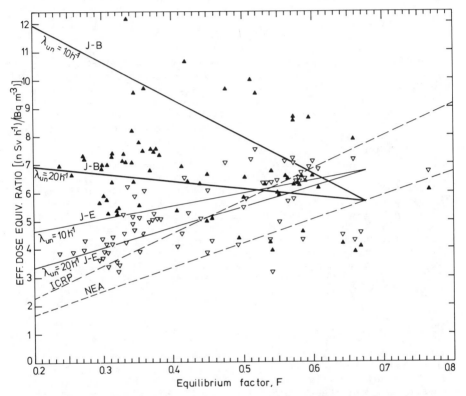

Figure 7. Effective dose equivalent per hour and per unit radon concentration (▲ J-B, ▽ J-E) as a function of the equilibrium factor. The full lines are calculated with the mean values of the 72 measurements (λ_a = .37/h, λ_{vent} = .41/h, P = .53, A.M.D. = .15 µm) and changing attachment rates.

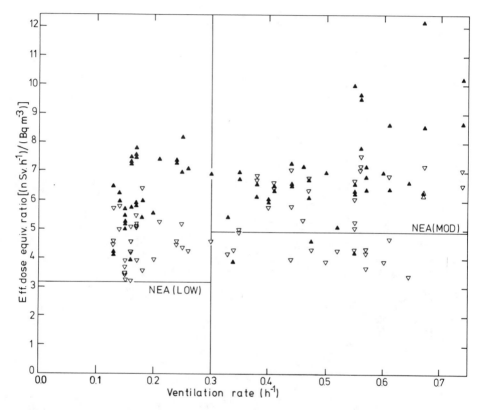

Figure 8. Effective dose equivalent per hour and per unit radon concentration (▲ J-B, ▽ J-E) versus ventilation rate. The NEA conversion factor is multiplied by the mean equilibrium factor of the measurements indicated in the ventilation interval.

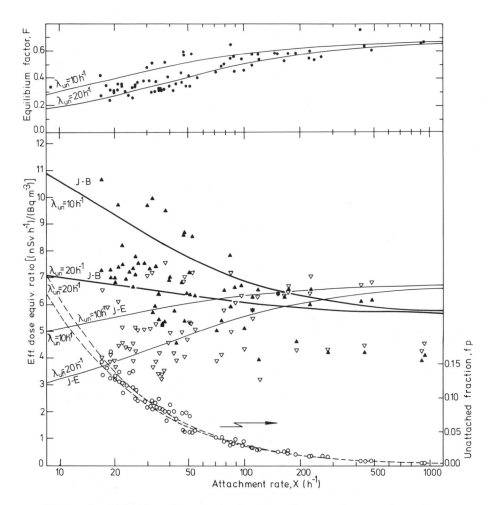

Figure 9. Effective dose equivalent per hour and per unit radon concentration (▲ J-B, ▽ J-E), equilibrium factor (●) and unattached fraction (o, right ordinate) versus the attachment rate. The curves are calculated as in Figure 7.

Table IV. Results of the thoron measurements

Loc. : No	date	λ_{vent} 1/h	λ_a 1/h	EEC s.d. $\mu J/m^3$		He $\mu Sv/h$	EEC $\frac{Th}{Rn}$	He $\frac{Th}{Rn}$
1 :	22/01	.17	.38	.053	.009	.032	.34	.14
1 :	30/01	.14	.24	.092	.004	.055	.33	.13
1 :	31/01	.60	.38	.026	.001	.016	.63	.25
1 :	01/12	.26	.29	.050	.002	.030	.43	.17
2 :	10/06	.61	.31	.038	.001	.023	.12	.05
2 :	11/06	.74	.35	.028	.001	.017	.13	.05
2 :	12/06	.56	.48	.031	.001	.18	.17	.07
2 :	13/06	.47	.24	.040	.002	.024	.12	.05
3 :	18/11	.15	.49	.028	.001	.017	.38	.15
3 :	19/11	.13	.38	.038	.002	.023	.44	.18
3 :	20/11	.17	.50	.033	.001	.020	.38	.14
3 :	21/11	.15	.41	.024	.001	.014	.36	.15
4 :	09/12	.38	.36	.022	.005	.013	.65	.26
4 :	10/12	.40	.36	.026	.001	.016	.82	.33
4 :	11/12	.67	.40	.018	.001	.011	.84	.34

Discussion and conclusion

The deposition rate and the corresponding deposition velocity of the unattached daughters is found to have a value of 18/h and .19 cm/s in the rooms and 8/h and .08 cm/s in the cellar. The latter could be corrected to .12 cm/s applying the same decrease in deposition velocity to the attached daughters as was found for the unattached daughters. The remaining difference is probably due to a smaller air velocity in the cellar and to a difference in roughness of the surface (concrete instead of wall paper and carpets).

Our values agree, within the systematic uncertainty of about 40%, with most of the values reported in the literature (see Table V) (Wicke et al.,1982; Bruno,1983; Scott,1983; Toohey et al.,1984; Offermann et al.,1984; Porstendörfer,1984; Bigu,1985; Israeli,1985; Reineking,1985). However, they are, on the one hand, an order of magnitude lower than those initially reported by Jacobi (1972) and Porstendörfer (1978) and, on the other hand, significantly higher than the recent estimations of Knutson (1983) and Schiller (1984).

Our analysis shows that the unattached fraction in the domestic environment is between .05 and .15 without any aerosol sources in the room and can decrease below .05 in the presence of aerosol sources. These values are much larger than assumed by James (1984) (fp=3%) and by the NEA-report (1983) (fp=2%). However the few experimental results reported in the literature agree with our findings. Bruno (1983) found an unattached fraction of .07 Reineking (1985), Shimo (1984) and Duggan (1969) measured about .10. The last two results are calculated by means of the room model from the reported unattached Po-218 concentrations.

Table V. Summary of the deposition velocities of the attached and
         unattached radon daughters and the corresponding deposi-
         tion rates from various researchers.

Researcher	:	place	λ_{un}	λ_a	S/V	v_{un}	v_a
	:		1/h	1/h	1/m	cm/s	cm/s
Jacobi	1972 :	dwel.	72	.72	2	1.0	.010
Porstendörfer	1978 :	dwel.	72	.072	2	1.0	.001
Wicke	1982 :	dwel.	30	.2	2	.42	.003
Bruno	1983 :	dwel.	8	-	-	-	-
Knutson	1983 :	Rn chamb.	3.6	.054	2	.05	.00075
Scott	1983 :	dwel.	-	-	-	.2	.005
Toohey	1983 :	dwel.	-	-	-	.36	-
Offermann	1984 :	Rn chamb.	15	.16	1.9	.22	.0023
Porstendörfer	1985 :	dwel.	10	.1	5	.05	.0005
Schiller	1984 :	dwel.	3.6	.004	2	.05	.00006
Bigu	1985 :	Rn chamb.	-	-	-	.2	-
Israeli	1985 :	dwel.	8	1	2	.1	.015
Reineking	1985 :	dwel.	30	.7	-	-	-
This paper	1986 :	Rn chamb.	19	-	6.6	.08	-
This paper	1986 :	dwel.	18	.37	2.6	.19	.0040

Due to the higher conversion factor of the unattached
fraction in the lung dosimetry models and the inverse relationship
between the unattached fraction and the equilibrium factor, it was
found that in the domestic environment the effective dose equivalent
is more related to the radon concentration than to the equilibrium
equivalent radon concentration. The same conclusion can be drawn
from results reported by Nero (1985, Figure 3) and by Jonassen
(1985, Figure 5). On the basis of our analysis a conversion factor
per unit radon concentration of 5.6 $(nSv/h)/(Bq/m^3)$ or 50
$(\mu Sv/y)/(Bq/m^3)$ can be put forward. It is the mean value of the
computed J-B and J-E doses in the two bedrooms and in the living
room. This conversion factor is to some extend correlated to the
deposition rate of the unattached daughters (see Figure 7 and Figure
9). Applying the conversion factor to the average radon
concentration in Belgium, which is 50 Bq/m^3 (Poffijn et al.,1985,
Figure 1) and assuming an occupancy factor of 80% gives an average
dose equivalent to the population of 2 mSv/y.

Acknowledgment
_____

This work was performed under contract no BI6-112-B with the
Commission of the European Communities, with support of the
Interuniversity Institute for Nuclear Sciences.

Literature Cited

Bigu, J., Radon daughter and thoron daughter deposition velocity and unattached fraction under laboratory-controlled conditions and in underground uranium mines, J.Aerosol Sci. 16 : 157-165 (1985)

Bruno, R.C., Verifying a model of radon decay product behaviour indoors, Health Phys. 45 : 471-480(1983)

Cliff, K.D., Assessment of airborne radon daughter concentrations in dwellings in Great Britain, Phys. Med. Biol. 23 : 1206-1217 (1978)

Duggan, M.J. and Howell D.M., The measurement of the unattached fraction of airborne RaA, Health Phys. 17 : 423-427 (1969)

ICRP Publication 39, Principles for limiting exposure of the public to natural sources of radiation, Oxford, Pergamon Press (1984)

Israeli, M., Deposition rates of Rn progeny in houses, Health Phys. 49 : 1069-1083 (1985)

Jacobi, W., Activity and potential α-energy of Rn-222 and Rn-220 daughters in different air atmospheres, Health Phys. 22 : 441- 450 (1972)

James, A.C., Dosimetric approaches to risk assessment for indoor exposure to radon daughters, Rad. Prot. Dos. 7 : 353-366 (1984)

Jonassen, N. and McLaughlin, J.P., The reduction of indoor air concentrations of radon daughters without the use of ventilation, The Science of the Tot. Env. 45 : 485-492 (1985)

Knutson, E.O., George, A.C., Frey, J.J. and Koh, B.R., Radon daughter plateout II : Prediction model, Health Phys. 45 : 445-452 (1983)

Mercer, T.T., The effect of particle size on the escape of recoiling RaB atoms from particulate surfaces, Health Phys. 31 : 173-175 (1976)

Miles, J.C.H. and Sinnaeve, J., Results of the second CEC intercomparison of active and passive dosemeters for the measurement of radon and radon decay products (1984)

Nazaroff, W.W., Optimizing the total-alpha three-count technique for measuring concentrations of radon progeny in residences, Health Phys. 46 : 395-405 (1984)

NEA Experts Report, Dosimetry aspects of exposure to radon and thoron daughter products, OECD-NEA (1983)

Nero, A.V., Sextro, R.G., Doyle, S.M., Moed, B.A., Nazaroff, W.W., Revzan, K.L. and Schwehr, M.B., Characterizing the sources, range, and environmental influences of Radon-222 and its decay products, The Science of the Tot. Env. 45 : 233-244 (1985)

Offermann, F.J., Sextro, R.G., Fisk, W.J., Nazaroff, W.W., Nero, A.V., Revzan,K.L. and Yater, J., Control of respirable particles and radon progeny with portable air cleaners, Report No. LBL- 16659 (1984)

Poffijn, A., Marijns, R., Vanmarcke, H. and Uyttenhove, J., Results of a preliminary survey on radon in Belgium, The Science of the Tot. Env. 45 : 335-342 (1985)

Porstendörfer, J., Wicke, A. and Schraub, A., The influence of exhalation, ventilation and deposition processes upon the concentration of Rn-222, Rn-220 and their decay products in room air, Health Phys. 34 : 465-473 (1978)

Porstendörfer, J., Behaviour of radon daughter products in indoor air, Rad. Prot. Dos. 7 : 107-113 (1984)

Raabe, O.G., Concerning the interactions that occur between radon decay products and aerosols, Health Phys. 17 : 177-185 (1969)

Raes, F., Janssens, A., De Clercq, A. and Vanmarcke, H., Investigation of the indoor aerosol and its effect on the attachment of radon daughters, Rad. Prot. Dos. 7 : 127-131 (1984)

Raes, F., Janssens, A. and Vanmarcke, H., A closer look at the behaviour of radioactive decay products in air, The Science of the Tot. Env. 45 : 205-218 (1985a)

Raes, F., Description of the properties of unattached RaA and ThB particles by means of the classical theory of cluster formation, Health Phys. 49 : 1177-1187 (1985b)

Reineking, A., Becker, K.H. and Postendörfer, J., Measurements of the unattached fractions of radon daughters in houses, The Science of the Tot. Env. 45 : 261-270 (1985)

Schery, S.D., Measurements of airborne Pb-212 and Rn-220 at varied indoor locations within the United States, Health Phys. 49 : 1061-1067 (1985)

Schiller, G.E., A theoretical convective-transport model of indoor radon decay products, Ph. D. Thesis, University of California, Berkely (1984)

Scott, A.G., Radon daughter deposition velocities estimated from field measurements, Health Phys. 45 : 481-485 (1983)

Shimo, M. and Ikebe, Y., Measurements of radon and its short-lived decay products and unattached fraction in air, Rad. Prot. Dos. 8 : 209-214 (1984)

Stranden, E., Thoron and radon daughters in different atmospheres, Health Phys. 38 : 777-785 (1980)

Toohey, R.E., Essling, M.A., Rundo, J. and Wang Hengde, Measurements of the deposition rates of radon daughters on indoor surfaces, Rad. Prot. Dos. 7 : 143-146 (1984)

Unscear, Ionizing radiation : sources and biological effects, Report to the General Assembly, United Nations, New York (1982)

Vanmarcke, H., Calibration of a radon daughter detection system, Report Nucl. Phys. Lab. (1984)

Vanmarcke, H., Janssens, A. and Raes, F., The equilibrium of attached and unattached radon daughters in the domestic environment, The Science of the Tot. Env. 45 : 251-260 (1985)

Wicke, A. and Porstendörfer, J., Radon daughter equilibrium in dwellings, in Natural Radiation Environment (K.G.Vohra, ed.), Wiley Eastern Limited, 481-488 (1982)

RECEIVED August 4, 1986

Chapter 24

A Model for Size Distributions of Radon Decay Products in Realistic Environments

F. Raes, A. Janssens, and H. Vanmarcke

Nuclear Physics Laboratory, State University of Gent, Proeftuinstraat 86, B-9000 Gent, Belgium

A model has been developed to calculate the size distributions of the short lived decay products of radon in the indoor environment. In addition to the classical processes like attachment, plate out and ventilation, clustering of condensable species around the radioactive ions, and the neutralization of these ions by recombination and charge transfer are also taken into account. Some examples are presented showing that the latter processes may affect considerably the appearance and amount of the so called unattached fraction, as well as the equilibrium factor.

This paper is the third chapter of our theoretical investigations of the size distribution of radon decay products. In all of the work we concentrated more particularly on the behavior of the so called unattached fraction under different environmental conditions. In a first paper (1), we have applied the theory describing the clustering of mixtures of condensable vapors around ions, offering a conceptual framework to cope with the broad spectrum of experimental data on the size of unattached RaA and ThB particles. It was shown that the observations could largely be explained by ion cluster formation and growth, looking at the radioactive ions as though they were normal positively charged ions. As a second step we investigated the impact of ion clustering and growth on the form of the active size distribution in different environmental conditions (2). The key question was whether the description of airborne activity in terms of an "unattached fraction" with a single diffusion coefficient and an "attached fraction" described by a unimodal lognormal distribution, is appropriate in all conditions. We showed that the composition of the atmosphere determines the physical appearance of radioactive particles in two ways: 1) the presence of condensable products may broaden the size distribution of the free radioactive fraction either by clustering or by forming an ultra fine aerosol to which the free fraction can become attached; and 2) the neutralization of

0097-6156/87/0331-0324$06.00/0

radioactive ions, mainly by charge transfer with appropriate gases quenches clustering and influences the deposition of airborne activity, especially in rooms with dielectric walls. Whereas in the second paper, we considered a rather general system of ions that were continuously being formed by radioactive decay of a gaseous mother isotope, we will now consider radon and all of its short lived decay products, and we will calculate size distributions for each decay product.

Ion Cluster Formation and Growth

The classical theory of charged cluster formation, and its application to radon decay products is given extensively in (1). The theory demonstrates that when an ion (in our case the radioactive decay product) is exposed to a super-saturated vapor, or to some binary mixture of condensable vapors (both under-saturated), the ion may start growing spontaneously, when it passes some critical size. The main conclusions regarding the size of the unattached fraction, and the comparison with the experimental findings are summarized in Table I. It is shown that the consideration of the clustering of binary mixtures of condensable products (H_2SO_4 and H_2O in our case) rather than the clustering of H_2O alone is essential to account for the observed range of diffusion coefficients of radon decay products and for the observed growth of these products.

Competitive Removal Processes

It takes some time for an ion to grow from the molecular size to the critical cluster size (1). When the ion becomes neutralized or is scavenged from the gas phase within that period of time, obviously no growth will occur. The ion removal processes are neutralization by recombination or charge transfer, deposition on surfaces, attachment to aerosol particles, ventilation and radioactive decay. The description of these processes is given in (2). Table II summarizes the mean ion lifetime corresponding to the mentioned removal processes and compares them with the time an ion needs to become a critical $H_2O-H_2SO_4$ cluster. The growth time is calculated for two different H_2SO_4 concentrations at a relative humidity of 75%, considering the kinetics of H_2SO_4 collisions with (subcritical) $H_2O-H_2SO_4$ ion clusters (1). Table II shows that ion growth will be inhibited in the most stringent conditions of neutralization, deposition or attachment, but also that there may be situations where the time an ion needs to become a stable particle is comparable to or smaller than the ion lifetime characteristic for each removal process.

The Numerical Model AERO1A

The model used here is similar to the one used in (2), except that it is now extended to calculate the evolution in time of RaA, RaB and RaC containing particles. The core of the model is the code AERO1, which was developed for the description of photolytic and radiolytic

Table I. Comparison of the experimental findings and theoretical
          predictions of the size of free radon decay products

	Experimental Findings	Theory : Clustering of H_2O	Theory : Clustering of H_2O and H_2SO_4
Diffusion Coefficient $(cm^2 s^{-1})$	$(0.55 \leftarrow)$ $0.08 - 0.03$ $(\rightarrow 0.003)$	$0.1 - 0.08$	$0.1 - 0.03 \rightarrow$
Occurence of Growth	on some occasions	impossible	possible in "polluted" atmospheres
Effect of Relative Humidity on Diffusion Coefficient	inconsistent	$D \downarrow$ when r.h.$\uparrow$	$D \downarrow$ when r.h.$\uparrow$

Table II. Comparison of the time an ion cluster needs to reach the critical size with the characteristic lifetimes corresponding with different competitive removal processes

Process	Conditions	Time of Reach the Critical Size (s)
ion clustering	H_2SO_4 = O	∞
	H_2SO_4 = 10^8 molec cm^{-3}	300
	H_2SO_4 = 10^{10} molec cm^{-3}	3

		ion mean lifetime (s)
recombination	Rn = 37 Bq m^{-3}	260
	Rn = 3.7 10^4 Bq m^{-3}	10
	Rn = 3.7 10^{10} Bq m^{-3}	0.06
charge transfer	NO_2 = O molec cm^{-3}	∞
	NO_2 = 10^{11} molec cm^{-3}	0.2
deposition	S/V = 0.02 cm^{-1},	
	D = 0.04 $cm^2 s^{-1}$	
	conductive wall	2500
	dielectric	
	(surface charge density	
	= 8 10^{-13} C cm^{-2})	7
attachment	aged aerosol	420
	fresh nucleation aerosol	2
ventilation	1 h^{-1}	3600
radioactive decay	RaA	265
	RaB	2320
	RaC	1710

formation of H_2O-H_2SO_4 aerosol from the gas phase. The full
description of AERO1 is given in (3).

The Size Distribution of Stable Particles

This size distribution is approximated by two modes: a nucleation
mode containing particles with a diameter between the critical size
($\sim$ 1 nm, depending on the conditions) and 0.4 µm; and an accumulation
mode between 0.01 and 1 µm. The accumulation mode is described by a
lognormal distribution with a fixed median diameter, geometric
standard deviation and total number concentration and is considered
here as a background aerosol that does not change in size itself, but
plays a role in scavenging radon daughters, ions and H_2SO_4 molecules.
H_2O-H_2SO_4 particles enter the nucleation mode at the critical size by
homogeneous or ion-induced aerosol formation from the gas phase, and
they further grow by condensation. The rate of aerosol formation is
very much dependent on the H_2O and H_2SO_4 concentratons in the gas
phase. The H_2O concentration is held constant at 75% r.h. at 25°C.
Contrary to our calculations in (2), where the H_2SO_4 concentration
was fixed in the course of the calculation, we now have solved the
balance equation for H_2SO_4, simultaneously with the balance equations
for the particles. Hence we are able to keep track of variations in
particle formation and growth rates due to variations in the H_2SO_4
concentration. It is assumed that the charge of the particles in
both modes is given by the Fuchs equilibrium charge distribution (5),
and all particles are subject to deposition and ventilation.

The Size Distribution of RaA Containing Particles

This size distribution is approximated by three modes: a free
activity mode; a nucleation mode; and an accumulation mode. The free
activity mode contains the subcritical clusters and is confined
between the diameter of the pure water ion cluster and the critical
size. Free RaA ions are formed continuously by radioactive decay of
radon. Each radioactive decay will produce 3.4 10^5 gaseous ions
(positive and negative) that will play a role in the recombination
process, or may induce aerosol formation. Free radioactive ions may
become uncharged by recombination or charge transfer, but only the
charged ones are able to grow from the free activitiy mode to the
nucleation mode at a rate given by ion induced nucleation theory (3).
Both charged and uncharged free RaA particles may become attached to
stable particles in the nucleation and accumulation mode. For the
nucleation and accumulation mode, Fuch's equilibrium charge
distribution is assumed.

The Size Distribution of RaB and RaC Containing Particles

RaB and RaC size distributions are calculated in the same way.
Recoil of RaB ions from aerosol particles is taken into account in
the way proposed by Mercer (6). If RaA ions or molecules are
attached to aerosol particles in the accumulation mode, it is assumed
that they remain on the surface of the particles; for RaA

attached to or grown into the nucleation mode, it is assumed that the RaA is submersed in the middle of the $H_2O-H_2SO_4$ droplet.

Comments

The decay of RaA, RaB and RaC will also contribute to the formation rate of positive and negative gaseous ions, and this contribution will change in time along with the concentration of the airborne radon daughters. Although it is straightforward to incorporate balance equations for the gaseous ions, this was not done in the present version of AERO1A. The ionization rate is calculated considering only the radon concentration, and this will lead to a slight underestimation of the growth process.

Although AERO1A contains the necessary coagulation algorithms, coagulation was not taken into account in the present calculations, because of the resulting excessive calculation times. Hence, radon daughters that enter the nucleation mode by growth or attachment are not allowed to interact with the accumulation mode any more. However, since in our examples the particle concentrations are rather low, it is not expected that omission of coagulation will induce large errors. One test calculation showed that the attached fraction is underestimated by 1% in conditions similar to Case 1 (see below). In cases of high particle concentrations no nucleation mode of radioactive particles will exist and the problem becomes irrelevant (see e.g. Case 5 below).

By using the classical theory of ion induced nucleation to describe the growth of radon daughters from the free activity mode to the nucleation mode, we loose information about the size of the subcritical clusters. These clusters are all lumped together between the size of a pure H_2O ion cluster at 75% r.h. and the size of the critical $H_2O-H_2SO_4$ cluster. The model only does keep track of the growth by condensation of the radon daughters once they arrived in the nucleation mode.

The prediction made by the model calculations should be taken with some care for two reasons: 1) H_2O and H_2SO_4 are considered to be the condensing species, whereas other species may be active in experimental or domestic environments; 2) the model uses classical nucleation theory, which is the only workable theory, but which is also to be criticized because it applies macroscopic entities to clusters that contain only a few molecules (3).

Results and Discussions

The environmental conditions for each of the cases considered below are summarized in Table III; all these parameters are constant in time. The build up of the nucleation mode of the stable particles and the build up of both the nucleation and accumulation modes of the radon decay products is calculated, and the results are given after a process time of one hour. Figures 1 to 5 show the size distributions of stable and radioactive particles, and Table IV gives the disequilibrium, the equilibrium factor F, the "unattached fraction" f_p and the plate-out rates for the different daughters.

Table III   Description of the conditions for which the size distri-
             butions were calculated

	NO_2 molec cm^{-3}	wall characteristics	N_{accum} cm^{-3}
Case 1	0	conductive	100
Case 2	10^{11}	conductive	100
Case 3	0	dielectric	100
Case 4	10^{11}	dielectric	100
Case 5	10^{11}	dielectric	10.000

In all of the cases :   Rn = 12 Bq m^{-3} ; H_2SO_4 formation rate = 5 10^5
molec $cm^{-3}s^{-1}$; ventilation rate = 3 $10^{-4}s^{-1}$; surface to volume ratio = 0.02 cm^{-1}. The accumulation mode is lognormal with a median
diameter of 0.1 μm and a geometric standard deviation of 1.8.

Table IV   Disequilibrium between radon and its daughter products,
            equilibrium factor (F) and unattached fraction (f_p) with
            respect to the potential α energy and plate out rates of
            the unattached RaA, RaB and RaC fractions for the different
            cases considered. The values prevail after one hour

Case	disequilibrium (Rn;RaA;RaB;RaC)	F	f_p	plate out[f] (h^{-1}) RaA	RaB	RaC
1	1;0.89;0.44;0.23	0.40	0.76	0.61	0.31	0.11
2	1;0.86;0.36;0.17	0.34	0.46	1.47	1.47	1.47
3	1;0.02;0.00;0.00	0.003	0.93	520.	520.	520.
4	1;0.54;0.07;0.02	0.10	0.85	2.1	1.8	1.8
5	1;0.81;0.37;0.20	0.35	0.06	4.0	3.7	3.7

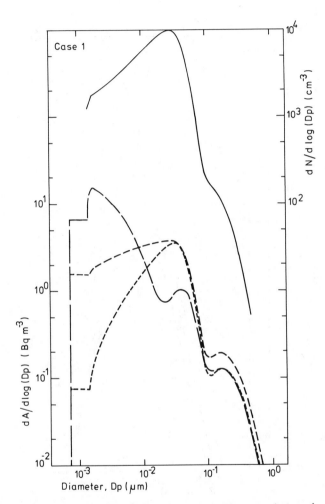

Figure 1. Size distributions of the stable particles (————) of RaA (——— ———), RaB (— — —) and RaC (--------) containing particles, after 1 hour, Case 1.

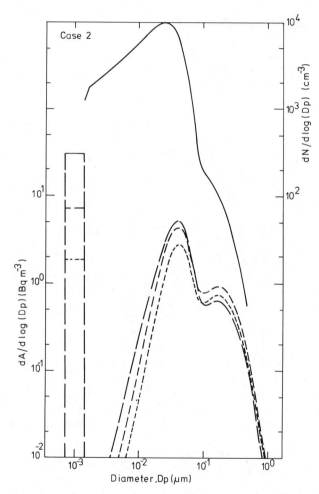

Figure 2. Size distributions of the stable particles (————)
of RaA (—— ——), RaB (— — —) and RaC (--------) containing
particles, after 1 hour, Case 2.

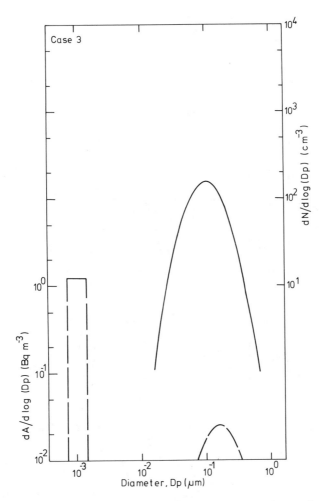

Figure 3. Size distributions of the stable particles (————) of RaA (—— ——), RaB (— — —) and RaC (--------) containing particles, after 1 hour, Case 3.

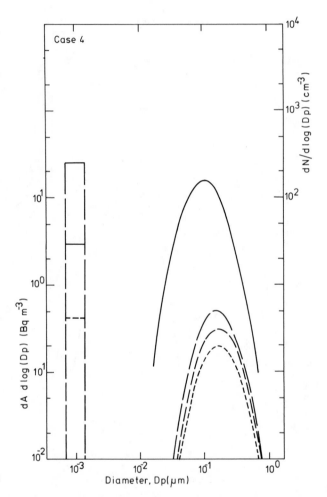

Figure 4. Size distributions of the stable particles (————)
of RaA (——— ———), RaB (— — —) and RaC (--------) containing
particles, after 1 hour, Case 4.

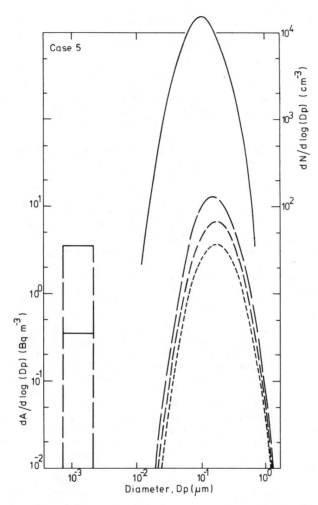

Figure 5. Size distributions of the stable particles (————)
of RaA (—— ——), RaB (— — —) and RaC (--------) containing
particles, after 1 hour, Case 5.

From Table I we can infer that growth of radioactive ions may
be expected in a steel room operated at low radon concentrations, low
concentrations of primary aerosol particles, low concentrations of
gases that induce charge transfer (like NO_2,NO), and with some
formation of condensable products. These conditions are simulated in
Case 1 (Figure 1). The stable aerosol size distribution is bimodal.
The RaA size distribution turns out to be trimodal: the mode at the
smallest sizes contains the free and nucleated RaA particles and can
be called the unattached fraction; the second and third mode contain
the RaA that is attached to the stable nucleation and accumulation
mode respectively. It should be noticed that the activity contained
in the attached fraction is more than 10 times smaller than in the
unattached fraction, such that only one mode, the unattached
fraction, will probably be detectable. The mean diffusion
coefficient of that fraction is 0.013 cm^2s^{-1}. The RaB and RaC size
distributions are bimodal, again only the mode in the lower size
range will probably be detectable. Strictly speaking, this mode
cannot be called "unattached fractions", since they contain both
nucleated RaB (or RaC) and RaB (or RaC) that is attached to the
stable nucleation model. The mean diffusion coefficients of these
modes are 0.007 and 0.002 $cm^{-2}s^{-1}$ for RaB and RaC respectively.

In Cases 2, 3 and 4, we focus attention on the effect of
electrical parameters like the charge of the radon daughter and the
conductivity of the walls. Both a high radon concentration (i.e.,
high ionization rate of the air) and a high NO_2 concentration can
affect the charge of the radon daughters; however, since high NO_2
concentrations will more readily occur, and since neutralization by
NO_2 goes linear with the NO_2 concentration whereas neutralization by
recombination goes only with the square root of the radon
concentration (2) we changed only the NO_2 concentrations in our
calculations. Calculations were performed considering both
conducting and dielectric walls, the latter having a surface charge
density of 8 10^{-13} C cm^{-2} (2). The size distributions for Case 2 are
shown in Figure 2. In this case, NO_2 is present at 4 ppb; the radon
daughters get uncharged by charge transfer, hence are not available
for nucleation anymore, and we end up with a well defined unattached
fraction. Charge transfer does not affect the amount of gaseous ions
in the air, such that nucleation still occurs around these ions, and
a stable nucleation mode develops Radon daughters attach to these
particles as well as to the preexisting accumulation mode, and a
broad bimodal attached fraction is formed.

In the following cases we deal with dielectric walls, which
largely enhance the deposition of ions and charged particles by
electrophoresis (2). In Case 3 no NO_2 is present and the radon
daughters remain charged, but all ions, stable and radioactive, are
scavenged by the walls before they can grow or even attach to the
preexisting aerosol. The nucleation modes collapse and virtually no
airborne activity is left (Fig. 3). In Case 4, NO_2 is present; most
parts of the radon decay products become neutral, such that they are
not susceptible to electorphoretic deposition anymore. They remain
airborne for a longer time and can attach to the preexisting aerosol
(Fig. 4). The plate out rates of the unattached fraction of the
different daughters are slightly different (Table IV). This is
because a different fraction of each of the daughters remains

charged, and this in turn is due to their different radioactive lifetime, hence a different exposure time to NO_2.

Case 5, in which the concentration of the preexisting aerosol is raised to 10,000 cm^{-3}, can be considered as a simulation of a domestic environment. The resulting size distributions are given in Figure 5. The increase of the aerosol concentration drastically affects the partitioning of the radioactivity between the attached and unattached fraction, as was to be expected. However, the increase in the aerosol concentration also results in an increase of the plate out rate (Table IV). Normally this is attributed to the smaller residence time of the unattached fraction, resulting in smaller clusters. However, this cannot be the reason here since the model predicts the absence of growth, and cannot keep track of changes in size of subcritical clusters (see above). Again the reason is that the ratio of charged to uncharged unattached daughters is higher in Case 5 than in Case 4. This time this is due to the shorter residence time of the unattached daughters (controlled by the attachment process in this case), such that neutralization by charge transfer is not as complete as in Case. 4

Conclusions

We have developed a model that allows calculation of the evolution in time of the size distributions of the short lived decay products of radon. All relevant processes (including clustering and growth, and neutralization of the decay products) were taken into account, except different patterns of air circulation and different degrees of turbulence which may also affect the deposition of airborne material. Application of the model in some selected conditions showed that the active size distributions and the amount of airborne activity is largely affected not only by the aerosol content of the atmosphere, but also by its chemical composition, as well as by the dielectric/conductive characteristics of the surfaces in the room. These factors have not been taken into account in previous models (7,8). Case 1 showed that clustering and growth of radon decay products is to be considered especially in experimental environments (i.e., steel rooms and partly treated atmospheres). From Cases 2 to 5 we may infer that clustering and growth is not too relevant in domestic environments because of important competitive removal processes like neutralization by charge transfer and/or attachment and/or deposition by electrophoresis. On the other hand, the charge properties of the radon daughters seem to be very important in the domestic environment.

We believe that the calculations presented here give a better understanding of the many factors that determine the behavior of radon decay products, and that they explain why such a large range of values is being found of diffusion coefficients of the unattached fraction, of equilibrium constants, plate out rates, etc. (see (1) for a review, (9) for experiments in steel rooms and (10), (11), (12) for field studies in domestic environments).

It is difficult to use the model as a predictive model, since this would imply the knowledge of the chemical composition of the indoor atmosphere, as well as the formation rates of some relevant condensable products, which is hard to achieve. Linking

the radon problem with the general problem of indoor pollution
becomes necessary here. However, the model is a powerful tool in the
assessment of the risk caused by radon and in the assessment of the
effectiveness of remedial actions in widely different environments.

Acknowledgments

This work is supported by the Communion of the European Committees,
contract number B10-F-496 and by the Interuniversity Institute for
Nuclear Physics. R.F. greatly acknowldeges the School of Engineering
and Applied Science of the University of California, Los Angeles, for
the many C.P.U. hours that were made available for this work, and
Maureen Kronish for typing the manuscript.

Literature Cited

Fuchs, N.A., On the stationary charge distribution on aerosol
particles in a bipolar ionic atmosphere, Pure and Applied Geophysics
56 : 185 (1963)

George, A.C., Knutson, E.O. and Tu, K.W., Radon daughter plateout-I,
Health Physics 45 : 439 (1983)

Israeli, M., Deposition rates of Rm progeny in houses, Health Physics
49 : 1069 (1985)

Jacobi, W., Activity and potential α-energy of Rm-222- and Rm-220
daughters in different air atmospheres, Health Physics 22 : 441
(1972)

Mercer, T.T., The effect of particle size on the escape of recoiling
RaB atoms from particulate surfaces, Health Physics 31 : 173 (1976)

Porstendorfer, J., Wicke, A. and Schraub, A., The influence of
exhalation, ventilation and deposition processes upon the
concentration of Rn-222, Rn-220 and their decay products in room air,
Health Physics 34 : 465 (1978)

Raes, F., Description of the properties of unattached RaA and ThB
particles by means of the classical theory of cluster formation,
Health Physics 49 : 1177 (1985a)

Raes, F., Janssens, A. and Vanmarcke, H., A closer look at the
behaviour of radioactive decay products in air, The Science of the
Total Env. 45 : 205 (1985b)

Raes, F. and Janssens, A., Ion induced nucleation in a $H_2O-H_2SO_4$
system : extension of the classical theory and search for
experimental evidence, J. Aerosol Sci. 16 : 217 (1985c)

Raes, F. and Janssens, A., Ion induced aerosol formation in a H_2O-H_2SO_4 system : numerical calculations and conclusions, J. Aerosol Sci. (1986), to be published

Reinking, A. Becker, K.H. and Porstendorfer, J., Measurements of the unattached fractions of radon daughters in houses, The Science of the Total Env. 45 : 261 (1985)

Vanmarcke, H., Janssens, A. and Raes, F., The equilibrium of attached and unattached radon daughters in the domestic environment, The Science of the Total Env. 45 : 251 (1985)

RECEIVED August 20, 1986

Chapter 25

Measuring Polonium-218 Diffusion-Coefficient Spectra Using Multiple Wire Screens

R. F. Holub[1] and E. O. Knutson[2]

[1]Geophysics Division, Denver Research Center, U.S. Bureau of Mines, Department of the Interior, Denver, CO 80225
[2]Environmental Measurement Laboratory, Department of Energy, New York, NY 10014

The Bureau of Mines, Denver Research Center and the Department of Energy, Environmental Measurement Laboratory, developed through parallel efforts, two closely related techniques for the measurement of ^{218}Po (RaA) diffusion coefficient spectra. This work was prompted by reports in the past 5 years indicating that the diffusion coefficient of unattached ^{218}Po may vary due to various physical and chemical factors in different environments. The diffusion coefficient is important because it affects the amount and site of ^{218}Po deposition in the respiratory tract.

Our multiple screen techniques are adapted from George's 1972 single screen technique for measuring the unattached ^{218}Po activity. The Environmental Measurements Laboratory technique uses three screens in parallel, the Bureau of Mines three (or four, in a later version) in series. After drawing a sample, the alpha activity on the front face of each screen is counted with a 2-pi scintillation counter and a factor we have determined by experiment is applied to determine the total activity on each screen. The diffusion coefficient spectrum is then determined by applying a variant of the theory by Cheng and Yeh. Results to date indicate that, except in situations where there is intense aerosol formation, the spectrum is peaked at about $0.08 \pm .01 cm^2/sec$.

Each liter of air normally contains a few atoms each of ^{218}Po, ^{214}Pb, ^{214}Bi and ^{214}Po, which are the short-lived decay products of the radioactive noble gas radon. When inhaled, these atoms can be deposited on the lining of the respiratory tract, causing irradiation of the tissue due to further radioactive decay. This irradiation accounts for about one half of the average persons dose

from natural radiation, and it is believed that this does contribute
to the incidence of lung cancer (Sinnaeve, 1984). In some
underground mines and in some homes, the concentration of radon and
its decay products can become high enough to constitute a serious
health concern. This concern motivated our research.

Both polonium nuclides are alpha emitters and therefore of
particular concern. In health physics it is customary to
differentiate between attached and unattached ^{218}Po: the former,
usually the larger of the two consists of ^{218}Po atoms attached to
airborne particles which are copiously present in virtually every
atmosphere; the latter consists of a ^{218}Po atom or ion, frequently
surrounded by several dozen molecules of a condensible species
present in the air. The purpose of this paper is to present a new
method for measuring the size properties of these unattached ^{218}Po
clusters.

In 1984 our two laboratories began to develop a multiple screen
technique for making amount and size measurements on ^{218}Po. The
work proceeded independently at first, leading to different
configurations of screens. Increasingly, the value of collaboration
became apparent and this led to a one-week collaborative experiment
at the BOM laboratory, September 23-27, 1985. This paper covers
essentially all of our work on screens to date.

Prior Work

Prior Work on Diffusion Coefficients. Beginning with Chamberlain
and Dyson (1956), several investigators have measured and reported
diffusion coefficients for radioactive heavy metal atoms or ions in
air. The Chamberlain - Dyson value of 0.054 cm^2/sec, measured for
^{212}Pb, is still widely used for both ^{212}Pb and ^{218}Po. Busigin et.
al (1979), who have reviewed the existing data and reported new
measurements, contend that the diffusion coefficient of ^{218}Po can
vary from 0.001 to 0.1 cm^2/sec. Possible gas-phase chemical
reactions were mentioned that might cause these large changes.
Goldstein and Hopke (1985) found that the diffusion coefficient of
^{218}Po could be "adjusted" in the range 0.03 to 0.08 cm^2/sec by
controlling the admixed trace gases. The diffusion coefficient was
used as an index of the degree of electrical neutralization of ^{218}Po
ions.

Approaching from another direction, Sinclair et al (1978) and
Knutson et al (1984) report that diffusion battery measurements of
radon daughter aerosol-size distributions often show a small peak
which could be interpreted as the unattached fraction. Its position
would indicate diffusion coefficients from 0.0005 to 0.05 cm^2/sec.

Quite recently Raes (1985) applied the classical theory of
homogenous nucleation originally developed by Bricard et al (1972)
to atmospheres containing SO_2, H_2O and ^{218}Po ions. Depending on the
H_2O and SO_2 concentrations, ions could grow to a quasi-stable
cluster which would evaporate upon electrical neutralization, or to
a larger size which would survive neutralization.

Implicit in the results discussed above, especially the
Goldstein and Hopke and the Raes papers, is the fact that the ^{218}Po
diffusion coefficient can be multi-valued. Therefore,

experimentalists should be prepared to deal with a spectrum rather than a single discrete value for the diffusion coefficients.

Prior Work on Wire Screens. The concept of using wire screens in the measurement of radioactive airborne atoms or ions dates back at least to Barry (1968), who developed a method for measuring the amount of radioiodine by drawing air through a screen, and then performing a radio assay on the screen. Iodine atoms, having a relatively high diffusion coefficient, diffuse readily from the airstream to the surface of the wires. Iodine atoms that impinge on the wires have a high probability of sticking due to their chemical activity. Barry presented data on the collection efficiency of copper screens for iodine.

Following Barry, James et al (1972), and Thomas and Hinchliffe (1972) investigated the use of wire screens for collecting ^{218}Po atoms or ions. Experiments were done in the absense of aerosol particles, yielding collection efficiency as a function of screen dimensions and face velocity. Information was developed on the fraction of deposited α-activity that could be counted from the front and back sides of the screens.

George (1972) settled on a 60 mesh stainless steel wire screen operated at a face velocity of 10-20 cm/sec as a means of measuring the amount of unattached ^{218}Po. Using fresh ^{218}Po and no aerosol, he found that α-counting the front face of the screen yielded 50% of the counts obtained by α-counting a filter sample. In tests with a condensation nucleus counter he found that the screen collected only a few percent of aerosol particles. This led to a standard technique for measuring the unattached fraction: to sample simultaneously on a 60 mesh screen and a filter and α-count both; extract ^{218}Po data from both by modified Tsovoglou or equivalent technique; multiply the screen result by 2 and divide by the result from the filter.

In 1975 there was a new development in the use of wire screens; Sinclair and Hoopes (1975) described a diffusion battery (for measuring the particle size of aerosols) made of very fine 635-mesh stainless steel screen. An empirical equation was developed for the collection efficiency. This diffusion battery has become one of the standard techniques in aerosol measurements. Later, Sinclair et al (1978) described a screen diffusion battery configuration suited for measuring the activity - weighted size distribution of radon daughter aerosols.

Cheng and Yeh (1980) greatly facilitated the use of wire screens by showing that a theory developed for fiber filters also worked for predicting aerosol penetration through 635-mesh wire screens. The equation is:

$$P = \exp\left(-A \ Pe^{-2/3}\right) \tag{1}$$

where $\quad$ P = penetration of screen
$\qquad\qquad$ Pe = 2a U/D
$\qquad\qquad$ a = wire radius (cm)
$\qquad\qquad$ U = flow velocity (cm/sec)
$\qquad\qquad$ D = diffusion coefficient of particle (cm^2/sec)

$$A = \frac{4B\alpha h}{\pi (1-\alpha)\alpha} = \begin{cases} 10.56 & \text{for parallel staggered cylinder model,} \\ 4.52 & \text{for fan model,} \\ 2.51\text{-}3.35 & \text{for real filter.} \end{cases}$$

h = thickness of screen (cm)

$$\alpha = \frac{\text{volume of solid}}{\text{total volume}} = \frac{4m_s}{\pi d_s^2 h \rho_s}$$

m_s = mass of screen (g)
d_s = screen diameter (cm)
ρ_s = screen density (g cm^{-3})

The constant B, which is a function of geometric arrangement, is given by

$$BC = 2.9(-0.5 \ln \alpha + \alpha - 0.25\alpha^2 - 0.75)^{-1/3}$$
for parallel staggered cylinder model,

$$BF = 2.7 \qquad \text{for fan model,}$$
$$B = 1.5\text{-}2.0 \qquad \text{for real filter.}$$

Using this equation, one could calculate the effect of changing the screen geometry or the face velocity. Later papers by the same authors showed by experiment that the theory works well for a variety of screens and face velocities. Particle sizes from .015 to 0.5µ were tested. Scheibel and Porstendorfer (1984) have shown independently that the Cheng-Yeh equation works well down to particle size = 4 nm.

Samuelsson (1984) reintroduced wire screens as a means of measuring the amount of unattached ^{218}Po. He assumed 0.05 cm^2/sec for the diffusion coefficient, then applied the Cheng-Yeh equation to calculate the amount caught on the screen. An experiment like that of James et al (1972) was done to determine α-activity countable from the front face of the screen. For a 200 mesh screen at face velocity comparable to our face velocities, Samuelsson found this amount to be 80% of that for the same amount of material deposited on a filter. This agreed with the James et al (1972) result for comparable face velocity. In conclusion, it is clear that no size spectra could be measured by the above techniques for unattached daughters and that a better technique must be developed.

New Work On Screens

Test of the Cheng-Yeh Equation Against Old Data. As already mentioned, the Cheng-Yeh equation seems to be well established for aerosol particles in the size range 4 to 200 nm. (A modification, the Cheng-Keating-Kanapilly equation, is valid for larger sizes.) As a preliminary step towards smaller size, we have tested the equation against the existing data of Barry (1968), James et al (1972) and Thomas and Hinchliffe (1972). One of the authors, EOK, is fortunate to have access to the original Thomas-Hinchliffe data books; other data points were read off graphs appearing in the cited papers. Samples of the original Thomas-Hinchliffe screens were weighed and measured to calculate the solid fraction, α. For the other authors, we estimated α from the information given, mainly wire diameter and mesh size.

The best overall fit between the Cheng-Yeh equation and pooled data was obtained for a diffusion coefficient of 0.05 cm^2/sec for ^{218}Po (other values were tried). The fit, although not perfect, was remarkably good in view of the very broad range of conditions (10-fold range in both face velocity and mesh size) covered by the pooled data. We conclude that the existing data generally support the Cheng-Yeh equation.

Elementary Theory of the Front-to-Back α-count Ratio for Wire Screens. The use of wire screens for measurement of unattached fraction of ^{218}Po hinges on two factors. It is necessary to know: how much ^{218}Po deposits on the screen (as a function of diffusion coefficient and other parameters) and how this deposit is distributed around each wire. The latter point is important because α-particles from the decay of ^{218}Po on the back of a wire cannot penetrate the wire to be counted from the front. Experiments to investigate these are discussed later.

Although the Cheng-Yeh equation appears to be adequate for describing total deposition on a screen, it gives no information about the exact site of deposition. To get some insight into the site of deposition, we make use of an equation developed by Friedlander (1977):

$$\left(\frac{\partial n}{\partial y_1}\right)_{y_1 = 0} = \frac{(APe)^{1/3} \, n_\infty \, \sin^{1/2} x_1}{1.45 \, \chi^{1/3}}$$

where the left hand is the gradient of concentration (to which deposition rate is proportional) at the point x on the surface of the wire, and

A = a constant (value not needed in our calculation)
Pe = the Peclet number, defined earlier
n_∞ = concentration far from the cylinder surface

x_1 = x/a, distance of the point on the surface from the front of the cylinder, measured in units of cylinder radius, a

$$\chi = \int \sin^{1/2} x_1 \, . dx_1 \tag{3}$$

Figure 1 shows the geometry. Strictly speaking, equation 2 applies only to single cylinders in slow cross flow, for point particles with negligible deposition and with the thickness of boundary layer much less than the radius of the cylinder.

In order to study the implications of Equation 2, it was evaluated at 80 points in the range $x_1 = 0$ to π. At $x_1 = 0$, L'Hospital's rule from calculus was needed. For larger x_1, Equation 3 was evaluated for each x_1 using trapezoid rule numerical integration, yielding values for use in Equation 2. It was found that the rate of deposition is the highest for x_1 near zero, diminishing to zero at $x_1 = \pi$.

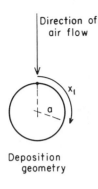
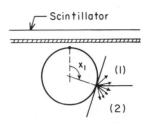

Direction of
air flow

Scintillator

(1)

(2)

Deposition
geometry

COUNTING   GEOMETRY

(1) Region of forward directed
alpha's. angle = $\pi - x_1$ radians

(2) Region of backward directed
alpha's. angle = x_1 radians.

Figure 1. Illustration of the geometry aspects for calculating the front-to-back activity ratio.

To calculate counting efficiencies for alpha particles, a counting geometry factor is needed as a function of x_1. Figure 1 illustrates geometry factors based on a simple, single-cylinder model for the screen. We assume that alpha particle emission is isotropic and that all alpha particles emitted with a forward component of velocity will reach a scintillator placed (after sampling) on the front of the screen. From this it follows that the geometry factor for counting from the front of the screen is $(x_1 - \pi\phi)/\pi$. By the same reasoning, the geometry factor for a scintillator placed on the back side of the screen is x_1/a. By combining the counting geometry factors with the deposition patterns (from Equation 2), using an 80 point trapezoid rule, and integrating these two geometry factors over the deposit on the cylinder yields 0.59 and 0.41 as the alpha activity countable from the front and back, respectively. The theoretical front-to-back ratio is therefore 1.44.

In fact, there is usually a substantial drop in ²¹⁸Po concentration on passing through the screen, so the deposition on the wires of a screen is usually not negligible. One would expect that the back-side deposition would be reduced, leading to a larger front-to-back ratio. This is partially borne out by the experiments, discussed next.

New Measurement of Front-to-Back Ratio and Losses. Table 1 shows the physical chararacteristics of the screens used in this work – two sets of three screens for BOM and one set for EML. (BOM has recently added a fourth screen to each set.) Note that the two BOM sets have widely different characteristics.

We have found that it is essential to measure physical properties carefully. The solid fraction was determined in each case from the mass and thickness of circles cut to an accurately known size (using a lathe in the case of BOM). The thickness was measured with a micrometer. The mass was determined with standard deviation about 0.5% (from three weighings). The wire diameter and screens' mesh dimensions were measured from photomicrographs and from the specifications given by manufacturers. The front-to-back measurement will be described below.

Table 1. Screen Characteristics

Numerical Mesh (in^{-1})	Wire Diameter (cm)	Measured Thickness (cm)	Mass per Unit area g/cm^2	Solid Fraction α	Counting Characteristics F/B*
BOM - Set 1					
40 (ss)**	.0229	.052	.121	.298	2.45
60 (c)	.0191	.0431	.0994	.258	2.55
80 (ss)	.0140	.0296	.0750	.319	2.62
BOM - Set 2					
120 (c)	.0089	.0188	.0569	.342	2.9
325 (ss)	.0036	.0067	.0152	.287	4.0
635 (ss)	.0020	.0050	.0131	.346	7.2
EML Set (ss)					
60	.016	.0356	.102	.36	2.55
100	.010	.0259	.063	.308	2.88
200	.004	.0135	.029	.275	3.35

*At face velocity = 18-21 cm/sec, SL = 1.07 ± 0.09% per screen, using the first method (p. 7)

*F/B = 2.34 ± .32 and SL = 1.22 ± 0.08 per three screens, using the second method (p. 7)

**ss = stainless steel, c = copper

Unlike the previous BOM work, where many screens were sampled consecutively (Holub, 1984) the two sets of screens were exposed and measured simultaneously in series. The screens were mounted on brass rings 1 mm thick and the whole stack loaded into an adapted filter holder. The screens were placed going from lower coarsest to finest mesh so that first catches the smallest size clusters and so on, as shown in figure 2. Correct mounting of the filter (the last in the stack) was checked by measuring a pressure drop across the probe to detect any leak. In order to test the reliability of the method the sets consisted of 40, 60, 80 and 120, 325, 635 mesh screens with the filters at the end of each set. The face velocity was kept around 18 cm/sec and the radioactivity on the screens and the filters were measured using standard equipment (ZnS scintillator, PM tube, automatic counters) by means of the modified Tsivoglou method. The radon daughters measured were generated from about 1000 pci/l of radon with various levels of the condensation nuclei (CN), humidities and trace gas concentrations. In performing the measurements of one set of screens after another, the usual assumption of constancy of the main parameters of the chamber was made (Holub, 1984; Droullard et al, 1984) so that the results from appropriate consecutive measurements can be compared. This assumption was proven correct by reproducibility of the results.

In order to determine the efficiency of the screens, the front-to-back ratio (F/B) of activities and a loss factor (SL) by which one multiplies to correct for loss in the screens (the alpha particles absorbed in a screen that cannot be detected either during the front or back measurements, see figure 1) have to be determined. Experimentally, this was accomplished by two methods:

The first method consisted of two measurements made simultaneously side-by-side, one with one screen wrong side out. The results for the F/B ratios are shown in figure 3. The error bars are that from the counting statistics only because the other errors are negligible. Again assuming the constancy of all relevant parameters in the chamber during the experiment the loss factors on one screen plus filter system were SL = 1.07 ± 0.09. Any mesh size dependency of the loss factors could not be ascertained because of the large error. The dependency of F/B on mesh size, however, is explicit and caused by the fact that each screen is sampling the same (small) size clusters. These results, in fact are not in disagreement with the results of method #2 indicating their size dependency. The reasons for F/B 500 mesh being higher than 635 are not known.

The second method on which most data in figures 4a and 4b are based, consists of making two consecutive measurements, one of which had the full screens set (3 in these experiments) right-side-out with respect to the flow direction through the probe (the front measurement) and the other had the screen set wrong-side-out with respect to the flow direction. Repeated pairs of measurements (about 20) were made and the ratio found to be F/B = 2.34 ± .32 and the losses SL = 1.22 ± .08. Note that 1.22 is approximately equal to 1.07^3, where 1.07 is the single screen SL factor. The apparent

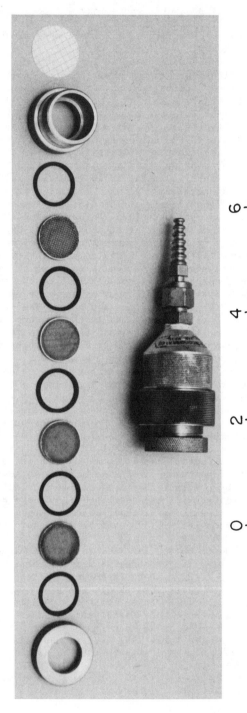

Figure 2. BOM stacked screens probe. The rubber rings between screens are used for tight fit.

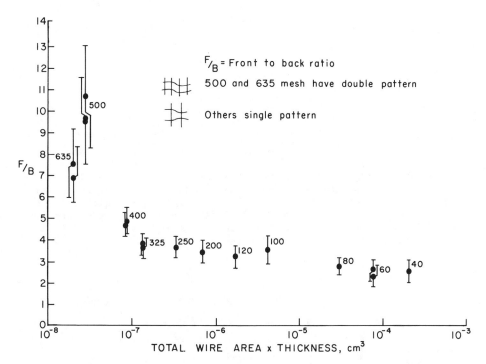

Figure 3. The front-to-back activity ratio as measured by method 1 as function of the total wire surface area times the thickness of the screen. The numbers by the points are mesh size per inch. The error bars are calculated from counting statistics. The reason for 500 mesh having higher F/B than 635 mesh is not understood.

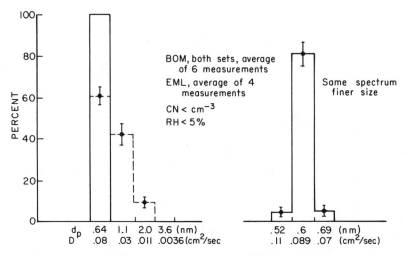

Figure 4a. Typical size distribution of ^{218}Po, full line BOM
results (both screen sets), dashed line EML results, for no growth
regime. CN levels below 70 per cm^3, no SO_2 added, humidity 2-5% or
75-90% made no difference. The difference in width is probably due
to better resolution for series than for parallel method. Error
bars are standard deviations for the measurements.

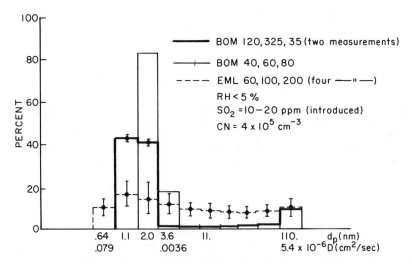

Figure 4b. Typical size distribution of ^{218}Po, full line BOM,
dashed line EML, for growth regime. CN level was in general
around 4×10^5 per cm^3, humidity 3-5%, SO_2 introduced concentration
10-20 ppm. The presence of a peak at size $.075\mu$ to $.11\mu$ is not
fully understood and may be an artifact of the deconvolution
program. Error bars are standard deviations for the measurements.

independence of these results on the size of the clusters
(particles) implies their dependency on the size because the
individual screens are catching consecutively greater size particles
in view of the fact that the smaller sizes are preferentially
removed by the preceding screens. To quantify this dependency,
however, is not a trivial problem and we plan to pursue it in the
future. In view of the difficulties, the assumption of constancy is
an acceptable first approximation. Note that the size dependency
does not agree with the results of the Friedlander's equation
discussed above because our particles are not point particles.

**First Measurements Of ²¹⁸Po Diffusion Coefficient Spectra and
Discussion.** A thorough test of the BOM and EML diffusion
coefficient measurement techniques was conducted during a one week
experiment at BOM on September 23-27, 1985. The BOM radon chamber
(Droullard et al, 1984), in which the atmosphere can be rigorously
controlled, was essential to this experiment. The strategy was to
test conditions which might yield differing growth of ²¹⁸Po. In
this way the two laboratories' techniques could be tested against
each other, and it could be determined if both methods could detect
changes.

The two atmospheres we decided to use to test our methods were
growth and no-growth regimes. The first one was accomplished by
introducing SO_2 at 10-20 ppm levels, the second by using outside air
where SO_2 on NO_x concentrations rarely exceed few ppb levels. The
detailed discussion of these results will be given elsewhere in
these proceedings.

The calculation of diffusion coefficient spectra using modified
Twomey method (1975) are based on these premises and assumptions:
The highest diffusion coefficient is equal to 0.12 cm^2/second for
radon atom (Hirst and Harrison, 1939) and this anchors the Twomey-
type fitting calculations at one end. In order to achieve good
agreement between the standard formula relating size and diffusion
coefficient at Rn atom sizes, the Cunningham slip correction has to
be modified by using a sum of the particle diameter and the air
molecules diameter, $(d_p + d_{air})$, rather than the usual d_p. This is
not surprising when one realizes that d_{Rn} = .48 nm which is
comparable to d_{air} = .37 nm. The low diffusion coefficient end is
set to coincide with the end of a measured CN counter/ diffusion
battery particle size distribution (usually at $D=10^{-4}$, size .11μ).
The low diffusion coefficient end is not, at present, of prime
interest; the main interest is to determine the cluster size
spectrum and be able to separate it from the large-size (attached)
part of the distribution.

Consecutive iterations in the Twomey code were stopped when the
results started to diverge, reached a very good fit (less than
10^{-6}%), or reached a built-in limit of 5000 iterations. Variations
between different calculations were not uniform; there seems to be
great sensitivity to certain changes in the data while other
variations do not change the results significantly. The most
reassuring feature of our results is the basic agreement between 40,
60, 80 and 120, 325, 635 sets (see figure 4a) and the agreement
between BOM and EML data.

The distribution of the activities for the growth regime is more favorable for the second set (all activities not much different from each other) and it is this set that gives a peak at larger sizes. The EML set gives better data for the no-growth regime. Three or even four screens (at present we are temporarily limited by the number of counters) cannot determine the exact size of the large-size ("attached") component. Even though there appears to be a small peak at that size in the diffusion battery data (see fig. 4b and 5), it would be premature to speculate that about 20% activity has grown to the size of $.075\mu$.

Figure 4a shows the results for CN levels at 70 cm^{-3} and below. As pointed out earlier (Holub, 1984), there are no attached radon daughters because the attachment rate is negligible compared to ^{218}Po half life. There is no observable growth and the clusters are very close to the oxide of ^{218}Po. All sets agree reasonably well. Error bars are from repeated measurements.

Figure 4b shows unmistakably a growth mode occurring at SO_2 concentration 10-20 ppm injected into the mainstream at the point where the CN were injected (Droullard et al, 1984). The condensation nuclei level, as measured continuously by a TSI counter, increased to about 4×10^5 cm^{-3} level during the time when measurements in figure 3b were taken. Note that at size 4.2nm the counter efficiency is only 8% (Agarwal and Sem, 1980). The value 4×10^5 is corrected for that.

The diffusion battery results are plotted together with the above results in figure 5. Note that the CN counter cannot detect particles below 4.2nm . The underlying assumption in this comparison is that regardless of whether the condensation nucleus has radioactive core or whether it comes from the nucleation process due to the α particle irradiation, the same growth would apply to both. Of great potential interest might be the presence of a small peak around $.075\mu$ size. This peak could be the result of runaway growth as suggested by Raes (1985). Another possibility is that it is an artifact of the Twomey code.

Using the B constant for cylinder arrangement, BC, rather than the fan constant, BF, in the Cheng and Yeh equation (1980), leads to lower diffusion coefficients (greater sizes) that never appear at the .64 nm class interval. In other words, regardless of humidity or SO_2 concentration there would always be growth to a size of at least 1.1nm − a highly unlikely case. As discussed previously and in view of the above, the BF Chen-Yeh results again proved to be preferable even though a screen might appear more like "staggered cylinders" than "fans".

The n-screen method (requiring n + 1 counters) to measure the ultrafine radon daughters distributions can be used to test Bricard-Raes theory. Most of the results reported here could be interpreted using this theory, however, more tests of the computational part will have to be done before calculating and analyzing the individual results. Also a new method, recently developed by Maher and Laird (1985), will be compared to Twomey's code.

A very approximate mass balance calculation for the whole range of employed SO_2 concentrations based on known amount of SO_2

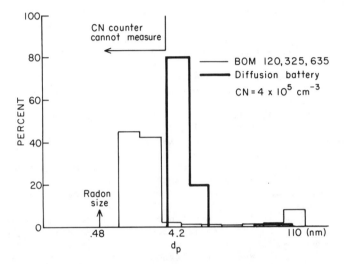

Figure 5. The same data as in figure 4b plotted together with the diffusion battery data. Note, the CN counter cannot see any particle below 4.2 nm. According to rough mass balance estimate 10^6 – 10^8 small clusters per cm^3 could be present.

introduced upstream into the chamber, the number of CN measured in the chamber (see fig. 5) and on the assumption there is no sink for SO_2 except the H_2SO_4 formation, indicate there might be about 10^6-10^8/cm^3 droplets of H_2SO_4 below the CN counter threshold of detection.

Summary and Conclusions

1. A study of published data indicates that the Cheng–Yeh theory reasonably well describes the collection efficiency of single wire screens, for both atoms and particles, and for a wide variety of screen geometries, using the BF constant.

2. Our data show that the front-to-back ratio (F/B) for α-counts after sampling unattached ^{218}Po through a screen is about 2.5 for coarse screens (40 to 80 mesh), increasing gradually to 7 to 11 for finer screens for small sizes. For larger sizes the ratio is closer to 2.5.

3. Our data show that the unattached ^{218}Po, the sum of front and back α-counts, is about 7% less than the average of the count from a reference filter.

4. For ^{218}Po sampled from a radon chamber free of particles (CN < 70 cm^{-3}) and with no introduced SO_2 concentration, our two multiple-screen methods yield a unimodal spectrum of diffusion coefficients, with the mode at .08 ± .01 cm^2/sec.

5. The distribution of diffusion coefficients in item 4 above is virtually monodispersed; a more exact value cannot be given due to limits in resolution and uncertainties in the mathematics of deconvolution.

6. For ^{218}Po sampled from a radon chamber containing radiolysis aerosol (CN ≈ 4 X 10^5/cm^3, formed from SO_2), the series-arranged screens (BOM) yielded a unimodal spectrum with mode at .02 cm^2/sec (1.5 nm) with geometric standard deviation, σg = 1.4. The parallel-arranged screens (EML) nucleus counter response in this size gives results that are in basic agreement with the BOM data.

Acknowledgments

Valuable discussions with R.F. Droullard and T.H. Davis are gratefully acknowledged. Thanks are extended to W.P. Stroud for help in setting up the SO_2 experiments and to K.W. Tu for help in operating the EML screen equipment.

Literature Cited

Ararwal, J.K. and G.J. Sem, Continuous Flow, Single-Particle-Counting Condensation Nuclei Counter, J. Aerosol Sci. 11:343-358 (1980).

Barry, P.J., Sampling for Airborne Radioiodine by Copper Screens, Health Phys. 15:243-250 (1986).

Bricard, J., M. Cabone, G. Madeleine, and D. Vigla, Formation and Properties of Neutral Ultrafine Particles and Small Ions, in Aerosols and Atmospheric Chemistry (G.M. Hidy, ed.) pp. 27-43 Academic Press (1972).

Busigin, A., A.W. van der Vooren, and C.R. Phillips, Attached and Unattached Daughters; Measurements and Measurement Techniques in Uranium Mines, in Proceedings of the Bureau of Mines Conf. on Radon Daughter Plateout Phenomena, (P.K. Hopke, ed.) pp. 99-101, Univ. of Illinois, Urbana, IL, April 16-18 (1980).

Chamberlain, A.C. and E.D. Dyson, The Dose to the Trachea and Bronchi from the Decay Products of Radon and Thoron, Brit. J. Radiol. 29:317-325 (1956).

Cheng, Y.S. and H.C. Yeh, Theory of Screen Type Diffusion Battery, J. Aerosol Sci. 11:313-319 (1980).

Droullard, R.F., T.H. Davis, E.E. Smith, and R.F. Holub, Radiation Hazard Test Facilities at Denver Research Center, Bureau of Mines IC 8965:1-22 (1984).

Friedlander, S.K., Smoke, Dust and Haze, John Wiley & Sons, Inc., New York (1977).

George, A.C., Measurement of the Uncombined Fraction of Radon Daughters with Wire Screens, Health Phys. 23:390-392 (1972).

Goldstein, S.D. and P.K. Hopke, Environmental Neutralization of Polonium-218, Environ. Sci. Technol. 19:146-150 (1985).

Hirst, B.W. and G.E. Harrison, The Diffusion of Radon Gas Mixtures, Proc. Roy. Soc. Lond. A169:573-586 (1939).

Holub, R.F., Turbulent Plateout of Radon Daughters, Rad. Prot. Dos. 7:155-158 (1984).

James, A.C., G.F. Bradford, and D.M. Howell, Collection of Unattached RaA Atoms Using Wire Gause, J. Aerosol Sci. 3:243-250 (1972).

Knutson, E.O., A.C. George, R.H. Knuth, and B.R. Koh, Measurement of Radon Daughter Particle Size, Rad. Prot. Dos. 7:121-125 (1984).

Maher, E.F. and N.M. Laird, E.M. Algorithm Reconstruction of Particle Size Distribution from Diffusion Battery Data, J. Aerosol Sci. 16:557-570 (1985).

Raes, F., Description of Properties of Unattached ²¹⁸Po and ²¹²Pb Particles by Means of Classical Theory of Cluster Formation, Health Phys. 49:1177-1187 (1985).

Samuelsson, C., private communication (1984).

Scheibel, H.G. and J. Porstendorfer, Penetration Measurements for Tube and Screen Type Diffusion Battery in Ultrafine Particle Size Range, J. Aerosol Sci. 15:673-679 (1984).

Sinclair, D. and G.S. Hoopes, A Novel Form of Duffision Battery, ATHA 36, pp. 39-42 (1975).

Sinclair, D., A.C. George, and E.O. Knutson, Application of
Diffusion Batteries to Measurement of Submicron Radioactive
Aerosols, in Airborne Radioactivity (D.T. Shaw, ed.) American
Nuclear Society, La Grange Park, IL, pp. 103-114 (1978).

Sinnaeve, J., Opening Address, Rad. Prot. Dos. 7:13-14 (1984).

Thomas, J.W. and L.E. Hinchliffe, Filtration of 0.001   Particles
with Wire Screens, J. Aerosol Sci. 3:387-393 (1972).

Twomey, S., Comparison of Constrained Linear Inversion and an
Iterative Nonlinear Algorithm Applied to the Indirect Estimation
of the Particle Size Distribution, J. Comp. Phys. 18:188-200 (1975).

RECEIVED November 13, 1986

Chapter 26

Development of a Mobility Analyzer for Studying the Particle-Producing Phenomena Related to Radon Progeny

Lisa M. Kulju[1], Kai-Dee Chu[2], and Philip K. Hopke[3]

[1]Department of Civil Engineering and Institute for Environmental Studies,
University of Illinois, Urbana, IL 61801
[2]Department of Nuclear Engineering and Institute for Environmental Studies,
University of Illinois, Urbana, IL 61801
[3]Departments of Civil Engineering and Nuclear Engineering and Institute for
Environmental Studies, University of Illinois, Urbana, IL 61801

The determination of the activity size distribution of
the ultrafine ions is of particular interest due to
their influence on the movement and deposition of Po-218.
These ultrafine ions are the result of radiolysis and
their rate of formation is a function of radon concentration, the energy associated with the recoil path of
Po-218, and the presence of H_2O vapor and trace gases
such as SO_2. A joint series of experiments utilizing a
mobility analyzer, the separate single screen method,
and the stacked screen method were conducted to examine
the activity size distribution of the ultrafine mode.
The results obtained from the mobility analyzer are discussed in this paper.

Radon-222, a decay product of the naturally occuring radioactive
element uranium-238, emanates from soil and masonry materials and is
released from coal-fired power plants. Even though Rn-222 is an
inert gas, its decay products are chemically active. Rn-222 has a
a half-life of 3.825 days and undergoes four succesive alpha and/or
beta decays to Po-218 (RaA), Pb-214 (RaB), Bi-214 (RaC), and Po-214
(RaC'). These four decay products have short half-lifes and thus
decay to 22.3 year Pb-210 (RaD). The radioactive decays products of
Rn-222 have a tendency to attach to ambient aerosol particles. The
size of the resulting radioactive particle depends on the available
aerosol. The attachment of these radionuclides to small, respirable
particles is an important mechanism for the retention of activity in
air and the transport to people.

The ionic charge, diffusivity, and electrical mobility
associated with these small radioactive particles are three parameters controlling plateout. A particle can acquire an electrical
charge by a number of mechanisms which promote the transfer of
electrons to and from the particle surface, therefore producing a
negatively and positively charged particle, respectively. With the
decay of Rn-222, an alpha particle and Po-218 are formed. As these

0097-6156/87/0331-0357$06.00/0

two particles move in opposite directions, electrons are stripped
from the Po-218 by the departing alpha particle or by the recoil
motion. The ion regains most of its electrons as it slows to rest,
such that 88% of the time, Po-218 exists as a singly charged,
positive ion. The remaining 12% of the time, Po-218 occurs as the
neutral species (Wellisch, 1913; Porstendorfer and Mercer, 1979).

These radon daughter ions have electrical mobilities in the same
range as ordinary atmospheric ions and the ions can be divided into a
number of groups where each group has a particular mobility (Nolan,
1916; McClelland and Nolan, 1926). Bricard et al. (1966) reports
measurements for five distinct mobility groups with values in the
range of 0.4-2.2 $cm^2s^{-1}V^{-1}$ for small radioactive ions in air.

Electrical Mobility

In past studies, the mobility spectrum of small radioactive ions in
air was determined by the utilization of two different spectrometers:
a Zeleny spectrometer and an Erikson spectrometer. In the Zeleny
spectrometer, air enriched with radon or thoron is drawn through a
chamber and the mobility of the resulting ions is determined by
drawing the ions from the chamber into a Zeleny-tube, a cylinder with
a negatively charged electrode positioned down the center. The ions
are separated by their electrical mobilities and the distribution of
radioactivity attached to the central wire is used to deduce the
mobility spectrum of the small radioactive ions. The alpha activity
is measured by the Renoux Method (Renoux, 1961) where the wire is cut
into equal length segments that are placed between two ZnS scintillators. In studies by Blanc et al. (1963), two mobility groups in air
were observed by the utilization of a Zeleny spectrometer suggesting
the existence of two types of ions. The mobility spectrum of Po-216
and Pb-212 ions as determined by Blanc (1963) suggests that the
mobility peaks at 0.82 $\pm$ 0.03 $cm^2s^{-1}V^{-1}$ and 1.65 $\pm$ 0.05 $cm^2s^{-1}V^{-1}$
correspond to aerosols that carry one and two electrical charges.

In the Erikson spectrometer (Erikson, 1922), the mobility is
determined from the distribution of activity deposited on a nuclear
emulsion in a parallel plate configuration rather than onto a collecting electrode as in the Zeleny spectrometer. The two parallel,
nuclear photographic plates are positioned such that the sensitive
surface is oriented toward the interior. Experiments performed by
Bricard et al. (1966) utilized air enriched in radon or thoron to
insure a significant number of tracks on the photographic plate. The
existence of two mobility peaks was observed where the broad band had
mobilities in the range of 0.3 to 1.1 $cm^2s^{-1}V^{-1}$ and the narrow band
had a mobility peak centered around 2.1 $cm^2s^{-1}V^{-1}$. Mobility spectra
using an Erikson spectrometer and a Zeleny spectrometer show four
mobility peaks which indicate four different groups of ions (Fontan
et al., 1969). The majority of these experiments were performed
under unknown conditions in terms of the concentrations of trace
gases in the air and humidity levels. With unknown reaction conditions, the true chemical nature of the molecules is hard to
elucidate. Therefore, there is a need for a well controlled system.

Design

A diffusion chamber (Figure 1) originally designed for neutralization
rate studies by Chu and Hopke (1985) has been modified to be an

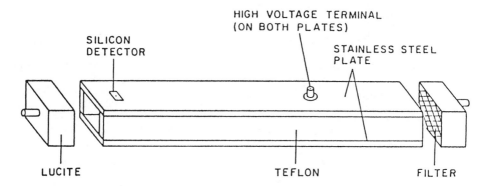

Figure 1. The diffusion chamber.

Erikson-type spectrometer. The system is composed of two parallel
stainless steel plates and two Teflon plates that form a rectangular
chamber. Both ends of the chamber are designed to hold filters which
are able to create a homogeneous flow for the gas as it enters the
chamber. A uniform electric field is obtained by connection to a DC
high voltage. A continuous monitoring process is feasible because a
custom fabricated rectangular silicon detector is mounted into the
ground plate. This detector has been fabricated to be as flat as
possible across its face and to have its surface at ground potential.
Thus, its presence in the ground plate does not perturb the electric
field lines.

The gas is drawn through one opening followed by two glass fiber
filters thus permitting only the radon and test gas into the chamber.
Part of the Po-218 formed in the chamber will deposit on the walls by
diffusion. To insure the absence of plated-out Po-218 alpha
particles in the detector, the distance between the two metal plates
was designed to be 6.0 cm. This distance was chosen because Po-218
alpha particles have a range of 4.6 cm in air and nitrogen. Thus, a
6.0 cm distance is large enough so that no counts can come from the
bottom plate.

This diffusion chamber was modified to provide a uniform flow
from two channels at the entrance, one for the filtered room air and
the other for the gas from the radon chamber. This modified mobility
analyzer is schematically shown in Figure 2. The pressure heads are
adjusted so that the gas velocities, v, are the same in both channels. An adjustable vertical electric field, E, is provided through
the analyzer so that charged particles are drawn toward the detector
located at x cm from the entrance. With the known distance, d,
between the radon-laden gas channel and the detector implanted plate,
the mobility can then be determined from

$$K = (vd)/(Ex) \qquad (1)$$

Results and Discussion

As previously mentioned, past studies used non-filtered air with
unknown concentrations of trace gases at unknown relative humidities.
Also, many of the studies used plastic aging chambers that may have
introduced volatile monomers into the air. These unknown factors are
important to determine in order to fully understand the nature of the
ultrafine particle mode. According to the classical thermodynamic
theory of ion cluster formation (Coghlan and Scott, 1983), the
relative humidity and trace gases will affect the existence of
condensation nuclei. Megaw and Wiffen (1961) observed an increase in
nuclei formation with the presence of sulfur dioxide.

In order to examine the process of ultrafine particle formation,
a joint series of experiments were conducted at the Denver Research
Center of the U.S. Bureau of Mines. In the Denver radon chamber, the
activity size distribution of the ultrafine mode was measured using
the mobility analyzer designed by Chu and Hopke (1985), the separate
single screen method (Holub and Knutson, 1987), and the stacked
single screen method (Holub and Knutson, 1987) for various relative
humidities and for various concentrations of SO_2. The results

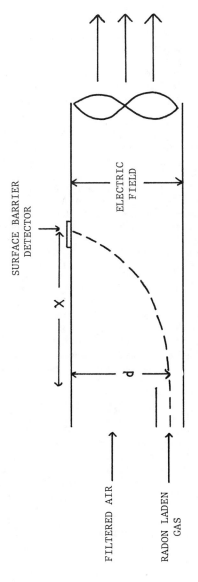

SURFACE BARRIER
DETECTOR

ELECTRIC
FIELD

X

d

FILTERED AIR

RADON LADEN
GAS

Figure 2. The mobility analyzer modified from the Erikson spectrometer.

obtained from the mobility analyzer will be discussed in some detail in this paper. For additional information concerning this joint project, the reader is referred to Chu et al. (1987).

From the data obtained by the utilization of the mobility analyzer it was observed that H_2O decreased the amount of Po-218 ions and the addition of SO_2 increased particle formation. From Figure 3, it can be seen that with an increase in the relative humidities and absence of SO_2, there is a corresponding decrease in the number of counts. This decrease in the number of counts recorded is a result of an increase in neutralization of the Po-218 ions by water vapor. Since the mobility analyzer is only capable of detecting ions, the mobility spectra obtained in the presence of water are of a different type than those spectra obtained in the absence of water.

Figure 3 also represents the affect of SO_2 and H_2O on particle formation. High concentrations of SO_2 (10-20 ppm) in the presence of H_2O result in a decrease in the total number of ions but an increase in the number of ions with lower mobilities. This is especially significant for Po-218 ions with mobilities centered around $2.0 \text{ cm}^2\text{V}^{-1}\text{s}^{-1}$. At low humidity and high SO_2 concentrations there is an increase in particle formation at higher mobilities. Under this set of conditions the examination of the activity size distribution of the finest mode of the ultrafine ions is possible.

There are some problems associated with this experiment in terms of low statistical precision, uncertainity in SO_2 concentrations, and the presence of butanol in the system. Statistical precision is low because of the small number of counts and low number of data points taken. Thus, the multiple peaks observed in Figure 3 may actually be only 2 distinct peaks where the broad peak has mobilities ranging from $0.4 \text{ cm}^2\text{V}^{-1}\text{s}^{-1}$ to $1.6 \text{ cm}^2\text{V}^{-1}\text{s}^{-1}$ and the narrow peak has mobilities centered around $2.0 \text{ cm}^2\text{V}^{-1}\text{s}^{-1}$. These peaks thus correspond to those reported by Bricard et al. (1966).

Another problem associated with this experiment is the uncertainity in the SO_2 concentrations. The high SO_2 concentration of 10 to 20 ppm is only an approximation. Because of the lack of precise control or measurement of the SO_2 concentrations, there is some ambiguity in interpreting the results. Lastly, reevaluation of the apparatus set-up revealed the presence of butanol from the TSI CNC in the chamber. From the literature it is known that the presence of radical scavengers, such as straight chain alcohols, may reduce the production of condensation nuclei by ionizing radiation (Coghlan and Scott, 1983). Therefore, butanol in the system suppresses particle formation. Follow-up experiments performed as part of the joint series conducted at the Denver Research Center have addressed for these problems. Thus, in conjunction with the separate single screen method (Holub and Knutson, 1987) and stacked single screen method (Holub and Knutson, 1987), the mobility analyzer is a useful device for determining the activity size distribution of the ultrafine charged particles.

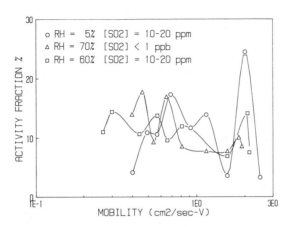

Figure 3. Mobility spectra of Po-218 ions at various relative humidities and SO_2 concentrations.

Acknowledgments

This work was funded in part by the US Department of Energy under contract number DE AC02 83ER60186.

Literature Cited

Blanc, D., J. Fontan, and M. Juan, On the Mobilities of Natural Ions Obtained in Filtered Air, Compte Rendus de l'Academie des Sciences de Paris 260:2099-2100 (1963).

Bricard, J., F. Billard, D. Blanc, M. Cabane, and J. Fontan, Detailed Structure of the Mobility Spectrum of the Small Radioactive Ions in the Air, Compte Rendus de l'Academie des Sciences de Paris 263:761-764, Series B, (1966).

Chu, K.D. and P.K. Hopke, Continuous Monitoring Method of the Neutralizing Phenomena of Polonium-218, Presented at the 78th Annual Meeting of the Air Pollution Control Association, Detroit, Michigan, (June 1985).

Chu, K.D., P.K. Hopke, E.O. Knutson, K.W. Tu, and R.F. Holub, The Induction of an Ultrafine Aerosol by Radon Radiolysis. This volume.

Coghlan, M. and J.A. Scott, A Study of the Formation of Radiolytic Condensation Nuclei, Proc. in Atmospheric Electricity (1983).

Erikson, H.A. On the Nature of the Negative and Positive Ions in Air, Oxygen, and Nitrogen, Physical Review Second Series, XX:117-126 (1922).

Fontan, J., D. Blanc, M.L. Hubertas, and A.M. Mart, Mesure de la Mobilite et der Coefficient de Diffusion des Particules Radioactives, in Elementary Electrodynamics (S.C. Coronitte and J. Hughes, ed) pp. 257-267, Gordon and Breach Science Publishers, New York, (1969).

Holub, R.F. and E.O. Knutson, Measurement of Po-218 Diffusion Coefficient Spectra Using Multiple Wire Screens. This volume.

McClelland, J.A. and P.J. Nolan, The Nature of Ions Produced by Bubbling Air Through Mercury, Proc. Roy Irish Acad. 23A:24 (1926).

Megaw, W.J. and R.D. Wiffen, The Generation of Condensation Nuclei by Ionizing Radiation, Geofisica pura e applicata 50:118-12 (1961).

Nolan, J.J. The Mobilities of Ions Produced by Spraying Distilled H_2O, Proc. Roy Irish Acad. 23A:9-23 (1916).

Porstendorfer, J. and T.T. Mercer, Influence of Electric Charge and Humidity upon the Diffusion Coefficient of Radon Decay Product, Health Physics 15:191-199 (1979).

Renoux, A. These de Specialite, Paris (1961).

Wellisch, E.M. The Distribution of Active Deposit of Radium in an Electric Field, Philosophical Magazine 26:623-635 (1913).

RECEIVED August 20, 1986

Chapter 27

Induction of an Ultrafine Aerosol by Radon Radiolysis

Kai-Dee Chu[1], Philip K. Hopke[2], E. O. Knutson[3], K. W. Tu[3], and R. F. Holub[4]

[1]Department of Nuclear Engineering and Institute for Environmental Studies,
University of Illinois, Urbana, IL 61801
[2]Departments of Civil Engineering and Nuclear Engineering and Institute for
Environmental Studies, University of Illinois, Urbana, IL 61801
[3]Environmental Measurement Laboratory, Department of Energy, New York, NY 10014
[4]Geophysics Division, Denver Research Center, U.S. Bureau of Mines, Department of the
Interior, Denver, CO 80225

Recent advances in measurement methods have allowed
for the more complete determination of the aerosol
size distribution for particles incorporating Po-218.
It has been found that the "unattached" fraction is
an ultrafine particle aerosol with a size range of
0.5 to 3 nm. In order to initiate studies on the
formation mechanism for these ultrafine particles, a
series of experiments were made in the U.S. Bureau of
Mines' radon chamber. By introducing SO_2 into the
chamber, particles were produced with an ultrafine
size distribution. It has been found that the particle
formation mechanism is supressed by the presence of
radical scavengers. These experiments suggest that
radiolysis following the decay of Rn-222 gives rise
to the observed aerosol and the properties of the
resulting aerosol are dependent on the nature and the
amount of reactive gas present.

It has been reported for many years that condensation nuclei can
be produced by ionizing radiation. Recent studies have improved
the measurement of the activity size distribution of these ultrafine
particles produced by radon and its daughters (Reineking, et al.,
1985; Knutson, et al., 1985). It seems that the Po-218 ion is formed
by the radon decay, is neutralized within a few tens of milliseconds,
and then attached to an ultrafine particle formed by the radiolysis
generated by the polonium ion recoil. Although there will be
radiolysis along the alpha track, those reactions will be very
far away (several centimeters) from the polonium nucleus when it
reaches thermal velocity. The recoil path radiolysis therefore
seems to be the more likely source of the ultrafine particles
near enough to the polonium atom to rapidly incorporate it.
These ultrafine particles have high mobility and can coagulate

0097-6156/87/0331-0365$06.00/0

rapidly with the preexisting aerosol. These ultrafine particles
are what had been termed the "unattached fraction". It will be
important to measure the full range of particle activity size
distribution if the properties of these radionuclides are to be
understood.

Background

Chamberlain, Megaw and Wiffen (1957) and Megaw and Wiffen (1961)
presented early reports of the ability of ionizing radiation to
induce the formation of condensation nuclei in laboratory air.
Megaw and Wiffen (1961) as well as Burke and Scott (1973) suggested
that there was a minimum dose that must be deposited in a volume
of air before the appearance of condensation nuclei. Similar
results were obtained in our laboratory where the effects of
ionization sources used to neutralize generated particles could
themselves produce particles when a sufficiently high ionization
density occured in a particular volume of air (Leong, et al., 1983).
        However, these results are probably the result of the method
of detecting particles and thereby provide an operational definition
of "particle". All of these studies use a condensation nuclei
counter to determine the presence of "particles". These devices
have been found to have sharply decreasing efficiency for detecting
particles less than 0.01 μm (Leong, et al., 1983, for example).
Thus, there is a size threshold that must be reached before a
cluster of atoms becomes big enough to be detected and turns
into a "condensation nuclei". Recent work by Madelaine and
coworkers (Perrin, et al, 1978; Madelaine, et al., 1980) have
extended the size of measurable ultrafine particles to the order
of 0.003 μm. They find rapid coagulation of this ultrafine aerosol
to a larger average diameter one that is easily observable.
Martell (1977) has suggested the importance of the microphysical
and chemical processes that take place in the alpha track may
lead to the formation of nuclei from the alpha radiolysis. A
similar process may then occur in the shorter but potentially
more ionized region of the polonium recoil track. Perrin et al.
(1978) used short-lived radon-219 to create their ultrafine
particles. With a half-life of 3.96 seconds, there is a pulse
of radiolysis occurring after the injection of the radon into
particle-free ambient air resulting in a rapid increase in observable
nuclei and a longer decrease as they coagulate. The qualitative
features of this process could be explained with a simple theoretical
coagulation calculation. Thus, it appears that there may be a
continuous size range of molecular clusters up to the size at
which they become "particles". It appears that the rate at
which the small, highly mobile clusters are formed is dose-dependent
and thus at a sufficiently high dose rate, the clusters rapidly
coagulate to condensation nuclei as has been observed. In this
paper, the word "condensation nuclei" means the particles that
can be detected by a CN counter, and the word "particles" means
both condensation nuclei and ultrafine particles.
        There has been a study by Vohra, Subbarama and Rao (1966)
in which they use ethanol and an electron source to directly

produce clusters to which radon (Rn-222) decay products rapidly attach. In the absence of the electron source, they find little cluster formation. This result is not surprising since small chain alcohol molecules represent extremely good free radical scavengers that could inhibit the formation of molecular clusters (Vohra, et al., 1966; Gusten, et al., 1981). However, for more normal atmospheric constituents like SO_2, ionizing radiation has been found to play a significant role in SO_2 to sulfate conversion under laboratory conditions (Vohra, 1975). In fact, Megaw and Wiffen (1961) in their work observed "some proportionality between the sulfur dioxide content of the air and the nucleus concentration produced by a given radiation dose". However, they also found that the rate of nuclei formation was greater than could be explained by SO_2 alone. It has been suggested that the primary mechanism of particle formation is radiolysis of water to form hydroxyl radicals (Coghlan and Scott, 1983) and that the presence of NO suppresses the formation of nuclei filtered laboratory air presumably by scavenging oxidizing free radicals (Chamberlain, et al., 1969). Coghlan and Scott (1983) have shown that hydroxyl radical scavengers like short chain alcohols suppress the formation of particles with the suppression correlated with the rate constant for reaction between hydroxyl radical and the radical scavenger. Thus, there is literature showing the formation of clusters and some of the potential interactions that occur in the air. However, there has not been a systematic study of the mechanisms of cluster formation and the possible interactions of various common atmospheric species in that process.

There is very recent evidence that there is an extremely fine mode in the particle activity size distribution. Reineking, Becker and Porstendorfer (1985) used several diffusion batteries to determine a Po-218 activity peak in the 1-3 nm diameter range. The Po-218 activity was also attached to particles in the accumulation mode peak in the 0.1 to 1.0 μm range. The Po-214 (RaC') activity was only observed in the accumulation mode and not associated with the ultrafine particles. Thus, the initial motion and deposition of much of the polonium-218 may be related to the transport by these ultrafine clusters.

Similar results were reported by Knutson, George, Hinchliffe and Sextro (1985). In this study, single screens were used to improve the size resolution for determining the activity size distribution in the 1-4 nm range. For low and moderate condensation nuclei concentrations (<1000/cm³), a distinct peak is observed in the size distribution in the 1-2 nm range. They have also shown that it is almost impossible to accurately determine the size of this ultrafine mode with a conventional screen type diffusion battery. In the presence of high ambient particle concentration (≥90,000/cm³), they find that the ultrafine mode coagulates too rapidly with the preexisting aerosol to be observed.

Experimental Systems

In order to examine the process of ultrafine particle formation, a joint series of experiments were conducted at the Denver Research

Center of the U.S. Bureau of Mines. The radon chamber in Denver
Research Center is designed to provide adjustable humidities and
a well controlled, monitored radon and CN concentration. It has
a length of 213 cm, a diameter of 152 cm, and a volume of about
3.89 m^3. The walls are made of 0.5-cm rolled steel with welded
seams. Figure 1 shows the general scheme of the test chamber
excluding the transducers and data aquisition system.

Rn-222 is produced from a dry Ra-226 source with an activity
of about 2 mCi housed in a lead shield. Radon is carried from
the source by means of compressed air with a regulated flow rate
of 200 cm^3/min. The test chamber is operated with a 75 l/min
air pump located downstream (Figure 1)(Droullard, et al., 1984)
that exhausts to the outside of the building. The flow rate of
this air pump is monitored by a mass flow transducer whose signals
are converted to volumetric flow rates.

A TSI Condensation Nuclei Counter model 3020 is used to
continuously monitor the aerosol concentration in the chamber
atmosphere. The chamber air is drawn through a port in the
chamber wall into the counter and returned to the chamber through
another port. The chamber interior temperature and humidity are
monitored with a commercial hygrometer system.

The activity level of Rn-222 and its daughters in the test
chamber is measured. Radon activity levels are occasionally
measured by grab sampling using scintillation cells. Radon-
daughter measurements for control or calibration purpose are
made by the modified Tsivoglou method (Thomas, 1976). Samples
can be collected through a sampling wand with a filter holder
being put into the chamber through a sampling port. The air in
the chamber is pulled through the filter and then recycled back
into chamber by a pump. Working levels are continuously monitored
by a continuous working-level detector developed by the Bureau
of Mines (Droullard and holub, 1977; Droullard and Holub, 1980).

The activity size distribution of the ultrafine mode was
measured using the BuMines stacked single screen method (Holub
and Knutson, 1986), the EML separate single screens approach
(Holub and Knutson, 1986), and the UI mobility analyzer (Kulju,
et al., 1986) for various relative humidities and the presence or
absence of added SO$_2$.

Results and Discussion

A possible particle formation mechanism is as follows:

$$H_2O \rightarrow H + OH$$

$$SO_2 + OH \rightarrow SO_3 + H$$

$$H_2O + SO_3 \rightarrow H_2SO_4$$

The hydroxyl radicals produced by water molecule radiolysis
react promptly with oxidizable species such as SO$_2$ in air and
form a condensed phase. These molecules further coagulate and
become ultrafine particles. The radon concentration in the

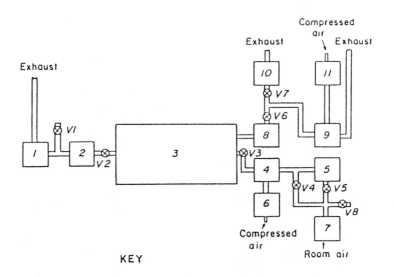

KEY

1	Air pump	*10*	Exhaust radon flowmeter
2	Mass flowmeter	*11*	Dry radon source
3	Radon test chamber	*VI*	Air control valve
4	Mixing chamber	*V2*	Chamber exhaust valve
5	Humidifier	*V3*	Chamber input valve
6	Aerosol generator	*V4*	Bypass valve
7	Air filter	*V5*	Humidifier valve
8	Chamber radon flowmeter	*V6*	Chamber radon control valve
9	Radon source control	*V7*	Exhaust radon control valve
		V8	Room air valve

Figure 1. Scheme of radon test chamber.

chamber was kept at 1000 pCi/l. The experimental conditions
were set to observe particle formation at both high and low
humidities combined with high and low SO_2 concentrations (Table
I). However, in the first set of experiments, the SO_2 concentrations
were only poorly controlled. We could detect only background
condensation nuclei (1-80 /cm^3) present when no additional SO_2
was injected into the chamber at both high and low humidity. A
large number of condensation nuclei were formed at about 10 ppm
SO_2 concentration and 6% relative humidity. However, there was
no increase in condensation nuclei formed in higher humidities.

Table I. Observed Condensation Nuclei Counts in Different
Atmospheric Conditions, September Experiments, 1985

SO_2	RH	Condensation Nuclei
background	10%	70 /cm^3
background	60%	50 /cm^3
10-20 ppm	6%	26000 /cm^3
(refill butanol in CNC)		
10-20 ppm	60%	80 /cm^3

Figure 2 and Figure 3 show the size distribution of these
ultrafine particles measured by separate single screens (EML),
stacked single screen (USBM), and the mobility analyzer (UI).
Since the mobility analyzer collects only charged particles, the
size distribution derived from the mobility spectrum is only for
the charged particles. There are some correlations among these
three different methods at both high and low SO_2 concentrations.
At higher humidities, more neutralization occurs leading to
fewer polonium ions that can be observed.

Because of the uncertainty in the SO_2 concentrations in these
experiments, another set of experiments were performed in March
1986 using only stacked single screen method. Similar combinations
of SO_2 concentrations and relative humidities were used (Table
II). The same lack of particle formation was observed when
there were only a few condensation nuclei formed at 1.1 ppm SO_2
concentration and 58% relative humidities. In reviewing the
system, we found that the design of the system was to recycle
the sample gas for CNC back into the chamber. The TSI CNC uses
butanol as its condensable fluid and the butanol is a radical
scavenger that removes hydroxyl radical rapidly enough to supress
particle formation. When radical scavengers such as butanol or
nitric oxide are present, a competing reaction occurs, i.e.

$$ROH + OH \longrightarrow RO + H_2O$$

and consequently supresses the particle formation process. This
reaction may explain the observation of particles in the earlier
experiments when the butanol in CNC was depleted, and after

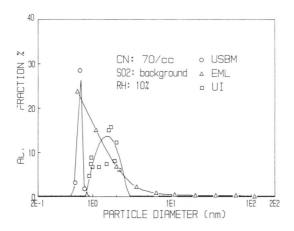

Figure 2. Size distributions measured with separated single screen method (EML), stacked single screen method (USBM), and mobility analyzer (UI).

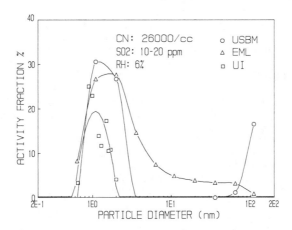

Figure 3. Size distributions measured with separated single screen method (EML), stacked single screen method (USBM), and mobility analyzer (UI).

refilling butanol in CNC, the particle counts dropped drastically
(Table I).

Table II. Observed Condensation Nuclei Counts in
Different Atmospheric Conditions

SO_2	RH	Condensation Nuclei
11 ppb	73%	21 /cm^3
110 ppb	68%	24 /cm^3
1.1 ppm	63%	400 /cm^3
(stop recycling CNC effluent back into chamber)		
11 ppb	5%	40 /cm^3
110 ppb	<2%	38 /cm^3
1.1 ppm	<2%	4000 /cm^3
11 ppm	<2%	27000 /cm^3
(injection of 92 ppb NO)		
110 ppb	<2%	9 /cm^3

When the flow through the CNC was exhausted outside of the
laboratory, we observed particle formation at higher SO_2
concentrations as expected (Table II). To prove that the radical
scavenger effect is reproducible, another radical scavenger (92
ppb nitric oxide) was used in the presence of 110 ppb SO_2
concentration and 2% humidity, and the supression in particle
formation was observed. Another possible mechanism that supressed
the particle formation is that more neutralization of polonium
ions occurred at the higher humidities and thus ion-induced
nucleation would be suppressed.

Figure 4 shows the size distribution of these ultrafine
particles at low humidities being shifted up in size by increasing
SO_2 concentration. There is an overlap of spectra at 11 ppb and
110 ppb of SO_2 concentrations. Figure 5 reveals the same results
at higher humidities, but the overlapped spectra were separated
with a smaller size distribution at 110 ppb in the the presence
of additional water vapor. We also obtained two different size
distributions at 1.1 ppm in two sets of experiments. Thus, there
are still some uncertainties in understanding the size distribution
of ultrafine particles.

In addition, an interesting phenomenon was observed when
pumping the sample gas through the sampling probe. The chamber
air is pulled through the screens and filter, and through a dry
test meter with a Metal Bellows pump. When pump is on, a steadily
increasing number of detectable condensation nuclei is found.
When the pump is turned off, the condensation nuclei count decreases
by ventilation and plate-out to that observed before the sampling
process. Thus, there must be a process during the sampling
period that shifts the particle size to a size detectable by the
CNC. The "pump effect" has further been found to be proportional
to the SO_2 concentrations (Figure 6). It was not observed when

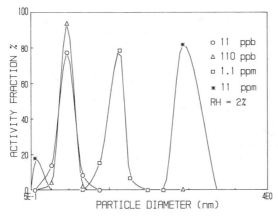

Figure 4. Size distributions measured with stacked single screen method.

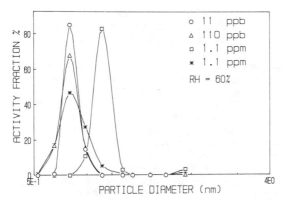

Figure 5. Size distributions measured with stacked single screen method.

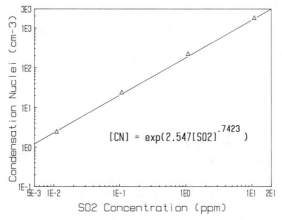

Figure 6. Proportionality between the "pump effect" and the concentrations of SO_2.

clean filtered, radon and SO_2-free laboratory air was drawn
through the pump. The "pump effect" can also be supressed by
the presence of radical scavengers such as butanol or nitric
oxide. The presence of 50 ppb nitric oxide supressed the pump
effect at 110 ppb.

Since the sample air was drawn through a filter by the pump
and then recycled back into the chamber, the air that flows into
the pump should be particle-free. A possible cause to the particle
formation phenomenon is that there is a build-up pressure required
before the release valve in the pump opens to return the gas
back into the chamber. Thus, during the pressurized period,
more collisions can occur and result in the increase of observable
condensation nuclei.

Summary

It has been found that the activity which is conventionally
referred to as the "unattached" fraction is actually an ultrafine
particle aerosol with a size range of 0.5 to 3 nm. The hydroxyl
radical from water molecule radiolysis is a key element to the
particle formation mechanism. By injecting different concentrations
of SO_2 into the test chamber, a possible particle formation
mechanism has been suggested as follows: Oxidizable species such
as SO_2 reacts promptly with hydroxyl radicals and form a condensed
phase. These molecules coagulate and become ultrafine particles.
The size distribution of these ultrafine particles can be shifted
upwards with the increase of SO_2 concentrations.

The compression of sampling air, the "pump effect", appears
to cause particle size changes resulting in an increase in
condensation nuclei counts. The increase in counts is related
to the concentration of SO_2. Radical scavengers supress particle
formation. They also supress the "pump effect" to different
degrees in according to the chemical composition of the radical
scavengers.

Acknowledgment

The work at the University of Illinois was funded in part by the
Department of Energy under contract DE ACO2 83ER60186. We would
also like to express our thanks to Ted Davis of the Denver Research
Center for his assistance in performing the experiments in the
Denver radon chamber.

References

Burke, T.P. and Scott, J.A., The Production of Condensation
Nuclei by Alpha Radiation, Proc. Roy. Irish Acad., 73:151-158 (1973).

Chamberlain, A.C., Heard, M.J., Penkett, S.A. and Wells, A.C.,
Supression of Radiolytic Nuclei in Air by Nitric Oxide, Health
Physics, 37:706-707 (1969).

Chamberlain, A.C., Megaw, M.J. and Wiffen, R.D., Role of Condensation Nuclei as Carrier of Radioactive Particles, Geofisica pura e applicata, 36:233-242 (1957).

Coghlan, M. and Scott, J.A., A Study of the Formation of Radiolytic Condensation Nuclei, Proc. Atmospheric Electricity (1983).

Droullard, R.F., Davis, T.H., Smith, E.E. and Holub, R.F., Radiation Hazard Test Facilities at the Denver Research Center, Bureau of Mines Information Circular (1984).

Droullard, R.F. and Holub, R.F., Continuous Working Level Measurements Using Alpha or Beta Detectors, BuMines, RI 8237, p 14 (1977).

Droullard, R.F. and Holub, R.F., Method of Continuously Determining Radiation Working Level Exposure, U.S. Pat. 4,185,199, Jan. 22 (1980).

Gusten, H., Filby, W.G. and Schoaf, S., Prediction of Hydroxyl Radical Reaction Rates with Organic Compounds in the Gas Phase, Atmospheric Environ., 15:1763-1765 (1981).

Holub, R.F. and Knutson, E.O., Measurement of Po-218 Diffusion Coefficient Spectra Using Multiple Wire Screens, this volume (1986).

Knutson, E.O., George, A.C., Hinchliffe, L. and Sextro, R., Single Screen and Screen Diffusion Battery Method for Measuring Radon Progeny Size Distributions, 1-500 nm, Presented to the 1985 Annual Meeting of the American Association for Aerosol Research, Albuquerque, NM, Nov. (1985).

Kulju, L.M., Chu, K.D. and Hopke, P.K., The Development of a Mobility Analyzer for the Studying of Neutralization and Particle-Production Phenomena Related to Radon Progeny, this volume (1986).

Leong, K.H., Hopke, P.K., Stukel, J.J. and Wang, H.C., Radiolytic Condensation Nuclei in Aerosol Neutralizers, J. Aerosol Science, 14:23-27 (1983).

Madelaine, G.J., Perrin, M.L. and Renoux, A., Formation and Evolution of Ultrafine Particles Produced by Radiolysis and Photolysis, J. Geophys. Res., 85:7471-7474 (1980).

Martell, E.A., Radioactive Emanations and Formation and Composition of Aerosols, Proc. Garmisch-partenkirchen Symposium on Radiation in the Atmosphere 44, Science Press., Princeton, NJ (1977).

Megaw, W.J. and Wiffen, R.D., The Generation of Condensation Nuclei by Ionizing Radiation, Geofisica pura e applicata, 50:118-128 (1961).

Perrin, M.L., Maigne, J.P. and Madelaine, G.J., Etude experimentale et theorique de l'evolution d'un aerosol de dimension inferieure a 0.02 μm, J. Aerosol Science, 9:429-433 (1978).

Reineking, A., Becker K.H. and Porstendorfer, J. ,Measurement of the Unattached Fractions of Radon Daughters in Houses, Presented to the Seminar on Exposure to Enhanced Natural Radiation and its Regulatory Implications, Maastricht, The Netherlands (1985).

Thomas, J.W., Measurement of Radon Daughters in Air, Health Physics, 23:782-789 (1976).

Vohra, K.G., Gas-to-Particle Conversion in the Atmospheric Environment by Radiation-Induced and Photochemical Reactions, Radiation Research, Acad. Press, New York, 1314-1325 (1975).

Vohra, K.G., Subbaramu, M.C. and Mohan Rao, A.M., A Study of the Mechanism of Formation of Radon Daughter Aerosols, Tellus, 18:672-678 (1966).

RECEIVED October 3, 1986

Chapter 28

Aerodynamic Size Associations of Natural Radioactivity with Ambient Aerosols

E. A. Bondietti[1], C. Papastefanou[2], and C. Rangarajan[3]

[1]Environmental Sciences Division, Oak Ridge National Laboratory, Oak Ridge, TN 37831
[2]Nuclear Physics Department, Aristotle University of Thessaloniki, Thessaloniki 540 06, Greece
[3]Division of Radiological Protection, Bhabha Atomic Energy Research Centre, Trombay, Bombay 400 085, India

The aerodynamic size distributions of Pb-214, Pb-212, Pb-210, Be-7, P-32, S-35 (as SO_4^{2-}), and stable SO_4^{2-} were measured using cascade impactors. The activity distribution of Pb-212 and Pb-214 was largely associated with aerosols smaller than 0.52 μm. Based on 46 measurements, the activity median aerodynamic diameter of Pb-212 averaged 0.13 μm (σ_g = 2.97), while Pb-214 averaged 0.16 μm (σ_g = 2.86). The larger median size of Pb-214 was attributed to α-recoil depletion of smaller aerosols following decay of aerosol-associated Po-218. Subsequent Pb-214 condensation on all aerosols effectively enriches larger aerosols. Pb-212 does not undergo this recoil-driven redistribution. Low-pressure impactor measurements indicated that the mass median aerodynamic diameter of SO_4^{2-} was about three times larger than the activity median diameter of Pb-212, reflecting differences in atmospheric residence times as well as the differences in surface area (Pb-212) and volume (SO_4^{2-}) distributions of the atmospheric aerosol. Cosmogenic radionuclides, especially Be-7, were associated with smaller aerosols than SO_4^{2-} regardless of season, while Pb-210 distributions in summer measurements were similar to sulfate but smaller in winter measurements. Even considering recoil following Po-214 α-decay, the average Pb-210-labeled aerosol grows by about a factor of two during its atmospheric lifetime. The presence of 5 to 10% of the Be-7 on aerosols greater than 1 μm was indicative of post-condensation growth, probably either in the upper atmosphere or after mixing into the boundary layer.

The decay of Rn-222 and Rn-220 in the atmosphere produces low vapor pressure progeny that coagulate with other nuclei or condense on existing aerosols. These progeny include 3.0-min (radioactive half-life) Po-218, 26.8-min Pb-214, and 10.6-h Pb-212. A

0097-6156/87/0331-0377$06.00/0

long-lived daughter in the Rn-222 decay chain, 22-year Pb-210, is
produced about an hour after attachment of Po-218. Cosmic ray
bombardment of gases in the atmosphere also produces condensable
radionuclides. For example, N_2 and O_2 are precursors of 53.3-d
Be-7, and Ar-40 gives rise to 87-d S-35 and 14-d P-32. The
atmospheric residence times of these radionuclides are controlled
either by radiodecay or aerosol removal.

In addition to these nuclear reactions, myriads of other
gas-phase transformations produce low-vapor pressure species, with
the oxidation of SO_2 and other reduced sulfur species dominating
aerosol formation and growth. Oxidation of SO_2 in the gas phase
produces H_2SO_4, a readily condensable species that either
combines with other molecules (new particle formation) or condenses
on existing aerosols.

While condensational growth is principally responsible for
aerosol growth up to several tenths of a micron (Hering and
Friedlander, 1982; McMurray and Wilson, 1983), oxidation of SO_2
also occurs in the aerosol phase through chemical reactions. This
mode of SO_2 oxidation is believed to be responsible for growth of
aerosols above several tenths of a micron (Hering and Friedlander,
1982; McMurray and Wilson, 1983), because growth rates increase
with aerosol diameter, in contrast to condensational growth, where
diameter growth decreases as diameter increases (Friedlander, 1977).

Aerodynamic Size Distributions of Naturally-Radioactive Aerosols.

Measurements of radionuclide distributions using cascade
impactors indicate that Be-7 and Pb-210 are associated with larger
aerosols than Pb-212 and Pb-214 (Röbig et al., 1980; Papastefanou
and Bondietti, 1986). Measurements of Pb-210 associations over
oceans indicated activity median aerodynamic diameters (AMAD) near
0.6 μm (Sanak et al., 1981). The impactor measurements of Moore
et al. (1980) on Pb-210, Bi-210, and Sr-90 sizes in continental air
indicated that about 80% of the activity from all three nuclides
was associated with aerosols below 0.3 μm. That work also
determined that the mean age of aerosol Pb-210 was about a week.
Knuth et al. (1983) compared Pb-210 and stable Pb sizes at a
continental location and found that 78% of the Pb-210 found below
1.73 μm was smaller than 0.58 μm. Young (1974) reported that
the most of the Be-7 in the atmosphere was associated with
submicron aerosols.

This paper summarizes part of the results of an investigation
designed to characterize the aerodynamic size distributions of
natural radioactivity and to evaluate the results in the context of
sulfate distributions and recent advances in the understanding of
aerosol growth mechanisms. This paper, while emphasizing our
results on Pb-212 and Pb-214, also summarizes our initial data for
longer-lived radionuclides.

Experimental Methods

Measurements on aerodynamic sizes of atmospheric aerosols and
associated radionuclides were carried out with Anderson 2000, Inc.,
1-ACFM Ambient Impactors with or without the Anderson low-pressure
modification, as well as with Sierra model 236 (six-stage)
high-volume impactors (HVI). The 1-ACFM design operated at 28

L min^{-1} (1 ft^3 min^{-1}). The stages had effective cutoff
diameters (ECDs) of 0.4, 0.7, 1.1, 2.1, 3.3, 4.7, 7.0, and
11.0 μm. The low-pressure modification, which alters the
impactor's operation by increasing the resolution in the submicron
region, involves a regulated flow rate of 3 L min^{-1}, five
low-pressure (114 mm Hg) stages for the submicron region and eight
atmospheric pressure stages for separating aerosols above
1.4 μm. The ECDs of the low-pressure stages were 0.08, 0.11,
0.23, 0.52, and 0.90 μm, whereas for the upper stages they were
1.4, 2.0, 3.3, 6.6, 10.5, 15.7, 21.7, and 35.0 μm. The stainless
steel plates supplied by the manufacturer were used for aerosol
collection. Either polycarbonate or glass-fiber backup filters
were used to collect all particles below the 0.08-μm collection
plate. The Sierra HVI had ECDs of 0.41, 0.73, 1.4, 2.1, 4.2, and
10.2 μm. No collection substrates were used since prelimary
comparisons of ^{7}Be sizes using two impactors operated in parallel
indicated that the use of glass fiber substrates shifts the median
size of ^{7}Be from about 0.40 μm (bare impactor surface) to about
0.6 μm (glass fiber). All impactors were operated at the flow
rates and pressures specified in the manufacturer's operating
manuals. The ECDs used for reporting distributions were those
reported by the manufacturer as we had no method for independently
calibrating the impactors using aerosols of the same geometry and
density as the ambient aerosol.

The length of each collection period varied from about 3 h
for Pb-214, to 30 or 40 h for Pb-212 and 1 to 14 d for Be-7,
Pb-210, S-35, and P-32, depending on impactor and objective. The
amount of air sampled with the LPI was usually less than 10 m^3,
with less than 0.1 mg total SO$_4^{2-}$ found to have deposited. The
samples were collected 13 m above the ground on the roof of the
Environmental Sciences Division building, Oak Ridge National
Laboratory.

The deposits on the stainless-steel collection plates of the
low-pressure impactor (LPI) were leached with a solution of 1 M
HNO$_3$ and the leachate rapidly evaporated on 5.08-cm stainless
steel plates (using a hot plate). The concentrations of Pb-212 and
Pb-214 in the impactor plate leachates were measured using seven
ZnS(Ag) alpha scintillation counters.

The HVI plates were leached with 0.1 M HCl. Be-7 was
measured using intrinsic germanium coaxial and well detectors.
Lead-210 was determined 30 days after collection stopped by
separating and measuring Bi-210 (Poet et al., 1972). When Pb-210
was measured, the upper two HVI stages were coated with a thin
layer of petroleum jelly to minimize soil particle bounce.

Sulfur-35 was measured by purifying the sulfate from the HVI
leachates using an alumina column (Veljkovic and Milenkovic,
1958). After eluting the sulfate from the alumina column with
NH$_4$OH, it was converted to H$_2$SO$_4$ using a small Dowex-50(H$^+$)
column. The resulting solution of dilute sulfuric acid was
concentrated and the β-activity determined using liquid
scintillation techniques. The purity of S-35 was checked by
following the decay rate of the isolated β-activity. Phosphorus-32
was isolated from 1 M HCl by precipitation of zirconium phosphate
(Mullins and Leddicotte, 1962). Nonradioactive sulfate was
determined by Dionex anion-exchange chromatography methods.

While nuclepore polycarbonate membranes (0.4 μm) were
preferred as backup filters when sulfate was to be measured,
glass-fiber filters had to be used in the HVI measurements.
Because of potential glass fiber sulfate artifacts, a separate
sampling for total sulfate was made using a polycarbonate filter.

Results

Pb-214 and Pb-212 Distributions. The α-activity recovered from
the LPI stages initially decayed at a 26.8-min half-life (i.e.,
Pb-214); after three hours, the rate approached that of 10.6-h
Pb-212 (Papastefanou and Bondietti, 1986). The measured alphas
were actually derived from the daughters of these nuclides. The
counting rate data were analyzed by differential decay rate
analysis after background correction to determine the Pb-214 and
Pb-212 activity of the samples at the stop sampling time. The
1 σ counting uncertainties were 5% or less for stages
corresponding to aerosols below 0.52 μm and <15% for the stages
collecting aerosols above 0.52 μm. Backgrounds averaged 0.025
cpm for the α-detectors. The calculated activities of Pb-214 and
Pb-212 were then used to determine the activity size distribution
of each radionuclide. Po-218 does not directly contribute alpha
particles during the measurements because of the long sampling time
(>3 h) and the 15-min minimum delay between the time sampling
stopped and the time counting began. Alphas from Pb-210 and other
nuclides were negligible as confirmed by repeated examination of
the count rates after the normal 24-h recording period.
        Examples of the observed activity size distributions of
Pb-214 and Pb-212 vs aerodynamic diameter (D_p) are presented by
the four subplots in Figure 1. These examples were selected from
46 measurements made over a ten-month period (Papastefanou and
Bondietti, 1986). About 46% of the measurements showed a
radioactivity peak in the 0.11- to 0.23-μm region (subplot a),
while 39% showed a peak in the 0.23- to 0.52 μm region (subplot
b). The remaining 15% of the measurements resulted in
distributions similar to those in subplot c (8.7%) or d (6.5%),
where Pb-214 and Pb-212 activities were highest in different size
ranges in the same spectrum. On the average about 76% of the
Pb-214 activity and 67% of the Pb-212 activity was found to be
associated with aerosols in the 0.08- to 1.4-μm size range. The
activity associated with aerosols smaller than 0.8-μm can also be
substantial, as indicated in Figure 1.
        Table I presents the average aerodynamic distributions of
Pb-212 and Pb-214, as well as the frequency with which Pb-214 or
Pb-212 was the dominant isotope in each size range. The Aitken
nuclei fraction (below 0.08 μm) contained a higher percentage of
Pb-212 activity compared with Pb-214 in 69.6% of the measurements.
The predominance of Pb-212 in this fraction is also illustrated by
the distributions reported in Figure 1. In the remaining
measurements, where Pb-214 was fractionally more abundant below
0.08 μm, the disparity between the relative amounts of each
isotope was not nearly as dramatic. Conversely, Figure 1 and
Table I illustrate that Pb-214 is generally enriched in the
accumulation mode aerosol, particularly between 0.11 and 0.52 μm,
where most of the surface area and mass occurs.

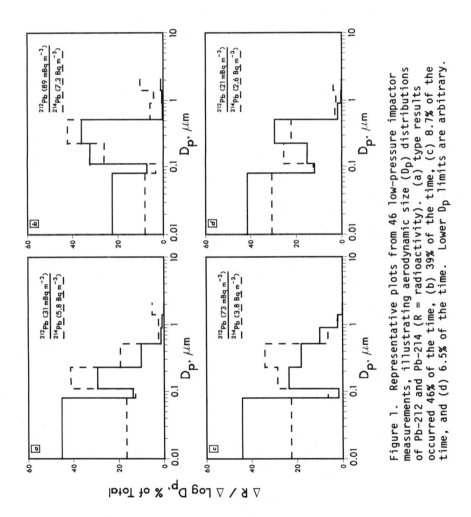

Figure 1. Representative plots from 46 low-pressure impactor measurements, illustrating aerodynamic size (D_p) distributions of Pb-212 and Pb-214 (R = radioactivity). (a) type results occurred 46% of the time, (b) 39% of the time, (c) 8.7% of the time, and (d) 6.5% of the time. Lower D_p limits are arbitrary.

Table I. Mean Observed Pb-212 and Pb-214 Distributions and Frequency of Dominance

	≤ 0.08 μm	0.08-0.11 μm	0.11-0.23 μm	0.23-0.52 μm	0.52-0.9 μm	0.09-1.4 μm
			Distribution of 44 measurements (%)			
Pb-212	32.6	9.9	25.9	27.0	3.02	1.58
Pb-214	21.3	11.4	27.4	30.5	4.12	2.63
			Frequency of dominance (%)			
Pb-214	28.2	52.2	65.2	58.7	73.9	71.7
Pb-212	69.6	47.8	34.8	41.3	23.9	23.9
Equal	2.2	0	0	0	2.2	4.4

The shift of Pb-214 to a slightly higher size distribution
compared to Pb-212 was also found using 1-ACFM and HVI impactors
(Fig. 2). The higher flow rates of these impactors, as well as the
ability to measure HVI activity by gamma spectroscopy, made us
confident that this shift was real and not a data analysis artifact.
The Pb-214 AMADs, determined with the LPI, varied from 0.10
to 0.37 μm (mean value, 0.16 μm). For Pb-212, they varied from
0.07 to 0.245 μm (mean value 0.12 μm) (Papastefanou and
Bondietti, 1986). These AMAD calculations were made assuming
lognormal distributions and using the manufacturer's cutoff
points. An abbreviated version of these results is presented in
Table II.

Pb-212 vs SO_4^{2-} LPI Distributions. Figure 3 presents a summary
of the average Pb-212 AMADs and SO_4^{2-} MADs (mass median
aerodynamic diameters) determined from a series of LPI measurements
made during the period January to October, 1985. The Pb-212 data
were derived from collections made at the same time as SO_4^{2-}
and from measurements made to compare Pb-212 vs Pb-214. The mean
aerodynamic diameter of Pb-212 was about three times smaller than
SO_4^{2-}. Much less sulfate was found in the aerosol fraction
below 0.08 μm, compared with Pb-212. While Pb-212 was largely
absent above 0.52 μm, about 20% of the SO_4^{2-} occurred above
this size.

Be-7, Pb-210, S-35, P-32 and SO_4^{2-}. The longer-lived
radionuclides were associated with larger aerosols than Pb-214 or
Pb-212. An example of these differences is presented by Figure 4,
which compares the distribution of Pb-212, Pb-210, SO_4^{2-}, Be-7,
S-35, and P-32 found on two-six stage HVIs operated continuously
for one week in February, 1985. Figure 4 illustrates that Pb-210,
the cosmogenic radionuclides, and SO_4^{2-} were associated with
larger aerosols than Pb-212. In most of our analyses, the fraction
of Be-7 associated with aerosols above 1.4 μm was usually between
5 and 10%; in this measurement 4.5% was found in the 1.4 to 2.1
μm size range, 1.1% in the 2.1 to 4.2 μm size range, and only
0.2% in sizes greater than 4.2 μm. Figure 4 also shows that
cosmogenic S-35, measured as SO_4^{2-}, does not have the same
aerosol distribution as stable SO_4^{2-}; this same result occurred
in two other measurements. The abundance of S-35 above 1.4 μm is
unknown; the S-35 and P-32 distributions presented in Figure 4b (as
well as the Pb-210 distributions in Figure 4a) represent only the
three collection stages below 1.4 μm, due to a current limitation
in the detection of low concentrations of these beta-emitters.
Table III summarizes the median aerodynamic diameters of
Pb-210, Be-7, and SO_4^{2-} found in measurements made through
March 1986. Be-7 distributions are substantially smaller than
SO_4^{2-}, regardless of the time of the year. The Pb-210 data,
while limited, suggests that summer sizes are larger than winter
sizes.

Pb-210 vs Pb-214 Distributions. Pb-210 is produced from the
α-decay of Po-214, the event used to quantify Pb-214
distributions on the LPI impactors. While the relationship between
the aerodynamic sizes of Po-214 and Pb-210 is complicated because

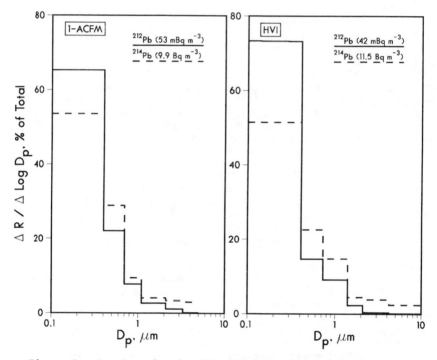

Figure 2. Aerodynamic size (D_p) distributions of Pb-212 and
Pb-214 activity (R) found with 1-ACFM and high-volume (HVI)
impactors illustrating the particle shift of Pb-214. Lower D_p
limits are arbitrary.

Table II. Summary of Mean Monthly Activity Median Aerodynamic
Diameters (AMAD) and Geometric Standard Deviations
(σ_g) of Radon and Thoron Daughter Size Distributions in
Ambient Aerosols

Period (month)	Number of samples	Pb–214		Pb–212	
		AMAD (μm)	σ_g	AMAD (μm)	σ_g
December 1984	2	0.16	2.35	0.15	2.80
January 1985	2	0.17	2.46	0.09	2.96
March	5	0.17	2.33	0.12	2.46
April	5	0.18	2.08	0.17	2.25
May	3	0.20	2.28	0.20	2.35
June 7	4	0.21	2.98	0.17	2.38
July	6	0.16	3.27	0.23	3.52
August	6	0.12	32.51	0.10	3.11
September	10	0.13	3.68	0.11	3.44
October	2	0.09	3.13	0.09	2.85
November	1	0.07	3.20	0.12	6.70
December	2	NR*	NR*	0.13	4.35

*NR = not recorded.

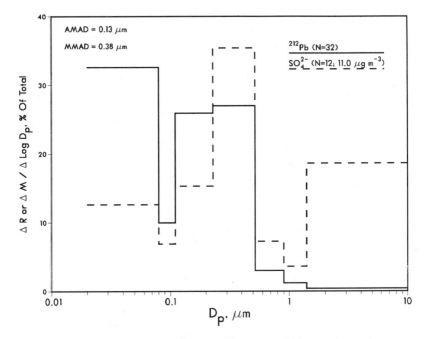

Figure 3. A comparison of mean Pb-212 activity (R), and
SO_4^{2-} mass (M) aerodynamic size (D_p) distributions from
low-pressure impactor measurements made between Feb. and Sept.
1985. Lower D_p limits are arbitrary.

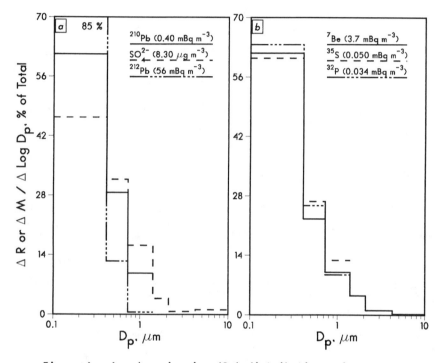

Figure 4. Aerodynamic size (D_p) distributions of radionuclides (R) and SO_4^{2-} (M) derived from a 7-d sampling (Feb. 1986) made using two high-volume cascade impactors. Pb-212, S-35, P-32, and Pb-210 were only measured on three stages (< 1.4 μm). Lower D_p limits are arbitrary.

Table III. Median Aerodynamic Diameters (MAD) and Geometric Standard
Deviations (σ_g) of Pb-210 and Be-7 Based on
Radioactivity and SO_4^{2-} Measured with
High-Volume Cascade Impactors

Month	MAD, μm (±σ_g)		
	Pb-210	Be-7	SO_4^{2-}
June 1985	–	0.50 (2.3)	0.50 (1.80)
July #1	0.49 (1.8)	0.30 (3.5)	0.49 (1.8)
July #2	–	0.31 (2.7)	0.38 (2.0)
July #3	–	0.48 (2.1)	0.48 (2.1)
August #1	0.40 (2.0)	0.36 (2.5)	0.41 (2.2)
August #2	0.40 (1.9)	0.30 (2.2)	0.40 (1.9)
September	–	0.34 (2.2)	0.45 (2.3)
November	–	0.34 (2.2)	0.45 (2.3)
December	–	0.32 (2.5)	0.58 (2.5)
January #1, 1986	–	0.32 (2.6)	0.42 (2.3)
January #2	0.32 (2.1)	–	–
February #1	0.32 (2.0)	–	–
February #2	0.36 (1.8)	0.34 (2.3)	0.43 (2.1)
March #1	0.28 (1.6)	0.32 (2.5)	–
March #2	–	0.29 (2.2)	0.41 (2.1)

of the large differences in their atmospheric lifetimes, Pb-210 has
always been found associated with aerosols larger than Po-214, as
indicated by the differences in AMADs reported in Tables II and III.

Discussion

Alpha-Recoil: An Explanation for the Pb-214 Distributions. The
longer half-life of Pb-212 compared to Pb-214 might favor the
presence of larger aerosol associations of Pb-212 if coagulation
rates are fast relative to radioactive decay rates, although
impactor measurements may not be sensitive enough to record this
effect.   Instead, the measurements indicated that the shorter-lived
chain (Po-218, Pb-214, etc.) was more often associated with larger
aerosol sizes than the longer-lived chain (Po-216, Pb-212, etc.).
The HVI measurements reported by Röbig et al. (1980) also indicated
a large particle shift of Pb-214 relative to Pb-212, although they
did not offer an explanation for the observation.

The Pb-214 shift might be explained by the fact that a
significant fraction of the 3.05-min parent of Pb-214 should attach
to an existing aerosol or coagulate with other nuclei during its
lifetime, as mean attachment half-lives are the order of a minute
less (Pörstendorfer and Mercer, 1980).   When this attached Po-218
α-decays, the recoiling Pb-214 daughter can escape the aerosol.
Complete recoil loss would occur if the diameter of the aerosol
were smaller than the range of the recoiling nucleus.   In water
this recoil range is 0.13 μm (Mercer, 1976) and should be
somewhat less in the atmospheric aerosol that has a density closer
to 1.5 g cm^{-3} (Herring and Friedlander, 1982).   By contrast, very
little of the 0.146 μs Po-216 would attach before decaying to
Pb-212 because of its short life relative to attachment times.   A
considerable fraction of the Pb-214 should undergo recoil
detachment, particularly from aerosols with diameters smaller than
0.1 μm (diameters approximating the recoil range).   The
probability of loss would decrease with increasing radius (Mercer,
1976).   If the recondensing Pb-214 behaves like the original
Po-214, the net effect would be a shift of Pb-214 to a larger size
distribution.

Since the subsequent β-decays of Pb-212 and Pb-214 + Bi-214
do not result in significant recoil (Mercer, 1976), the alpha
measurement of Po-214 and the Pb-212 daughters is in reality
tracing the aerosol distribution of a Rn-220 daughter atom (Pb-212)
which has condensed only once and a Rn-222 daughter atom (Pb-214)
that has probably condensed more than once.   This stability of the
Pb isotopes is the basis for our generic reference to Pb-212 and
Pb-214 distributions.

A Recoil Model That Accounts for The Pb-214 Shift.   The following
equations describe an empirical model that outlines the hypothesis
more formally.   First, the radon daughters are defined in terms of
their atmospheric distributions at some time during their life:

A = the fraction of the total Po-218 that attaches
         to any aerosol before it decays to Pb-214;
$1 - A$ = the fraction of the total Po-218 that decays
         before aerosol attachment;

R_i = the fraction of the Pb-214 that, through recoil, is lost from aerosol size interval i following Po-218 decay;

$1 - R_i$ = the fraction of Pb-214 that remains with aerosol size interval i following Po-218 decay;

f_i = fraction of the total unattached atoms (Po-218 or Pb-214) that condense on size interval i.

Then, it is assumed that f_i, the fractional distribution of condensing isotopes on size interval i, can be derived from the measured Pb-212 distributions (i.e., the half-life of Po-218 is short enough that the distribution of daughter Pb-212 represents the initial fate of condensing species).

For A, the fraction of the total Po-218 that attaches before decay, the recoiling daughter Pb-214 produced in any size interval can fractionate as follows:

$$Fe_i = Af_i R_i \tag{1}$$

where Fe_i is the fraction of the total Po-218 atoms that result in recoiling Pb-214 atoms that escape size interval i, and

$$Fn_i = Af_i (1 - R_i) \tag{2}$$

where Fn_i is the fraction of the total attached Po-218 atoms that result in recoiling Pb-214 atoms that do not escape size interval i. Summing all size intervals in terms of Equation (1) gives:

$$Fe_s = \sum_i Af_i R_i = A\sum_i f_i R_i \tag{3}$$

where Fe_s represents the sum of all Po-218 atoms that condense and then escape (as Pb-214) the atmospheric aerosol after decay.

Condensing Pb-214, while assumed to follow the same size distribution as observed for Pb-212, is derived from two sources which must be considered separately. First,

$$Fu_i = f_i (1-A) \tag{4}$$

where Fu_i is the fraction of the Pb-214 in size interval i that originated by condensation of Pb-214 that originated from the decay of Po-218 before condensation; and second,

$$Fd_i = f_i Fe_s \tag{5}$$

where Fd_i is the fraction of the total Pb-214 associated with size interval i that originated from the deposition of the Pb-214 that had escaped from all aerosol fractions (Fe_s) following decay of attached Po-218. These Pb-214 atoms, which originated by recoil detachment from the general aerosol population, are condensing for the second time. Because the probability of recoil detachment

decreases with increasing aerosol diameter, the greater depletion
of the smaller aerosols effectively results in a shift of the total
Pb-214 atoms to a higher size distribution.

From the above definitions the total Pb-214 atoms can be
accounted for as follows:

$$Ft_i = Fu_i + Fd_i + Fn_i \tag{6}$$

where Ft_i is the fraction of the total measured Pb-214 that
occurs in size interval i. Summing Equation (6) for all size
intervals gives:

$$Ft_s = \sum_i Ft_i \tag{7}$$

Equations 6 and 7 were solved for values of R_i and A that
gave the best fit between observed and calculated Pb-214
distributions. The various impactor stages were used for deriving
the size intervals. An example of the results of this model is
presented in Figure 5. In this calculation, A was assumed to be
0.963 (i.e., 96.3% of the Po-218 was assumed to attach to ambient
aerosols before decay to Pb-214). The best agreement between
calculated and measured Pb-214 distributions was found for the case
when calculated recoil losses (R_i) of 100, 70, 65, and 35%
occurred from the < 0.11-, 0.11- to 0.23-, 0.23- to 0.52-, and
0.52- to 0.90-μm size ranges, respectively. In this example, the
percentage of the total Pb-214 that underwent recoil detachment was
calculated to be 79.6%, very similar to values predicted in a
theoretical analysis (Mercer, 1976) and calculated from
experimental data obtained when atmospheric aerosols were exposed
to an enriched radon atmosphere (Mercer and Stowe, 1971). In
solving for the best fit between calculated and observed Pb-214
distributions, the <0.08-μm and 0.08- to 0.11-μm measurements
were combined to improve the fit. This simplification was made
because of (1) the low fractional contribution of the 0.08- to
0.11-μm region to the total activity, (2) the likelihood that
recoil losses will approach 100% in this size region (Mercer,
1976), and (3) the possibility that coagulative growth of Pb-212
out of the < 0.08-μm region might be significant during its
lifetime.

This model does not explicitly consider that a fraction of the
measured Pb-214 actually deposits in the impactor as
particle-associated Po-218. The Pb-214 daughters produced under
this condition would either not recoil off the plate or, if they
did, they might end up associated with a smaller size fraction on a
lower stage. In terms of both the model and the measurements, this
fraction of the total Po-218 is not operationally different from
the fraction which decays before attachment (1-A) or is not lost
following recoil; both represent Po-218 which does not undergo
recoil redistribution.

Estimating Recoil Redistribution of Pb-210 Following Po-214
a-Decay. In addition to Pb-214, Pb-210 would also undergo
α-recoil following decay of Po-214. Since the Po-214 is
separated from Pb-214 by only β-decays, the model parameters

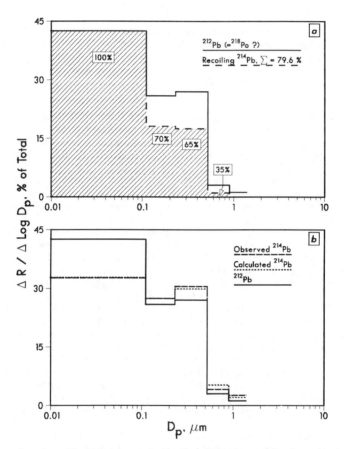

Figure 5. Results of an empirical model that calculated, for each low-pressure impactor size range, the percentage recoil losses of Pb-214 [subplot (a)] necessary to produce a calculated Pb-214 distribution is best agreement with the observed distribution [subplot (b)]. Lower D_p limits are arbitrary.

derived above can be used to calculate the aerodynamic size
distribution of Pb-210 that would result if condensation processes
(and recoil) alone affect radioactivity sizes.

For the values of R_i derived previously (Figure 5), the
amount of recoiling Pb-210 can be calculated using Equation 3,
substituting the measured Pb-214 distribution (Table I) for f_i.
The distribution of condensing Pb-210 can be derived from
Equation 5, where Fd_i now represents the fraction of the total
Pb-210 associated with size interval i that originated from Pb-210,
which had escaped from all aerosol fractions following Po-214
decay. The calculated Pb-210 distributions, using this model, are:
<0.11 μm, 31.7% (vs 32.7% for Pb-214; 0.11 to 0.23 μm, 26.2%
(vs 27.4% for Pb-214); 0.23 to 0.52 μm, 30.8% (vs 30.5% for
Pb-214; 0.53 to 0.9 μm, 5.1% (vs 4.1% for Pb-214); and 0.9 to 1.4
μm, 3.5% (vs 2.6% for Pb-214). The AMAD derived from this new
Pb-210 distribution is 0.18 μm, substantially lower than actually
measured, indicating that the presence of Pb-210 in larger aerosol
sizes must result from post-condensation growth.

Pb-212 and SO_4^{2-} Distributions. The aerodynamic size
distributions of Pb-212 and SO_4^{2-} were quite different,
reflecting the different dependencies of surface area and volume on
aerosol diameter (Friedlander, 1977). Pb-212, like the other
radionuclides, becomes associated with the atmospheric aerosol by
condensation or coagulation processes that are surface-area
related. Sulfate, on the other hand, is the main solute in the
accumulation mode aerosol so that its steady-state distribution is
proportional to volume, even though condensation of H_2SO_4 may
dominate its initial aerosol association. The difference also
reflects residence time: 16.7 h for Pb-212 + Bi-212 and about a
week for sulfate.

Pb-210 and the Cosmogenic Radionuclides. We noted earlier that our
measurements of Pb-214 were really measurements of Po-214 decay,
that is, the production of Pb-210. The mean AMAD of these
measurements was about 0.16 μm, with the AMAD of Pb-210 predicted
to be ~0.18 μm after recoil. However, the summer AMAD of
Pb-210, after aging in the atmosphere for about a week (Moore et
al., 1980), was closer to 0.4 μm, indicating that Pb-210's AMAD
approximately doubles during its lifetime in the atmosphere. The
limited measurements reported here suggest that the AMAD of Pb-210
is smaller in winter than summer, possibly reflecting differences
in aerosol growth rates. The summer measurements were also not
different from simultaneous SO_4^{2-} measurements.

The cosmogenic radionuclides, particularly Be-7, are
associated with slightly smaller aerosols than sulfate, both in
winter and summer. The reasons for this difference may be either
(1) faster growth rates of secondary SO_4^{2-} aerosols in plumes
from sources or (2) the direct contribution of primary sulfate
aerosols from these sources. Coarse aerosol sulfate (above several
μm) may be a factor in some cases. By contrast, the cosmogenic
nuclides are produced in the stratosphere and upper troposphere and
mix into the boundary layer with considerable week-to-week
fluctuations (Rangarajan and Gopalakrishnan, 1970). While sizes
distributions in the upper atmosphere have not been measured,

artificial radioactivity from high-altitude nuclear tests was found
associated with aerosols about 0.1 to 0.2 μm in size (Martell,
1962).

Radionuclide Sizes Relative to Growth Mechanisms
of Sulfate Aerosols.

Hering and Friedlander (1982), in a study of
the Los Angeles Basin sulfate aerosol, observed the occurrence of
two categories of sulfate distributions: one with a mass median
diameter of 0.54 $\pm$ 0.07 μm and one with a mass median diameter of
0.20 $\pm$ 0.02 μm. Based on a growth model, they concluded that
condensational processes dominated the formation of the small
distributions. At much higher sulfate loadings than found in our
measurements (i.e., 11 μg m^{-3}, Figure 4), they concluded that
coagulation times of several weeks were required to increase
aerosol volume distributions from 0.25 to 0.50 μm. Because air
masses have much shorter residence times in the Basin, they
concluded that distributions near 0.5 μm were the result of rapid
growth in the liquid phase.

McMurray and Wilson (1982; 1983) examined aerosol growth in
urban plumes and remote locations and concluded that both
condensation and droplet-phase reactions contributed to aerosol
growth, with humidity and sunlight important variables. For
example, an analysis of measurements on the St. Louis, Missouri,
urban plume indicated that 75% of secondary aerosol volume
formation was attributed to condensation and 25% to droplet-phase
reactions. They further concluded that in humid climates,
droplet-phase growth was responsible for the presence of submicron
volume distributions peaking in the 0.3- to 0.5 μm-dia range; in
arid climates the presence of submicron volume distributions, which
peaked near 0.2 μm, was considered to be due to
condensation-dominated growth.

Our measurements on radionuclide sizes can be evaluated in
terms of these mechanisms, although the significance of
droplet-phase growth remains uncertain. Lead-212 distributions
reflect, of course, condensational growth. The rate at which mass
(as Pb-212 atoms) deposits to the aerosol decreases with increasing
diameter, as predicted by the growth law (McMurray and Wilson,
1982). The Pb-214 distributions also reflect condensational growth
but with the added complication of recoil. The oxidation rate of
SO_2 is at a minimum during winter, affecting both the rate of new
particle production and the growth of aerosols by chemical
reactions in the liquid phase. Therefore, the growth of Pb-210
from less than 0.2 μm to slightly over 0.3 μm during the winter
may establish a lower limit for growth by coagulation during the
approximately one week Pb-210 resides in the atmosphere.

McMurray and Wilson (1982) calculated aerosol growth rates by
determining the change in diameter (D_p) with respect to time. By
analogy, the growth of Pb-210 in the atmosphere can be calculated
by dividing the ~0.2 μm change in AMAD by the mean residence
time of Pb-210. The resulting growth rate is approximately 0.001
μm h^{-1}, indicating growth by condensation or coagulation
(McMurray and Wilson, 1982; 1983). We also conclude that the upper
limit of aerosol growth is between 2 and 4 μm, as indicated by
the Be-7 results. Additional insights into aerosol growth rates
may be gained by evaluating the ages of Pb-210 on the 0.73- to

2.1-μm size range (stages 5 and 4 of the Sierra 236 HVI) using the Bi-210 in-growth method (Poet et al., 1972; Moore et al., 1980). Very little Pb-214 condenses in this size range so that the age of those fractions should reflect post-condensation growth aging. This approach may be especially useful during the summer when growth rates are fast. Measurements must be done in climates where soil resuspension is minimal, however, and the fractional distribution of Be-7 should be used as a guide for evaluating if Pb-210 distributions can be safely attributed to a condensation origin.

The rates and mechanisms of aerosol growth in the atmosphere are obviously quite complicated. Natural radionuclides are unique tracers in that they can show (1) where condensational growth is occurring (Pb-212), (2) the average growth rate that occurs from 0.2 to 0.4 μm during (Pb-210), and (3), how growth rates differ with respect to season and entry point into the mixed layer (Be-7 and Pb-210). The presence of Be-7 as a global tracer of aerosols mixing into the boundary layer also suggests that the influences of humidity, aerosol precursor concentrations, and other variables can be compared geographically. For example, we observed much larger Pb-210 aerosol sizes than reported by Moore et al. (1980) in Colorado. Is this a humidity or pollution effect or a methodology difference? Sanak et al. (1981) found larger Pb-210 aerosols in the marine boundary layer. Why? Was it because they used glass-fiber substrates on each impactor stage or do aerosols attain different sizes over the oceans? More systematic comparisons are obviously needed.

Summary

The aerodynamic size distributions of Pb-214, Pb-212, Pb-210, Be-7, P-32, S-35-SO_4^{2-}, and stable SO_4 were measured using cascade impactors. Pb-212 and Pb-214, measured by alpha spectroscopy, were largely associated with aerosols small than 0.52 μm. Based on over 46 low-pressure impactor measurements, the mean activity median aerodynamic diameter (AMAD) of Pb-212 was found to be 0.13 μm, while for Pb-214 the AMAD was larger—0.16 μm. The slightly larger size of Pb-214, confirmed with operationally different impactors, was attributed to α-recoil-driven redistribution of Pb-214 following decay of aerosol-associated Po-218. A recoil model was presented that explained this redistribution. Low-pressure impactor measurements indicated that the mass median aerodynamic diameter of SO_4^{2-} (0.38 μm) was about three times larger than the activity median diameter of Pb-212, reflecting differences in atmospheric residence times as well as the differences in surface area (Pb-212 and volume (SO_4^{2-}) distributions of the atmospheric aerosol. Cosmogenic radionuclides, especially Be-7 and S-35, were associated with smaller aerosols than SO_4^{2-} regardless of the time of year. Lead-210 distributions were similar to sulfate in summer measurements but smaller in winter measurements. Even considering recoil following Po-214 alpha decay, the average Pb-210-labelled aerosol grows by about a factor of two during its atmospheric lifetime. Five to ten percent of the Be-7 occurred on aerosols larger than 1 μm, indicating that post-condensational growth is significant.

Acknowledgments

Research sponsored by the Office of Health and Environmental
Research, U.S. Department of Energy, under Contract No.
DE-AC05-840R21400 with Martin Marietta Energy Systems, Inc.
Environmental Sciences Division, Publication 2770.

References

Friedlander, S. K., Smoke, Dust, and Haze, John Wiley & Sons, Inc.,
New York, New York (1977).

Hering, S. V., and S. K. Friedlander, Origins of Aerosol Sulfur
Size Distributions in the Los Angeles Basin, Atmos. Environ.
11:2647-2656 (1982).

Knuth, R. H., E. O. Knutson, H. W. Feely, and H. L. Volchock, Size
Distributions of Atmospheric Pb and Pb-210 in Rural New Jersey:
Implications for Wet and Dry Deposition, in Precipitation
Scavenging, Dry Deposition, and Resuspension, Vol. 2, pp.
1325-1336, Elsevier Science Publishers (1983).

Martell, E. A., The Size Distribution and Interaction of
Radioactive and Natural Aerosols, Tellus, 18:486-498 (1962).

McMurray, P. H. and J. C. Wilson, Droplet Phase (Heterogeneous) and
Gas Phase (Homogeneous) Contributions to Secondary Ambient Aerosol
Formation As Functions of Relative Humidity, Atmos. Environ.,
16(1):121-134 (1982).

McMurray, P. H. and J. C. Wilson, Growth Laws for the Formation of
Secondary Ambient Aerosols: Implications for Chemical Conversion
Mechanisms, J. Geophy. Res., 88(C9):5101-5108 (1983).

Mercer, T., The Effect of Particle Size on the Escape of Recoiling
RaB Atoms from Particulate Surfaces, Health Phys. 31:173-175 (1976).

Mercer, T. and W. A. Stowe, Radioactive Aerosols Produced by Radon
in Room Air, in Inhaled Particles II, The Gresham Press, Surry,
England (1971).

Moore, H. E., S. E. Poet and E. A. Martell, Size Distributions and
Origin of Pb-210, Bi-210, and Po-210 on Airborne Particles in the
Troposphere, in Natural Radiation Environment III, (T. F. Gesell
and W. M. Lowder, eds.), CONF-780422, Vol, 1, pp. 415-429, National
Technical Information Service, Springfield, Virginia (1980).

Mullins, W. T. and G. W. Leddicotte, The Radiochemistry of
Phosphorus, NAS-NS 3056, National Technical Information Service,
Springfield, Virginia (1962).

Papastefanou, C. and E. A. Bondietti, Aerodynamic Size Associations
of Pb-212 and Pb-214 in Ambient Aerosols, submitted to Health Phys.
(1986).

Poet, S. E., H. E. Moore and E. A. Martell, Lead-210, Bi-210, and Polonium-210 in the Atmosphere: Accurate Ratio and Application to Aerosol Residence Time Determination, J. Geophys. Res., 77:6515-6527 (1972).

Porstendörfer, J. and T. Mercer, Diffusion Coefficient of Radon Decay Products and their Attachment Rate to the Atmosphere Aerosol, in Natural Radiation Environment III, (T. F. Gesell and W. M. Lowder, eds.), CONF-780422, Vol. 1, pp. 281-293, National Technical Information Service, Springfield, Virginia (1980).

Rangarajan, C. and S. S. Gopalakrishnan, Seasonal Variation of Beryllium-7 Relative to Cesium-137 in Surface Air at Tropical and Subtropical Latitudes, Tellus 22:115-120 (1970).

Röbig G., K. H. Becker, A. Hessin and J. Porstendörfer in Proc. 8th Conference on Aerosol Sci., pp. 96-102, George-August-University, Gottingen, West Germany (1980).

Sanak, J., A. Gaurdy and G. Lambert, Size Distributions of Pb-210 Aerosols over Oceans, Geophys. Res. Lett., 8(10):1067-1070 (1981).

Veljkovic, S. R., and S. M. Milenkovic, Concentration of Carrier-free Radioisotopes by Adsorption on Alumina, in Proc. of the 2nd UN Conference on Peaceful Uses of Atomic Energy. Vol. 20, pp. 45-49, United Nations (1958).

Young, J. A., The Particle Size Distribution of Man-made and Natural Radionuclides, in Pacific Northwest Laboratory Annual Report for 1973, Part 3, pp. 16-17 (atmospheric sciences), Richland, Washington (1974).

RECEIVED September 18, 1986

HEALTH EFFECTS

Chapter 29

A Reconsideration of Cells at Risk and Other Key Factors in Radon Daughter Dosimetry

A. C. James

National Radiological Protection Board, Chilton, Didcot, Oxfordshire OX11 0RQ, United Kingdom

In assessing dose to lung from radon daughter exposure, it is appropriate to consider secretory cells as well as basal cells as sensitive targets; hence, it is more defensible to average dose over the whole thickness of bronchial epithelium. The mean bronchial absorbed dose calculated in this way is higher by about 60% for the unattached fraction of potential alpha-energy and about 30% for the attached aerosol than the dose to basal cells alone.

The data on bronchial dimensions, airway deposition and clearance of radon daughters are reviewed and a model consistent with these data is formulated. This gives dose conversion factors of 130 mGy per WLM exposure for unattached radon daughters and approximately 8 mGy per WLM for the aerosol fraction. The latter value is sensitive to the particle size distribution, varying inversely with aerosol AMD. Breathing rate and age are shown to be minor factors in determining lung dose. Assuming that radon daughter aerosols do not grow in size at physiological humidity, one finds that typical exposure conditions in the home may be assessed by an effective dose equivalent of 15 mSv per WLM.

Consideration of the inverse variation of equilibrium factor and unattached fraction in room air leads to a relatively constant dose per unit concentration of radon gas, which facilitates the interpretation of monitoring data.

The risk of lung cancer from exposure to radon daughters in homes is derived by assessing lung dose, either absolutely by evaluating an effective dose equivalent (UNSCEAR, 1982; NEA, 1983) or by scaling the

0097-6156/87/0331-0400$06.00/0
Published 1987 American Chemical Society

excess rate of lung cancer incidence in uranium miners in proportion to dose rates in mines and homes (NCRP, 1984; Jacobi and Paretzke, 1985; Ellett and Nelson, 1985).

Lung dose must be calculated by modelling the sequence of events involved in inhalation, deposition, clearance and decay of radon daughters within the bronchial airways. Various models have shown that dose for a given exposure depends on environmental and personal factors, principally the aerosol size distribution, breathing rate and lung size. An improved understanding of the physical behaviour and size distribution of radon daughter aerosols in room air is emerging (Porstendorfer, 1984), supported by an increasing body of data. It is therefore worthwhile reconsidering the influence of environmental conditions on lung dose and taking this opportunity to incorporate improved knowledge of the parameters and processes involved in modelling that results from recent research.

Realistic models of lung dose proceed by defining the tissue at risk and the location of sensitive cells. The 'basal cells' lining the basement membrane of bronchial epithelium are generally assumed to be targets for cancer induction, but a different consensus is now emerging. There are also differences of opinion on the extent of sensitive bronchial tissue. The concept of an effective dose equivalent calls for dose to be averaged over the whole tissue (NEA, 1983), whereas some researchers prefer to focus attention on dose to a few larger airways where the majority of cancers are thought to occur (NCRP, 1984; Ellett and Nelson, 1985). These fundamental issues are being considered by a Task Group set up by the ICRP to review the current lung model (Bair, 1985). In this paper I shall attempt to formulate a model for radon daughter dosimetry consistent with the Task Group's present views and to examine the practical import.

Cells at risk

The histological types of lung cancer seen to excess in uranium miners reflect those in the population at large (Masse, 1984). These occur almost entirely in bronchial airways. Approximately 20% are adenocarcinomas which occur in peripheral bronchioles (Spencer, 1977) where there are no basal cells. Squamous cell cancers predominate in miners exposed early in life to relatively low concentrations of radon daughters (Saccomanno et al., 1982). These are considered likely to arise from the secretory small mucous granular cells which undergo cell division and extend to the epithelial surface (Masse, personal communication). Division of these cells is accelerated after irritation by toxicants such as cigarette smoke or infectious diseases (Trump et al., 1978).

It is reasonable to conclude that dose to cells throughout the bronchial tree may contribute to the risk of lung cancer and not just the dose received by certain cells in the large central airways. It is probably also appropriate to evaluate the dose absorbed by cells throughout the depth of bronchial epithelium, i.e. the mean dose,

rather   than   focus   on   basal   cells   alone   (Masse,   1984).   These
conclusions underlie the revised dosimetry developed below.

Dose per disintegration

The dose received by cells in the bronchial epithelium from decay of
the alpha-emitting radon daughters at the airway surface decreases
rapidly with depth in tissue. The data available show that the
thickness of bronchial epithelium is highly variable (Gastineau et
al., 1972) but the distribution of values can be formalised (Wise,
1982) by the summary shown in Table I.

Table I. Normal Distribution of Epithelial Thickness (μm)

Airways	Mean	Standard Deviation
Main bronchi	80	6
Lobar bronchi	50	12
Segmental bronchi	50	18
Transitional bronchi	20	5
Bronchioles	15	5

Figure 1 shows the range of doses to basal cells calculated for
one alpha-decay of RaA or RaC' per cm^2 of airway surface in each
bronchial generation. The hatched areas show the effect of considering
radon daughter activity in the mucosa, i.e. the airway wall, rather
than entirely in mucus (minimum values) (James et al., 1980; NEA,
1983). The location of radon daughter decays relative to basal cells
is clearly important in the upper airways, including the segmental
bronchi (generations 3-5) which are commonly regarded as a major site
of lung cancer (NCRP, 1984; Ellett and Nelson, 1985). The Figure also
shows the values adopted by the NCRP (Harley and Pasternack, 1982)
which assume a target cell depth of 22 μm in the upper airways and 10
um in the bronchioles and the range of values used in other dosimetric
models (Jacobi and Eisfeld, 1980; Hofmann, 1982). These different
dosimetric factors, arising from different assumptions about target
cell depth, should be taken into account when comparing the results of
various models.

The conversion factor varies much less when the mean dose to all
epithelial cells is evaluated (Figure 2). This is especially marked
for RaC' decays which contribute most of the dose. In this case, very
similar doses are calculated if the complex depth distributions of
Table I are represented by a single epithelial thickness of 50 μm in
the bronchi, i.e. generations 1-10, and 15 μm in bronchioles.

Lung models

Most dosimetry models have incorporated the so-called Weibel 'A'
airway dimensions (Weibel, 1963) in order to calculate aerosol
deposition, clearance and the density of alpha-decays per unit surface

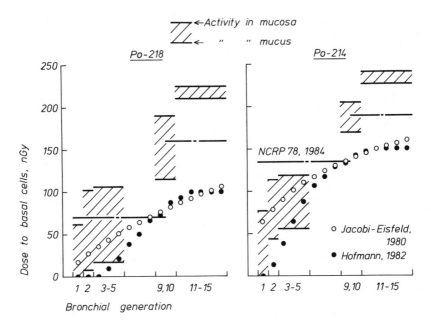

Figure 1. Doses to basal cells in each bronchial generation
from 1 disintegration of RaA or RaC' per cm^2 airway surface.

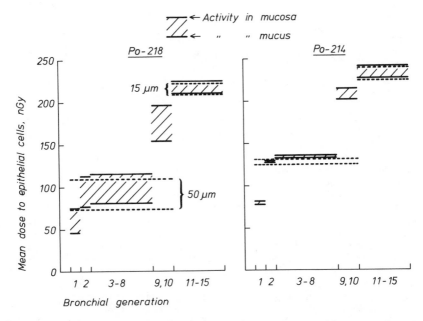

Figure 2. Doses averaged over all epithelial cells in each
bronchial generation from 1 disintegration of RaA or RaC'
per cm^2 airway surface.

area. The Yeh-Schum airway model (Yeh and Schum, 1980) has also been used (James et al., 1980; Harley and Pasternack, 1982) and dosimetric results from both models have been compared (NEA, 1983; NCRP, 1984). A further model of the bronchial tree, referred to here as 'UCI', has been reported more recently (Phalen et al., in press). This includes novel data for children's lungs at various ages.

Both the Weibel 'A' and Yeh-Schum models need to be reduced in scale to represent adult human lung at a normal level of inflation corresponding to 3000 ml functional residual capacity (FRC) (Yu and Diu, 1982). Partial scaling has been included in some dosimetric models (NEA, 1983; James, 1984) but not in others. In all cases the airway sizes used to represent adult lung correspond to a higher level of inflation than the standard FRC, leading to general but relatively small underestimates of bronchial dose.

The volumes and surface areas of airways in each generation that result from scaling the Weibel 'A', Yeh-Schum and UCI lung models to the standard FRC (Yu and Diu, 1982), are shown in Figure 3. The residual differences in airway size are appreciable, but there is no overriding reason to prefer a particular model. Dosimetric results are therefore evaluated below for all three.

Deposition

Diffusion is the dominant mechanism of lung deposition for radon daughter aerosols. It is generally assumed that airflow is laminar in the smaller airways and that deposition in each airway generation can be calculated adequately (Chamberlain and Dyson, 1956; Ingham, 1975). However, there is no such consensus on the treatment of deposition in the upper bronchi. Some authors (Jacobi and Eisfeld, 1980; NCRP, 1984) have considered deposition to be enhanced by secondary flow, on the basis of experimental results (Martin and Jacobi, 1972). It has been shown that this assumption reduces the calculated dose from unattached radon daughters by a factor of two (James, 1985).

The deposition of sub-micron aerosols in a hollow cast of human bronchi has recently been measured under realistic conditions (Cohen et al., in press). Typical data are shown in Figure 4. These are inconsistent with convective enhancement of deposition but support the classical treatment of deposition by diffusion (Chamberlain and Dyson, 1956).

The deposition model used here includes expressions for diffusion (Ingham, 1975) sedimentation (Pich, 1972) and impaction (Egan and Nixon, 1985) and a realistic treatment of lung ventilation. It can be shown that this predicts the aerosol deposition measured in the lungs of human subjects (summarised by Rudolf (1986)) over the range of aerosol size from 5 nm to 5 μm diameter, and for all breathing conditions tested, to within 20% of measured values.

Loss by filtration in the nose or mouth of radon daughter activity attached to aerosols can be neglected. The few data available for unattached daughters (George and Breslin, 1969) indicate that about 50% of activity inhaled through the nose is lost but only a negligible amount during mouth breathing. These values are consistent with human data obtained recently for a range of particle sizes down to 5 nm diameter (Schiller, 1986) which suggest that nasal deposition efficiency increases with the square root of the particle diffusion

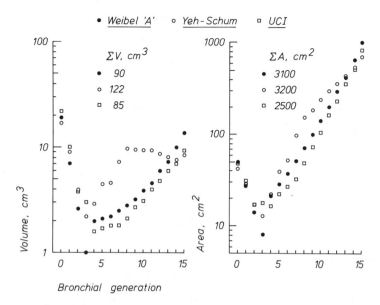

Figure 3.   Volumes and surface areas of airways in each
bronchial generation of the Weibel 'A', Yeh-Schum and UCI lung
models, with total volumes ΣV and areas ΣA.

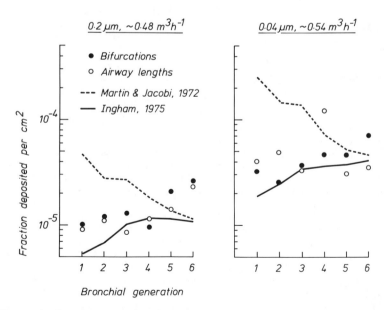

Figure 4. Deposition of submicrometer aerosols measured for
cyclic flow through a hollow cast of the human bronchial
tree, compared with calculated values.

coefficient, under constant breathing conditions. However, the
estimation of nasal filtration of unattached radon daughters for
different breathing rates and ages remains uncertain. Knowledge of the
mechanism involved is needed.

Clearance

Radon daughters are deposited on the surface of mucus lining the
bronchi. It is generally assumed that the daughter nuclides, i.e.
polonium-218 (RaA), lead-214 (RaB) and bismuth-214 (RaC), remain in
the mucus and are transported towards the head. However, one
dosimetric model assumes that unattached radon daughters are rapidly
absorbed into the blood (Jacobi and Eisfeld, 1980). This has the
effect of reducing dose by about a factor of two. Experiments in which
lead-212 was instilled as free ions onto nasal epithelium in rats have
shown that only a minor fraction is absorbed rapidly into the blood
(Greenhalgh et al., 1982). Most of the lead remained in the mucus but
about 30% was not cleared in mucus and probably transferred to the
epithelium.
    The absorption characteristics of radon daughters remain somewhat
uncertain, as do the rates of mucous clearance at various levels in
the bronchial tree. Accordingly, the effect on calculated doses of a
range of assumed clearance behaviour is examined below. It is
considered that the following postulates determine the possible range
of doses:

    (i)   insoluble daughters totally transported in mucus
    (ii)   insoluble daughters with no clearance
    (iii) partially soluble daughters with 30% transferred to
           mucosal tissue.

Dose per unit exposure

Figure 5 shows, on logarithmic scales, absorbed doses averaged over
all epithelial cells in the bronchial and alveolar regions of the
lung, for the wide range of aerosol size associated with radon
daughters in room air. The size range can be divided into two bands,
corresponding to unattached and attached daughters. The aerosol
parameter determining dose is the AMD (activity median diameter). It
is assumed that free atoms and small ion clusters are 1 nm and 3 nm in
diameter, respectively. Larger attached species are assumed to be
distributed in particle size, with geometric standard deviations of
1.5 at an AMD of 0.01 µm and 2.0 for AMDs larger than 0.05 µm. A
breathing rate of 0.75 m³ per h is assumed to characterise domestic
exposure of adult males (UNSCEAR, 1982; NEA, 1983).
    Doses calculated using the Weibel 'A' or UCI lung dimensions are
uniformly about 30% higher than values obtained with the Yeh-Schum
model (Figure 5). The bands of dose plotted here show variation due to
the three different assumptions of clearance behaviour. The effect of
clearance is entirely negligible for radon daughters attached to
aerosols, but for unattached daughters, variability in bronchial dose
due to clearance amounts to about ± 30%. Larger uncertainty could
possibly be introduced by habitual mouth breathing. Conversely,

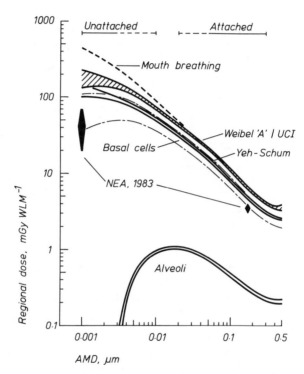

Figure 5.   Doses averaged over all epithelial cells in the bronchial and alveolar regions of the lung per unit exposure to potential alpha-energy as a function of aerosol size, compared with doses to basal cells; for several models of airway size and clearance behaviour.

bronchial dose from unattached daughters is relatively independent of particle size, but the dose decreases markedly as the aerosol AMD becomes larger. Dose to alveoli is always less than 10% of that to the bronchi and can be safely neglected.

The range of doses calculated when only basal cells are assumed at risk is also shown in Figure 5. For unattached daughters, doses are approximately one half and for attached daughters three quarters of values derived by averaging over all cells. These doses are to be compared with the range derived by the NEA (NEA, 1983). The reference values recommended by the NEA and adopted by UNSCEAR (UNSCEAR, 1982) lie at the bottom of the range of doses to basal cells derived here.

Table II lists values of mean bronchial dose obtained by weighting equally the results from all three lung models and clearance assumptions. I propose these as reference values.

Table II. Reference Values of Mean Bronchial Dose from Exposure to 1 WLM Potential Alpha-energy in Homes

Aerosol size (AMD, μm)	Absorbed dose (mGy per WLM)
Unattached	130
0.05	20
0.1	10
0.15	7
0.2	5
0.3	4
0.4 - 0.5	3

It can be concluded that mean bronchial dose is inversely proportional to the AMD of the inspired aerosol over the size range 0.05 - 0.2 μm. This corresponds to the range reported for room air under normal conditions (Reineking et al., 1985). However, it is not certain that the size of radon daughter aerosols is stable once inspired. It is quite possible that they could grow rapidly to roughly double their ambient size by absorbing water (Martonen and Patel, 1981). In absolute terms, therefore, bronchial dose from the attached fraction of radon daughter activity could be as low as half that indicated by environmental measurements. It would be useful to resolve this uncertainty experimentally.

Figure 6 shows doses averaged over epithelial cells in the segmental bronchi, i.e. generations 3 - 5. In this case the Yeh-Schum and UCI lung dimensions give similar results, whereas the Weibel 'A' model alone gives results about 30% higher. The dose to shallow basal cells assumed by the NCRP (NCRP, 1984) to represent exposure to the domestic radon daughter aerosol (0.125 μm diameter) lies within the range of values calculated here. Doses averaged over all epithelial cells in segmental bronchi are seen in Figure 6 to be uniformly double those to basal cells only. In the case of unattached daughters, the latter are similar to the value of about 140 mGy per WLM assumed by the NCRP.

It can be seen by comparing Figures 5 and 6 that doses averaged over all cells or to basal cells alone are within a factor two of each other. Thus, irrespective of aerosol size, doses to segmental bronchi are similar to the average value for the whole bronchial region. Within certain bounds, the effect of increasing the size of carrier aerosol particles, i.e. the attached fraction of radon daughter activity, is to reduce lung dose proportionally. The position is more complex for unattached daughters. Increasing aerosol size in this range serves to reduce proportionally dose to the upper airways, i.e. segmental bronchi. However, this has the effect of increasing deposition and dose in smaller airways, maintaining the mean bronchial dose more or less constant.

Influence of breathing rate. Figure 7 relates average doses to epithelial cells at various breathing rates. Doses are normalised to values calculated at each aerosol size for an adult male breathing at the reference rate of 1.2 m^3 per h (c.f. a miner) (ICRP, 1981). Apart from the values at unusually large aerosol sizes, results for both the bronchial region as a whole and for segmental bronchi are remarkably constant at each breathing rate. The latter are represented well by the fine broken lines which show values proportional to the square root of breathing rate. This relationship results from the combination of intake rate (simply proportional to breathing rate) and deposition probability, which has been shown in human subjects to be proportional to the square root of residence time in the lung when deposition occurs predominantly by diffusion (Gebhart and Heider, 1985).

It can be concluded that uncertainties in breathing rate are relatively unimportant in determining lung dose and that dose varies with particle size in much the same manner over the whole physiological range of breathing rates.

Age dependence

The main parameters influencing lung dose as a function of age are the breathing rate and lung dimensions. Breathing rates have been assessed from data on dietary intake as a function of age (Adams, 1981). It is assumed that oxygen consumption is proportional to energy expenditure and thus dietary intake. The following expression has been developed (Adams, 1981) to relate the energy expenditure rate R in a child of age t (years) and mass M(t) (kg) to that of an adult:

$$R = (M(t)/70).\exp[0.047(21 - t)]. \qquad t \leq 21$$

Using this expression, with data on body weight as a function of age (Altman and Dittmer, 1972), one gets the breathing rates relative to adult values shown in Figure 8. These correspond to rates of 0.095, 0.34, 0.48 and 0.61 m^3 per h at birth, 2, 6 and 10 years, respectively, relative to an adult rate of 0.75 m^3 per h. The NEA and NCRP have assumed generally lower rates of intake by children (NEA, 1983; NCRP, 1984).

The UCI lung model (Phalen et al., in press) includes for the first time measurements of airway diameter and length throughout the bronchial tree in 20 lungs taken from children or young adults at various ages, including neo-natal specimens. Linear regressions of

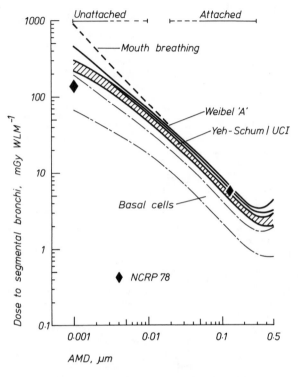

Figure 6.   Doses averaged over epithelial cells in segmental bronchi per unit exposure to potential alpha-energy.

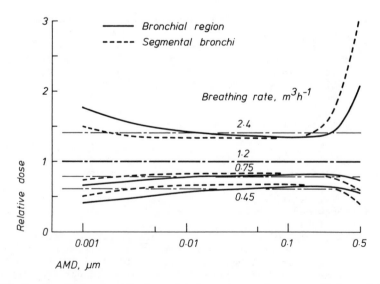

Figure 7.   Doses to the bronchial region and segmental bronchi as a function of breathing rate, relative to values for miners breathing at 1.2 $m^3 h^{-1}$.

both diameter and length were reported for the trachea and 15 bronchial generations with respect to height. These functions can be used to scale for age the airway dimensions in each of the three models of adult lung discussed earlier. The resulting scaling factors for diameter and area are shown in Figure 9.

It is seen that the diameters of bronchioles (averaged over generations 11 - 15) vary little with age. The increase in bronchial size is greater, but still less than might be expected if airways are simply scaled for overall body dimensions (illustrated by the dashed curves in Figure 9, which are functions of body weight W). Since bronchiolar diameter does not change much with age it is likely that the thickness of bronchiolar epithelium is also relatively constant. However, in the case of the bronchi, it is reasonable to assume that epithelial thickness is proportional to bronchial diameter. Thus, it is necessary to use age dependent conversion factors between the surface density of alpha-decays and dose to cells.

The effect of these considerations on mean dose to epithelial cells at various ages relative to adult values is shown in Figure 10. It is seen that the mean bronchial dose is only marginally increased in young children. This effect is smaller than the age dependence considered earlier (NEA, 1983; NCRP, 1984) and can surely be regarded as insignificant. A greater age dependence of dose to segmental bronchi is shown in Figure 10, but this is still not large except for neonates.

Dose conversion factors

It has been shown elsewhere that the individual radon daughters, ie RaA, RaB and RaC, contribute to lung dose strictly in proportion to the amount of potential alpha-energy associated with each radionuclide and not in proportion to their individual activities (Jacobi and Eisfeld, 1980; James, 1984). This leads to the theoretical expectation that risk is proportional to the special quantity 'exposure to potential alpha-energy', under otherwise similar environmental conditions (NEA, 1983; ICRP, 1981).

In practice, it is often neither possible nor convenient to monitor 'exposure to potential alpha-energy', whereas monitoring of exposure to radon gas by means of the time-integral of the radon concentration in air is relatively straightforward (NEA, 1985). Hence, the additional parameter F, known as the equilibrium factor, has been devised for practical application (UNSCEAR, 1982). F expresses the airborne concentration of potential alpha-energy as a fraction of the highest possible value achieved when all the daughters have the same activity as the measured radon gas. Thus, the potential alpha-energy concentration is 1 WL when the radon concentration is 3700 Bq per m^3 and F = 1, i.e. radon and daughters are in radioactive equilibrium. The annual exposure to potential alpha-energy, E_p, is then related to the average radon concentration, C_{Rn} by:

$$E_p \text{ [WLM]} = F \times C_{Rn} \text{ [Bq m}^{-3}\text{]} \times n \times 8760 / 170 / 3700$$

where 8760 = number of hours per year
     170 = number of hours per working month
       n = fraction of time spent indoors (occupancy)

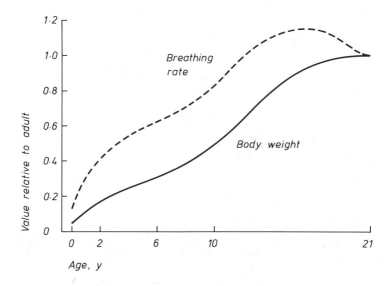

Figure 8.   Breathing rate and body weight as a function of age.

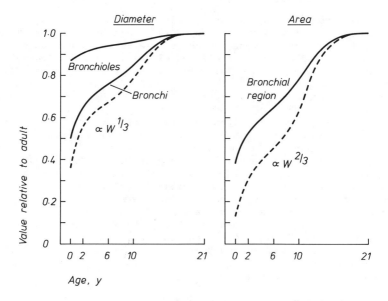

Figure 9.   Diameters and surface areas of bronchial airways
as a function of age, according to the UCI lung model.

Taking n as 0.8 (UNSCEAR, 1982; NEA, 1983), the annual exposure to potential alpha-energy is given by F x 0.011 (WLM per y per Bq m^{-3}).

This conversion procedure can be used with dosimetric factors chosen from here or elsewhere to estimate dose rates as a function of the average radon gas concentration for different environmental conditions. A conversion factor of 130 mGy per WLM is recommended here (Table II) for that part of the exposure associated with unattached daughters, i.e. the unattached fraction. However, for the attached fraction, it is necessary to specify the aerosol AMD, bearing in mind that dose is inversely proportional to AMD. A value of 0.12 µm may well be typical of domestic rooms. In this case, a dose conversion of about 8 mGy per WLM is recommended (Table II). It should be noted here that the numerical factor could be as high as 20 under unusual conditions (Reineking et al., 1985), or as low as 4 if aerosol growth in the humid airways of the lung is in fact significant.

Figure 11 shows the annual absorbed doses, D_B, calculated by applying these recommended dose conversions to the range of unattached fractions and equilibrium factors that might normally be encountered in rooms (Porstendorfer, 1984). The equilibrium factor, F, generally increases with aerosol concentration. This will lead to a proportionally higher total exposure to potential alpha-energy for a given concentration of radon gas in room air. However, the fraction of the exposure associated with unattached daughters, f_p, decreases dramatically for high F. The net effect of these changes is that the dose rate for a given radon concentration is relatively independent of room conditions. On this basis, a single conversion factor could be recommended, with a value somewhere between 50 and 100 uGy per y per Bq m^{-3} radon gas concentration.

It is to be noted that the conversion factor of 20 mSv effective dose equivalent per 200 Bq m^{-3} equilibrium equivalent radon concentration, considered by the ICRP (ICRP, 1984) in recommending an 'Action Level' for remedying high indoor radon concentrations, corresponds to a dose rate of about 40 µGy per y per Bq m^{-3} radon gas concentration.

Effective dose equivalent. If it is assumed that the weighting factor for bronchial dose equivalent is 0.06, the unattached fraction of potential alpha-energy in room air is typically about 5%, and that the aerosol AMD is typically 0.12 µm (Reineking et al., 1985), the dosimetry developed here gives a conversion factor to effective dose equivalent of approximately 15 mSv per WLM exposure in homes. This is three times higher than the value commonly used (UNSCEAR, 1982).

Conclusions

(i) In order to assess risk, it now seems more defensible to consider the average dose to all epithelial cells than just the dose to deep-lying basal cells. A dosimetric model formulated in this way gives mean bronchial doses higher than heretofore by about 60% for the unattached fraction of potential alpha-energy and about 30% for radon daughters attached to room aerosols.

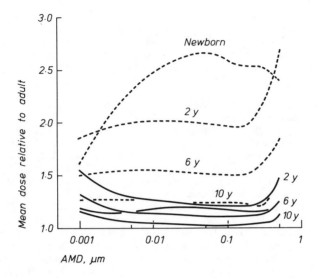

Figure 10. Doses to the bronchial region (solid curves) and to segmental bronchi in newborn infants and children of various ages, relative to values for adults.

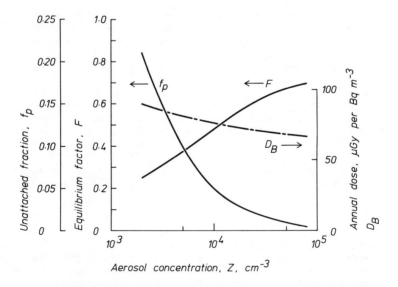

Figure 11. Variation of unattached fraction of potential alpha-energy and equilibrium factor according to a model of room aerosol behaviour and the effect on bronchial dose rate per unit radon gas concentration.

(ii) Refinement of the assumptions made in modelling lung deposition and clearance, based on recent experimental data, leads to an estimate of absorbed dose per unit exposure to unattached daughters three times higher than that adopted by UNSCEAR (UNSCEAR, 1982). The new value of 130 mGy per WLM is similar to that recommended by NCRP (NCRP, 1984).

(iii) This pointer to higher dose conversion factors is reinforced by an emerging congruence of experimental results, indicating that exposure to unattached daughters is higher in homes than previously assumed and that the attached aerosol is typically smaller. The combined effect of these factors suggests an increase in the estimate of conversion factor to effective dose equivalent in homes by a factor of three, to about 15 mSv per WLM.

(iv) It has been shown that both breathing rate and age are minor factors in determining dose per unit exposure.

(v) According to the dose conversion factors developed here and current understanding of the inverse relationship between equilibrium factor and unattached fraction, lung dose rate is simply proportional to the radon gas concentration over a wide range of conditions. This has beneficial implications for domestic and occupational monitoring schemes.

(vi) In order to reinforce or refute these conclusions priority should be given to experimental studies and predictive modelling of aerosol size and unattached fraction in room air. It is important also to know if radon daughter aerosols grow significantly in size at physiological humidity and to refine our understanding of nasal deposition of unattached daughters.

Acknowledgment

This work was partly funded by the Commission of the European Communities, under contract BI6-116-UK.

References

Adams, N., Dependence on Age at Intake of Committed Dose Equivalents from Radionuclides, Phys. Med. Biol. 26:1019-1034 (1981).

Altman, P.L. and D.S. Dittmer, Biology Data Book, Vol 1, pp. 195-201, Federation of American Societies for Experimental Biology, Bethesda, MD (1972).

Bair, W.J., ICRP Work in Progress: Task Group to Review Models of the Respiratory Tract, Radiol. Prot. Bull. 63:5-6 (1985).

Chamberlain, A.C. and E.D. Dyson, The Dose to the Trachea and Bronchi from the Decay Products of Radon and Thoron, Br. J. Radiol. 29:317-325 (1956).

Cohen, B.S., N.H. Harley, R.B. Schlesinger and M. Lippmann, Nonuniform Particle Deposition on Tracheobronchial Airways: Implications for Lung Dosimetry, Ann. occup. Hyg. in press.

Egan, M.J. and W. Nixon, A Model of Aerosol Deposition in the Lung for Use in Inhalation Dose Assessments, Radiat. Prot. Dosim. 11:5-17 (1985).

Ellett, W.H. and N.S. Nelson, Epidemiology and Risk Assessment, in Indoor Air and Human Health (R.B. Gammage and S.V. Kaye, eds) pp. 79-107, Lewis, Chelsea, MI (1985).

Gastineau, R.M., P.J. Walsh and N. Underwood, Thickness of Bronchial Epithelium with Relation to Exposure to Radon, Health Phys. 23:857-860 (1972).

Gebhart, J. and J. Heyder, Removal of Aerosol Particles from Stationary Air within Porous Media, J. Aerosol Sci. 16:175-187 (1985).

George, A.C. and A.J. Breslin, Deposition of Radon Daughters in Humans Exposed to Uranium Mine Atmospheres, Health Phys. 17:115-124 (1969).

Greenhalgh, J.R., A. Birchall, A.C. James, H. Smith and A. Hodgson, Differential Retention of Pb-212 Ions and Insoluble Particles in Nasal Mucosa of the Rat, Phys. Med. Biol. 27:837-851 (1982).

Harley, N.H. and B.S. Pasternack, Environmental Radon Daughter Alpha Dose Factors in a Five-lobed Human Lung, Health Phys. 42:789-799 (1982).

Hofmann, W., Dose Calculations for the Respiratory Tract from Inhaled Natural Radionuclides as a Function of Age - II. Basal Cell Dose Distributions and Associated Lung Cancer Risk, Health Phys. 43:31-44 (1982).

ICRP; International Commission on Radiological Protection, Limits for Inhalation of Radon Daughters by Workers, Publication 32, Ann. of ICRP 6:1-24 (1981).

ICRP; International Commission on Radiological Protection, Principles for Limiting Exposure of the Public to Natural Sources of Radiation, Publication 39, Ann. of ICRP 14:1-8 (1984).

Ingham, D.B., Diffusion of Aerosols from a Stream Flowing through a Cylindrical Tube, Aerosol Sci. 6:125-132 (1975).

Jacobi, W. and K. Eisfeld, Dose to Tissues and Effective Dose Equivalent by Inhalation of Radon-222, Radon-220 and their Short-lived Daughters, GSF Report S-626, Gesellschaft fur Strahlen-und Umweltforschung, Munich-Neuherberg (1980).

Jacobi, W. and H.G. Paretzke, Risk Assessment for Indoor Exposure to Radon Daughters, Sci. Total Environ. 45:551-562 (1985).

James, A.C., J.R. Greenhalgh and A. Birchall, A Dosimetric Model for Tissues of the Human Respiratory Tract at Risk from Inhaled Radon and Thoron Daughters, in Radiation Protection. A Systematic Approach to Safety, Vol 2, pp. 1045-1048, Pergamon, Oxford (1980).

James, A.C., Dosimetric Approaches to Risk Assessment for Indoor Exposure to Radon Daughters, Radiat. Prot. Dosim. 7:353-366 (1984).

James, A.C., Dosimetric Assessment of Risk from Exposure to Radioactivity in Mine Air, in Occupational Radiation Safety in Mining (H. Stocker, ed) Vol 2, pp. 415-425, Canadian Nuclear Association, Toronto (1985).

Martin, D. and W. Jacobi, Diffusion Deposition of Small-sized Particles in the Bronchial Tree, Health Phys. 23:23-29 (1972).

Martonen, T.B. and M. Patel, Modelling the Dose Distribution of H_2SO_4 Aerosols in the Human Tracheobronchial Tree, Am. Ind. Hyg. Assoc. J. 42:453-460 (1981).

Masse, R., Cells at Risk, in Lung Modelling for Inhalation of Radioactive Materials (H. Smith and G. Gerber, eds) pp. 227-246, EUR 9384 EN, CEC, Brussels (1984).

NCRP; National Council on Radiological Protection, Evaluation of Occupational and Environmental Exposures to Radon and Radon Daughters in the United States, NCRP Report No. 78, Bethesda, MD (1984).

NEA; Nuclear Energy Agency Group of Experts, Dosimetry Aspects of Exposure to Radon and Thoron Daughter Products, OECD Nuclear Energy Agency, Paris (1983).

NEA; Nuclear Energy Agency Group of Experts, Metrology and Monitoring of Radon, Thoron and their Daughter Products, OECD Nuclear Energy Agency, Paris (1985).

Phalen, R.F., M.J. Oldham, C.B. Beaucage, T.T. Crocker and J.D. Mortensen, Postnatal Enlargement of Human Tracheobronchial Airways and Implications for Particle Deposition, Anat. Rec. in press.

Pich, J., Theory of Gravitational Deposition of Particles from Laminar Flow in Channels, Aerosol Sci. 3:351-361 (1972).

Porstendorfer, J., Behaviour of Radon Daughter Products in Indoor Air, Radiat. Prot. Dosim. 7:107-113 (1984).

Reineking, A., K.H. Becker and J. Porstendorfer, Measurements of the Unattached Fractions of Radon Daughters in Houses, Sci. Total Environ. 45:261-270 (1985).

Rudolf, G., J. Gebhart, J. Heyder, C. Schiller and W. Stahlhofen, An Empirical Formula Describing Aerosol Deposition in Man for Any Particle Size, J. Aerosol Sci. 17:350-355 (1986).

Saccomanno, G., V.E. Archer, O. Auerbach, M. Kuschner, M. Egger, S. Wood and R. Mick, Age Factor in Histological Type of Lung Cancer in Uranium Miners, a Preliminary Report, in Radiation Hazards in Mining (M. Gomez, ed) pp. 675-679, Society of Mining Engineers, New York (1982).

Schiller, Ch.F., Diffusionsabscheidung von Aerosolteilchen im
Atemtrakt des Menshen. PhD Thesis, J.W. Goethe-Universitat,
Frankfurt/Main (1986).

Spencer, H., Carcinoma of the Lung, in Pathology of the Lung, Vol 2,
pp. 773-859, Pergamon, Oxford (1977).

Trump, B.E., E.M. McDowell, F. Glavin, L.A. Barrett, P. Becci,
W. Schurch, H.C. Kaiser and C.C. Harris, The Respiratory Epithelium
IV. Histogenesis of Epidermoid Metaplasia and Carcinoma in Situ in the
Human, J. Natl. Cancer Inst. 61:563-575 (1978).

UNSCEAR; United Nations Scientific Committee on the Effects of Atomic
Radiation, Ionizing Radiations: Sources and Biological Effects, Report
to the General Assembly, Annex D, United Nations, New York (1982).

Weibel, E.R., Morphometry of the Human Lung, Academic Press, New York
(1963).

Wise, K.N., Dose Conversion Factors for Radon Daughters in Underground
and Open-cut Mine Atmospheres, Health Phys. 43:53-64 (1982).

Yeh, H.C. and G.M. Schum, Models of Human Lung Airways and Their
Application to Inhaled Particle Deposition, Bull. Math. Biol.
42:461-480 (1980).

Yu, C.P. and C.K. Diu, A Comparative Study of Aerosol Deposition in
Different Lung Models, Am. Ind. Hyg. Assoc. J. 43:54-65 (1982).

RECEIVED November 13, 1986

Chapter 30

Updating Radon Daughter Bronchial Dosimetry

Naomi H. Harley and Beverly S. Cohen

Institute of Environmental Medicine, New York University Medical Center, New York, NY 10016

The lung cancer risk from radon daughter
exposure is known only for occupationally
exposed males. In order to determine the
risk in environmental situations it is
necessary to determine whether the
bronchial alpha dose, which confers the
risk, is similar to that in mines.
Particle size is a major factor which
determines the alpha dose conversion factor
for radon daughters (mGy/WLM). Data on
indoor environments are emerging and
indicate that a variety of specific
conditions exist. For example, a dose
factor four times that for a nominal
occupational or environmental exposure
exists if kerosene heater particles
dominate the indoor aerosol and four times
smaller if a hygroscopic particle
dominates.

Many states in the U.S. are currently involved in large scale
surveys to measure radon levels in homes in an attempt to assess
the environmental risk from radon and radon daughter exposure.
Radon daughters deliver the largest radiation exposure to the
population and it is estimated that 0.01% of the U.S. population
(23,000 persons) are exposed from natural sources to greater than
those levels allowed occupationally (4 WLM/yr) (NCRP, 1984).
  In order to estimate the risk from environmental exposure it
is necessary to utilize the underground miner epidemiology which
relies exclusively upon data from occupationally exposed males.
It is possible to infer cross population risk if the detailed
dose delivered to target cells in bronchial epithelium is
calculated for the various population groups since it is the dose
which confers the risk.

0097-6156/87/0331-0419$06.00/0
© 1987 American Chemical Society

Several models have been published concerning the bronchial
dosimetry for radon daughters (Harley, 1984; ICRP, 1981; James,
1984) and these are in general agreement provided that the
reference atmosphere is chosen to be similar.
It is important to update the bronchial dosimetry for
radon daughters as new information becomes available. It is the
purpose of this study to show that there is a potential for
either significantly increased bronchial dose in the home per
unit exposure if the ambient particle size is artificially
reduced due, for example, to open-flame burning or use of
kerosene heaters, or a decreased dose if hygroscopic particles
dominate the indoor aerosol.

Indoor Characteristics

Recent measurements have indicated that the indoor aerosol is
generated primarily indoors (Tu et al., 1985) unless the dwelling
is, for example, a public building with high ventilation rates
(Walker et al., 1980). Spengler (Spengler et. al., 1984) has
shown that a single smoker in a home can double the normal
particle loading from about 20 to 40 $\mu g/m^3$. Measurements made in
an energy efficient test home (0.1 air changes/hr) showed that
one cigarette dominated the indoor aerosol characteristics for
many hours (Offerman et al., 1985).
Although many factors need to be considered in calculating
the alpha dose to target cells in bronchial epithelium, many of
the factors do not produce a large change in the dose conversion
factor (mGy/WLM) (Harley, 1984). The particle size distribution,
however, is a sensitive factor in the calculation of bronchial
dose and smaller or larger particle sizes than usually assumed
for the indoor aerosol can produce a substantial increase or
decrease in the bronchial dose. The particle size chosen in past
radon daughter dose estimates for occupational and environmental
exposure was from the work of George et al. (1975) and George and
Breslin (1980) who reported mines to have particles with activity
median diameter of 0.18 μm, and homes, 0.12 μm. Extensive
measurements of the indoor environmental particle size
distribution are not available, but existing data indicate a
large potential for differences from this value.
Tu et al. (1985) measured several types of indoor
environmental aerosols and found that with open flame burning or
with the use of kerosene heaters, the resulting hydrophobic
particles dominated the indoor aerosol. Their measurements also
show that cigarette smoke can dominate the indoor aerosol. It is
assumed here, for purposes of calculating a range of dose factors
that a hygroscopic particle can also be present in the
environment. Tu (1986) has calculated the activity median
diameter for radon daughter attachment in two houses from his
measurements for these differing situations and these are shown in
Table I.

Table I. Calculated activity median diameter for attached radon daughters from measured aerosol characteristics (Tu, 1986).

Site   Activity median diameter(μm)   geometric standard deviation
----

House A
normal room air   0.075   1.8
cigarette smoked   0.123   1.79

House B
normal room air   0.086   1.8
kerosene heater on   0.030   1.92

Tu and Knutson (1984) also measured the particle deposition of hydrophobic and hygroscopic particles in the human respiratory tract. They showed that the hygroscopic particles grow by a factor of 3.5 to 4.5 at the saturated humidity present in the lung. For the purpose of calculating bronchial deposition for a hygroscopic aerosol we assume an increase in size by a factor of 4 upon entry into the bronchial tree.

Tracheobronchial Dose

The tracheobronchial dose has been calculated previously using average reported values for the various factors needed (Harley, 1984). The following list of factors involved in the dose calculation indicates where differences in the modeling occurs for this update.

Physical characteristics

Unattached fraction of daughters
The same. Mines ^{218}Po/^{222}Rn = 4%; indoors 7%. ^{218}Po/^{214}Pb=10. (NCRP, 1984a). Diffusion coefficient after Knutson et al. (1983) of 0.0025 cm^2/sec.

Radon daughter disequilibrium
    Different. ^{222}Rn/^{218}Po/^{214}Pb/^{214}Bi, nominal mine 1/.61/.29/.21 (George et al., 1975); nominal home 1/.9/.6/.4 (Fisenne and Harley, 1974). High particle home 1/.9/.7/.7., assumed similar to outdoors (Fisenne and Harley, 1974)

Particle deposition models
    Different. Ingham (1975) diffusion deposition but corrected by turbulent diffusion factors in airways 0 to 6 as measured by Cohen (1986).

Particle size distribution
    Different. Nominal calculation for mine and home but adding calculations which include kerosene heater particles and hygroscopic particles. (Tu et al., 1985)

Physical dose calculation
Same as Harley (1984).

Biological characteristics

Breathing pattern/nasal deposition
Same as Harley (1984).

Bronchial morphometry
Same as in Harley (1984) from Yeh and Schum (1980).

Mucociliary clearance rates
Same as Harley (1984).

Location of target cells
Same as Harley (1984).

Mucus thickness
Same as Harley (1984).

      The specific values of the physical and biological
characteristics such as breathing patterns for occupational
exposure and active and resting patterns for environmental
exposure are given in the footnotes to the Tables. A few remarks
should be made concerning the parameters used which affect the
dose calculation significantly.
      In the present study the bronchial morphometry of Yeh and
Schum (1980) is utilized instead of the Weibel model. This is a
more accurate description of the bronchial airway lengths,
diameters and branching pattern. It does not assume dichotomous
branching and therefore does not suffer from the problem of the
artificially high surface area leading to low alpha dose in the
more distal airways common to other models of the human airways.
      It is well known that enhanced deposition in the first few
airways occurs due to the turbulence produced. Turbulent
diffusion is accounted for by using factors (ratio of observed
deposition to calculated diffusion deposition) to correct the
diffusion deposition. These had formerly been measured by Martin
and Jacobi (1972) in a dichotomous plastic model of the upper
airways. The data used here are from measurements performed by
Cohen (1986) using hollow casts of the upper bronchial tree which
included a larynx. This cast was tested using cyclic flow with
deposition measured for 0.03, 0.15 and 0.20 um diameter
particles. Her turbulent diffusion factors are used in the
calculation here (14 for generation 0, and 2 for generations 1 to
6).
      There are only scant data on nasal deposition. The
available studies reported utilized micron sized particles and
the dominant mode of deposition is impaction. This is not the
case for the particles considered here and diffusion and
turbulent diffusion are the mechanisms of interest. George and

Breslin (1969) measured nasal deposition for both the attached
and unattached fraction of radon daughters in a laboratory
atmosphere and found it to be about 1% for the attached and 60%
for the unattached fraction. The size characteristics of the
unattached fraction is assumed to be that found by Knutson et al.
(1983) for older air of 0.005 μm (D=0.0025 cm^2/sec). No data are
available for nasal deposition of the small sized kerosene heater
particles, 0.030 μm, and the value for the attached radon
daughters of 1.3% is used.

Results and Conclusions

The calculations are shown for the adult male breathing under
normal environmental conditions. The alpha dose combines the
daytime and nighttime breathing patterns.
    Table II shows the nominal alpha dose factors for
occupational mining exposure. Table III shows the alpha dose
factors for the nominal environmental situation. Table IV shows
the bronchial dose factors for the smallest sized particles, that
dominated by the kerosene heater or 0.03 μm. particles. The
radon daughter equilibrium was shifted to a somewhat higher value
in this calculation because this source of particles generally
elevates the particle concentration markedly with consequent
increase in the daughter equilibrium. Table V shows the alpha
dose for a 0.12 μm particle, the same as the nominal indoor
aerosol particle, but for a particle which is assumed to be
hygroscopic and grows by a factor of 4, to 0.5 μm, once in the
bronchial tree.
    The calculation of effective dose equivalent is sometimes
used even when reporting values for natural radioactivity. The
concept of effective dose equivalent was developed for
occupational exposures so that different types of exposure to
various organs could be unified in terms of cancer risk. It is
highly unlikely that the general population would require
summation of risks from several sources of radiation exposure.
The normal or average risk from whole body gamma-ray exposure in
the environment is only about 10% of that from average radon
daughter exposure and much less in elevated indoor environments.
Considering that the radon daughter lung cancer risk can be
derived directly from exposure in most cases, effective dose
equivalent is an unnecessary step.
    The results in Tables II to V indicate that a range of alpha
dose factors for radon daughters of about a factor of ten exists
and that they are site specific depending primarily on the indoor
aerosol. The potential for delivering an alpha dose four times
greater or smaller than that in a nominal atmosphere exists if
there is either a small sized or hygroscopic aerosol. The
calculations have been performed here for males, but it has been
shown that this model is easily adapted to calculating the alpha
dose to women, children and infants (Harley, 1984).

TABLE II. Alpha Dose in mGy/WLM for Occupational Mining
Atmosphere. Breathing Pattern and Specific Conditions,
See Footnote

```
----------------- LOBE --------------------
```

GEN	RIGHT UPPER	RIGHT MIDDLE	RIGHT LOWER	LEFT UPPER	LEFT LOWER
0	5.4	5.4	5.4	5.4	5.4
1	3.5	3.5	3.5	3.3	3.3
2	4.5	3.1	3.1	3.9	3.2
3	4.5	3.8	3.2	3.9	3.2
4	4.0	4.0	3.1	3.7	3.1
5	3.8	3.8	3.0	3.5	3.0
6	3.9	4.3	2.9	3.4	2.8
7	2.8	3.5	2.7	2.8	2.3
8	3.6	4.8	2.7	2.9	2.3
9	5.5	7.1	2.8	4.2	2.4
10	4.7	6.0	1.7	3.1	1.6
11	5.2	7.3	2.1	3.4	1.8
12	6.2	7.5	2.3	3.5	2.1
13	6.8	7.9	2.5	3.9	2.6
14	0.0	10.0	2.7	0.0	3.2
15	0.0	0.0	3.1	0.0	0.0
16	0.0	0.0	3.5	0.0	0.0

Unattached daughters Po-218/Rn, 4.0%; Pb-214/Rn, 0.4%.
Daughter Ratio Rn/Po-218/Pb-214/Po-214 1/.61/.29/.21.
Breathing Pattern 18 lpm, 15 Breaths/min.
Breathing cycle-inspiration, pause, expiration, pause
(3/8, 1/8, 3/8, 1/8)
Diffusion Coefficient for Unattached 0.0025 cm²/sec.
Particle diameter 0.17 μm (AMD) for Attached Daughters
Nasal Deposition 1.3% for Attached and 60% for Unattached Daughters
Alpha dose to cells at 22 μm depth below epithelial surface
Twenty percent alveolar deposition prior to expiration

TABLE III. Alpha Dose in mGy/WLM for Males. Environmental Aerosol. Breathing Pattern and Specific Conditions, See Footnote.

------------------ LOBE ------------------

GEN	RIGHT UPPER	RIGHT MIDDLE	RIGHT LOWER	LEFT UPPER	LEFT LOWER
0	6.1	6.1	6.1	6.1	6.1
1	4.1	4.1	4.1	3.8	3.8
2	5.2	3.6	3.6	4.5	3.7
3	5.1	4.5	3.7	4.5	3.7
4	4.6	4.6	3.6	4.3	3.6
5	4.4	4.4	3.4	4.0	3.5
6	4.5	5.0	3.4	3.9	3.2
7	3.3	4.1	3.2	3.2	2.7
8	4.2	5.7	3.2	3.4	2.7
9	6.4	8.4	3.3	4.9	2.9
10	5.5	7.1	2.0	3.6	1.9
11	6.1	8.7	2.4	4.0	2.1
12	7.3	8.9	2.7	4.1	2.5
13	8.1	9.5	3.0	4.6	3.1
14	0.0	12.0	3.3	0.0	3.9
15	0.0	0.0	3.8	0.0	0.0
16	0.0	0.0	4.2	0.0	0.0

Unattached Daughters. Po-218/Rn, 7.0%, Pb-214/Rn, 0.7%
Daughter Ratio Rn/Po-218/Pb-214/Po-214 1/.9/.6/.4.
Breathing Pattern 18 lpm active, 15 Breaths/min.
Breathing Pattern 9 lpm resting, 12 Breaths/min.
Breathing Pattern, 16 Hours Active, 8 Hours Resting
Diffusion Coefficient for Unattached 0.0025 cm^2/sec.
Particle diameter 0.12 μm (AMD) for Attached Daughters
Nasal Deposition 1.3% for Attached and 60% for Unattached Daughters
Alpha dose to cells at 22 μm depth below epithelial surface

TABLE IV. Alpha Dose in mGy/WLM for Males. Kerosene Heater
Aerosol.   Breathing Pattern and Specific Conditions,
             See Footnote.

------------------ LOBE ------------------

GEN	RIGHT UPPER	RIGHT MIDDLE	RIGHT LOWER	LEFT UPPER	LEFT LOWER
0	26.7	26.7	26.7	26.7	26.7
1	17.9	17.9	17.9	17.0	17.0
2	22.7	15.9	15.9	19.7	16.5
3	22.2	19.3	16.1	19.7	16.3
4	20.0	20.1	15.5	18.9	16.0
5	19.3	19.4	14.9	17.8	15.3
6	20.0	21.9	14.6	17.2	14.2
7	14.4	17.9	13.7	14.2	12.1
8	18.7	24.8	14.0	15.1	12.2
9	28.7	37.3	14.5	22.2	12.9
10	25.0	32.3	9.1	16.5	8.7
11	28.2	39.8	11.0	18.4	9.7
12	33.7	41.0	12.5	19.1	11.7
13	37.0	43.6	13.7	21.1	14.2
14	0.0	54.7	15.0	0.0	17.6
15	0.0	0.0	17.2	0.0	0.0
16	0.0	0.0	19.1	0.0	0.0

Unattached Daughters,Po-218/Rn 7.0%, Pb-214/Rn, 0.7%
Daughter Ratio Rn/Po-218/Pb-214/Po-214 1/.9/.7/.7.
Breathing Pattern 18 lpm active, 15 Breaths/min.
Breathing Pattern 9 lpm resting, 12 Breaths/min.
Breathing Pattern, 16 Hours Active, 8 Hours Resting
Diffusion Coefficient for Unattached 0.0025 cm²/sec.
Particle diameter 0.03 μm (AMD) for Attached Daughters
Nasal Deposition 1.3% for Attached and 60% for Unattached Daughters
Alpha dose to cells at 22 μm depth below epithelial surface

TABLE V. Alpha Dose in mGy/WLM for Males. Hygroscopic
Aerosol. Breathing Pattern and Specific Conditions,
See Footnote.


```
------------------- LOBE -------------------

        RIGHT    RIGHT    RIGHT    LEFT     LEFT
GEN     UPPER    MIDDLE   LOWER    UPPER    LOWER

 0       2.7      2.7      2.7      2.7      2.7
 1       1.7      1.7      1.7      1.6      1.6
 2       2.2      1.5      1.5      1.9      1.5
 3       2.2      1.9      1.5      1.9      1.5
 4       1.9      1.9      1.5      1.8      1.5
 5       1.8      1.8      1.4      1.6      1.4
 6       1.9      2.1      1.4      1.6      1.3
 7       1.3      1.7      1.3      1.3      1.1
 8       1.7      2.2      1.3      1.3      1.1
 9       2.5      3.2      1.3      1.9      1.1
10       2.1      2.7      0.8      1.3      0.7
11       2.3      3.2      0.9      1.5      0.8
12       2.7      3.2      1.0      1.5      0.9
13       3.0      3.4      1.1      1.6      1.1
14       0.0      4.2      1.1      0.0      1.3
15       0.0      0.0      1.3      0.0      0.0
16       0.0      0.0      1.4      0.0      0.0
```

Unattached Daughters. RaA/Rn, 7.0%, RaB/Rn, 0.7%
Radon Daughter Ratio Rn/RaA/RaB/RaC' 1/.9/.6/.4.
Breathing Pattern 18 lpm active, 15 Breaths/min.
Breathing Pattern 9 lpm resting, 12 Breaths/min.
Breathing Pattern, 16 Hours Active, 8 Hours Resting
Diffusion Coefficient for Unattached 0.0025 cm²/sec.
Particle diameter 0.12 µm (AMD) for Attached Daughters
Assume hygroscopic with 4x growth in airways to 0.5 µm
Nasal Deposition 1.3% for Attached and 60% for Unattached Daughters
Alpha dose to cells at 22 µm depth below epithelial surface

Acknowledgments

The authors would like to thank Dr. Keng Tu of the USDOE
Environmental Measurements Laboratory for allowing us to use his
unpublished calculations. This work was performed with support
from USDOE Contract DE AC02 80 EV 10374 and in part by Center
Grants ES00260 and CA13343 from NIEHS and the National Cancer
Institute and Special Emphasis Research Career Award Grant No. OH
00022 from NIOSH. All support is gratefully acknowledged.

Literature Cited

Cohen, B.S., Deposition of Ultrafine Particles in the Human
Tracheobronchial Tree: A Determinant of the Dose from Radon
Daughters, Proc. Am. Chem. Soc. Annual Meeting 1986, New York.

Fisenne, I.M. and N.H. Harley, Lung Dose Estimates from Natural
Radioactivity Measured in Urban Air, USAEC Report HASL-TM-74-7
New York, NY (1974).

George, A.C. and A.J. Breslin, Deposition of Radon Daughters in
Humans Exposed to Uranium Mine Atmospheres, Health Phys. 17:115-
124 (1969).

George, A.C., L. Hinchliffe, and R. Sladowski, Size Distribution
of Radon Daughter Particles in Uranium Mine Atmospheres, Am. Ind.
Hyg. J., 36:484-490 (1975).

George, A.C. and A.J. Breslin, The Distribution of Ambient Radon
and Radon Daughters in Residential Buildings in the New Jersey-New
York Area, in Natural Radiation Environment
III; (Gesell, T.and W. Lowder, Eds.); National Technical
Information Service, Springfield, VA. (1980).

Harley, N.H., Comparing Radon Daughter Dose:Environmental versus
Underground Exposure, Rad. Protection Dosimetry 7:371-375 (1984).

ICRP, "Limits for Inhalation of Radon Daughters by Workers,"
International Commission on Radiation Protection
Report 32, Pergamon Press, New York, 1981.

Ingham, D.B., Diffusion of Aerosols from a Stream Flowing
Through a Cylindrical Tube, J. Aero. Sci., 6: 125-132 (1975).

James, A. C., Dosimetric Assessment of Risk from Exposure to
Radioactivity in Mine Air Proc. Occupational Radiation Safety in
Mining, (H. Stocker, ed) pp415-426 Canadian Nuclear
Assn.,Toronto, Canada (1984).

Knutson, E.O., A.C. George, R.H. Knuth and B.R. Koh, Radon
Daughter Plateout II. Prediction Model, Health Phys. 45:445-452
(1983).

Martin, D. and W. Jacobi, Diffusion Deposition of Small-sized Particles in the Bronchial Tree,Health Phys. 23: 23-29 (1972).

NCRP, "Exposures from the Uranium Series with Emphasis on Radon and its Daughters," National Council on Radiation Protection and Measurements Report No. 77. Bethesda, MD. (1984)

NCRP, "Evaluation of Occupational and Environmental Exposures to Radon and Radon Daughters in the United States," National Council on Radiation Protection and Measurements Report No. 78 Bethesda, MD (1984a).

Offermann, F.J., R.G. Sextro, W.J. Fisk, D.T. Grimsrud, W.W. Nazaroff, A.V. Nero, K.L. Revzan, K.L. and J. Yater, Control of Respirable Particles in Indoor Air with Portable Air Cleaners, Atmo. Env. 19: 1761-1775 (1985).

Spengler, J.D. and K. Sexton, Indoor Air Pollution: A Public Health Perspective, Science, 221: 9-17 (1983).

Tu, K.W. and E.O. Knutson, Total Deposition of Ultrafine Hydrophobic and Hygroscopic Aerosols in the Human Respiratory System, Aerosol Sci. and Technology, 453-465 (1984).

Tu, K.W., E.O. Knutson and H. Franklin, Aerosol Measurements in Residential Buildings in New Jersey, Abstract. Am. Assoc. for Aerosol Res. Annual Meeting. Albuquerque, NM. (1985).

Tu, K.W. Private communication. 1986.

Walker, M.V. and C.J. Weschler, Water-Soluble Components of Size-Fractionated Aerosols Collected after Hours in a Modern Office Building, Env. Sci and Technology 14:594-597 (1980).

Yeh, H.C. and M. Schum, Models of Human Lung Airways and their Application to Inhaled Particle Deposition, Bull. Math Biol. 42: 461-480 (1980).

RECEIVED October 3, 1986

Chapter 31

The Validity of Risk Assessments for Lung Cancer Induced by Radon Daughters

F. Steinhäusler

Division of Biophysics, University of Salzburg, A-5020 Salzburg, Austria

Available input data for the risk assessment from low
level radon daughter (Rn-d)exposure are mostly either
of low quality, partially contradicting or simply
"guesstimates". Therefore at present only the upper
limit of this risk can be estimated. Results of epidemiological studies amongst miners are associated
with large uncertainties with regard to the assessment of past radiation exposure, lung cancer diagnos-
tic and/or classification and synergistic effects due
to smoking and dust exposure. An alternative approach
uses dosimetric modelling for Rn-d inhalation to obtain Rn-d exposure-dose conversion factors. Large uncertainty is caused by individual variability due the
influence of life style, physical and biological parameters. It is concluded that for "normal" indoor
Rn-d exposure the resulting risk is neglegible compared to other risks "accepted" by society.

Lung cancer due to inhalation of radon decay products (Rn-d)
represents the single most significant risk from all natural and
non-made radiation sources. There is growing concern about its
impact on society for occupationally and non-occupationally exposed
persons. Increased lung cancer incidence amongst miners has already
led to the request by labour unions for a lowering of the presently
recommended Rn-d exposure limits (Williams, 1985). Trends in energy
conservation efforts and recycling of unsuitable industrial wastes
as construction material (Steinhäusler and Pohl, 1983) indicate an
increase of a Rn-d induced lung cancer risk for members of the
public. In view of the significant socio-economic consequences of
these issues it is important to ensure that any input data used for
risk assessments are subjected to stringent quality control. Such
data can be derived from epidemiological studies, experimental
animal inhalation research and theroretical dosimetric modelling.
All three data pools are associated with considerable uncertainties.
In this paper the following aspects are critically reviewed with

0097-6156/87/0331-0430$06.00/0
© 1987 American Chemical Society

regard to the applicability to the risk estimation for low-level
Rn-d exposure:
- limitations of Rn-d related epidemiological studies
- suitability of extrapolating from results obtained
  from animal inhalation studies to man
- inherent uncertainties of presently used lung models.

Data Needed for Risk Assessment

Epidemiology. Using the epidemiological approach the following
preconditions have to be met in order to establish a dose-effect
relationship:

a)   individual doses due to Rn-d inhalation in the exposed group
     have to be determined with a high degree of reliability ("dose-
     data");
b)   in order to discriminate from lung cancer cases due to carcino-
     gens other than Rn-d, ideally a specific type of lung cancer
     should be identified which is induced by Rn-d only ("effect-
     data");
c)   a control population has to be defined which is comparable to
     the exposed group under investigation in all aspects,
     except for its lack of Rn-d exposure.

Individual dose assessment requires radiological data on all
external and internal sources contributing to occupational and non-
occupational radiation exposure (Steinhäusler and Pohl, 1983). This
is of particular importance in the case of low level Rn-d exposure,
as man is always exposed to Rn-d at varying levels through all
stages of life, e.g. at school, home or work. The resulting lifetime
risk from this chronic exposure is influenced by the latent
period between the initial Rn-d exposure inducing the cellular
transformation and the subsequent macroscopic expression as a
pulmonary cancer. The latent period varies from less than ten years
up to over fifty years, depending e.g. on smoking characteristics
(Härting and Hesse, 1879; Radford et al., 1984). As the latent
period is about inversely proportional to the lifetime risk,
continous low level Rn-d exposure may lead to a lengthening of the
latent period, even in excess of the individual lifetime.

     Subsequently, individual data on exposure are converted to dose
by using conversion factors (OECD/NEA, 1983). The choice of the
appropriate numerical value depends on physiological parameters
(e.g. respiratory minute volume) as well as physical characteristics
of the inhaled aerosol (e.g. particle size). Mean values range
typically from about 5 mSv/WLM (non-occupational exposure) to about
10 mSv/WLM (occupational exposure).

     Critical analysis of the data on lung cancer is equally
important but frequently overlooked and assumed as largely error-
free. This group of "effect-data" comprises medical data on lung
cancer diagnosis and classification, demographical information on
the population group under investigation and the quantitative
assessment of smoking characteristics or exposure to other carcinogens in the atmosphere.

All this has to be seen against a mostly increasing background of "spontaneous" lung cancer cases in the general population. Members of this population are themselves exposed to a wide range of Rn-d levels which varies typically by an order of magnitude even in a relatively small city (Steinhäusler et al., 1983).

Animal Inhalation Experiments. Such experiments overcome at least partially one of the major problems inherent to human Rn-d epidemiological studies, i.e. the identification of a homogenous control group. Although the control group itself is still exposed to Rn-d, this exposure can occur at low environmental levels under controlled laboratory conditions. Furthermore this type of study offers the possibility of either simultaneous or alternative exposure to atmospheric carcinogens other than Rn-d, e.g. smoke, diesel fumes or ore dust. In this manner the question of synergistic superposition of different lung carcinogens can be addressed. The possibility of achieving true randomization represents a further advantage over epidemiology with respect of obtaining unbiased lung cancer incidence data. Since the number of Rn-d exposed animals is at least theoretically limited only by financial constraints, such inhalation experiments can provide statistically significant numbers on lung cancer induction even at low Rn-d levels.

However, in order to be able to use the information from animal studies for Rn-d related dose-effect studies and risk assessments for humans the following conditions should be met:

a)   in view of the uncertainties associated with high to low dose extrapolations in radiobiological studies, animal Rn-d exposure levels should be of a similar magnitude as those occuring in human occupational and non-occupational conditions;

b)   exposure to carcinogens other than Rn-d (e.g. tobacco smoke) should be physiologically realistic, simulating those of human exposure conditions;

c)   dosimetric models developed for the human respiratory tract should be adapted for the specific test animal used, taking into account differences in lung physiology and -geometry;

d)   radiation sensitivity and cancer induction mechanisms in test animals should be similar to that in humans.

Dosimetric Studies. The main objective of Rn-d dosimetry is to enable the assessment of the observed dose from Rn-d deposited in the respiratory tract from measured Rn-d concentration in the atmosphere. As a consequence of an inhomogeneous Rn-d deposition within the lung, different dosimetric concepts have been developed to describe either regional mean dose values (e.g. for tracheobronchial or pulmonary region) or microdosimetric dose calculations (e.g. for the basal cell layer). Using Monte Carlo calculation methods it is also possible to account for the random nature of cellular hits by deposited alpha particles. The results of such dosimetric calculations should provide the following information needed for Rn-d risk assessment:

a)   which biological targets (e.g. bifurcations, epithelial cells)
     within the respiratory tract have the highest hit probability
     by alpha particles.
b)   which physiological parameters (e.g. age, physical activity,
     sex) have the most influence on the individual Rn-d exposure-
     dose conversion;
c)   which physical parameters (e.g. particle size, unattached
     fraction, disequilibrium factor) significantly effect the
     conversion of exposure to dose.

Data Available for Risk Assessment

Epidemiology. Hitherto, epidemiologic research has been carried out
mainly amongst mining populations. The authors of many of the
original studies have been fully aware of the severe inherent
shortcomings of their investigations, mostly due to unavailable
individual exposure data. Therefore they stated caveats concerning
the potential misuse of their data at a later stage, e.g. for the
deduction of a quantitative dose-effect relationship:
"...this analysis is most emphatically not offered as a basis for
any estimate of risk per unit dose..."(Royal Commission, 1976)
"...the study was not designed... to demonstrate that low levels of
exposure cause any detectable increase in the lung cancer incidence
rate..."(Congress of the United States, 1967)
"...it is therefore not considered appropriate to extrapolate at
present from the truncated period of observation to lifetime
risk..."(Muller et al., 1983)
"...it is not possible to draw any conclusions about dose and effect
below about 100 WLM due the considerable uncertainty in exposure and
statistical error..."(Snihs, 1973)
     In general the uncertainties associated with these epidemiolo-
gical studies can be grouped into three categories:

a)   uncertainties associated with the reconstruction of past
     individual Rn-d exposure, mostly due to the lack of measure-
     ments and high job mobility
b)   uncertainties of diagnosis and classification of lung cancer
     data
c)   uncertainty about synergistic effects between Rn-d and
     other carcinogens present in the inhaled atmosphere.

In order to establish a causal relationship rather than a pure
association between the suspected carcinogen Rn-d and the induced
effect "lung cancer " it is most important to know the individual
Rn-d exposure. This is the particular weakness of all these studies:
since these measurements were not carried out fifteen to thirty
years ago, this essential information cannot be provided by any of
the available retrospective studies with a sufficient degree of
reliability and accuracy (Steinhäusler, in press). Therefore they
have to rely on indirect estimation of past Rn-d exposure levels,
using e.g. inference from uranium ore content.
     Low level Rn-d exposure-effect studies that may be relevant to
the general indoor exposure situation, at least as far as lifetime
exposure is concerned, are available for miners in CSSR, US, Canada,

Newfoundland and Norway (Figure 1). At these low exposure levels, it can be seen that in the case of the Czechoslovakian and Norwegian study significantly more lung cancer cases were observed than expected. However, in both cases lung cancer incidence amongst miners was compared to the relevant data for non-mining male members of the general population. In the Newfoundland study, the noted relatively small increase cannot be regarded as statistically significant due to the large uncertainties associated with the retrospective exposure estimates, since Rn-d exposure assessments are available only for the post-1960 period. In the case of the US- and Canadian miners the expected number of cases even exceeded that actually observed.

As these studies used non-miners as controls there is an additional inherent "safety" margin, since the frequency of smokers amongst miners is generally higher by a factor 2 to 3 compared to members of the public, i.e. all these assumed Rn-d related lung cancer cases also have a strong association with exposure to smoke.

Medical data on lung cancer diagnosis and classification are subject to errors and uncertainties just like any other scientifi- cally derived information. At present, lung cancer induced by low level Rn-d exposure cannot be distinguished with histological- cytological methods from lung cancer initiated by other carcinogens, e.g. tobacco smoke. It has been argued repeatedly that Rn-d induced lung cancer is associated with an unusual cell type frequency distribution, i.e. high frequency of small cell. However, a review of quality control aspects of the medical procedures and data used for the quantification of this supposedly Rn-d induced "effect" revealed that this is questionable for the following reasons (Steinhäusler, in press):

a) there is a high degree of intra- and interobserver variability in lung cancer tissue classification by pathologists, causing large differences in the numbers classified as a particular cancer type;

b) the source term, i.e. biopsy or autopsy, is of significance for the result of potentially misclassified lung tumors, e.g. crush artifacts are categorized as "small cell carcinomas" in 25% of all biopsies;

c) in post-mortem examinations compared to anti-mortem diagnosis 40% of primary lung cancer are missed altogether, 30% are false-positive.

d) histo-cytological typing of lung cancer is difficult in prin- ciple because these tumors are generally heterogeneous, i.e. intermixing of squamous, adeno-, large- and small cell carcinoma cells is frequent.

In view of these fundamental difficulties it is understandable that lung cancer diagnosis and classification represents a diffifult task, resulting frequently in data of low reliability.

Dosimetry. Dosimetry for inhaled Rn-d particles has made significant advances over the past twenty years. Lung modelling, based initially to a large extent on the WEIBEL'A' -symmetrical lung model, has been

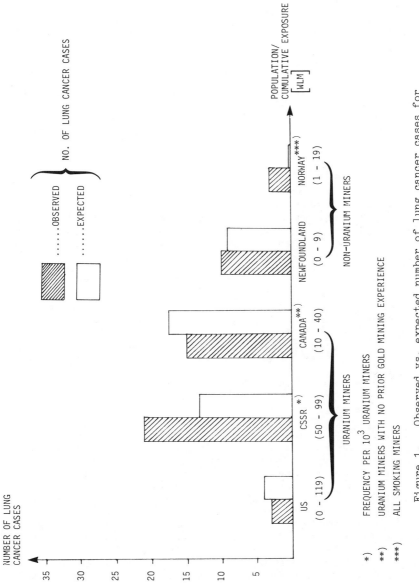

Figure 1. Observed vs. expected number of lung cancer cases for miners exposed to low-levels of radon daughters.

refined to present models based on thorax-casts made from human
lungs (Weibel, 1963; Yeh and Schum, 1980). Thereby potential
artefacts are minimized and the information on in vivo three-
dimensional lung geometry (e.g. branching angles, bronchial inclina-
tion) is maintained.

However, results obtained from lung dose modelling still show a
large range of values for the conversion of Rn-d exposure to dose
for the following reasons:

a)   target geometry:
     basal cells in the bronchial region are generally considered to
     be the sensitive target for lung cancer induction due to Rn-d
     deposited in the airways. The assumed depth of this cell layer
     below the bronchial surface determines to a large extent the
     dose received by these cells from inhaled Rn-d. Physiological
     information on the depth distribution is scarce.

b)   unattached/attached Rn-d:
     The nature and behavior of "free" Rn-d ions or atoms, i.e. not
     attached to atmospheric condensation nuclei, is still subject
     of controversy, particularly with regard to the influence of
     environmental atmospheric conditions, such as humidity and
     presence of other gases (Busigin et al., 1981). Free Rn-d atoms
     are one of the most critical parameters for the exposure-dose
     conversion. This can be of particular importance in indoor
     exposure situations with a large ratio of unattached to
     attached Rn-d.

Figure 2 shows the spread of results for two different physiological
and dosimetric models, even though a standardized exposure situation
is assumed (OECD/NEA, 1983).

Due to the superposition of various other biological, physiological and physical parameters used in modelling, the published
exposure-dose conversion factors range from 2 to 120 mGy per WLM.
However, a sensitivity analysis indicated that for most indoor
exposure situations compensatory effects can reduce this range to
about 5 to 10 mGy/WLM for the indoor situations occurring most
frequently (OECD/NEA, 1983).

Animal inhalation studies. Experimental Rn-d inhalation studies
have been carried out on dogs, hamsters, rats and mice. The two most
extensive and comparable investigations were carried out on rats,
using mean Rn-d exposure levels from 20 to 8 000 WLM (Chameaud and
Lafuma, 1984) and from 640 to 2 560 WLM (Cross et al., 1984),
respectively. Only the low-level exposure up to about 100 WLM is
relevant to the indoor exposure situation experienced during the
lifetime of a member of the public. Figure 3 shows the lung tumor
rate obtained from these studies as a function of Rn-d exposure. It
can be seen that the tumor rate for rats is between 2% and 4% at
about 100 WLM. These data have not been corrected for shortening of
the lifespan as the result of causes other than lung cancer from
Rn-d. However, in order to extrapolate these findings from rat to
man it has to be assumed that the susceptibility to Rn-d induced
lung cancer is similar for the two species. Furthermore, it is

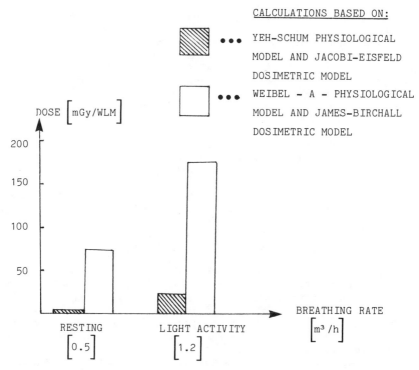

Figure 2. Mean bronchial dose to basal cells, standardized for 1 WLM exposure to free radon 222 daughter atoms for different physical activities (assuming typical indoor exposure with equilibrium factor F = 0.4).

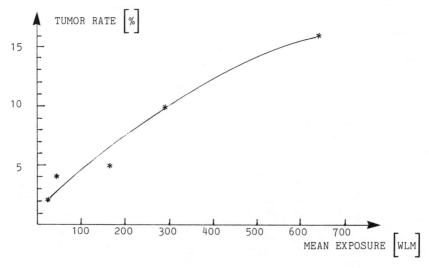

Figure 3. Primary epithelial lung tumor rate in rats as a function of Rn-d exposure.

questionable whether the same relationship between exposure and dose applies in animals and man. From the animals study carried out so far it has been concluded that the lung cancer risk (LC) per WLM is of a similar magnitude for Rn-d exposed miners and animals inhaling Rn-d (Cross et al., 1982). An analysis of the relevance of animal models for Rn-d inhalation in man demonstrated the following (Hofmann and Daschil, 1983): if it is justified to compare only radiation doses in both cases, i.e. neglecting all other cocarcinogenic effects, then only the dose in the critical region is effective in tumor induction.

Animal studies have also been used to study the simultaneous influence of Rn-d and ore dust, diesel fumes and cigarette smoke on lung cancer induction. However, results do not clearly indicate a potentially synergistic effect. Whilst ore dust and diesel fumes showed no effect on the lung cancer rate, exposure to smoke after initial Rn-d exposure increases the lung cancer rate; this is in contrast to results obtained from alternating smoke and Rn-d exposure, where the lung cancer induction rate was found to be decreasing (Cross et al., 1978; Chameaud et al., 1982).

Risk Assessment for Indoor Exposure. The assessment of the lung cancer risk resulting from a chronic low-level lifetime exposure to Rn-d is largely determined by the assumption on the length of the latent period, followed by a period of tumor expression. This necessitates the follow-up period to be sufficiently long to include all Rn-d induced lung cancer cases. Therefore all epidemiological studies are truncated studies which do not fulfill this important criterion. Non-occupational Rn-d indoor exposure occurs at typically low exposure rates. Therefore it is likely that the latent period is increased up to 40 years as compared to the 10 to 20 years observed with uranium miners, who most probably received their Rn-d exposures at high exposure rates over a period of a few years during the early phase of uranium mining (Steinhäusler, in press).

The choice of mathematical risk model used to assess the lifetime risk, i.e. relative vs. absolute risk model, is often overrated in its importance because the input data used are associated with large and often unquantifiable uncertainties. Furthermore, in case of complete follow-up of the Rn-d exposed population groups, both models will give the same result (BEIR-II, 1980). The multiple uncertainties in the data used to assess the lung cancer risk from Rn-d inhalation is reflected in the wide range of values published, differing by more than an order of magnitude (Figure 4).

The information provided by epidemiology, dosimetry and animal studies with regard to indoor Rn-d exposure can be summarized as follows:

1) Physical characteristics of the air inhaled (e.g. fraction of unattached Rn-d), ventilation of the room and physiological parameters (e.g. breathing rate) can influence significantly the exposure-dose conversion for the individual inhabitant, with the most frequently occurring value for indoor Rn-d exposure of about 5 mSv/WLM.
2) Cigarette smoking may act as a promotor for lung cancer induction.

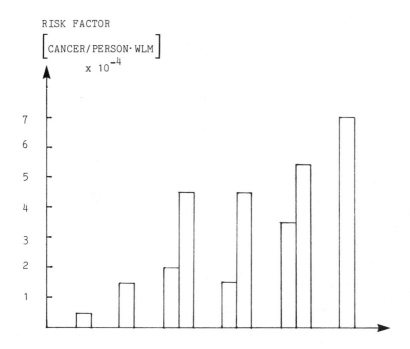

Figure 4. Published lifetime risk factors for Rn-d related lung cancer risk.

3) Up to 100 WLM of cumulative exposure, lung cancer risk shows
    linear dependence on Rn-d exposure; this is followed by a
    plateau region up to about 500 WLM.

Due to the various sources of uncertainties associated with both
groups of input data (dose and effect) needed for risk assessment,
at present it is only possible to give an upper limit for the pure
Rn-d related cancer risk.

Using the miner data for this purpose may be overestimating the
risk due to indoor exposure, since it includes risks from simultaneous exposure to external gamma radiation, long lived alpha
emitters. This effect, however, may be outweighed   by the fact
that smokers amongst miners are generally twice as frequent as the
number of smokers among members of the public. Therefore risk
estimates derived from miners may only reflect the risk of male
smokers exposed to Rn-d but may not be valid for children and women.
Altogether it can be assumed that such a risk factor obtained from
miners represents an upper limit for the risk associated with indoor
exposure.

Another possibility is the use of lung cancer incidence amongst
non-smoking members of the public, representing the maximum value
that can be attributable to Rn-d only. Because of statistical
problems with a rare event, such as lung cancer amongst non-smokers,
estimates of annual rates range from about 20 to 300 cases per 10
men depending on the age (Steinhäusler, in press).

A third possibility consists of comparing the theoretically
calculated lung cancer rate based on risk coefficients derived from
miners with the actual cancer occurrence among non-miners, derived
from Rn-d exposure assessment in dwellings and using appropriate exposure-dose conversion factors (Steinhäusler et al., 1983;
Edling, 1983).

From the above it can be concluded that the risk for lung
cancer induction from chronic indoor exposure to Rn-d is
unlikely to be higher than 1.10^{-4}/mSv. In order to understand the
magnitude of this risk it has to be emphasized that man can be
exposed to a multitude of different hazardous materials in the
indoor atmosphere besides Rn-d, such as formaldehyde, nitrogen
dioxide, carbon monoxide, nitrosamines, polyaromatic hydrocarbons,
volatile organic compounds, asbestos and pesticides (Gammage and
Kaye, 1985).

Literature Cited

Busigin,A., Van der Vooren,A.W., Babcoock, J.C. and C.R. Phillips,
The Nature of Unattached RaA (Po 218) Particles, Health Phys.40: pp
333-344 (1981).

Chameaud,J., Perraud,R., Lafuma,J. and R. Masse, Cancers Induced by
Rn-222 in the Rat, In: Proc. of Spec.Meet. on Ass. of Radon and
Radon Daughter Exposure and Related Biological Effects (G.F. Clemen-
te, A.V. Nero, F. Steinhäusler and M. R. Wrenn, eds) p. 198,
R.D. Press, Salt Lake City (1982).

Chameaud,J., and J. Lafuma, Experimental Study of Cancer Induction
with Rn 222 Daughters, Rad.Prot.Dosimetry 7: 385 (1984).

Congress of the United States,Joint Committee on Atomic Energy, Part I and II (1967).

Cross,F.T., Palmer,R.F., Filipy,R.E., Busch,R.H. and B.O. Stuart, Study on the Combined Effects of Smoking and Inhalation of Uranium Ore Dust, Radon Daughters and Diesel Oil Exhaust Fumes in Hamsters and Dogs, Pacific Northwest Laboratory Rep. No. PNL-2744, Richland, National Technical Information Service, Springfield, VA., USA (1978).

Cross,F.T., Palmer,R.F., Dagle,G.E., Busch,R.H., and R.L. Buschbom, Influence of Radon Daughter Exposure Rate, Unattached Fraction and Disequilibrium on Occurrence of Lung Tumors, Rad.Prot.Dosimetry 7: 381 (1984).

Cross,F.T., Palmer,R.F., Busch,R.H., Filipy,R.E., Dagle,G.E. and B.O. Stuart, Carcinogenic Effects of Radon Daughters, Uranium Ore Dust and Cigarette Smoke in Beagle Dogs, Health Phys. 42: 33 (1982).

Edling,C., Lung Cancer and Radon Daughter Exposure in Mines and Dwellings: study no. V. Linköping University, Medical Dissertation No. 157, Dept. of Occup. Med., Linköping, Sweden (1983).

Evans,R.D., Harley,J.H., Jacobi,W., McLean,A.S., Mills,W.A. and C.G. Stewart, Estimate of Risk from Environmental Exposure to Radon-222 and its Decay Products, Nature 290: 98 (1981).

Gammage,R.B.and S.V. Kaye (eds), Indoor Air and Human Health, Lewis Publ. Inc., Chelsea, MI., USA (1985).

Härting,F.H., und W. Hesse, Der Lungenkrebs, die Bergkrankheit in den Schneeberger Gruben, Teil I, Eulenbergs Vierteljahrschriftf., Gerichtliche Medizin und öffentliches Gesundheitswesen, neue Folge 30: 296 (1879).

Hofmann,W., and F. Daschil, The Relevance of Animal Models for Radionuclide Inhalation in Man, in Current Concepts in Lung Dosimetry, Proc. of the 30th Annual Meeting of the Radiation Research Society at Salt Lake City 1982 (D.R. Fisher, ed) pp 95-102, Utah (1983).

International Commission on Radiological Protection, Limits for Inhalation of Radon Daughters by Workers, ICRP Publication, No.32, Pergamon Press (1981).

Lundin,F.E., Wagoner,J.K. and V.E. Archer, Radon Daughter Exposure and Respiratory Cancer, Quantitative and Temporal Aspects, Nat. Inst. Occup. Safety and Health/Nat. Inst. Env.Sciences, Joint Monograph No.1, US Dept. of Health, Education and Welfare, Public Health Service (NTIS, No. PB 204871), Washington, D.C. (1971).

Masse,R., Histiogenesis of Lung Tumors Induced in Rats by Inhalation of Alpha Emitters: an Overview, in Pulmonary Toxicology of Respirable Particles (C.L. Sanders et al., eds) p. 498, CONF - 791002, National Technical Information Service, Springfield, USA (1980).

Muller, J., Wheeler, W.C., Gentleman, J.F. Suranyi, G. and R. Kusiak, Study of Mortality of Ontario Miners, Part I, Rep. Ontario Ministry of Labour,Ontario Workers' Compensation Board, Atomic Energy Control Board of Canada (1983).

OECD - Nuclear Energy Agency, Dosimetry Aspects of Exposure to Radon and Thoron Daughter Products, NEA Experts Report ISBN 92-64-12520-5, OECD, Paris (1983).

Radford,E.P., and K.G. St.Clair Renard, Lung Cancer in Swedish Iron Miners Exposed to Low Doses of Radon Daughters, The New England Journal of Medicine, 310: 1485 (1984).

Royal Commission,Report of the Royal Commission on the Health and Safety of Workers in Mines (J.M. Ham, F.R. Hume, C.C. Gray, A.L. Gladstone, J. Beaudry, E.A. Perry, R.P. Riggin, eds), Ministry of the Attorney General, Ontario, Canada (1976).

Snihs,J.O., The Approach to Radon Problems in Non-Uranium Mines in Sweden, in Proc. Third Int. Congr. Prot. Assoc. p 909, (IRPA), U.S.A.E.C. Conf. 73097-P.2. (1973).

Steinhäusler,F. and E. Pohl, Lung Cancer Risk for Miners and Atomic Bomb Survivors and its Relevance to Indoor Radon Exposure, Radiation Protection Dosimetry: Vol.7, No.1-4: 389-394 (1983).

Steinhäusler,F., and Pohl,E., Lung Cancer as a Result of Energy Conservation and Waste Recycling, Proc. XI. Regional Congress of IRPA, Vol.I: pp 258-262, Vienna, Austria (1983).

Steinhäusler,F., Hofmann,W., Pohl,E. and J. Pohl-Rüling, Radiation Exposure of the Respiratory Tract and Associated Carcinogenic Risk due to Inhaled Radon Daughters, Health Physics Vol.45, No.2: 331-337 (1983).

Steinhäusler,F., The Epidemiological Evidence for Health Risks, in Radon and its progeny in indoor air, (A. Nero et al., eds), CRC-Press, (in press).

United Nations Scientific Committee on the Effects of Atomic Radiation (UNSCEAR), Sources and Effects of Ionizing Radiation, Report to the General Assembly with Annexes, United Nations, N.Y., USA, (1977).

US-Environmental Protection Agency, Draft Environmental Impact Statement for Remedial Action Standards for Inactive Uranium Processing Sites (40 CFR 192), US-EPA Rep. No. EPA 520/4-80-011 (1980).

US-National Academy of Sciences, National Research Council BEIR-II, The Effects on Populations of Exposure to Low Levels of Ionizing Radiation, Washington, DC, USA (1980).

US-National Academy of Sciences, National Research Council BEIR-II: Biological Effects of Ionizing Radiation, Washington, D.C., USA (1972).

Weibel E.R.,   Morphometry of the Human Lung, Springer Verlag, Berlin (1963).

Williams,   L.   R. ,   Labour   viewpoint,   in   <u>Proc.Int.Conf.</u> <u>on</u> <u>Occupat.</u> <u>Rad.</u>   <u>Safety</u> <u>in</u> <u>Mining</u> (H.   Stocker,   ed) 29, Canadian Nucl. Assoc., Toronto, Canada (1985).

Yeh,H.C.,   Schum,   G.M.,   Models of the Human Lung Airways and their Application to Inhaled Particle Deposition,   <u>Bull.Math.Biol.</u> 42:   pp 461-480 (1980).

RECEIVED September 9, 1986

Chapter 32

Critique of Current Lung Dosimetry Models for Radon Progeny Exposure

Edward A. Martell

National Center for Atmospheric Research, Boulder, CO 80307

Based on studies of the uranium miners, elaborate models have been developed for estimating the risks of lung cancer to non-smokers in the general population from exposure to indoor radon progeny. These models can be faulted on mechanistic, dosimetric, and epidemiological grounds. Even the basic model assumption, that the relevant lung dose is the alpha radiation dose to basal cells of the bronchial epithelium, is no longer valid. These models are based on a simplistic correlation of risk versus radiation dose, with no consideration of multistage processes which may explain latent-period variations and the age-incidence of lung cancer. Questions of considerable relevance–such as the minimum tissue volume for tumor induction, or synergism between radon progeny and tobacco smoke–have not been taken into account. Because they are not applicable to groups at highest risk–i.e., smokers and passive smokers–such models, even after improvement, will yield qualitative risk estimates of marginal value.

Elaborate radiation dosimetry models have been developed for the purpose of estimating risks of lung cancer to nonsmokers in the general population due to exposure to indoor radon progeny (United Nations, 1977; National Academy of Sciences, 1980; Harley and Pasternack, 1981; ICRP, 1981; NCRP, 1984). These models involve assumptions that can be seriously questioned on the basis of epidemiological, dosimetric and mechanistic considerations. Some alternative dosimetric concepts have been discussed by Hofmann (1983). Other serious shortcomings of these models have been identified (Ellett and Nelson, 1985). One of the basic model assumptions–that basal cells of the bronchial epithelium are the target cells for induction of bronchial cancer–must now give way to impressive evidence that secretory cells are the primary progenitor cells for malignant transformations (Keenan *et al.*, 1982; McDowell and Trump, 1983). Questions of major importance–e.g., the minimum cell population for bronchial cancer induction, the nature of multistage processes that may explain latent period variations and the age-incidence of lung cancer, and the relative merits of alpha and β^- radiation interactions

0097-6156/87/0331-0444$06.00/0
© 1987 American Chemical Society

as initiators and promoters in the progression of malignancy–have not even been considered.

It is widely recognized that failure to assess the effect of smoking on risks of radon exposure, a common failure of all current models, may result in a serious overestimation of the lung cancer risks to nonsmokers. Thus, for example, it is acknowledged (National Academy of Sciences, 1980, p. 328), "If the lung-cancer risk after radiation exposure is proportional to the usual age-specific rates for smokers and nonsmokers, then the estimates of excess risk should be increased by about 50% to apply to smokers, and reduced by a factor of about 6 for nonsmokers, as well as delayed in time." The epidemiological evidence for lung cancer in passive smokers, plus other considerations discussed in this paper, indicate that lung cancer in nonsmokers resulting from indoor radon progeny exposure alone may be a relatively rare disease of old age, with an incidence that is far lower than current model predictions.

In this paper I review epidemiological and experimental evidence which has considerable relevance to these issues. The age-incidence of lung cancer vs smoking rate and gender clearly indicates a multiplicative effect due to synergistic interactions between smoking and radon progeny exposure. Serious shortcomings of the limited evidence for an additive effect (Radford and Renard, 1984) are pointed out. Several lines of evidence indicate that bronchial cancer can be induced by irradiation of as little as 1 mg of epithelium at bifurcations, and that such "hot spots" may be attributable mainly to the selective deposition of mainstream cigarette smoke. Possible complementary roles of β^- and alpha radiation interactions as initiators and promoters in the multistage process of bronchial cancer induction also are discussed.

Epidemiological Evidence

The observed high incidence of lung cancer in uranium miners provided the first evidence that clearly implicated inhaled radon progeny in the etiology of human lung cancer (Altshuler et al., 1964). Respiratory cancer deaths in uranium miners not only increase with cumulative radon progeny exposure but also were observed to be about ten times higher in smoking uranium miners than in nonsmokers (Lundin et al., 1969). Commenting on these observations, Doll (1971) concluded: "The data fit the hypothesis that the agents interact to produce their effects by multiplication and the hypothesis that they act independently is hardly tenable (P < 0.001)." A more comprehensive follow-up of uranium miners (Archer et al., 1976) showed that, over a wide range of cumulative radon progeny exposures, miners who smoked more than 20 cigarettes per day had seven to nine times higher incidence of lung cancer deaths than nonsmokers–clearly indicating a multiplicative effect. It is particularly noteworthy that approximately the same ratio holds for the lifetime incidence of lung cancer deaths in male smokers and nonsmokers of the general population (Kahn, 1966), summarized in Figure 1. The implications are obvious–synergistic interactions between indoor radon progeny exposure and cigarette smoking also may explain the high incidence of lung cancer in smokers in the general population. This important possibility is reinforced by the experimental evidence reviewed below.

The age-specific incidence of lung cancer deaths in male smokers and nonsmokers (Figure 1) reflects the dominant influence of smoking rate and duration of smoking in years. When smoking rates are not taken into account, the age-specific incidence of lung cancer deaths in the general population appears to decrease markedly after age 65 (Kohn, 1978; Ellett and Nelson,

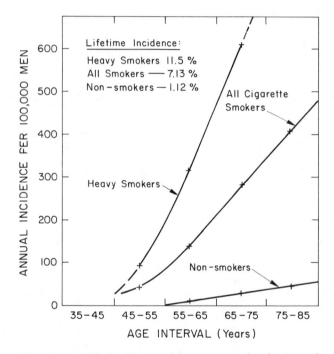

Figure 1. The age-specific incidence of lung cancer deaths in male cigarette smokers and nonsmokers (Kahn, 1966, Appendix Table A). The lifetime incidence for nonsmokers includes lung cancers attributable to passive smoking, asbestos inhalation, and other occupational exposures.

1985), a trend which has been attributed to competing risks. However, an alternative explanation is provided by the fact that smokers have a higher and earlier incidence of all common causes of death (Kahn, 1966) and thus a shorter life expectancy, as shown in Figure 2. The decrease in the incidence of lung cancer after age 65 is more logically attributed to the earlier depletion in the ranks of those at highest risk. The age-incidence gradually drops off from that influenced by heavy smokers to that for nonsmoking females. However, within each group classified by sex and smoking rate, the age-specific incidence of lung cancer increases markedly with age, as shown in Figure 1 for males and elsewhere for females (Hammond, 1966; Enstrom and Godley, 1980).

Below age 75, the incidence of lung cancer in nonsmoking females is only about one-third that for nonsmoking males (Enstrom and Godley, 1980). This sex difference can, in part, be attributed to higher levels of exposure to occupational factors and promoters in the work place for males. In addition, many of the lung cancers in nonsmokers can be attributed to passive smoking effects. Published evidence does not provide an adequate assessment of the lung cancer risk for passive smokers; however, the positive studies suggest a modest risk of about two to three times that of other nonsmokers (Samet, 1985). The prevalence of passive smoking among nonsmokers, approximately 63% in the United States (Friedman *et al.*, 1983), suggests that passive smoking may account for 50% to 80% of the lung cancers in nonsmokers. On this basis the lifetime incidence of lung cancer in nonsmokers who are not passive smokers would be about 0.2% for males and 0.1% for females.

One recent study of lung cancer mortality in Swedish iron miners (Radford and Renard, 1984) yielded results which suggest that the effects of smoking and exposure to radon progeny may be nearly additive and not synergistic. Based on these results, Radford and Renard propose that exposure to indoor radon progeny alone accounts for an appreciable number of lung cancers in the general population. These limited results are not very convincing because they are at odds with results for United States uranium miner studies (Lundin *et al.*, 1969; Archer *et al.*, 1976; Whittemore and McMillan, 1983) and for animal experiments (Chameaud *et al.*, 1982) which clearly indicate marked synergistic effects. Other recent studies of lung cancer in iron ore miners in Northern Sweden (Damber and Larsson, 1985; Jorgensen, 1984) show marked, multiplicative effects of smoking and underground mining, in disagreement with the Radford-Renard results. Radford and Renard included long-term ex-smokers with "nonsmokers" and included pipe and cigar smokers as well as short-term ex-smokers with "smokers." This is a highly questionable procedure, particularly in view of the results of Archer *et al.* (1976) who observed a systematic increase in lung cancers with cumulative radon progeny exposure only for smoking rates of 20 cigarettes per day or more. Light smokers had fewer lung cancers and more excess deaths from other respiratory disease at lower and intermediate cumulative radiation exposures, indicating that the promotion of lung cancer is more effective at higher smoking rates. Archer *et al.* (1976) also show a much higher incidence of lung cancers in ex-smokers than in nonsmokers over a wide range of cumulative radiation exposures. Thus, the best way to test the multiplicative effects of smoking and radon progeny exposure would be to compare lung cancer deaths vs cumulative radon progeny exposure only between smokers of one pack per day or more and true nonsmokers who never smoked. The mortality ratios for current cigar and pipe smokers, current cigarette smokers, and heavy smokers (>39 cigarettes per day) are 1.7,

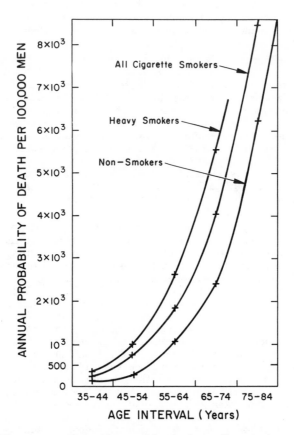

Figure 2. The age-specific incidence of deaths from all causes in male cigarette smokers and nonsmokers (Kahn, 1966, Appendix Table A).

10.9, and 23.6, respectively (Kahn, 1966). Clearly, pipe and cigar smokers should not be included in epidemiological studies of multiplicative effects.

Radford and Renard also assumed that for the miners, the relative risks of lung cancer for equal current smoking status were independent of age. On the contrary, there is an extraordinary dependence of lung cancer risk on duration of smoking in years (Peto and Doll, 1985) and thus with age. For a given smoking rate, the annual excess incidence of lung cancer is about 20 times higher after 30 years of smoking than it is after 15 years. The age-specific incidence of lung cancer in smokers, and of other common human cancers, is the basis for proposed multistage models of carcinogenesis (Armitage and Doll, 1961; Doll, 1971).

Synergistic Interactions

The observed lung cancer mortality in smoking and nonsmoking miners (Lundin *et al.*, 1969; Archer *et al.*, 1976; Whittemore and McMillan, 1983; Damber and Larsson, 1985; Jorgensen, 1984) clearly indicates marked multiplicative effects. This implies a synergism for exposure to both radon decay products and cigarette smoke. Such synergism has been confirmed in animal experiments. Large numbers of mice, exposed continuously to inhalation of high concentrations of radon and radon decay products attached to natural indoor aerosols, experienced substantial life shortening and weight loss compared to controls, but did not develop lung cancer (Morken and Scott, 1966; Morken, 1973). Experiments in which rats inhaled very high concentrations of cigarette smoke for prolonged periods resulted in no malignant lung tumors (Chameaud *et al.*, 1982). Rats exposed to very large cumulative doses of radon decay products–500 and 4,000 WLM (working level months)–developed some lung tumors and both the incidence and the degree of malignancy were increased by exposure to cigarette smoke (Chameaud *et al.*, 1982), clearly demonstrating synergism at these high exposure levels. These authors pointed out that the combined effect of cigarette smoke and radon is not additive because no cancers were found in rats exposed only to smoke, thus indicating only a promoting effect for cigarette smoke.

Synergistic interaction mechanisms which explain how cigarette smoking may enhance the risk of lung cancer from indoor radon progeny exposure have been identified. For a given indoor radon concentration and ventilation rate, the concentration of airborne radon progeny is substantially enhanced in smoke-filled rooms compared to that in clean, filtered indoor air (Martell, 1983; Bergman and Axelson, 1983). Inhaled smoke tar particles and attached radioisotopes are selectively deposited and retained at respiratory tumor sites. Because smoke tars are highly resistant to dissolution in lung fluid (Ermala and Holsti, 1955), the radiation insult is localized at tissue sites with concentrated smoke tar deposits, due to a high degree of radioactive decay of short-lived radon decay products at these sites before clearance. These processes may explain the surprising differences in patterns of cancer incidence in smokers and passive smokers (Higgins, 1985), discussed below. Experimental evidence which indicates that bronchial tumors are induced by radiation in small volumes of bronchial epithelium and that radioisotopes in mainstream cigarette smoke contribute substantially to the high risk of lung cancer in smokers is briefly reviewed in the next two sections of this paper.

Target Cell Population

Current lung dosimetry models are based on the assumption that basal cells of the bronchial epithelium are the critical target cells for malignant transformation and that the alpha dose to these cells is the relevant radiation dose.

Because the alpha particle range is short, the thickness of the bronchial epithelium and mucus layers and the variable depth of basal cells within the bronchial tree are considered to be important dose-determining factors. The alpha dose is averaged over large tissue volumes, ranging from 45 grams of the upper bronchial tree (ICRP, 1981) to the regional dose in a few grams of a single bronchial generation (NCRP, 1984). However, there is now compelling evidence indicating that secretory cells rather than basal cells may be the primary targets for malignant transformation and that tumors may be induced focally in much smaller volumes of tissue.

A comprehensive study of the repair of injured tracheal epithelium of hamsters, in which tritiated thymidine labelling was used to identify mitotically active cells, demonstrated the dominant role of secretory cells in the regeneration of damaged epithelium (Keenan et al., 1982). Secretory cells outnumber basal cells by more than five to one and pass through DNA synthesis into mitosis at twice the rate of basal cells. The columnar secretory (mucus-secreting) cells of the bronchial epithelium extend all the way from the basal lamina to the surface of the airway. These numerous and rapidly dividing secretory cells are identified as the major progenitors for bronchiogenic neoplasms, regardless of phenotype (McDowell and Trump, 1983).

Evidence that alpha radiation doses from ^{210}Po (polonium-210) and from indoor radon decay products are implicated in the etiology of bronchial cancer in cigarette smokers is now substantial. Surprisingly high concentrations of ^{210}Po were found at individual bifurcations in seven of 37 cigarette smokers (Little et al., 1965). It was later pointed out that the observed high concentrations of ^{210}Po at the bifurcations of smokers could be explained by the presence of insoluble ^{210}Pb-enriched smoke particles produced by combustion of tobacco trichomes in burning cigarettes (Martell, 1974; 1975). This possibility has been confirmed by Radford and Martell (1977), who determined that high concentrations of ^{210}Po at bifurcations of smokers were invariably accompanied by higher concentrations of ^{210}Pb. Little et al. (1965) estimated that the alpha radiation dose due to ^{210}Po at "hot spots" in the bronchial epithelium of smokers would be about 20 rads in 25 years. This component of the alpha radiation dose at bronchial bifurcations of smokers can be attributed to the accumulation and persistence of ^{210}Pb-enriched particles in lesions with cilia absent and carcinoma in situ, lesions which increase with smoking rate and duration of smoking in years (Auerbach et al., 1961). However, during the earlier years of smoking, an important contribution to the alpha radiation dose at bronchial bifurcations of smokers can be attributed to the indoor radon decay products, ^{218}Po and ^{214}Po (Martell and Sweder, 1981, 1983; Martell, 1983a), discussed below.

Another issue of particular significance is that of the minimum, critical tissue volume or critical cell population for cancer induction. Lung dosimetry models are concerned with small differences in the alpha dose to basal cells within a relatively large volume of the upper bronchial tree. However, as Brues (1954) pointed out, radiation-induced tumors arise focally in small tissue volumes and tumors can be induced by β^- irradiation of 10^7 to 10^8 cells–i.e., only 10 to 100 mg of tissue. For chronic irradiation by alpha- and β^--emitting radon progeny at bronchial bifurcations, a cumulative total of 10^7 to 10^8 cells are irradiated in only ten cell generations within tissue volumes of 1.0 to 10 mg of bronchial epithelium. Thus it is reasonable to hypothesize that chronic irradiation of a dividing cell population in a few milligrams of bronchial epithelium may be adequate for the induction of malignant transformations. The minimum cell population for malignant transformation due to chronic irradiation by internal emitters is an unresolved question of great radiobiological significance.

Dilute Aerosols vs Mainstream Smoke

Current practice in lung dosimetry involves detailed calculations of patterns of lung deposition and clearance for dilute radon progeny aerosols in indoor air in the form of unattached ions and the fraction attached to natural, soluble aerosols. Consideration is given to the influence of the particle size spectrum, breathing patterns, and the degree of radon daughter disequilibrium, on deposition and clearance patterns and on bronchial airway dosimetry. Such model calculations provide estimates of the natural alpha radiation dose distribution for inhaled, natural radon progeny aerosols. However, exposure to this natural background radiation appears to have little direct relevance to the induction of human lung cancer. This possibility, first indicated by the experimental results of Morken and Scott (1966), is strongly reinforced by a consideration of the properties and distribution of mainstream cigarette smoke as well as the pattern of cancers in passive smokers.

Published results on the concentration and size distribution of small particles in mainstream smoke vary widely, with concentrations ranging from 10^7 to 10^{11} cm^{-3} and with NMAD (number median aerodynamic diameter) ranging from 0.2 to 0.7 μm (Ishizu *et al.*, 1978). The MMAD (mass median aerodynamic diameter) of undiluted mainstream smoke particles ranges between 0.93 and 1.00 μm (Langer and Fisher, 1956; Holmes *et al.*, 1959). Lower values of the MMAD for diluted mainstream smoke, which decreased with degree of dilution, are reported by Hinds (1978). However, the particle size distributions for mainstream smoke appear to have little relevance to its retention and distribution in the lung, for reasons discussed below. Note that the concentration of tars in mainstream smoke is about 1,000 times that of air in smoke-filled rooms.

The deposition and retention of dense, undiluted mainstream cigarette smoke is substantially greater than that expected for dilute aerosols of the same size distribution. Hinds *et al.* (1983) review several earlier studies which showed average depositions ranging from 82 to 97% of the mass of inhaled smoke. With improved techniques, Hinds *et al.* (1983) obtained results which varied widely for individual smokers, with an average deposition of 57% for male smokers–still about three times higher than expected for dilute particle behavior. By comparison, the retention of highly diluted mainstream cigarette smoke in the human lung was shown to be only 15-20% (Porstendorfer and Schraub, 1972), in good agreement with lung model predictions. It is evident that dilute aerosol behavior is not applicable to the deposition and retention of dense, undiluted mainstream smoke in the lungs of smokers.

This apparent contradiction is best explained by the ensemble behavior of dense mainstream smoke, described by Fuchs (1964). When undiluted mainstream tobacco smoke is passed horizontally into a chamber, the dense smoke column settles as an ensemble, with a gravitational settlement rate equivalent to that for individual particles of 47 μm diameter. Such behavior can explain the remarkable retention of undiluted smoke (Hinds *et al.*, 1983) as well as the highly nonuniform pattern of deposition for inhaled smoke tars observed by Ermala and Holsti (1955). The heaviest tar deposits occur where the dense smoke column strikes directly on projecting surfaces of the pharynx and the larynx, and at the tracheal and bronchial bifurcations (Ermala and Holsti, 1955). If gravitational sedimentation has a strong influence on the bronchial deposition of mainstream smoke, the orientation of bifurcations in the bronchial tree should influence deposition patterns, with highest deposition at the carina of bifurcations of large branching angle where the inspired air flow is directed downward. The actual pattern for the deposition of mainstream smoke in the bronchi of smokers should be determined in experiments in which ^{212}Pb-tagged mainstream smoke is passed through hollow casts of the tracheobronchial tree and the concentrations of ^{212}Pb-tagged smoke tar deposits at brochial bifurcations of various orientation are determined.

The possibility that it is the constituents of mainstream smoke in concentrated smoke tar deposits at tumor sites that account for the high incidence of respiratory cancers in smokers is further reinforced by the surprisingly different pattern of cancers in passive smokers. Passive smokers, who inhale the same dilute airborne particles in indoor air as that which smokers inhale between puffs and between cigarettes, have a surprisingly low incidence of lung cancer–no more than two to three times that of other nonsmokers (Samet, 1985). On the other hand, passive smokers have a relatively high incidence of paranasal sinus cancers (Hirayama, 1983), lymphomas, leukemia, and other cancers (Sandler et al., 1985a, b). This pattern of cancers in passive smokers may be attributed to their inhalation of dilute smoke tar particles and attached radon progeny in indoor air by normal nose breathing. A fraction of the inhaled particles is trapped on hairs in the nasal passages. Particles which are deposited in the normal, healthy lungs of passive smokers are readily cleared and, in part, accumulate at secondary tumor sites, including the lymph nodes and bone marrow. Thus, the otherwise surprising pattern of excess cancers in passive smokers appears to be explainable on the basis of differences in the tissue distribution for inhaled smoke tar particles and for the associated radiation insult.

Dosimetric Considerations

Based on the epidemiological and experimental evidence discussed in this paper, several basic assumptions common to all current lung dosimetry models are subject to question. There now is persuasive evidence that secretory cells rather than basal cells of the bronchial epithelium are the primary progenitor cells for malignant transformations (McDowell and Trump, 1983). On this basis the depth of the basal cell layer cannot be a significant factor in alpha radiation dosimetry. There also is good evidence which suggests that malignant transformations may be induced focally in as little as 1.0 to 10 mg of bronchial epithelium–some 10^6 to 10^7 cells.

Current models also fail to give adequate consideration to published evidence which shows the marked influence of solubility and other properties of alpha-emitting radioisotopes on their effectiveness as carcinogens. As Morken (1973) has pointed out, inhaled radon and its short-lived decay products on natural aerosols are particularly ineffective in the production of lung tumors in experimental animals, whereas inhaled polonium-210 and plutonium are highly effective. The effectiveness of low doses of ^{210}Po has been demonstrated in experiments with hamsters (Little et al., 1975). For the induction of lung cancer by plutonium, the RBE (relative biological effectiveness) ranges from about 23 for soluble compounds to about 85 for insoluble compounds compared with radiation from intratracheally administered beta-gamma emitters (ICRP, 1980). The effectiveness of both soluble and insoluble plutonium can be attributed to their persistence in tissue with a highly nonuniform distribution (ICRP, 1980, Figures 26-27), giving rise to a correspondingly nonuniform pattern of alpha interactions with cells. By contrast, the alpha radiation from the natural, soluble, short-lived radon decay products is distributed in a random, nonpersistent, relatively uniform pattern of cellular interactions. The implications of these differences for cancer in active smokers, passive smokers, and true nonsmokers are clear. In active smokers, the radiation insult is localized at tissue sites with concentrated smoke tar deposits–the respiratory tumor sites (Ermala and Holsti, 1955). In passive smokers, radon progeny associated with diluted smoke tar particles that are inhaled by normal nosebreathing give rise to a quite different tissue distribution of smoke tars and radioactivity, and a correspondingly different pattern of tumors. True nonsmokers exposed to indoor radon progeny on natural, soluble aerosols alone, should experience a negligible risk of lung cancer.

The special importance of radon progeny as well as polonium-210 in mainstream cigarette smoke for the induction of bronchial cancer in smokers has been overlooked because of the relatively small volume of inhaled air involved. A typical smoker will inhale only about 10 ℓ of air through burning cigarettes into mainstream smoke per day compared to a total volume of about 10^4 ℓ of air inhaled daily. However, the smoke tar concentration of mainstream smoke is about 1,000 times greater than that of indoor air which is inhaled between puffs and between cigarettes. The deposition and retention of dense, mainstream smoke also is several times higher than that for dilute aerosols of the same size distribution (Hinds et al., 1983). If this difference is explained by the ensemble behavior of the dense mainstream column described by Fuchs (1964), the smoke tars should be concentrated selectively at the carina of bronchial bifurcations, particularly at bifurcations of widest angle and where the inspired mainstream smoke column is directed downward. The more concentrated smoke tar deposits at bronchial bifurcations of active smokers will persist sufficiently long to allow almost complete radioactive decay of short-lived radon progeny, due to the far less effective clearance by ciliary streaming of mucous at bifurcations (Hilding, 1957). Clearance by the blood circulatory system is also very slow because smoke tars are highly resistant to dissolution (Ermala and Holsti, 1955). The comparative stasis of smoke tars, ^{210}Po, and radon progeny at bifurcations of smokers and the prolonged action of these agents on the epithelium at bifurcations can explain the progressive development of lesions with cilia absent and carcinoma *in situ*, which occur with a frequency that increases with smoking rate and duration of smoking in years (Auerbach et al., 1961). The presence of high concentrations of ^{210}Po in the bronchial epithelium at bifurcations of about 20% of older smokers (Little et al., 1965) can be explained by the accumulation and persistence of insoluble, ^{210}Pb-enriched particles present in mainstream smoke in lesions with cilia absent (Martell, 1975; Radford and Martell, 1977).

Radon progeny which pass from indoor air through burning cigarettes into mainstream smoke will contribute an alpha radiation dose comparable to that from ^{210}Po at smoke tar deposit sites during the early years of smoking. Recent surveys indicate that indoor radon levels in the United States exceed 4 pCi ℓ^{-1} in about 18.5% of homes and range up to more than ten times this level in homes with the highest readings (Hileman, 1983). In smoke-filled rooms with 4 pCi ℓ^{-1} of ^{222}Rn, the air concentrations of ^{218}Po, ^{214}Pb and ^{214}Bi will be about 4, 2, and 2 pCi ℓ^{-1}, respectively. Concentrations in mainstream smoke are about one half as high (Martell and Sweder, 1981). For ten draws of average puff volume of ~50 cm^3 per cigarette (Hinds *et al.*, 1983), the activity per cigarette in mainstream smoke is ~1.0 pCi of ^{218}Po and ~ 0.5 pCi each of ^{214}Pb and ^{214}Bi. There is ~0.11 pCi of ^{210}Po in the mainstream smoke of one nonfilter cigarette (Radford and Hunt, 1964). Assuming the same deposition pattern for mainstream smoke constituents and further assuming essentially complete decay of short-lived radon progeny at sites of concentrated smoke tar deposits, it is evident that the potential alpha dose from ^{218}Po and ^{214}Po approaches or exceeds that from ^{210}Po. The dose contribution of ^{218}Po and ^{214}Po will be relatively greater only during the early years of smoking, before the ^{210}Pb-enriched particles begin to accumulate in lesions at bifurcations. For smokers exposed to ≥ 4 pCi^{222}Rnℓ^{-1} the combined, 40 year cumulative alpha radiation dose-from ^{218}Po, ^{214}Po, ^{210}Po, thoron progeny, and ^{210}Po in ^{210}Pb-enriched particles in mainstream smoke, plus radon progeny inhaled between puffs and between cigarettes–is some 40 to 80 rads (800 to 1600 rem) in hot spots at bronchial bifurcations. This appears to be an effective carcinogenic dose of alpha radiation (Little *et al.*, 1975). Alpha particles also produce exceptionally high concentrations of ions, free radicals, and effective chemical mutagens within alpha particle tracks and inside the protective membrane of the cells which they traverse (see below).

Mechanisms

In all current lung dosimetry models the relevant radiation dose has been assumed to be the average alpha radiation dose to a large basal cell population. However there is no discussion of the specific mechanisms involved which can justify such an assumption. The fact that for lung cancer induction, the RBE for the alpha particle-emitting actinides is at least 20 times that for the same dose of X-rays (ICRP, 1980) appeared to provide some indirect support for it. It also should be noted that alpha particles are exceptionally effective in damaging or killing the cells whose nuclei they traverse (Goldman, 1976; Hofmann, 1983). However, premalignant and fully malignant cells are viable cells, with no serious loss in mitotic or metabolic capability, and with a proliferative advantage over normal diploid cells. In view of these apparent contradictions, how can alpha radiation be so effective a carcinogen? This dilemma can be resolved if, in the multistage process of malignant transformation (Wolman, 1983), one of the primary roles of alpha radiation is that of promotion by killing cells, a hypothesis which has been proposed elsewhere (Martell, 1983b).

A high frequency of lethal alpha interactions in small volumes of bronchial epithelium can promote by stimulating the mitotic activity of the surrounding dividing cell population. Dividing cells are far more vulnerable to mutagenic transformation than cells during interphase (Evans, 1962). Initiation can be attributed to a high frequency of non-lethal radiation interactions with this dividing cell population, by β^- interactions as well as by non-lethal alpha interactions. Tumor progression involves both the selective proliferation of premalignant cells in the mitotically active cell population and the induction of further chromosome changes in premalignant cells which enhance their proliferative advantage and give rise to progressively greater potential malignancy.

It is recognized that the proposed hypothesis is at odds with the view held by some radiobiologists that a more uniform distribution of alpha-emitting radionuclides is more carcinogenic than a highly nonuniform distribution (National Academy of Sciences, 1976). However, the latter viewpoint is based on a consideration of experiments with inhaled actinides which have a highly nonuniform distribution and does not explain why the RBE is several times higher for insoluble than for "soluble" actinides (ICRP, 1980). And this view is completely at odds with the important implications of animal experiments which demonstrate the ineffectiveness of natural, soluble radon progeny aerosol for lung cancer induction (Morken and Scott 1966; Morken, 1973). It appears that, for effective carcinogenic action, alpha particle-emitting radioisotopes must be present in a distinctly unnatural, nonuniform distribution and in a persistent form in tissue.

It has been widely assumed that radon progeny exposure and cigarette smoking are two distinctly separate and independent sources of lung cancer risk (Evans *et al.*, 1981; Harley and Pasternack, 1981) presumably because of early speculations that lung cancer in smokers can be attributed to suggested chemical carcinogens in cigarette smoke. However, the epidemiological and experimental evidence reviewed in this paper clearly establish that indoor radon progeny are implicated in the etiology of respiratory cancers in smokers. This possibility is strongly reinforced by a consideration of the exceptionally high concentrations of effective chemical mutagens produced inside the cells which are subjected to alpha interactions. Each alpha particle produces $\sim 2 \times 10^5$ ion pairs and $\sim 5 \times 10^6$ excited molecules and free radicals in a short, straight track through the several cells in its path. Depending on energy, alpha particle track lengths range from 47 to 88 μm in tissue of unit density. Alpha radiolysis products include O, O_3, OH, H_2O_2, CH_2O, and other highly reactive chemicals. Thus, alpha interactions with cells produce highly effective mutagens and carcinogens at concentrations which are orders of magnitude higher than that from other sources of radiation or chemical pollutants. Alpha radiation also is exceptionally effective in the production of DNA double-strand breaks, the most critical lesions leading to the types of chromosomal aberrations observed in malignancy (Leenhouts and Chadwick, 1978). By contrast, the suspected chemical carcinogens in tobacco smoke are present at such low concentrations that their effectiveness in contributing to malignant transformations is admittedly only a matter of conjecture (Wynder and Hofmann, 1976). Thus, although lung cancer has been causally associated with cigarette smoking, it is now evident that indoor radon progeny and ^{210}Po must be the primary carcinogenic agents for the induction of bronchial cancer in smokers.

Once it becomes recognized and accepted that radon progeny are the primary agents of lung cancer in smokers and that radiation-induced bronchial cancer is a multistage process, it becomes possible to design lung dosimetry models which may explain latent period variations and differences in the age-incidence of lung cancer vs gender and smoking rate. In the development of lung cancer in uranium miners it is observed that "the shortest latent periods are found among those men who are elderly at the start of mining, who smoke heavily, and who have the most intense (radiation) exposure" (NCRP, 1984, p. 111). This is readily explainable when we recognize that lung cancer in smokers is due to synergistic interactions with indoor radon progeny and the multistage process is dependent on smoking rate, cumulative indoor radon progeny exposure, and especially duration of smoking in years. Because the progress of lung cancer is already in an advanced stage in older smokers, it requires a much shorter period of exposure to higher radiation levels in underground mines to complete the process. Discussion of dosimetry models which take into consideration the important parameters that determine lung cancer risks is outside the scope of this paper. However, a mathematical model which describes a multistage process for radiation-induced bronchial cancer in smokers, in which both the initial and final stages of DNA transformation are dose-rate-dependent and in which premalignant cells undergo exponential proliferation, is described elsewhere (Martell, 1983b).

Summary Discussion

It is evident that current lung dosimetry models for exposure to indoor radon progeny have many serious shortcomings and grossly overestimate lung cancer risks for nonsmokers in the general population. The epidemiological evidence that takes into account the influence of gender and smoking rate shows large differences in the age-specific incidence which increases with smoking rate, especially with duration of smoking in years; is higher for males than for females; and increases with age–and thus with cumulative radon progeny exposure–in each group. The relatively low incidence of lung cancer in nonsmokers is in large part attributable to occupational factors and passive smoking effects. In female nonsmokers who are not passive smokers lung cancer is a relatively rare disease of old age, with cumulative lifetime incidence of about 0.1 percent. It is unlikely that a significant fraction of this low incidence is attributable to exposure to high indoor radon progeny alone.

Experiments in which mice, rats, and dogs were exposed to inhalation of high concentrations of radon progeny on natural aerosols for prolonged periods resulted in life shortening and weight loss but no bronchial tumors (Morken, 1973). Lung dosimetry modelers have chosen to disregard these experimental results and their important implications, ostensibly because other experiments with rats gave a few positive results (Chameaud et al., 1982). However, in the latter experiments radon progeny exposure levels and aerosol properties were inadequately described and no control animals were used. Positive results showed only two very small lung tumors (< 2mm diameter) in 28 rats exposed to an estimated 500 WLM of radon progeny exposure and a higher incidence in 50 rats exposed to the unrealistically high cumulative exposure of 4,000 WLM. This is hardly an adequate basis for rejecting results of a series of well designed experiments (Morken and Scott, 1966; Morken, 1973) in which much larger numbers of animals, including adequate numbers of controls, were used. Morken's findings are not inconsistent with the epidemiological evidence for true nonsmokers. They also suggest that mammalian cell populations have evolved to cope with the adverse effects of natural radon progeny exposure to a remarkable degree. These experimental results clearly imply that the inhalation of indoor radon progeny associated with natural, soluble aerosols should not be expected to give rise to a significant risk of lung cancer.

The epidemiological evidence clearly indicates the multiplicative effects of smoking and exposure to radon progeny. Results of animal experiments also show marked synergism for exposure to both radon progeny and cigarette smoke. Axelson (1984) concludes that "there is even some justification for believing that radon daughter exposure could be a main initiator of lung cancer, whereas smoking might merely act as a promoter." Evidence reviewed in this paper indicates that smoking promotes lung cancer by localizing the radiation insult in concentrated smoke tar deposits at respiratory tumor sites. It also is evident that concentrated smoke tar deposits at bronchial bifurcations and other tumor sites in smokers are due to the highly selective

tive deposition and retention of dense mainstream smoke–not to a selective deposition of dilute aerosols inhaled between puffs and between cigarettes. The alpha dose contributed by [214]Po due to short-lived radon progeny in mainstream smoke exceeds that from [210]Po at above average indoor radon progeny levels. This [214]Po dose contribution increases with smoking rate and indoor radon progeny level. The importance of mainstream smoke and its radioisotope constituents as agents of respiratory cancers in smokers is indirectly reinforced by the very different pattern of excess cancers in passive smokers. Passive smokers inhale diluted smoke tar particles and attached radon progeny by normal nose breathing. Excess cancers in passive smokers include a surprisingly low incidence of respiratory cancers (Samet, 1985) as well as an unexpectedly high incidence of lymphomas and leukemia (Sandler *et al.*, 1985a,b) and paranasal sinus cancers (Hirayama, 1983). This otherwise surprising pattern of cancers in passive smokers may be explained by differences in the pattern of deposition and clearance for inhaled smoke tar particles and the consequent differences in the pattern of radiation exposure.

A review of the relevant literature has revealed a number of other serious shortcomings of current lung models. Secretory cells-not basal cells-are the primary progenitors for bronchial cancers. Thus, the depth of basal cells is not an important factor. Tumors may be induced focally in small tissue volumes, possibly as little as 1.0 mg of tissue– $\sim 10^6$ cells. Tumors induced by inhaled plutonium are attributable to a highly nonuniform distribution of alpha interactions with target cell populations. In the multistage process of malignant transformation alpha interactions may play a dual role of (1) promotion by lethal interactions at hot spots, and (2) initiation by nonlethal interactions in nearby dividing cell populations. β^- radiation also may contribute significantly to the initiation and evolution of malignant stem cells. It should be possible to model the multistage processes which explain the age-specific incidence of lung cancer in smokers as well as the observed large variations in latent period for lung cancer development.

For those who might wish to preserve the speculative hypothesis that lung cancer in smokers is due to suspected chemical carcinogens rather than to radon progeny, the exceptional effectiveness of alpha particles as carcinogens has been pointed out. Alpha particles produce extremely high concentrations of ions, free radicals, and highly effective chemical mutagens inside the protective membranes of cells they traverse. By contrast, the effectiveness of suspected chemical carcinogens at the very low concentrations present in cigarette smoke has not been established (Wynder and Hofmann, 1976).

The long list of serious shortcomings and questionable assumptions which characterize current lung dosimetry models brings to mind a comment by Erwin Chargaff: "One of the most insidious and nefarious properties of scientific models is their tendency to take over, and sometimes supplant, reality." Undoubtedly, exposure to high levels of indoor radon and its decay products contributes significantly to cancer risks and other chronic health effects, particularly in smokers and passive smokers. However, there is no adequate basis for predicting a significant risk of lung cancer for true nonsmokers exposed to high indoor radon progeny alone.

Acknowledgments

The National Center for Atmospheric Research is sponsored by the National Science Foundation.

Literature Cited

Altshuler, B., M. Nelson, and M. Kushner, Estimation of lung tissue dose from the inhalation of radon and daughters, *Hlth. Phys. 10*: 1137-1164 (1964).

Archer, V. E., J. D. Gilliam, and J. K. Wagoner, Respiratory disease mortality among uranium miners, *Ann. N. Y. Acad. Sci. 271*: 280-293 (1976).

Armitage, P. and R. Doll, Stochastic models for carcinogenesis, in Proceedings, 4th Berkeley Symposium on Mathematical Statistics and Probability, Vol. 4, pp. 19-38, Univ. Calif. Press, Berkeley (1961).

Auerbach, O., A. P. Stout, E. C. Hammond, and L. Garfinkel, Changes in bronchial epithelium in relation to cigarette smoking and in relation to lung cancer, *New Engl. J. Med. 263*: 253-267 (1961).

Axelson, O, Room for a role for radon in lung cancer causation?, *Medical Hypotheses, 13*: 51-61 (1984).

Bergman, H., and O. Axelson, Passive smoking and indoor radon daughter concentrations, *Lancet*: 1308-1309 (1983).

Brues, A. M., Ionizing radiations and cancer, *Adv. Cancer Res., 2*: 177-195 (1954).

Chameaud, J., R. Perraud, J. Chrétian, R. Masse, and J. Lafuma, Lung carcinogenesis during in vivo smoking and radon daughter exposure in rats, in **Recent Results in Cancer Research, 82**, 11-20, Springer-Verlag, Heidelberg (1982).

Damber, L. and L-G. Larsson, Underground mining, smoking, and lung cancer: A case-control study in the iron ore municipalities in Northern Sweden, *JNCI, 74*: 1207-1213 (1985).

Doll, R., The age distribution of cancer: Implications for models of carcinogenesis, *J. Roy. Statistical Soc.*, Series A, Part 2:133-155 (1971).

Ellett, W. H. and N. S. Nelson, Epidemiology and risk assessment: Testing models for radon-induced lung cancer, in **Indoor Air and Human Health** (R. B. Gammage and S. V. Kaye, eds) pp. 79-107, Lewis Publishers, Inc., Chelsea, Michigan (1985).

Enstrom, J. E. and F. H. Godley, Cancer mortality among a representative sample of nonsmokers in the United States during 1966-68, *J. Natl. Cancer Inst.* 65:1175-1183 (1980).

Ermala, P., and L. R. Holsti, Distribution and adsorption of tobacco tar in the organs of the respiratory tract, *Cancer, 8*: 673-678 (1955).

Evans, H.J., Chromosome aberrations induced by ionizing radiations, *Int. Rev. Cytol., 13*:221-321 (1962).

Evans, R. D., J. H. Harley, W. Jacobi, A. S. McLean, W. A. Mills, and C. G. Stewart, Estimate of risk from environmental exposure to radon-222 and its decay products, *Nature, 290*: 98-100 (1981).

Friedman, G. D., D. B. Petitti and R. D. Bawol, Prevalence and correlates of passive smoking, *Am. J. Public Health,* 73:401-405 (1983).

Fuchs, N. A., **The Mechanics of Aerosols**, pp. 46-49, MacMillan, New York (1964).

Goldman, M., An overview of high LET radiation effects in cells, in **The Health Effects of Plutonium and Radium** (W. S. S. Jee, ed.), pp. 751-766, J. W. Press, Salt Lake City, Utah (1976).

Hammond. E. C., Smoking in relation to the death rates of one million men and women, *Natl. Cancer Inst. Monograph, 19* (W. Haenszel, ed) pp. 127-204, U. S. Department of Health, Education, and Welfare, Bethesda, MD, January 1966.

Harley, N. H., and B. S. Pasternack, A model for predicting lung cancer risks induced by environmental levels of radon daughters, *Hlth. Phys.,* 40:307-316 (1981).

Higgins, I., Lifetime passive smoking and cancer risk, *Lancet,* 866-867, April 13, 1985.

Hilding, A. C., Ciliary streaming in the bronchial tree and the time element in carcinogenesis, *New Engl. J. Med, 256,* 634-640 (1957).

Hileman, B., Indoor air pollution, *Environ. Sci. Technol., 17* :469A-427A (1983).

Hinds, W. C., Size characteristics of cigarette smoke, *Am. Ind. Hyg. Assoc. J.,* 39:48-54 (1978).

Hinds, W., M. W. First, G. L. Huber and J. W. Shea, A method for measuring respiratory deposition of cigarette smoke during smoking, *Am. Ind. Hyg. Assn. J.,* 44:113-118 (1983).

Hirayama, T., Passive smoking and lung cancer: Consistency of association, *Lancet,* 1425-1426, December 1983.

Hofmann, W., Dosimetric concepts for inhaled radon decay products in the human lung, in **Current Topics in Lung Dosimetry** (D. R. Fisher, ed) pp. 37-43, CONF-820492, NTIS, U. S. Department of Commerce, Springfield, Virginia (1983).

Holmes, J. C., J. E. Hardcastle and R. I. Mitchell, The determination of particle size and electric charge distribution in cigarette smoke, *Tobacco Science,* 3:148-153 (1959).

ICRP Publication 31, **Biological Effects of Inhaled Radionuclides,** International Commission on Radiological Protection, Pergamon Press, Oxford (1980).

ICRP Publication 32, **Limits for Inhalation of Radon Daughters by Worker's,** International Commission on Radiological Protection, Pergamon Press, Oxford (1981).

Ishizu, Y., K. Ohta and T. Okada, Changes in the particles size and the concentration of cigarette smoke through the column of a cigarette, *J. Aerosol Sci.,* 9:25-29 (1978).

Jorgensen, H. S., Lung cancer among underground workers in the iron ore mine of Kiruna based on thirty years of observation, *Annals. Acad. Med., Singapore, 13* (2, Supplement) 371-377, April 1984.

Kahn, H. A., The Dorn study of smoking and mortality among U. S. veterans: Report on eight and one-half years of observation, National Cancer Institute Monograph No. 19, pp. 1-125, H. E. W., Bethesda, Maryland (1966).

Keenan, K. P., J. W. Combs and E. M. McDowell, Regeneration of hamster tracheal epithelium after mechanical injury, *Virchows Arch.* [Cell Pathol.] 41, Parts I, II, and II: 193-252 (1982).

Kohn, R. B., **Principles of Mammalian Aging**, 2nd Edition, Prentice Hall, Inc. (1978).

Langer, G. and M. A. Fisher, Concentration and particle size of cigarette-smoke particles, *A. M. A. Arch. Ind. Hlth.*, 13:373-378 (1956).

Leenhouts, H. P. and K. H. Chadwick, An analysis of radiation induced malignancy based on somatic mutation, *Int. J. Radiat. Biol.* 33:357-370 (1978).

Little, J. B., E. P. Radford, H. L. McCombs, and V. R. Hunt, Distribution of polonium in pulmonary tissues of cigarette smokers. *New Engl. J. Med.*, 273:1343-51 (1965).

Little, J. B., A. R. Kennedy and R. B. McGandy, Lung cancer induced in hamsters by low-doses of alpha radiation from polonium-210, *Science* 188:737-738 (1975).

Lundin, F. E., J. W. Lloyd, E. M. Smith, V. E. Archer and D. A. Holaday, Mortality of uranium miners in relation to radiation exposure, hard-rock mining and cigarette smoking-1950 through September 1967, *Hlth. Phys.*, 16:571-578 (1969).

Martell, E. A., Radioactivity of tobacco trichomes and insoluble cigarette smoke particles, *Nature*, 249:215-217 (1974).

Martell, E. A., Tobacco radioactivity and cancer in smokers, *Amer. Sci.*, 63:404-412 (1975).

Martell, E. A., and K. S. Sweder, The roles of polonium isotopes in the etiology of lung cancer in cigarette smokers and uranium miners, in **Radiation Hazards in Mining**, (M. Gomez, ed.) pp. 383-389, New York, Society of Mining Engineers, A. I. M. E., Kingsport Press (1981).

Martell, E. A., α-Radiation dose at bronchial bifurcations of smokers from indoor exposure to radon progeny, *Proc. Natl. Acad. Sci. USA*, 80:1285-1289 (1983a).

Martell, E. A., Bronchial cancer induction by alpha radiation: A new hypothesis, Paper C6-11 in **Proceedings of the 7th International Congress of Radiation Research**, (J. J. Broerse *et al.*, eds), Martinus Nijhoff, Amsterdam (1983b).

Martell, E. A., and K. S. Sweder, Properties of radon progeny in mainstream cigarette smoke and the alpha dose at segmental bifurcations of smoke, in **Current Topics in Lung Dosimetry**, (D. R. Fisher, ed.) pp. 144-151, CONF-820492, NTIS, U. S. Department of Commerce, Springfield, Virginia (1983).

McDowell, E. M. and B. F. Trump. Histogenesis of preneoplastic and neoplastic lesions in tracheobronchial epithelium, *Surv. Synth. Path. Res.* 2:235-279 (1983).

Morken, D. A., and J. K. Scott, The effects on mice of continual exposure to radon and its decay products on dust, University of Rochester Atomic Energy Project Report UC-669, 134 pp, November 1966.

Morken, D. A., The biological effects of radon on the lung, in **Noble Gases** (R. E. Stanley and A., A. Moghissi, eds.) pp. 501-506, ERDA and TIC CONF-730915, Library of Congress No. 75-27055 (1973).

National Academy of Sciences-National Research Council, **Health Effects of Alpha-Emitting Particles in the Respiratory Tract**, Report of AD Hoc Committee on "Hot Particles," U. S. Environmental Protection Agency, EPA 520/4-76-013, October 1976.

National Academy of Sciences–National Research Council, **The Effects on Populations of Exposure to Low Levels of Ionizing Radiation**, National Academy Press, Washington, D. C. (1980).

NCRP Report No. 78, **Evaluation of Occupational and Environmental Exposures to Radon and Radon Daughters in the United States**, National Council on Radiation Protection and Measurements, Bethesda, Maryland (1984).

Peto, R. and R. Doll, The control of lung cancer, *New Scientist*, 26-30, 24 January 1985.

Porstendorfer, J. and A. Schraub, Concentration and mean particle size of the main and side stream of cigarette smoke, *Staub-Reinhalt. Luft*, 32:33-36 (1972).

Radford, E. P. and V. R. Hunt, Polonium-210: A volatile radioelement in cigarettes, *Science, 143*, 247-249 (1964).

Radford, E. P., and E. A. Martell, Polonium-210: lead-210 ratios as an index of residence times of insoluble particles from cigarette smoke in bronchial epithelium, in **Inhaled Particles IV, Part 2** (W. H. Walton, ed.) pp. 567-580, Pergamon Press, Oxford (1977).

Radford, E. P. and K. G. St. Clair Renard, Lung cancer in Swedish iron miners exposed to low doses of radon daughters, *New Engl. J. Med.* 310:1485-1494 (1984).

Samet, J. M., Relationship between passive exposure to cigarette smoke and cancer, in *Indoor Air and Human Health*, (R. B. Gammage and S. V. Kaye, eds) pp. 227-240 (1985).

Sandler, D. P., R. B. Everson and A. J. Wilcox, Passive smoking in adulthood and cancer risk, *Am. J. Epidemiol.*, 121:37-48 (1985a).

Sandler, D. P., R. B. Everson, A. J. Wilcox and J. P. Browder, Cancer risk in adulthood from early life exposure to parent's smoking, *Am. J. Publ. Hlth*, 75:487-492 (1985b).

United Nations Scientific Committee on the Effects of Atomic Radiation Sources and **Effects of Ionizating Radiation**, United Nations, New York (1977).

Whittemore, A. S. and A. McMillan, Lung cancer mortality among U. S. uranium miners: A reappraisal, *JNCI* 71:489-499 (1983).

Wolman, S. R., Karyotypic progression in human tumors, *Cancer Metastasis Reviews*, 2:257-293 (1983).

Wynder, E. L., and D. Hofmann, Tobacco and tobacco smoke, *Seminars in Oncology, 3*: 5-15 (1976).

RECEIVED November 13, 1986

Chapter 33

Surveys of Radon Levels in Homes in the United States
A Test of the Linear–No-Threshold Dose–Response Relationship for Radiation Carcinogenesis

Bernard L. Cohen

University of Pittsburgh, Pittsburgh, PA 15260

The University of Pittsburgh Radon Project for large
scale measurements of radon concentrations in homes
is described.   Its principal research is to test
the linear-no threshold dose-response relationship
for radiation carcinogenesis by determining average
radon levels in the 25 U.S. counties (within certain
population ranges) with highest and lowest lung cancer rates. The theory predicts that the former
should have about 3 times higher average radon levels
than the latter, under the assumption that any correlation between exposure to radon and exposure to other
causes of lung cancer is weak. The validity of this
assumption is tested with data on average radon level
vs replies to items on questionnaires; there is
little correlation between radon levels in houses
and smoking habits, educational attainment, or economic status of the occupants, or with urban vs rural
environs which is an indicator of exposure to air
pollution.

The University of Pittsburgh Radon Project makes measurements of
radon concentration in indoor air by use of diffusion barrier charcoal adsorption collectors (Cohen, 1986).   The measurements average
radon levels with an integration time constant of 3 days. Exposure
times are 7 days. All measurements are handled by mail. After exposure, collectors are returned to the Laboratory and their radon
content is measured by gamma ray counting for 40 minutes with 3-inch
diameter NaI (Tℓ) scintillation detectors to determine the number of
counts in the 295, 352, and 609 KeV gamma ray peaks. Currently, 20
of these measuring systems are in use, giving a capacity of about
650 measurements/day. At present, about 300 measurements per day
are being completed. About 13,000 collectors are in the field at
all times.

0097–6156/87/0331–0462$06.00/0
© 1987 American Chemical Society

The measurements fall into two categories:

1. "$12 measurements" provided to all requestors for a $12 charge. About 30,000 of these have been completed. They provide the financial support for the program and also provide a substantial amount of useful research information.

2. "Random" measurements made at no charge for willing participants selected randomly by mailing service computers. They are used for research purposes described below, and to study methods for making data from the $12 measurements useful.

The $12 measurements give higher average radon levels than the random measurements for several reasons, such as

a. measurements are more likely to be requested when a neighboring house is found to have a high level, when the house is in an area or on a geological formation with high radon levels, or when a house is recognized as having construction features associated with high radon levels

b. our service is known to the public principally through media coverage, and there is generally more media coverage in high radon areas

c. some measurements are for follow-up studies on houses already known to have high radon

d. people who purchase measurements are more likely to know that basements generally have higher levels and to test there, despite the advice in our instructions not to test in basements.

In order to overcome these problems, we include questions in the questionnaire as:

- Do you have any reason to believe that the radon level in your house is higher or lower than the average for your County?
- Do you know what factors affect radon levels in homes?

We form "modified sets" of measurements for which the answer to these two questions is NO, and for which measurements were not made in the basement.

As an example, our largest full data set — Nov. 1985 to Feb. 1986— includes 9882 measurements with an average radon level of 5.7 pCi/liter, whereas the modified set from it contains 1771 measurements with an average of 3.6 pCi/liter. Our random measurements during that period, discounting areas selected because they have high radon levels, give a modified set with an average 3.5 pCi/liter, in good agreement with the modified set from $12 measurements.

Data from the $12 measurement full set and modified set are shown in Table I. We see there that the full set exhibits the same general features as the modified set, except that the radon levels are somewhat higher. Since the former have far more measurements, they give more detailed information on geographic variations. The $12 measurements are also directly useful for studies of variations of average radon levels with age of house, socioeconomic factors, etc.

Test of the Linear-No Threshold Theory

The principal research goal of our project is to test the linear-no threshold dose-response theory for radon-induced lung cancer. This

Table I. Data from $12 Measurements, Nov. 1985 - Jan. 1986

Zip (x10^5)	0	0	0-1	1	2	2	3	4	5	6	7	8	9
States	ME NH VT MA CT RI	NJ	NY	PA	DC MD VA	WV NC SC	TN MS AL GA FL	MI OH IN KY	ND SD MN IA WI	IL MO NE KS	TX OK AR LA	AZ CO WY NY UT ID	WA OR CA
pCi per liter													
<10	370	3326	204	1029	179	50	124	219	95	67	44	134	126
10-20	18	137	6	139	28	2	8	20	11	8	0	13	1
20-30	7	48	3	51	5	1	3	3	1	2	0	4	4
30-50	2	45	0	39	3	1	2	1	1	0	0	0	1
50-100	0	22	2	27	5	0	2	2	0	0	0	0	1
>100	1	8	1	12	0	0	0	0	0	0	0	0	0
Aver.	3.6	4.3	3.7	12.0	6.5	3.5	5.1	4.6	4.9	4.0	2.0	5.0	3.5
MODIFIED SET													
Aver.	2.2	3.2	1.9	5.5	2.6	2.1	4.1	3.8	3.3	3.3	2.2	4.1	2.6
<10	44	793	41	290	42	16	29	69	31	13	16	193	32
10-20	1	30	2	24	3	0	0	6	1	1	0	8	0
20-50	0	17	0	12	0	0	2	1	0	0	0	3	0
50-100	0	3	0	2	0	0	0	0	0	0	0	0	0
>100	0	1	0	2	0	0	0	0	0	0	0	1	0

theory, in conjunction with data from studies of miners, makes definite predictions that are subject to experimental tests. We discuss one of these tests.

For each County in the United States in a certain population range, we define

M = total mortality rate from lung cancer

M_r = lung cancer mortality rate due to radon

M_n = lung cancer mortality rate due to non-radon causes including smoking and all other factors, known or unknown

r = average radon level in a county

Then, from the linear-no threshold theory,

$$M_r = kr \qquad (1)$$

where k is a constant determined from the studies of miners with due consideration for various other factors. From the definition of M_n,

$$M = M_r + M_n \qquad (2)$$

We also introduce three simplifying assumptions, the validity of which will be discussed below:

A. There is no correlation between M_r and M_n; this means, for example, that the average indoor concentration of radon in various counties is not correlated with the amount of cigarette smoking in those counties

B. There is no synergism between r and factors affecting M_n. Actually synergisms have been treated Cohen, 1986a), and were found not to affect the results appreciably

C. The distribution of r values for various counties is known.

Fig. 1 shows schematically the distributions of values of M, M_r and M_n for U.S. counties. They are related through (2), and with Assumptions A and B, any one of these can be mathematically calculated from the other two. For example, if the distribution of M and M_r are known, the distribution of M_n can be determined as follows:

(a) Assume that the distribution of M_n is gaussian centered at $\overline{M}_n$ and with standard deviation S_n. Pick trial values of $\overline{M}_n$ and S_n.

(b) Divide the M_n and M_r distribution in Figure 1 into 100 intervals of equal area under the curve (i.e. equal probability).

(c) For each of the (100 x 100 =) 10,000 combinations of M_n and M_r, calculate M from (2). Each of these 10,000 then has equal probability, and they form the calculated distribution of M values in Figure 1.

(d) Check this calculated M distribution against the known one. If they do not agree, pick different values of $\overline{M}_n$ and S_n and repeat (b), (c), (d).

(e) Repeat this process until agreement is obtained; the values of $\overline{M}_n$ and S_n that give agreement characterize the distribution of M_n we are seeking.

(f) In principle, one could then iterate the shape of the M_n distribution from gaussian to improve the detailed agreement with the known M distribution.

From Assumption C and Eq. (1), the distribution of M_r is accurately known. From statistics on lung cancer mortality, the distribution of M is accurately known. Thus, the distribution of M_r can be calculated mathematically, and the problem is completely solved, allowing us to derive predictions that can be tested. In particular,

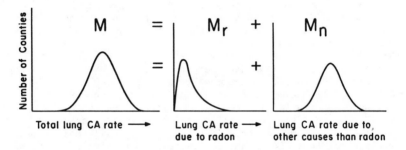

Figure 1. Relationship between frequency distribution of lung mortality rates for various U.S. counties. M is the total rate, M_r is the rate due to radon, and M_n is the rate due to causes other than radon. They are related by (2).

we can calculate the average values of r in the upper and lower ends of the distribution of M values — that is, in counties which have very high or very low lung cancer rates. This is a by-product of the calculation outlined above: as we know the M_r and M_n contribution to each of the 10,000 M values, we need only average the M_r contributions from the 100 highest and 100 lowest M values, and apply (2) to obtain the average r values for these. The results can then be tested experimentally by measuring r for those counties. If the experimental measurements do not agree with the results of the calculation, this can only mean either that our assumptions are not valid, or that Eq.(1) is not correct, which means that the linear-no threshold theory fails.

It is extremely important to note that this process requires no knowledge of smoking practices or of the other factors that influence M_n. If the linear-no threshold theory is correct and our assumptions A, B, C are valid, the distribution of M_n is calculated strictly mathematically from the known distributions of M and M_r, and there is no need to understand its causal factors. Since smoking cigarettes is an important contributor to lung cancer, counties with high lung cancer rates (large M-values) undoubtedly have a high incidence of smoking, and vice versa. But that is taken into account in the calculation. In fact if we assume some simple relationship between M_n and smoking frequency, we could do a completely analagous calculation to determine the relative smoking incidence in high lung cancer vs low lung cancer counties. But these matters are irrelevant. The theory makes definite predictions about radon levels in counties with large-M and small-M, and these predictions can be experimentally tested.

The above discussion depends on our three assumptions, A, B, and C. We next consider their validity:
- Assumption B, that there is no synergism, is not necessary. The calculation has been carried out with the largest synergism consistent with miner data, and the results are little changed.
- Assumption C, that the distribution of r values is known, is not literally correct, but there is enough information available to bound the types of distribution that are plausible, and calculations can be made for various distributions spanning these bounds. They then span the range of plausible results.
- Assumption A is the principal matter considered in this paper.

Results of Calculation and Comparison with Data

The test we are discussing is more sensitive if we go back to the time before lung cancer statistics were dominated by cigarette smoking. We therefore consider white females who died in 1950-1969. When the distribution of M values from that population is used in the above calculation, it is found (Cohen,1986a) that r should be about 3 times larger for the counties with the highest lung cancer rates than for those with the lowest lung cancer rates. This result is not much changed if we assume a strong synergism between smoking cigarettes and radon in causing lung cancer.

We have been testing this prediction by determining average radon levels in the 20 U.S. counties within certain population

restrictions with the highest and lowest lung cancer rates using random selection of invitations to participate in our measurement program at no charge. The studies to date have suffered from poor response and from marginal statistics, and the results cannot be taken
very seriously, but we exhibit them in Table II.

We see that these data suggest that average radon levels may be
lower in high lung cancer counties than in low lung cancer counties,
in gross disagreement with the prediction that they should be about
3 times higher! If this trend continues and is confirmed, there can
be two explanations:

(1)   Our Assumption A is false and there is a strong negative
      correlation between radon levels in houses and exposure
      to other things that cause lung cancer
(2)   Eq.(1) is not valid, which means that the linear-no
      threshold theory fails, grossly over-predicting the
      cancer risk at low doses.

Explanation (2) is what our project is intended to investigate, but
before that investigation becomes meaningful, the possibility of
explanation (1) must be explored. We next consider that problem.

Possible Negative Correlations Between Radon and Other Lung Carcinogens

Ab initio, there is little reason to expect a strong correlation between radon levels and exposure to other factors that cause lung
cancer; for example, it is difficult to see why houses of cigarette
smokers should have substantially lower radon levels than houses of
non-smokers. By far, the most important factor in determining radon
levels is geology, and it is difficult to see how that can correlate
with smoking or air pollution in any consistent way. The most
plausible explanations for a possible correlation are through ventilation rates (e.g. smokers may keep windows open more, which reduces radon), but radon levels have been found to have surprisingly
little correlation with ventilation rates (Nero, et al., 1983).

In this Section, we explore some possible correlations using
data from our studies.

1.   Houses of cigarette smokers may have lower radon levels than
     houses of non-smokers.

An item in our questionnaire asks how many cigarettes per day
are smoked in the house. Results to date are:

cig./day	$12 measurements		Random measurements	
	Number	Av. pCi/liter	Number	Av. pCi/liter
0	9484	4.6	838	4.0
1-5	467	3.7	39	2.0
6-19	623	3.9	109	2.4
$\geq$20	1014	3.7	167	2.6

Present indications are that there is some correlation which will
have to be taken into account in the analysis.

Table II. Average Radon Levels in Counties with Very High and Very Low Lung Cancer Rates from Random Studies in Various Time Periods

Low Lung Cancer Counties

County	CA rate	Summer '85	Jan-Feb '86	Mar-Apr '86
Salt Lake, UT	3.5	[2.3]		
Schuylkill, PA	3.6	1.8	5.5	2.0
Franklin, PA	3.7	1.8		5.4
Oneida, NY	4.0	0.7	2.1	5.2
Northumb'l'd, PA	4.1	1.4		1.6
Marion, OR	4.2	1.0	2.5	
Kalamazoo, MI	4.5			
Linn, IA	4.7	1.3		1.5
Chatauqua, NY	4.7	1.1	5.8	2.4
St. Clair, MI	4.7		0.9	1.1
Lycoming, PA	4.8	1.1		
Madison, IN	4.8	1.9		3.2
Cambria, PA	4.8		2.5	4.8
Washtenau, MI	4.8		4.4	2.8
Berrien, MI	4.9		1.6	1.5
Average		1.33	3.4	2.8
Hi lung CA aver.		0.90	1.9	1.2

High Lung Cancer Counties

County	CA rate	Summer '85	Jan-Feb '86	Mar-Apr '86
Charleston, SC	11.0	1.0	1.0	0.7
Clark, NV	9.6	1.1		
Newport, RI	9.2	0.7		
Solano, CA	9.1	0.9		2.1
Jefferson, MO	9.1		1.3	
Campbell, KY	9.0	1.1		
Humboldt, CA	8.8			0.8
Fairfax, VA	8.8	0.9	3.2	1.7
Harrison, MS	8.7	0.9		
Somerset, NJ	8.6	0.8		1.8
Monmouth, NJ	8.5	0.7		0.8
San Mateo, CA	8.5	1.1		
Ventura, CA	8.4	1.0		
Sarasota, FL	8.4	0.6		1.1
Atlantic, NJ	8.1	0.8		
Prince Geo.,MD	8.1	0.8		1.3
Contra Costa, CA	8.1	0.9		0.7
Sebastien, AR	8.1	0.6		0.5
Duval, FL	8.0	0.6		1.2
Tulsa, OK	8.0	1.2		1.1
Chatham, GA	8.0	1.1		1.2
Average		0.90	1.9	

2. Socioeconomic factors may correlate with both radon levels and
    lung cancer rates for unrelated reasons.

      Our questionnaire asks about dollar value of the house and
annual family income. Results to date for average radon levels
in pCi/liter are:

Value of house	$12 measurements		Random measurements	
	Number	pCi/liter	Number	pCi/liter
<$40K	159	2.2	152	2.2
$40K - $75K	1102	5.2	414	3.6
$75K-$130K	3639	4.3	278	3.5
>$130K	5398	4.4	223	3.6

Income/year	$12 measurements		Random measurements	
	Number	pCi/liter	Number	pCi/liter
<$15K	169	4.0	116	2.4
$15K-$25K	724	4.5	228	3.4
$25K-$45K	2789	4.0	369	3.3
>$45K	5991	4.5	310	3.0

One problem with these data is that they are not well distributed
geographically; over half of these measurements are from the Reading
Prong region of New Jersey, and it is quite possible that in that
region, poor people happen to live in a geological area with low
radon while wealthier people live in a geological area with high
radon. However, when large numbers of regions are considered, this
should average out.
      There is no indication in these data of a consistent monotonic
relationship between radon levels and wealth. There is a consistent
indication that very poor people have lower radon levels than others,
but this indication disappears rapidly for incomes above $15,000/yr
and for houses valued above $40,000. The data on very poor people
may be dominated by students and young people rather than by poor
families.

3. Education level may somehow correlate with both radon levels and
    lung cancer rates for unrelated reasons.

      Some data on this are listed below, as average Rn level vs years
of education beyond 8th grade for the head of household; in each
square, the upper figure is number of measurements and the lower
figure is average pCi/liter.

years	<0	1-3	4	5-7	8	>8
$12 measure- ments	1781 4.3	132 3.5	1349 4.4	1377 4.7	2599 4.3	4473 4.5
random mea- surements	153 6.3	48 4.0	320 3.0	192 3.2	177 2.4	263 3.8

There is little indication here of any correlation between radon exposure and educational level.

4.   There may be correlations, for some unknown reason, between radon levels and exposure to air pollution.

Air pollution is normally an urban problem, is less of a problem in suburban areas, and is almost never a problem in rural areas. Some data on this from a national survey of physics professors (Cohen, 1986b) follows:

Environs	number	Av. pCi/liter
urban	108	1.63
suburban	248	1.37
rural	84	1.53

There is no support here for the premise being studied.   Much more data on this question will soon be available.

Problems in Inferring Past Radon Exposure from Present Measurements

Any study of lung cancer-radon exposure correlations, including both case-control studies and the one considered here, must infer radon exposures many decades ago from measurements made at present.
    The most obvious reason why there might be differences are as follows:

1.   Differences in house construction practices

It is commonly believed that new houses are tighter and therefore have higher radon levels.   Our data on average radon level vs age of house are listed in Table III. They seem to indicate no very large effect, but some correction for this will be necessary.

2.   Recent weatherization activities may have increased radon levels

Nero et al. (Nero et al., 1983) report little correlation between radon level and ventilation rate.   Our data on this are listed in Table IV. The data in Table IVa indicate a negative correlation, but those in Table IVb and IVc indicate a rather strong positive correlation.   Hopefully this matter will be clarified when much more data become available.
    Typical estimates are that recent weatherization activities have reduced ventilation rates by about 20%, presumably increasing radon levels that much. Since only about half of all houses have been weatherized, this is only about a 10% correction.

3.   In earlier times windows were kept open in summer, whereas now many homes keep windows closed for air conditioning purposes.

Some of the data presented above indicate that average radon levels are now only half as large in summer as in winter.   Thus the summer air conditioning season contributes only 10-15% of our annual radon exposure, and the correction due to this factor is less than 10%.

Table III. Average Radon Level vs Age of House from Various Data Sets. Figures are the Number of measurements/Average pCi/liter.

SURVEY	AGE 1-9	AGE 10-19	AGE 20-29	AGE 30-49	AGE 50-79	AGE >80
$12-Winter '85-'86						
New Jersey	1325/5.6	1264/5.8	1236/3.7	1063/3.3	510/2.2	236/3.9
Pennsylvania	394/11.4	318/2.7	250/10.2	308/6.9	193/29.7	191/6.4
DC-MD-VA	198/5.5	70/5.8	108/2.5	106/2.5	18/7.2	11/1.6
Other	772/5.1	491/3.5	350/4.3	354/3.9	234/3.1	168/3.1
$12-Spring '86						
New Jersey	1204/4.8	1215/6.2	1080/4.9	946/2.8	437/2.1	317/4.9
Pennsylvania	408/8.4	266/7.7	195/10.5	184/5.1	105/3.7	150/5.3
DC-MD-VA	1036/3.7	613/3.6	474/2.8	465/2.2	119/2.4	68/17.6
Other	854/3.8	621/2.3	521/3.0	530/3.0	308/2.9	257/2.1
$12-Fall '85	189/12.4	185/7.5	168/15.1	169/10.1	127/14.3	100/6.7
Random-Summer '85	139/2.5	183/1.7	159/2.0	168/1.5	151/1.5	110/1.4
Random-Winter '85-'86	126/3.4	176/5.5	151/4.7	185/5.4	103/4.2	55/3.2
Random-Spring '86	188/5.1	259/3.0	267/3.5	267/5.6	179/2.1	101/2.8
Univ. Profs. - 1 yr	85/1.29	101/1.52	94/1.69	75/1.77	44/0.88	54/1.37
Pittsburgh - 1 yr	12/2.7	33/2.3	26/2.3	31/2.3	50/2.0	15/3.3
Cumberland-Winter '84-'85	25/7.7	43/10.8	40/9.8	35/11.1	13/3.9	9/4.0

Table IV. Replies to Questions about Weatherization from Various
Studies. Figures are Number of Measurements-Average
pCi/liter.

a. How much weatherization in past 10 yr?

	$12-Feb, Mar '86	Random, Summer '85
much	550-4.0 pCi/ℓ	458-1.6
little	687-4.9	408-1.9
nothing	226-4.5	

b. How much has been done in past 10 years?

		Random, Nov.85-Jan.86
Weatherstripping around doors and windows	much	249-6.2
	little	273-4.3
	nothing	147-3.3
Closing gaps under doors to outside	much	256-5.6
	little	263-5.1
	nothing	145-2.9
Caulking, glazing around windows	much	269-5.5
	little	220-4.1
	nothing	169-4.2

c. How much has been done to weatherize in past 10 years (March-
April 1986)?

	$12	Random
much	2270-6.0	422-3.7
little	2635-4.6	502-3.2
nothing	965-3.7	168-4.1

Questions on window opening practices have been added to our questionnaire. The results for winter indicate that nearly all windows are kept closed. Results for summer will give information on the question under discussion.

Literature Cited

Cohen, B.L., A Diffusion Barrier Charcoal Adsorption Collector for Measuring R_n Concentrations in Indoor Air, Health Phys. 50:457 (1986).

Cohen, B.L., Expected Indoor Radon Levels in Counties with Very High or Very Low Lung Cancer Rates, Health Phys. (submitted) (1986a).

Cohen, B.L., A National Survey of Radon in Homes and Correlating Factors, Health Phys. (in press) (1986b).

Nero, A.V., M.L. Baegel, C.D. Hollowell, J.G. Ingersoll, and W.W. Nazaroff, Radon Concentrations and Infiltration Rates Measured in Conventional and Energy-efficient Houses, Health Phys. 45:401 (1983).

RECEIVED August 4, 1986

Chapter 34

Deposition of Ultrafine Particles in the Human Tracheobronchial Tree
A Determinant of the Dose from Radon Daughters

Beverly S. Cohen

Institute of Environmental Medicine, New York University Medical Center, New York, NY 10016

The deposition of ultrafine particles has been measured in replicate hollow casts of the human tracheobronchial tree. The deposition pattern and efficiency are critical determinants of the radiation dose from the short lived decay products of Rn-222. The experimental deposition efficiency for the six airway generations just beyond the trachea was about twice the value calculated if uniform deposition from laminar flow is assumed. The measured deposition was greater at bifurcations than along the airway lengths for 0.2 and 0.15 µm diameter particles.

Exposure to alpha particle radiation from the short lived daughters of ^{222}Rn is a recognized cause of bronchogenic cancer in uranium and other underground miners. When ^{222}Rn decays, the ^{218}Po and succeeding progeny quickly attach to particles in the air. The activity median diameter of the attached particles varies with indoor, outdoor or underground mining atmospheres but ranges from about 0.1 to 0.4 µm (NCRP, 1984). When these particles are inhaled, a fraction deposit on the mucosal surface of the tracheobronchial tree. Subsequent radioactive decay will deliver the significant dose to the sensitive cells of the bronchial epithelium.

The radiation dose will depend critically on the efficiency with which the particles are deposited on the airway surfaces. In addition the pattern of deposition is important because substantial radioactive decay of the short lived radon daughters will take place before the initial particle deposit can be removed by normal clearance mechanisms (Cohen, et al., 1985).

Few data are available on the deposition of ultrafine particles (d $\leq$ 0.2 µm) on airway surfaces. Chamberlain and Dyson (1956) measured the deposition of unattached radon decay products in a rubber latex cast of a human windpipe which extended from the epiglottis to a few cm below the Carina. Martin and Jacobi (1972)

0097-6156/87/0331-0475$06.00/0
© 1987 American Chemical Society

examined the deposition of particles in the 0.2 to 0.4 µm range in a plastic dichotomous symmetrical branching model of the upper airways. The model was based on the morphometry of Weibel (1963). The test aerosol was produced by labeling the normal atmospheric aerosol with ^{212}Pb atoms, and had a geometric standard deviation (σ_g) of about 2.5. They found that the deposition probabilities exceeded those predicted for laminar flow in the model. A set of correction factors derived from that work have been used to predict the radiation dose to the bronchial epithelium from the short-lived radon daughters (NCRP 1984; Harley and Pasternack 1972, 1982). James (1977) measured the deposition of ^{212}Pb tagged condensation nuclei (AMD $\sim$ 0.14 µm, $\sigma_g \sim$ 1.3) in excised pig lungs ventilated in a cyclic manner. James used a form of the diffusion equation attributed to Davies (1973), which was subsequently shown to be in error by Ingham (1975). The error substantially affected the resulting ratios.

A model system which utilizes casts prepared from autopsy specimens of the human tracheobronchial tree for the study of particle deposition was developed by Schlesinger and Lippmann (1972). The particle sizes studied in the cast system were in the range where the dominant deposition mechanism is impaction. The initial studies demonstrated that deposition is not homogenous, but is enhanced at bifurcations. Additional studies showed that bifurcation deposition was enhanced at 0.3 µm as well as at larger particle sizes (Schlesinger, et al., 1977). Further measurements in the cast system using constant inspiratory flow have shown that particle deposition efficiency increases linearly with Stokes number (Chan and Lippmann, 1980). Subsequently, the deposition of 3 and 8 µm particles under cyclic inspiratory flow was compared with that under constant flow at equivalent mean flow rates (Schlesinger, et al., 1982; Gurman, et al., 1984). The bronchial deposition efficiency was shown to be greater under cyclic flow for both airways and bifurcations, however, tracheal deposition was generally greater for constant flow. The enhancement seen at bifurcations under constant inspiratory flow persisted under pulsatile flow.

This paper will present some results of a set of experiments carried out in the hollow airway cast system with ultrafine particles which are of particular interest for the calculation of the dose to the bronchial epithelium from the short lived radon daughters. Detailed deposition efficiencies and intrabronchial distributions are presented elsewhere (Cohen, et al., 1986).

Experimental Methods

Deposition experiments were carried out in replicate hollow casts of the upper airways of a human tracheobronchial tree using 0.2, 0.15 and 0.04 µm diameter particles and cyclic inspiratory flow rates of approximately 18 and 34 liters per minute (L/min) (see Table I). The replicates were produced from a single solid master airway cast prepared from the lungs of 34 year old male. The airway dimensions of the cast corresponded closely with the population mean of eight adult males as reported by Nikiforov and Schlesinger (1985). The airway diameters are somewhat larger and

the lengths somewhat shorter than those of Weibel (1963). Both
dimensions are smaller than those of the single pathway model of
Yeh and Schum (1980). The casting process and replicate production were done using methods described by Schlesinger, et al.
(1977). The casts, which were pruned to airways ≥ 3 mm, retained
141 airways which were demarcated into bifurcation and length segments as shown in Figure 1. The airways were assigned to generations according to the dichotomous branching model of Weibel
(1963). The trachea is considered to be generation 0, the major
bronchi comprise the first generation. Each bifurcation is
assigned to the generation of the distal pair of airways, thus,
the Carina, the bifurcation at which the trachea divides into the
right and left bronchi, is assigned to the first generation. In
this cast there were two trifurcations, one at the branch point of
the right upper lobar bronchus and the other in the left upper
lobe at the branching of the superior division bronchus. For
analysis of the deposition data the three airways distal to these
trifurcations were considered to be at the same branching level.
The velocity of the airstream exiting from each end airway was
measured with a hot wire anenometer probe (TSI Model 1260A-T15),
to determine the fraction of the flow in each airway for 15 and 30
L/min flowing into the trachea. A volumetric flow for each airway
was calculated based on these measurements and the measured cross
sectional area.

Table I. Tracheobronchial Cast Deposition Studies
Cyclic Inspiratory Flow

					Reynolds	Number
		Mean	Diffusion		Right	Left
Cast	D_p (μm)	Flow Rate (cm³/s)	Coefficient (cm²/s)	Trachea	Bronchus (.60)*	Bronchus (.40)*
II	0.20	265	2.22×10^{-6}	1060	850	580
VIII	0.15	325	3.92×10^{-6}	1300	1050	710
V	0.04	302	3.55×10^{-5}	1210	970	665
IV	0.20	625	2.22×10^{-6}	2510	2010	1370
VI	0.15	568	3.92×10^{-6}	2280	1830	1240
VII	0.04	547	3.55×10^{-5}	2190	1770	1210

*Fraction of flow.
$Re = d\overline{U}/\nu; \quad \nu = 0.167 \text{ cm}^2/\text{s}$

Particles were generated from a colloidal suspension of ^{99m}Tc
labeled ferric oxide using a Collison nebulizer, and were size
classified with an Electrostatic Classifier (ESC, TSI Model 3071).
The ESC initially charges the incoming aerosol particles to produce a bipolar charge distribution. The aerosol passes between
concentric cylinders across which a high electric field is
imposed. Only particles with a given mobility, or charge mass
combination, can exit through a slit at the end of the charged
central cylinder. Initial tests with NaCl nuclei demonstrated

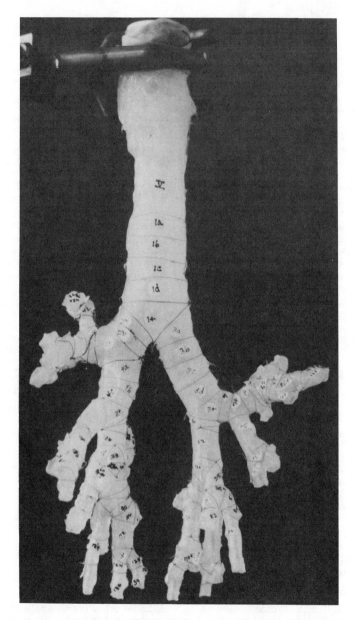

Figure 1. A replicate hollow airway cast showing demarcation of test sections.

that the particles penetrating had a σ_g of about 1.06. Samples of ^{99m}Tc-ferric oxide aerosols for the three particle sizes were collected onto Nuclepore filters following cast exposure. Scanning electron micrographs (SEM) revealed particle size distribution with a σ_g of about 1.3 for the 0.15 and 0.2 µm particles. This larger σ_g is attributed to the penetration of multiply charged particles. The 0.04 µm particles appeared very uniform on SEM but were too small to measure.

During each test, the cast was suspended from a mechanical larynx and enclosed in an acrylic thorax. The aerosol was charge neutralized to Boltzman equilibrium and mixed with clean dilution air, then drawn through the artificial thorax. A needle containing 0.76 mg of ^{226}Ra in equilibrium with its daughters was placed in the thorax for the duration of the experiment for additional charge neutralization. The flow through the system was controlled by the mechanical larynx-flow system developed by Gurman, et al. (1980). Flow was monitored with a pneumotachgraph, and the signal recorded on a strip chart. The average flow rate for each exposure was determined by cutting and weighing a section of the chart tracing. All radioactivity passing through the cast was collected on disposable thorax liners and by the final filter and electrostatic precipitator through which the airstream exited the thorax. The total amount of activity recovered was compared with the input as estimated by the filter samples taken at the ESC before and after each experiment. For the 0.2 and 0.15 µm particles, the agreement was within 15%. For the 0.04 µm particles, a series of filter samples indicated that the aerosol input was too variable to permit a satisfactory activity balance.

After exposure, the outside surface of the cast was cleansed until the activity of the washes was less than 10X the background of a gamma well scintillation counter. The cast was cut into separate bifurcations and airway sections and each section was counted to determine the amount of aerosol deposited. Some samples contained both airway and bifurcation sections because of the complex configuration of the cast. For combination samples, the total activity deposited was equally apportioned between each of the airways and bifurcations. End airways were included for determination or total deposition but not in any of the analyses because flow disturbances at open ends may have affected deposition. The surface area of each sample was measured separately. The surface density for each cast segment was calculated by dividing the activity measured in the sample by the interior surface area of that sample.

The deposition fraction on each of the airways was determined by dividing the measured activity in the segment by the fraction of the tracheal flow that passed through the segment and normalized for the total amount of activity that penetrated the cast. The same was done for each bifurcation except that the flow for the proximal airway was used. Thus, the deposition fraction for each segment is the fraction of the particles entering that segment which deposit there. The mean deposition efficiency measured for the generation for the airways and bifurcations were summed to determine the total deposition efficiency for each generation.

Deposition efficiency for ultrafine aerosols in the airways

is frequently calculated by assuming that the airflow pattern is laminar and then applying a correction factor. The deposition fraction F_d for laminar flow may be calculated from the Gormley-Kennedy (1949) equation for the penetration of diffusing particles through a cylindrical tube as formulated by Ingham (1975):

$$F_d = 1 - [0.819 \exp(-14.63\,\Delta) + 0.0976 \exp(-89.22\,\Delta) \\ + 0.0325 \exp(-228\,\Delta) + 0.0509 \exp(-125.9\,\Delta^{2/3})] \qquad (1)$$

where: $\Delta = \pi\,DL/4Q$
       D = particle diffusion coefficient
       L = tube length
       Q = mean flow rate through the tube

The solution to this equation was obtained for each airway in the cast for each set or experimental conditions using the measured values for length of the airway and measured fraction of the total flow which passes through the airway. The diffusion constants are shown in Table I. The mean deposition for each generation was obtained from the deposition calculated for all cast airways of the given generation.

Results

The mean measured activity per unit surface area are shown for airways and bifurcations separately in Table II. These data are for those segments which contained only airway lengths or bifurcations. The results are given as the number of particles which deposit per cm^2 for 10^6 particles which enter the trachea. This assumes that the particle and activity distributions are equivalent. For the 0.2 and 0.15 μm particles the surface density at the bifurcations is greater than that along the airway lengths at $p < .01$ when the paired data are compared by a one tailed t-test.

The total deposition in the airway cast was very low, ranging from 0.2% for the 0.15 μm particles at 20 L/min to 2.0% for the 0.04 μm particles at 18 L/min. In the trachea, if uniform deposition is assumed, the ratios of experimental to calculated deposition, are 14 ± 3 for mean flow rates of about 18 L/min and 4 ± 1 at about 34 L/min. The ratios of the calculated deposition in the trachea for low flow to that at high flow are 1.78, 1.45 and 1.49 for the 0.2, 0.15 and 0.04 μm particles, respectively. The measured ratios are 9, 7 and 4.

The ratios of the experimental to the calculated values for deposition efficiency for each generation are shown for each of the experiments in Figures 2-4. The error bars represent the standard error derived from the mean measured deposition fraction for the generation. A coefficient of variation was calculated for the measured deposition in each generation which was used to obtain an estimate of the standard error of the ratio. It was assumed that no additional variability in the ratio was introduced by the calculated deposition fraction. The mean ratio for all six sets of cast measurements is shown in Figure 5. The error bars in Figure 5 represent the standard error of the mean of the six experiments.

Table II.   Surface Density of Particles Deposited in the Hollow
Airway Cast[a]

Particle Diameter (μm)	Mean Flow Rate (L min⁻¹)	Group[b]	Number of Samples	Mean ± SE
0.20	16	B	21	17 ± 2
		A	27	13 ± 2
0.15	20	B	44	16 ± 2
		A	47	12 ± 1
0.04	18	B	25	52 ± 10
		A	35	58 ± 12
0.20	37	B	26	20 ± 3
		A	32	18 ± 3
0.15	34	B	33	10 ± 1
		A	46	8 ± 1
0.04	33	B	11	48 ± 16
		A	10	72 ± 21

[a] Number of particles which deposit per cm^2 for 10^6 particles entering the cast.
[b] B = bifurcation; A = airway lengths

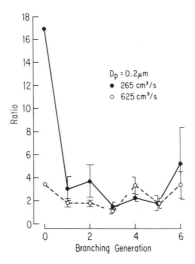

Figure 2. Ratio of measured to calculated deposition fraction
0.2 μm particles.

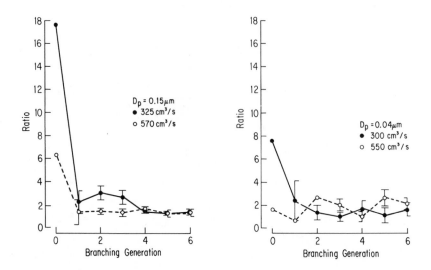

Figure 3 (left).   Ratio of measured to calculated deposition fraction 0.15 μm particles.

Figure 4 (right).   Ratio of measured to calculated deposition
fraction 0.04 μm particles.

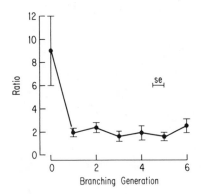

Figure 5. Mean ratio of measured to calculated deposition
fraction.

Discussion

The ratio of deposition at bifurcations to that on airway lengths
(1.25 $\pm$ .05 for 0.2 and 0.15 μm particles) is lower than that
measured in the tubular model of Martin and Jacobi (1972).   This
is primarily a result of the more realistic airway geometry of the
hollow cast model.   Chan, et al. (1980) pointed out the importance
of the change in shape and increase in cross sectional area in the
human tracheobronchial tree as the flow in an airway approaches a
branch point. The nonuniform deposition observed for bifurcations
as compared with airway lengths throughout the cast, and along the
major airways, has important implications for the dosimetry of the
inhaled submicrometer particles to which the radon daughters are
attached.   Implications of nonuniform particle deposition on the
tracheobronchial tree have been discussed by Cohen, et al. (1985).
The data suggest that the dose from ambient radon daughters will
be highest at the bifurcations and could be about 20% greater than
that estimated for uniform deposition.

    For each of the three particles sizes studied, the deposition
efficiency in the trachea was much greater than predicted if uniform deposition is assumed. In each data set the ratio at the
lower flow rate exceeded that at a higher flow. If the increase
results from turbulence introduced by the larynx it might be
expected to be more effective at the higher flow rate. The
discrepancy may result from the jet formed downstream of the larynx.   The center line velocity substantially exceeds the mean
velocity of the airstream and secondary circulation patterns are
set up near the wall which can act as dead zones (Ultman, 1985).

    The calculated deposition efficiency assumes laminar flow in
the airways, which is clearly not a good assumption. The Reynolds
numbers in the trachea and major bronchi range from about 600 to
2500 (see Table I), thus flow is frequently neither fully laminar
nor fully turbulent. As the total cross sectional area of the
airways increases, the Reynolds numbers decrease, and flow will be
more laminar. However, laminar flow may not be achieved in this
bifurcating system.   One measure of the distance needed for laminar flow to be established is the "entrance length".   If   the
length of a tube does not equal or exceed this measure, local turbulence will persist. The tube lengths in the human tracheobronchial tree are much less than the required entrance length down to
about the tenth generation (Olson, 1970).

    The airstream velocity profile downstream of a bifurcation is
asymmetrical. The peak velocity occurs near the inner wall of the
daughter branches in the plane of the bifurcation (Olson, et al.,
1973).   We observed this skewed distribution and unsteady flow
when the velocity was measured near the open end of recently
bifurcated airways for this model cast (Sussman, et al., 1985).

    In addition, the effects of pulsatile flow cannot be ignored.
One measure of the impact of oscillary flow is the Womersley
parameter (α); $\alpha = r\sqrt{2\pi f/\nu}$ where r is the tube radius, f the frequency of oscillation and ν is the kinematic viscosity of the
fluid (Womersley, 1955). The degree of departure from parabolic
flow increases with and frequency effects may become important
in straight tubes when $\alpha > 1$ (Ultman, 1985).   For conditions of
these experiments, α exceeds one to beyond the third generation.

It is clear that predictions based on uniform deposition from laminar flow are not satisfactory. Current models are not able to incorporate the effects of flow patterns in the upper airways which result in the inhomogeniety of deposition demonstrated in the cast system. It seems remarkable that predictions for diffusive deposition from laminar flow in cylindrical tubes are so close to those observed in the complex geometry and branching of the human tracheobronchial system. The results of the experiments reported here indicate that the deposition probability for 0.04 to 0.2 μm particles in generations 1-6 of the upper airways is greater than that predicted by equation (1) by a factor of about 2. Until better models are developed the correction factors shown in Figures 2 to 5 are useful for estimating deposition of inhaled paricles in the size range to which the decay products of radon are attached.

Acknowledgments

This work was supported by Grant No. ES 00881 from the National Institutes of Environmental Health Sciences (NIEHS) and Special Emphasis Research Career Award Grant No. OH 00022 from the National Institute for Occupational Safety and Health of the Centers for Disease Control. It is part of a Center program supported by Grant ES 00260 from NIEHS and Grant CA 13343 from the National Cancer Institute. The author wishes to thank Mr. Robert Sussman who assisted with all of the laboratory studies.

Literature Cited

Chamberlain, A.C. and E.D. Dyson, The Dose to the Trachea and Bronchi from the Decay Products of Radon and Thoron. Br. J. Radiol. 29: 317-325 (1956).

Chan, T.L. and M. Lippmann, Experimental Measurements and Empirical Modelling of the Regional Deposition of Inhaled Particles in Humans, Amer. Ind. Hyg. Assoc. J. 41:399-409 (1980).

Chan, T.L., R.M. Schreck and M. Lippmann, Effect of the Laryngeal Jet on Particle Deposition in the Human Trachea and Upper Bronchial Airways, J. Aerosol Sci. 11:447-459 (1980).

Cohen, B.S., N.H. Harley, R.B. Schlesinger and M. Lippmann, Nonuniform Particle Deposition on Tracheobronchial Airways: Implications for Lung Dosimetry. Ann. Occup. Hyg. (in press).

Cohen, B.S., R.G. Sussman and M. Lippmann, Deposition of Ultrafine Particles in Hollow Airway Casts of the Human Tracheobronchial Tree. In Preparation.

Gormley, P.C. and M. Kennedy, Diffusion from a Stream Through a Cylindrical Tube, Proc. R. Ir. Acad. Sect. A52:163-169 (1949).

Gurman, J.L., R.B. Schlesinger and M. Lippmann, A Variable-Opening Mechanical Larynx for Use in Aerosol Deposition Studies, Am. Ind. Hyg. Assoc. J., 41:678-680 (1980).

Gurman, J.L., M. Lippmann and R.B. Schlesinger, Particle Deposition in Replicate Casts of the Human Upper Tracheobronchial Tree Under Constant and Cyclic Inspiratory Flow. I. Experimental, Aerosol Science and Technology 3:245-252 (1984).

Harley, N.H. and B.S. Pasternack, Alpha Absorption Measurements Applied to Lung Dose from Radon Daughters, Health Phys. 23:771-782 (1972).

Harley, N.H. and B.S. Pasternack, Environmental Radon Daughter Alpha Dose Factors in a Five-Lobed Human Lung, Health Phys 42:789-799 (1982).

Ingham, D.B., Diffusion of Aerosols from a Stream Flowing Through a Cylindrical Tube, Aerosol Science, 6:125-132 (1975).

James, A.C., Bronchial Deposition of Free Ions and Submicron Particles Studies in Excised Lung, in Inhaled Particles and Vapours IV, (W.H. Walton, ed.), pp. 203-219, Pergamon Press, New York (1977).

Martin, D. and W. Jacobi, Diffusion Deposition of Small-Sized Particles in the Bronchial Tree, Health Physics, 23:23-29 (1972).

National Council on Radiation Protection and Measurements, Evaluation of Occupational and Environmental Exposures to Radon and Radon Daughters in the United States, NCRP Report No. 78, NCRP, Bethesda, MD 20814 (1984).

Nikiforov, A. and R.B. Schlesinger, Morphometric Variability of the Human Upper Bronchial Tree, Respiration Physiology, 59:289-299 (1985).

Olsen, D.E., G.A. Dart and G.F. Filley, Pressure Drop and Fluid Flow Regime of Air Inspired Into the Human Lung, J. Appl. Physiol. 28:482-494 (1970).

Olsen, D.E., M.F. Sudlow, K. Horsfield and G.F. Filley, Convective Patterns of Flow During Inspiration, Arch. Intern. Med. 131:51-57 (1973).

Schlesinger, R.B. and M. Lippmann, Particle Deposition in Casts of the Human Upper Tracheobronchial Tree, Am. Ind. Hyg. Assoc. J. 33:237-251 (1972).

Schlesinger, R.B., D.E. Bohning, T.L. Chan and M. Lippmnn, Particle Deposition in a Hollow Cast of the Human Tracheobronchial Tree. J. Aerosol Sci. 8:429-445 (1977).

Schlesinger, R.B., J.L. Gurman and M. Lippmann, Particle Deposition Within Bronchial Airways: Comparisons Using Constant and Cyclic Inspiratory Flows, Ann. Occup. Hyg. 26:47-64 (1982).

Sussman, R.G., B.S. Conen and M. Lippmann, The Distribution of Airflow in Casts of Human Lungs, Presented at the 1985 Annual Meeting of the American Association for Aerosol Research, Albuquerque, NM (November, 1985).

Ultman, J.S., Gas Transport in the Conducting Airways, in Gas Mixing and Distributions in the Lung (L.A. Engel and M. Paiva, eds.), pp. 63-136, Marcel Dekker, Inc., New York (1985).

Weibel, E.R. Morphometry of the Human Lung, Academic Press, New York (1963).

Womersley, J.R., Method for the Calculation of Velocity, Rate of Flow and Viscous Drag in Arteries when the Pressure Gradient is Known, J. Physiol. 127:553-563 (1955).

Yeh, H.C. and G.M. Schum, Models of Human Lung Airways and Their Application to Inhaled Particle Deposition, Bulletin of Mathematical Biology, 42: 461-480 (1980).

RECEIVED August 20, 1986

Chapter 35

Effect on Peripheral Blood Chromosomes

J. Pohl-Rüling, P. Fischer, and E. Pohl

Division of Biophysics, University of Salzburg, A-5020 Salzburg, Austria

A survey of investigations on chromosome aberrations
in peripheral blood lymphocytes of people exposed
to elevated levels of radon in the atmosphere shows
non linear dose relationship. At very low doses a
sharp increase occured, followed by a plateau. A
hypothesis involving DNA repair mechanism is given.
In vivo the doses are always not only due to internal
alpha but also due to the external gamma irradiation,
which must be taken into account, especially in the
low dose range. Therefore in vitro experiments with
alpha particles of radon daughters at doses comparable
to the in vivo investigations have been carried out
(0.05 to 3 mGy). The method is described. Preliminary
results show similar dose response as at our in vivo
studies.

Radon and daughters are causing a significant part of the normal
environmental irradiation of man. Furthermore they are involved
in several elevated natural and occupational radiation exposures.
Their main health hazard is the induction of lung cancer as the
lungs, getting the highest dose within the human organism, are
the critical organ.

All the other organs and tissues of the body, however, are
also exposed to a radiation burden due to the inhalation of radon
and daughters, as radon itself and about one third of the inhaled
decay products passes from the lungs into the blood stream (Pohl
and Pohl-Rüling, 1977, 1982).

The nucleus of all eucariotic cells contains the carrier
of the genetic information in the chromosomes. It is possible
to visualize the chromosomes and analyze their number and pattern
during a special period of cell division (the metaphase).
Alterations from their normal shapes are observed as structural
chromosome aberrations. These are chromosome type aberrations
(terminal and interstitial deletions, dicentrics and rings),
chromatid aberrations (gaps, breaks and exchanges) and sister
chromatid exchanges. Spontanous frequencies of such chromosome

anomalies occur normally in every person (2 to 8 chromosome type aberrations in 1000 metaphases of lymphocytes) and can be ascribed to physical (irradiation), chemical and biological agents. The induction of chromosome aberrations is a conseqence of a reaction of cells to induce and repair damages in their DNA.

One of the most sensitive biological effects of ionizing radiation is to increase the frequency of normally observed chromosome aberrations (but not to induce qualitatively special abnormalities). Peripheral blood lymphocytes are the most feasible cells for chromosome investigations, as blood samples are easy to obtain and the techniques to stimulate the lymphocytes to proliferate within a culture medium and to prepare suitable chromosome slides for microscopic analyzation have their routine protocoll (e. g. Yunis, 1965; Lloyd et al, 1982).

The main problems of the study of chromosome aberrations, caused by radon and daughters at their most frequently existing dose levels, i. e. boardering the natural burdens, are: (i) to get statistical significance at very low doses, and (ii) to study their induction by internal exposure to alpha emitters only.

Effects of Very Low Dose Levels.

A great many studies have been carried out on the induction of chromosome aberrations due to almost all kinds of radiations, with low and high LET, delivered acutely, fractionated or continously, in vitro as well as in vivo. To a certain degree the induction of chromosome aberrations has even been used as a biological dosimeter (e. g. Intern. Atomic Energy Agency, 1969; Lloyd and Purrott, 1981). The main part of these studies, however, dealt with medium or high doses (above 50 mGy). Effects at low doses are generally difficult to obtain, because of the large number of events needed to get statistical significance. Therefore, estimates of radiation effects and related risks at low levels are usually extrapolated from the effects at higher doses. Radiation induced effects are generally described by the relationship based on the theory of dual radiation action according to the equation

$$y = c + aD + bD^2 \qquad (1)$$

whereby y is the effect, D the dose, c, a and b are constants, (Kellerer and Rossi, 1972).

In radiation protection and risk assessment linear extrapolations are commonly used. Linearity, however, is not based on experimental evidence in the very low dose range and a departure from it would imply an over- or underestimation of risks. There are both theoretical and experimental indications wich contradict the linearity at doses considered here (e.g. Brown, 1977; Archer, 1984).

Burch (1983) suggests that repair mechanisms cause a non neglectible complication for extrapolation from high to low doses and presents a modification of the linear-quadratic formula given above. Katz and Hofmann (1982) carried out an analysis of particle tracks with the result that they find no basis for a linear or linear-quadratic extrapolation to low doses. Van Bekkum and Bentvelzen (1982) present a hypothesis of the gene transfer-

misrepair mechanisms of radiation carcinogenesis which implies
a non-linear extrapolation model for the calculation of cancer
risks caused by very low doses of low LET radiation. Baum (1982)
found in his theory of radiation carcinogenesis a dose response
for high LET radiation following a power function with an exponent
of 0.4. Marshall and Groer (1977) present a theory of the induction
of bone cancer by alpha radiation, in which the tumor rate is
proportional to D^2 at low doses. This latter result is
experimentally sustained by Kamiguchi and Mikamo (1982) and by
Frankenberg-Schwager et al. (1980).

A critical review of almost all experimental data up to 1978
on chromosome aberrations in human blood lymphocytes, induced
in vivo and in vitro, due to external and internal irradiation,
with low and high LET, at low and high dose levels, was made by
Pohl-Rüling et al. (1978). The results showed that chromosome
type aberrations rise at continous irradiation sharply in the
range boardering the natural environmental levels with neglectible
influence of LET. At additional doses of about 0.003 to 0.05 Gy,
delivered either acutely or accumulated over one year, the effects
are only weakly, if at all, dose dependent and therefore the dose
response curve reveals a plateau. At higher doses, above about
0.3 Gy, however, dose kinetics are followed which are best described
by the two component theory given above. A detailed discussion
regarding the comparability of the data is given by these authors.
They suggest the hypothesis that at natural levels of radioactivity
a basic amount of repair enzymes is present and with increasing
radiation also aberrations increase. After a certain level of
damage to the DNA, however, additional repair enzymes are induced,
resulting into a plateau of dose response. When this inducibility
is saturated at higher doses, aberration levels rise again.

In the following is shown that in vivo investigations on
chromosome aberrations in people exposed to an elevated level
of radon and daughters provide similar dose response.

In Vivo Investigations of Chromosomes Aberrations Due to Exposure of Radon and Daughters.

Investigations of chromosome aberrations induced by radon and
daughters can be carried out in population groups with an elevated
exposure to these radionuclides. These are mine workers and
inhabitants of areas with an elevated radon concentration in the
air: Radon is either excessively exhaled from ground or building
materials, or released from spring or well waters which are used
in dwellings or for therapeutical purposes (in radon spas). Such
population groups are, however, also exposed to an external
irradiation, more or less exceeding the normal natural levels.

Several papers carried out in various countries of the world
showed that aberration frequencies in people exposed to elevated
doses of radon and progenies are higher than in controls (Pohl-
Rüling and Fischer, 1983; Chen Deqing et al., 1985; The
Environmental Protection Agency, 1982; Nakai and Mifune, 1982).
These results, however, cannot be used to establish a dose-effect
curve, because of inadequate dose estimates. For this purpose
it is necessary to calculate the individual accumulated external
and internal doses of all persons involved into the investigation,

not only the exposed ones but also the control subjects. Even
in a normal radiation environment within small areas, such as
e. g. towns, the organ dose distribution within the population
has a wide range. This occurs due to differences in individual
living and working places, sex, age, and physical activity, which
influences the minute respiratory volume and consequently the
intake of radon and progenies (Steinhäusler et al., 1980).

Investigations in Badgastein, Austria. A study with individual
dose calculations was carried out with the residents of an area
in and around the spa resort Badgastein (Pohl-Rüling and Fischer,
1979, 1983). The radioactive environments of the entire region
outdoors and indoors have been measured for many years (Pohl et
al., 1976). These measurements included rooms with therapeutic
facilities, such as bath rooms, inhalations rooms and the interior
of the "Thermal Gallery" (a former gold mine with air temperatures
up to 41°C, high humidity and a mean radon concentration in the
air of 110 kBq/m^3). On the basis of these extended measurements
the radiation burdens from external gamma-and internal alpha-
irradiation were calculated. The exact dose to the lymphocytes
is not known, but since they continously circulate within the
body the blood dose was regarded to be the best approximation.
180 blood samples were taken from 122 persons. A dose-response
analysis was carried out whereby the age was taken into
consideration (not the smoking habits). This analysis showed that
the exposures accumulated over one month before blood sampling
have a greater significance than those accumulated over a larger
period of time (6 or 12 months). These blood doses of the
investigated persons ranged from 0.08 to 0.3 mGy for the external
gamma irradiation and from 0.001 to 2.7 mGy for the internal alpha
exposure.

Preliminary studies, already, resulted in a dose response
curve which rises sharply up to about 0.5 mGy/month and then
flattens into a plateau (Figure 1). In the dose range up to 0.3
mGy/month only a small part (about 0.014 mGy/month) was due to
internal alpha exposure, whereas at higher doses more and more
the exposure to alpha rays were dominant.

As this shape was not at all compatible with curves expected
from equation (1) a more sophisticated analysis of the data from
30770 cells was tried. Test persons were categorized into three
groups (I, II, III) according to their type of exposure pattern.
This was either only continous from the environment (group I)
or with addition of occupational burdens delivered fractionated
during two different periods of working times, either 8 hours
in therapeutical milieus (group II) or 2 to 4 hours within the
thermal gallery (group III). The dose relationships for persons
in the single groups followed the same shape as described in the
first analysis. Moreover it turned out that the amounts of the
aberrations at same dose decrease with decreasing fractionation
times, which also contradicts the usual results with in vitro
irradiation from other authors (Figure 2).

Investigations in Finland. The ground water of some areas has
been found to contain considerable amounts of Ra-226 and/or Rn-
222. The tap water in the dwellings of these regions acts as a

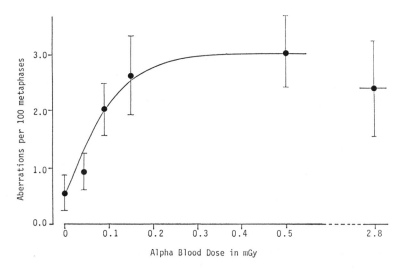

Figure 1. Total chromosome type aberration frequencies (sum of dicentrics, terminal and interstitial deletions) in the population of Badgastein.

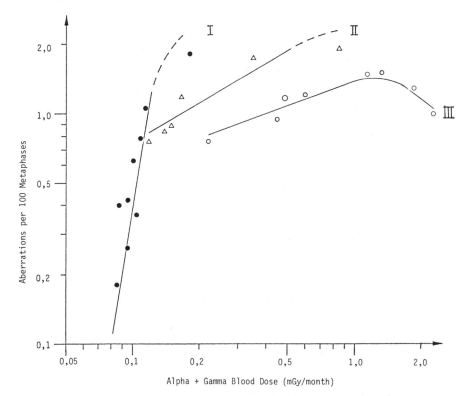

Figure 2. Total chromosome type aberration frequencies (sum of dicentrics, terminal and interstitial deletions) in the population of Badgastein, categorized into three groups.

radon source causing concentrations between 0.19 and 3.7 kBq/m   in
the indoor air (Stenstrand et al., 1979). In this work radon
concentrations in different dwellings are given. Therefore the
blood doses for three exposure categories could be assessed
according to our dose calculations. The aberration frequencies at
comparable mean doses fit well to the results obtained in
Badgastein, Austria (Pohl-Rüling and Fischer, 1983, Table 25-12).

Investigations on Mine Workers. Such investigations have been
carried out in several countries. In a study in Yugoslavia
(Kilibarda et al., 1968) on 20 uranium miners no increase of
aberration frequencies were found against 7 control workers. In
some mines of Finland "with a high level of radon and daughters"
a small number of workers were investigated (Stenstrand, 1982).
An increase of aberration frequencies was found but no differences
occured between miners who worked for a short time only and those
who worked for more than 19 years. In 55 chinese uranium miners
an increase of the aberration frequencies has been found but it
was emphazised that no linear dose response was established (Cao
Shu-yuan et al., 1981). In another study in China uranium miners
with high gamma and low alpha exposure revealed significant greater
aberration frequencies than controls, whereas non-uranium miners
with low gamma and high radon burdens showed no remarkable increase
of aberrations compared to controls (Deng Zhi-cheng and Li Yuan-
hui, 1983).
     An investigation of Brandom et al. (1978) supplied results
showing non-linear dose relationships. They categorized 80 uranium
miners and 20 controls (balanced for age and smoking habits) into
6 groups according to their accumulated exposure in Working Level
Months (WLM). A total of 9849 metaphases showed increasing
aberration rates within 5 groups up to exposure levels between
1740 and 2890 WLM. The aberration frequencies of the highest burden
group, more than 3000 WLM, however, were lower than that of group
three. A direct comparision with our findings in the Badgastein
population, also showing a decrease of the aberration rates at
higher doses, is not possible. We used the blood doses accumulated
for one month before sampling, as we found this period to be the
most significant. In the paper of Brandom et al., however,
accumulated life time doses were given, with no information of
the radiation burden a short time before sampling.
     These results on mine workers showed that at a certain amount
of exposure the increase of chromosome aberration frequencies
is no longer linear. This could also be considered as a support
of the hypothesis of inducing repair enzymes.

Investigations with Animals. A further support of the hypothesis
described above can be found in investigations carried out with
animals: Leonard et al. (1979) found in an area with high natural
radioactivity in France a small but significant increase of
chromosome aberrations in the lymphocytes of rabbits. The rabbits
were kept in a hut for 12 months, and received up to 0.7 Gy/year
from gamma rays together with more than 6 Gy alpha doses from
radon and daugthers. Further experiments with rabbits at radon
exposure under controlled conditions have shown that the chromosome

aberrations increased rapidly with dose but then attained a plateau.
It was concluded that the aberrations are not due to radon exposure
but essentially to gamma irradiation (Leonard et al., 1981).
   Chinese experiments with rats, inhaling radon and daughters
for several weeks with accumulated doses up to 3500 WLM, did not
show any increase in their chromosome aberration frequencies over
that of the control group (Deng Zhi-cheng and Li Yuan-huy, 1983).
Similar results on rats have been obtained in an unpublished
investigation by Pohl-Rüling and Pohl.

In Vitro Investigations of Human Blood Chromosome Aberrations
Due to Alpha Irradiation.

Such studies could contribute to the knowledge of the effect of
radon and daughters on chromosomes without any external irradiation.
There is only a small number of studies dealing with in vitro
alpha irradiation.

   Vulpis (1973) added pure natural boron as H_3BO_3, which contains
about 19% B-10, to the blood culture. It was then irradiated with
thermal neutrons from a reactor. The reaction of B-10(n,α)Li-7
within the culture served as alpha source. The dose range was
35 to 195 mGy according to her dose estimation which was complicated
by subtracting the dose due to the reactor radiation alone. The
dicentric chromosomes appeared to follow a linear response up
to about 0.18 Gy and then leveled off to a plateau due to a
"saturation" effect. The relative biological efficiency (RBE)
with respect to X-rays with the doses up to 5,1 Gy was found to
be 23.
   DuFrain et al. (1979) used the alpha particles from Am-241.
This was added to lymphocyte-rich blood plasmas and then removed
after about 2 hours exposure. They found dose varying RBEs of
119 to 45 for doses of 8.5 to 70 mGy. Their dose estimation,
however, was critizised by Fisher and Harty (1982) according to
microdosimetric considerations. Therefore also the over tenfold
discrepancy of their RBEs for alpha particles to the other studies
was attributed to erraneous dose estimations.
   Purrott et al. (1980) used a method similar to DuFrain et
al., see above. They added Pu-239 and Am-241 nitrate solutions
to agitated whole blood samples. A linear dose relationship was
observed over the dose range of 0.13 to 1.6 Gy and the RBE compared
to Co-60 gamma rays was 4.6.
   Edwards et al. (1980) used a Cm-242 source to irradiate
externally a thin film of blood. The energy of 4.9 MeV of the
alpha particles were almost entirely absorbed by the blood. The
dicentrics yield was linear from 0.11 to 4.2 Gy. From this resulted
a RBE of 17.9 with respect to Co-60 gamma rays. It was, however,
only 6.0 at the initial slope. To explain this RBE, which was
low compared to that expected from neutron experiments, a model
is presented taking into account cell killing and mitotic delay.
   All these investigations with internal and external alpha
sources are not directly comparable with the alpha irradiation
due to radon and decay products directly solved within the blood.
An external alpha source yields certainly another dose distribution
within the blood. Internal alpha sources, as described above,

were provided by long-lived radionuclides in concentrations at
which their chemotoxicity certainly must be taken into account.
It is unknown whether and to what amount these substances would
induce chromosome aberrations according to their toxicity without
the effect of irradiation. Therefore the knowledge of the effect
of the alpha rays of radon and daughters on chromosomes is still
missing.

To avoid all these problems we developed a method, described
below, for internal irradiation of blood with short-lived Rn-222
decay products. These radionuclides are always present in the
entire human body, according to the inhalation of the normal
atmosphere. Therefore no substances are brought into the blood
which could have additional toxic effects. Furthermore the
concentration of these nuclides in the blood is always extremely
small, because of their short half-life, so that any chemotoxic
effect can be excluded.

In Vitro Irradiation of Blood Cultures with Short-lived Rn-222
Decay Products.

Method. Figure 3 shows the equipment used by us for loading a
culture medium with radon and decay products. Air was circulated
in a closed system, driven by a membran pump (MP). The system
consisted of a Ra-226 solution (Ra), a security bubble flask with
water (H_2O), a membran bacteria filter (MF) and a second bubble
flask containing 100 ml RPMI 1640 culture medium (CM). This medium
contains 100 IE/ml penicilin and streptomycin and 0.75% L-glutamin.
Foetal calf serum, an essential part of blood cultures, must
not be added, else the airstream would develop foam. Furthermore
we added a small amount of $Pb(NO_3)_2$ and $Bi(NO_3)_2$, about 10 ng
of each, as "carriers" for the radon decay products to avoid a
"wall effect".

The air circulation must continue at least three hours. It
is not only necessary to distribute the radon accumulated in the
radium solution bottle into the entire system (for this purpose
a shorter time would be sufficient) but it takes three hours to
establish the equilibrium between radon and its short-lived
daughters in the medium.

After loading the medium with radon and daughters as described
above it is disconnected from the circuit. Now clean air is bubbled
through the medium for some minutes to remove the radon. The solid
decay products Po-218, Pb-214 and Bi-214, being in equilibrium
at this time, remain in the solution. After 30 minutes Po-218
(half-life 3.05 min.) has completely decayed.This time is taken
as zero time, t=0, for the dose calculation and for adding the
loaded medium to the whole blood culture. The blood culture have
been set up with the necessary amount of unloaded medium and the
required agents according to routine methods. Part of the loaded
medium is measured with a gamma spectrometer from which it is
possible to determine the Pb-214 and Bi-214 content at t=0.

In the blood cultures Pb-214 and Bi-214 decay completely
within four hours. The dose due to their disintegrations can be
calculated according to the following equations. Only the alpha
rays of Bi-214/Po-214 have to be taken into consideration, because
the doses due to beta and gamma rays are neglectible.

If a sample contains at the time t=0 the activities A_B^o and A_C^o (in Bq) of Pb-214 (RaB) and Bi-214 (RaC) then the activity of Bi-214 for any successive time, t, is A_C^t and can be calculated according to equation 2:

$$A_C^t = \frac{\lambda_C}{\lambda_B - \lambda_C} (e^{-\lambda_B t} - e^{-\lambda_C t}) A_B^o + e^{-\lambda_C t} A_C^o \tag{2}$$

λ_B and λ_C are the decay constants for Pb-214 and Bi-214, resp. From this the entire number of alpha particles, N_α, emitted in the blood culture is

$$N_\alpha = \int_0^\infty A_C^t \, dt \tag{3}$$

with the values for λ_B and λ_C the N_α becomes

$$N_\alpha = 2317 \, A_B^o + 1707 \, A_C^o \tag{4}$$

N_α and the energy of the Bi-214/Po-214 alpha rays (7.69 MeV = 1.232×10^{-12} Joule) yield the alpha dose D_α (in Gy):

$$D_\alpha = \frac{10^{-9}}{m} (2.855 \, A_B^o + 2.103 \, A_C^o) \tag{5}$$

whereby m is the mass of the blood culture in kg.

To get a series of various doses for one experiment we added different amounts of the loaded medium to the various blood cultures (e.g. 0.1, 0.2, 0.5, 1.0, 2.0 ml of the loaded medium to the blood cultures of 20 ml each). Therefore the A_C^o and A_B^o are different for the single blood cultures.

The orders of magnitude of doses for various series of experiments can be selected by using Ra-226 sources with different activities.

Preliminary results: The goal of the first step of our investigation was to study the effect of very low alpha doses which could be compared with the dose of alpha and gamma rays, accumulated over one month, of the people in Badgastein, see above. Because of the great statistical errors at low incident rates an amount of cells of the order of magnitude of 10000 should be analysed for one single dose. But by our small group it was hitherto only possible to score altogether about 5000 metaphases. Therefore the result can only be regarded to show the trend of the dose effect curve. Figure 4 shows the frequencies of the total number of chromosom type aberrations. In spite of the big statistical errors it can be seen that the dose effect increases rapidly with very small doses and then flattens into a plateau. This corresponds to the dose response which we found in our in vivo study, see Figure 1. In that work we believed that the sharp increase of the aberration frequencies is mainly caused by gamma irradiation (Pohl-Rüling and Fischer, 1979). Our in vitro investigation, however, indicates that the increase must be caused by alpha rays. The plateau could not be caused by a decreased rate of mitosis due to affected or damaged cells at our low doses, between 0.3 and 3 mGy! They are three to four orders of magnitude lower than those at which other authors found still a linear rise of aberrations with dose.

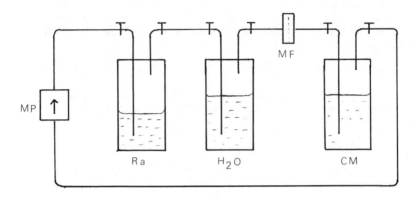

Figure 3. Equipment for loading a culture medium (CM) with Rn-222 decay products.

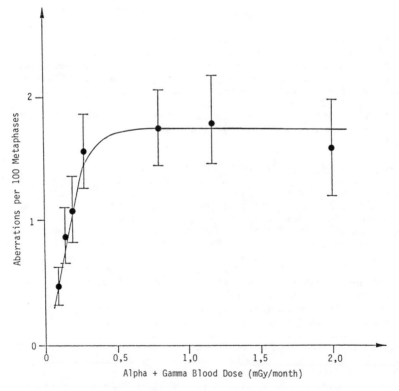

Figure 4. Total chromosome type aberration frequencies (sum of dicentrics, rings, terminal and interstitial deletions) of in vitro blood samples, irradiated with Rn-222 decay products.

No calculation of the RBE of the alpha rays at this very low dose range can be made, as no in vitro study was carried out with x- or gamma rays at such low dose levels. The lowest dose of x-rays with which a chromosome aberration study has been carried out was 4 mGy (Pohl-Rüling et al., 1983). A very rough estimation of the RBE at doses between 2 and 4 mGy yield a value between 2 and 3.

To get a significant value it would be necessary to irradiate blood samples from the same donor at the same time with the same doses of alpha and gamma rays, and to score a great amount of metaphases.

Conclusions.

In vivo as well as in vitro investigations have shown that very low doses result into a sharp increase of chromosome aberration frequencies, followed by a plateau. Although the statistical errors in these investigations are large it can be excluded that the dose relationship is represented by a linear function in the whole dose range considered. To explain this effect we tried the hypothesis that after a certain amount of damage to the DNA additional repair enzymes are produced. The implications of this hypothesis is that protective mechanisms within the organisms are capable of controling the degree of radiation damage upon reaching a certain level. The production of repair enzymes can be triggered as it is the case also for other inducible enzymes. This would result in a plateau of the dose response curve. If the additional repair enzyme concentration reaches the point of saturation, the damage, here the chromosome aberration frequencies, again becomes strongly dose dependent, i. e. they rise with dose.

As radon and decay products are always present in our environment the effect of their alpha rays on chromosomes should be more extensively investigated. This is not possible with our small group therefore we suggest a cooperation of many laboratories. A model for such an investigation is given by the coordinated research programmes, sponsored by the International Atomic Energy Agency, Vienna, on acute in vitro irradiation of blood lymphocytes with x-rays and neutrons at dose levels between 4 and 300 mGy, (Pohl-Rüling et al., 1983a, 1983b, 1986).

Acknowledgments

We want to thank Dr. O. Haas (St. Anna Children's Hospital, Vienna) for testing the mitosis rates and the stimulation effect of PHA in whole blood cultures containing 10 ng/ml of each $Pb(NO_3)_2$ and $Bi(NO_3)_2$. He confirmed that the addition of these substances at the given amount are not toxic at all for the process of culturing lymphocytes and preparing chromosome slides. Furthermore we thank him, Mrs. T. Ferstel (St. Anna Children's Hospital, Vienna) and Mr. Chr. Atzmüller (Division of Biophysics, University of Salzburg) for scoring part of the slides for this study.

Literature cited

Archer, V.J., Oncology Overview, Selected Abstracts on Risk of
Cancer from Exposure to Low Level Ionizing Radiation, PB84-922906,
International Cancer Research Data Bank Program, National Cancer
Institute, U.S. Department of Health and Human Services, Salt
Lake City, Utah, (August 1984).

Baum, J.W. Clonal Theory of Radiation Carcinogenesis, Proceedings of
the Symposium on Microdosimetry pp. 12, National Technical
Information Service, Springfield, VA, DE 82022044, (1982).

Brandom, W.F., G. Saccomanno, V.E. Archer, P.G. Archer, and A.D.
Bloom, Chromosome Aberrations as a Biological Dose-Response Indicator
of Radiation Exposure in Uranium Mines, Radiation Research 76:159-
171 (1978).

Brown, J.M., The Shape of the Dose-Response Curve for Radiation Carcinogenesis: Extrapolation to Low Doses, Radiation Research
71:34-50 (1977).

Burch, P.R.J., Problems With the Linear-Quadratic Dose-Response
Relationship, Health Physics 44:411-413 (1983)

Cao Shu-yuan, Deng Zhicheng, Shou Zhen-ying, Li Yun-hua, and Yu
Cui-fang, Lymphocyte Chromosome Aberrations in Personnel occupationally Exposed to Low Levels of Radiation, Health Physics
41:586-587 (1981).

Chen Dequing, Zhang Chaiyang, and Yao Suyan, Further Investigations
on Chromosome Aberrations in Lymphocytes of Inhabitants in Heigh
Background Radiation Area in Yangjiang, Chinese Journ. Radiological
Medical Protection 5:116-119 (1985).

Deng Zhi-cheng and Li Yuan-huy, The Main Cause of Chromosome
Aberration Increase in Blood Lymphocytes of Uranium Miners, Private
Communication from Department of Radiation Medicine, North China,
Institute of Radiation Protection China (1983).

DuFrain, R.J., L.G. Littlefield, E.E. Joiner, and E.L. Frome,
Human Cytogenetic Dosimetry: A Dose-Response Relationship for
Alpha Particle Radiation from Am-241, Health Physics 37:279-289
(1979).

Edwards, A.A., R.J. Purrott, J.S. Prosser, and D.C. Lloyd, The
Induction of Chromosome Aberrations in Human Lymphocytes by Alpha-
Radiation, Internat. J. Radiation Biology 38:83-91 (1980).

Fisher, D.R. and R. Harty, The Microdosimetry of Lymphocytes
Irradiated by Alpha Particles, 29th Annual Scientific Meeting of the
Radiation Research Society, Minneapolis, Minnesota, (April, 1985).

Frankenberg-Schwager, M., D. Frankenberg, D. Blocher, and C. Adamczyk, The Linear Ralationship Between DNA Double-Strand Breaks and Radiation Dose (30 MeV Electrons) is Converted Into a Quadratic Function by Cellular Repair, Internat. J. Radiation Biology 37:207-212 (1980).

International Atomic Energy Agency, Routine Use of Chromosome Analysis in Radiation Protection, Proceedings of an Advisory Group Meeting, Fontenay-aux-Roses, France, TecDoc-224, Vienna (Dec. 1978).

Kamiguchi Y., and K. Mikamo, Dose Response Relationship for Induction of Structural Chromosome Aberrations in Chinese Hamster Oocytes After X-Irradiation, Mutation Research 103:33-37 (1982).

Katz, R. and W. Hofmann, Biological Effects of Low Doses of Ionizing Radiations: Particle Tracks in Radiobiology, Nuclear Instruments and Methods 203:433-442 (1982).

Kellerer, A.M. and H.H. Rossi, The Theory of Dual Radiation Action, Current Topics Radiation Research, Quarterly, 8:85-158 (1972).

Kilibarda, M., B. Marcovic, and D. Panov, Studies of Chromosome Aberrations in Persons Chronically Exposed to Radiation (X, Ra 226, and Rn 222), Studies Biophysics 6:179-186 (1968).

Léonard, A., M. Delpoux, J. Chameaud, G. Decat, and E.D. Léonard, Natural Radioactivity in Southwest France and Its Possible Genetic Consequences for Mammals, Radiation Research 77:170-181 (1979).

Léonard, A., M. Delpoux, J. Chameaud, G. Decat, and E.D. Léonard, Biological Effects Observed in Mammmals Maintained in an Area of Very High Natural Radioactivity, Canadian J. Genetic Cytology 23:321-326 (1981).

Lloyd, D.C., J.S. Prosser, and R.J. Purrott, The Study of Chromsome Aberrations Yield in Human Lymphocytes as an Indicator of Radiation Dose: Revised Techniques, NRPB-M70, National Radiological Protection Board, Chilton, Didcot, Oxon. OX11 0RQ, Great Britain, (November 1982).

Lloyd, D.C., and R.J. Purrott, Chromosome Aberration Analysis in Radiological Protection Dosimetry, Radiation Protection Dosimetry 1:19-29 (1981).

Marshall, J.H., and P.G.A. Groer, A Theory of the Induction of Bone Cancer by Alpha Radiation, Radiation Research 71:149-192 (1977).

Nakai, S. and K. Mifune, Personal Communication, (1982).

Pohl, E., and J. Pohl-Rüling, Dose Calculations Due to the Inhalation of Rn 222, Rn 220 and Their Daughters, Health Physics 32:552-555 (1977).

Pohl, E. and J. Pohl-Rüling, Dose Distribution in the Human Organism
Due to Incorporation of Radon and Decay Products as a Base for
Epidemiological Studies, in Proceedings of the International Radon
Specialist Meeting on the Assessment of Radon and Daughters
Exposures and Related Biological Effects, Rome, (RD Press) pp.
84-111, University of Utah, Salt Lake City, Utah, (1982).

Pohl, E., F. Steinhäusler, W. Hofmann, and J. Pohl-Rüling,
Methodology of Measurements and Statistical Evaluation of Radiation
Burden to Various Population Groups From All Internal and External
Natural Sources, in Proceedings of Biological and Environmental
Effects of Low-Level Radiation, (International Atomic Energy
Agency), Vol.II, pp. 305-315, Vienna, Austria (1976).

Pohl-Rüling, J., P. Fischer, and E. Pohl, The Low-Level Shape
of Dose Response for Chromosome Aberrations, in Proceedings of
the International Symposium of Late Biological Effects of Ionizing
Radiation (International Atomic Energy Agency) pp. 315-326, Vienna,
Austria, (1978).

Pohl-Rüling, J. and P. Fischer, The Dose Effect Relationship of
Chromosome Aberrations to Alpha and Gamma Irradiation in a
Population Subjected to an Increased Burden of Natural Radioacti-
vity, Radiation Research 80:61-81 (1979)

Pohl-Rüling, J. and P. Fischer, Chromosome Aberrations in Inhabitants of Areas With ELevated Natural Radioactivity, in Radiation-
Induced Chromosome Damage in Man (T. Ishihara and M.S. Sasaki,
ed) pp. 527-560, A.R. Liss, Inc., New York (1982).

Pohl-Rüling, J., P. Fischer, O. Haas, G.Obe, A.T. Natarajan, P.P.W.
van Buul, K.E. Buckton, N.O. Bianchi, M. Larramendy, M. Kučerová,
Z. Poliková, A. Léonard, L. Fabry, F. Palitti, T.Sharma, W. Binder,
R.N. Mukherjee, and U. Mukherjee, Effect of Low-Dose Acute X-
Irradiation on the Frequencies of Chromosomal Aberrations in Human
Peripheral Lymphocytes in Vitro, Mutation Research 110:7182 (1983).

Pohl-Rüling, J., P. Fischer, K.E. Buckton, R.N. Mukherjee, W.
Binder, R. Nowotny, W. Schmidt, N.O. Bianchi, P.P.W. van Buul, A.T.
Natarajan, L. Fabry, A. Léonard, M. Kučerová, D.C. Lloyd, U.
Mukherjee, G. Obe, F. Palitti, and T. Sharma, Comparison of Dose
Dependence of Chromosome Aberrations in Peripheral Lymphocytes
at Low Levels of Acute in Vitro Irradiation With 250 kV X-Rays
and 14 MeV Neutrons, in Proceedings of the International
Symposium on Biological Effects of Low-Level Radiation in Venice,
pp. 171-184, International Atomic Energy Agency, Vienna, Austria
(1983).

Pohl-Rüling, J., P. Fischer, D.C. Lloyd, A.A. Edwards, A.T. Natarajan, G. Obe, K.E. Buckton, N.O. Bianchi, P.P.W. van Buul, B.C. Das,
F. Daschil, L. Fabry, M. Kučerová, A. Léonard, R.N. Mukherjee,
U. Mukherjee, R. Nowotny, P. Palitti, Z. Poliková, T. Sharma,
and W. Schmidt, Chromosomal Damage Induced in Human Lymphocytes
by Low Doses of D-T Neutrons, Mutation Research 173:267-272 (1986).

Purrott, R.J., A.A. Edwards, D.C. Lloyd, and J.W. Stather, The Induction of Chromosome Aberrations in Human Lymphocytes by in Vitro Irradiation With Alpha-Particles From Pu 239, Internat. J. Radiation Biology 38:277-284 (1980).

Steinhäusler, F., W. Hofmann, E. Pohl, and J. Pohl-Rüling, Local and Temporal Distribution Pattern of Radon and Daughters in an Urban Environment and Determination of Organ Dose Frequency Distributions With Demoscopical Methods, in Proceedings of the Symposium on Natural Radiation Environment, III, Houston, Conf-780422, DOE Sym. Ser. 51, Vol. II, pp. 1145-1161, Houston NM, (1980).

Stenstrand, K., Personal Communication (1982).

Stenstrand, K., M. Annamäki, and T. Rytömaa, Cytogenetic Investigation of People in Finland Using Household Water with High Natural Radioactivity, Health Physics 36:441-444 (1979).

The Environmental Protection Agency, Office of Radiation Programs, Cytological Studies in Areas with Atypical Environmental Radon Concentrations: Radon Community Cytology Study, Final Report of Project No..Wa-78-C243, (September, 1984).

Van Bekkum, D.W. and P. Bentvelzen, The Concept of Gene Transfer-Misrepair Mechanism of Radiation Carcinogenesis May Challange the Risk Estimation For Low Radiation Doses, Health Physics 43:231-237 (1982).

Vulpis, N., Chromosome Aberrations Induced in Human Peripheral Blood Lymphocytes Using Heavy Particles From B10(n,α)Li7 Reaction, Mutation Research 18: 103-111 (1973).

Yunis, J.J., Human Chromosome Methology (J.J.Yunis ed) Academic Press, New York, London (1965).

RECEIVED September 9, 1986

Chapter 36

Biophysical Effects of Radon Exposure on Human Lung Cells

B. Reubel[1], C. Atzmüller[1], F. Steinhäusler[1], and W. Huber[2]

[1]University of Salzburg, Division of Biophysics, A-5020 Salzburg, Austria
[2]Lungenabteilung, Landeskrankenanstalten Salzburg, Austria

Radon and its decay products are suspected to be one of the strongest natural carcinogenesis in man's environment based on epidemiological evidence, animal inhalation experiments and dosimetric calculations. Presently no in vitro evidence of radon daughter carcinogenic potential is available. The aim of the present study is to investigate the reaction of human lung cells and tissue samples to different levels of exposure to radon and its daughters. Experimental methods consist of the use of different microelectrodes for intra- and extracellular recordings. Preliminary assessment of cellular dose due to radon exposure is based on exposure levels and cellular target geometry. Analysis of the investigation of the biophysical parameters revealed the following statistically significant results: a) metabolic changes as indicated by alteration of cellular oxygen consumption; b)alterations of membrane properties as measured by the transmembrane resting potential.

Radon 222 and its decay products, Rn-d, are well known as among the most powerful carcinogenic agents to which human beeings are exposed as members of the general public (e.g. indoor exposure), as a consequence of occupation (e.g. uranium miner) or as part of medical treatment (radon spa treatment).

Knowledge about the potential of Rn-d to cause lung cancer is presently derived from three sources of data:

1) Epidemiology of uranium miners exposed to high Rn-d levels .
2) Dosimetric calculations indicating that the basal cells of the bronchial epithelium are the sensitive target for lung cancer induction; whereby basal cell dose is the initiating agent and cell killing as well as other biological factors, such as smoking, are the promoting quantities (Hofmann, 1983).
3) Animal inhalation experiments: Experiments with rats showed

0097-6156/87/0331-0502$06.00/0

positive results of tumor induction after Rn-d exposure ranging
between 10^2 and 10^3 WLM (Chameaud and Perraud, 1980, Cross
et al, 1980). It could also be demonstrated that there is a
linear no-treshold relationship for lung cancer induction by
Rn-d between 0 and 50 WLM; the doubling dose for cancer
induction was 20 WLM (Chameaud et al, 1984; Chameaud et al,
1983).

Data quality underlying epidemiological and dosimetric assessment is
questionable for various reasons. Epidemiological investigations
suffer from the low quality of input data on Rn-d exposure and lung
cancer incidence (Steinhäusler and Pohl, 1983; Steinhäusler, in
press). Animal inhalation experiments allow conclusions on the
reaction of the whole body to Rn-d exposure, but can be extrapolated
to men only with caution due to physiological differences and
species-specific differences in the dose-effect relationship.
Besides these animal experiments and the chromosome aberration
studies no data exist for low level Rn-d induced in vitro effects
at the cellular or tissue level. Human lung cell/tissue
cultures are suitable objects for studies on the cancer induction
in the human lung due to Rn-d exposure because in vitro experiments
offer reproducible conditions with regard to characteristics of the
biological target as well as Rn-d exposure conditions. However,
results obtained from cell culture studies do not allow for regulatory effects present in in vivo-experiments.

Material and Methods

In the present investigation human embryonic lung cells (dimensions: 20 x 5 micrometer) and human lung tissue (obtained from
biopsy samples of patients) were exposed to two different levels of
radon and Rn-d atmosphere, generated by emanating radon-222 from an
open Ra-226-source into an incubator (Figure 1). Temperature, CO2
and O2 concentration in the incubator with Rn-d exposure and in the
reference-incubator with control cultures (without Rn-d exposure)
were maintained at constant levels. Irradiated and control cells in
log-phase were kept under standardized cultivation conditions in
order to ensure homogenous cell material of the same age. For
measurements of radon-222 and Rn-d in the air of the incubator an
ionization chamber, respectively a working-level-meter was used.
Assessment of the cellular dose was not possible at this stage.
Therefore atmospheric Rn-d levels, radon solubility in aqueous phase
and cellular target geometry as input parameters were used to
describe the Rn-d exposure of the cells. Further detailed microdosimetric calculations are in progress.
       Quantification of the cellular response to irradiation was
carried out using several biophysical parameters for in vitro-
cultured cells and tissue samples:

1)   Transmembrane resting potential (TMRP)
2)   Cellular oxygen uptake (pO2)
3)   Observation of the cell structure

The experimental set-up for TMRP measurements consists of following components: neuroprobe amplifier, chart recorder, camera-monitor-video system, remote controlled micromanipulator, inverted micros-cope, preamplifier, electrode holder with KCl-filled glass-micro-electrode and Ag/AgCl reference electrode.The electric circuit is closed with: medium - KCl-filled microelectrode - preamplifier - amplifier - reference electrode - medium. Optical control is pro-vided by inverted microscope, video camera and monitor. With the use of the micromanipulator the microelectrode is impaled into the cell under microscopical control to obtain a TMRP-recording. Figure 2 shows an intracellular measurement of a human cell in vitro.

The experimental set-up for cellular oxygen measurements (pO2) consists of following components: pO2 measuring micro chamber (volume 0.6 microliter), polarographic microelectrode , water-bath for constant temperature, chemical microsensor connected to a strip-chart recorder and gas calibration unit.

Cells, respectively human biopsies suspended in medium, are filled into the microchamber via a porthole. pO2 measurements are performed with polarographic microelectrodes, registering redox-processes at the electrode tip.

Cellular structures were investigated using standard light microscope techniques.

Radon/Radon Decay Products Exposure

Cell cultures. Cell cultures, covered with a 10 mm layer of medium in petri-dishes, were exposed for one week to an atmospheric radon concentration of 260 kBq/m^3 ("high level exposure"), resp. 37 kBq/m^3 ("low level exposure"). The Rn-decay product concentration was 6 WL, resp. 0,9 WL (RaA: 61, resp. 8,6 kBq/m^3 , RaB: 25, resp. 3,6 kBq/m^3 , RaC':13, resp. 1,8 kBq/m^3). After an exposure for one week in an elevated Rn/Rn-d atmosphere cell cultures were incu-bated in a normal atmosphere.

As a consequence of the high solubility of radon in an aqueous solution about 0,17 of the atmospheric Rn-activity is contained in the medium at 37° C, causing the alpha particle-exposure of the cells (Jennings and Russ, 1948).

Measurements of biophysical parameters were carried out three hours after the beginning of Rn/Rn-d exposure and then in 24 hour-intervals. Saturation of Rn/ Rn-d in the medium was reached after several hours.

Biopsies. Human lung biopsies, obtained from the lung department of Salzburg Clinics, were suspended in buffered phosphate saline. All samples were exposed to the high level Rn/Rn-d atmosphere for one day in order to measure the oxygen uptake following irradiation. Before irradiation the same biopsy served as control for unirradiated pO2 -measurements. Potential ageing effects of the biopsy, respectively bacterial infection showing a possible effect on oxygen consumption, were accounted for by repeating measurements with the same biopsy about 24 hours after the initial measurement.

Results

TMRP-measurements. Altogether 4100 single measurements have been carried out for the four series investigating the temporal changes of TMRP of irradiated and control cells.

Three days after the beginning of high level Rn/Rn-d exposure TMRP-values of the irradiated cells decreased significantly (level of significance: $\propto$ (two-sided) ≙ 0.05) below the control value : as an example series 2 is shown in Figure 3 . This is followed by a further decrease to a minimum on the 5th-6th day of exposure. After re-incubating the cells in a normal atmosphere for about two days, TMRP-values reached again control levels.

After low level Rn/Rn-d exposure the same pattern of time dependent TMRP-change was obtained: significantly decreasing TMRP-values in the time interval of 1-5 days after the onset of irradiation compared to the control values, followed by an increase back to control levels after re-incubation in normal atmosphere : as an example series 4 is shown in Figure 4. Mean TMRP-values and the results of the statistical u-test are shown in Table I,II.

Cellular oxygen consumption. In a preliminary test series 9 pO2 measurements have been analysed until now. Cellular oxygen measurements of human lung biopsies showed after Rn/Rn-d exposure a significantly increased (significance level: $\propto$ (two-sided) ≙ 0.05) oxygen consumption for tumor samples as compared to respective control value of the same tumor samples non-exposed to Rn/Rn-d (Figure 5).

It can also be seen that even 24 hours of storage of the unexposed biopsy samples resulted only in an insignificant change of the pO2 consumption. Normal lung tissue reacted to radon exposure with decreasing oxygen consumption (level of significance: $\propto$ (two-sided) ≙ 0.05).

Tables III,IV contain the results of the regression analysis for the pO2 measurements together with the statistical t-test.

Structural changes. At several loci cell membranes of Rn/Rn-d exposed cells appeared to be damaged similar to membrane damages observable after heat treatment, resulting in a change of the cell shape from originally spindle-shaped cells to spherically-shaped cells with membrane indentations (Figure 6).

Discussion

It was demonstrated in earlier investigations (Steinhäusler et al, 1985) that human cancer biopsies show a decreased TMRP compared to normal tissue by about 50%. The observed TMRP-values after exposure to radon and Rn-d with atmospheric activities of 6 WL and 0,9 WL indicate a decrease of a similar magnitude. This effect may be interpretated as an indicator of the initiation of a possible tumor transformation at the cellular level.

The frequency distribution of the TMRP measurements shows for both levels of exposure a bi-modal histogram - in contrary to the log-normal distribution of the control values (Figure 7). The bimodal distribution could be due to a reaction at the cellular level indicating a critical number of hits, i.e. peak A is representing the cell population with a TMRP unaffected by low level alpha exposure, while peak B represents cells reacting to alpha hits.

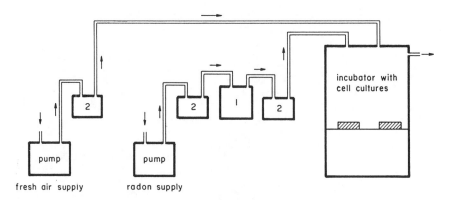

Figure 1.   In vitro radon-222-exposure of human cells: 1) Ra-226-
solution;   2) bubbler

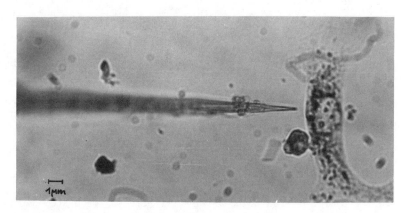

Figure 2.   TMRP measurement of a human lung cell using a microelectrode

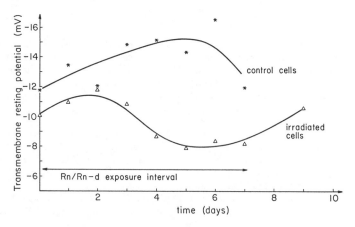

Figure 3.   Mean TMRP-values of WI-38 cells, exposed to an atmospheric concentration of radon daughters of 6 WL

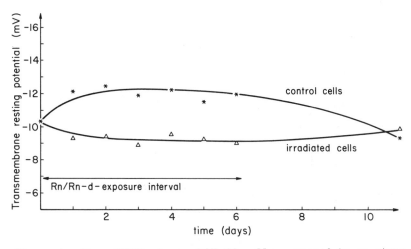

Figure 4. Mean TMRP-values of WI-38 cells, exposed to an atmospheric concentration of radon daughters of 0.9 WL

TABLE I. TMRP-Values of Lung Fibroblasts mean $(\bar{x}, \bar{c})$ $\mp$ Standard Deviation.

	Series 1 irradiated[1]/ control cells	Series 2 irradiated[1]/ control cells	Series 3 irradiated[1]/ control cells	Series 4 irradiated[2]/ control cells
before exposure	$\bar{x}_1 = 9.0 \mp 2.5$ $\bar{c}_1 = 8.9 \mp 3.7$	$\bar{x}_7 = 10.7 \mp 2.7$ $\bar{c}_7 = 11.8 \mp 3.0$	$\bar{x}_{16} = 11.9 \mp 2.5$ $\bar{c}_{16} = 12.4 \mp 2.4$	$\bar{x}_{22} = 10.2 \mp 2.9$ $\bar{c}_{22} = 10.4 \mp 2.8$
1d after exposure	$\bar{x}_2 = 9.8 \mp 4.0$ $\bar{c}_2 = 8.6 \mp 3.5$	$\bar{x}_8 = 10.9 \mp 2.9$ $\bar{c}_8 = 13.4 \mp 6.6$	$\bar{x}_{17} = 11.3 \mp 2.3$ $\bar{c}_{17} = 13.8 \mp 2.4$	$\bar{x}_{23} = 9.3 \mp 2.6$ $\bar{c}_{23} = 12.1 \mp 3.0$
2d after exposure	$\bar{x}_3 = 6.3 \mp 2.2$ $\bar{c}_3 = 9.1 \mp 3.7$	$\bar{x}_9 = 11.8 \mp 3.1$ $\bar{c}_9 = 12.1 \mp 4.1$	$\bar{x}_{18} = 11.9 \mp 2.5$ $\bar{c}_{18} = 13.7 \mp 2.2$	$\bar{x}_{24} = 9.4 \mp 4.8$ $\bar{c}_{24} = 12.4 \mp 3.1$
3d after exposure	$\bar{x}_4 = 6.1 \mp 2.0$ $\bar{c}_4 = 8.8 \mp 2.4$	$\bar{x}_{10} = 10.8 \mp 2.7$ $\bar{c}_{10} = 14.8 \mp 3.2$	$\bar{x}_{19} = 9.6 \mp 3.0$ $\bar{c}_{19} = 13.4 \mp 2.8$	$\bar{x}_{25} = 8.9 \mp 2.7$ $\bar{c}_{25} = 12.0 \mp 3.2$
4d after exposure	$\bar{x}_5 = 7.0 \mp 3.6$ $\bar{c}_5 = 8.8 \mp 3.3$	$\bar{x}_{11} = 8.7 \mp 2.9$ $\bar{c}_{11} = 15.1 \mp 4.5$	$\bar{x}_{20} = 9.1 \mp 2.9$ $\bar{c}_{20} = 12.0 \mp 2.5$	$\bar{x}_{26} = 9.5 \mp 2.6$ $\bar{c}_{26} = 12.2 \mp 3.2$
5d after exposure	$\bar{x}_6 = 6.5 \mp 2.7$ $\bar{c}_6 = 10.4 \mp 2.5$	$\bar{x}_{12} = 7.9 \mp 2.8$ $\bar{c}_{12} = 14.3 \mp 3.6$	$\bar{x}_{21} = 10.5 \mp 3.2$ $\bar{c}_{21} = 14.2 \mp 3.5$	$\bar{x}_{27} = 9.2 \mp 2.6$ $\bar{c}_{27} = 11.5 \mp 3.3$
6d after exposure		$\bar{x}_{13} = 8.3 \mp 3.0$ $\bar{c}_{13} = 16.5 \mp 4.2$		$\bar{x}_{28} = 9.0 \mp 3.0$ $\bar{c}_{28} = 12.0 \mp 3.2$
7d after exposure		$\bar{x}_{14} = 8.1 \mp 3.5$ $\bar{c}_{14} = 12.0 \mp 3.3$		
9d after exposure		$\bar{x}_{15} = 10.4 \mp 3.6$		
11d after exposure				$\bar{x}_{29} = 10.4 \mp 2.8$ $\bar{c}_{29} = 10.2 \mp 3.3$

[1]atmospheric Rn-d concentration: 6 WL
[2]atmospheric Rn-d concentration: 0.9 WL

TABLE II: Statistical Comparison of the TMRP-Mean-Values of Control and Irradiated Cells; α (two sided) = Level of Significance

	Series 1 irradiated[1]/ control cells α	Series 2 irradiated[1]/ control cells α	Series 3 irradiated[1]/ control cells α	Series 4 irradiated[2]/ control cells α
before exposure	$\bar{x}_1$ vs $\bar{c}_1$: 0.05	$\bar{x}_7$ vs $\bar{c}_7$: 0.05	$\bar{x}_{16}$ vs $\bar{c}_{16}$: 0.05	$\bar{x}_{22}$ vs $\bar{c}_{22}$: 0.05
1d after exposure	$\bar{x}_2$ vs $\bar{c}_2$: 0.05	$\bar{x}_8$ vs $\bar{c}_8$: 0.05	$\bar{x}_{17}$ vs $\bar{c}_{17}$: 0.001	$\bar{x}_{23}$ vs $\bar{c}_{23}$: 0.01
2d after exposure	$\bar{x}_3$ vs $\bar{c}_3$: 0.001	$\bar{x}_9$ vs $\bar{c}_9$: 0.05	$\bar{x}_{18}$ vs $\bar{c}_{18}$: 0.05	$\bar{x}_{24}$ vs $\bar{c}_{24}$: 0.01
3d after exposure	$\bar{x}_4$ vs $\bar{c}_4$: 0.001	$\bar{x}_{10}$ vs $\bar{c}_{10}$: 0.001	$\bar{x}_{19}$ vs $\bar{c}_{19}$: 0.001	$\bar{x}_{25}$ vs $\bar{c}_{25}$: 0.05
4d after exposure	$\bar{x}_5$ vs $\bar{c}_5$: 0.05	$\bar{x}_{11}$ vs $\bar{c}_{11}$: 0.001	$\bar{x}_{20}$ vs $\bar{c}_{20}$: 0.001	$\bar{x}_{26}$ vs $\bar{c}_{26}$: 0.01
5d after exposure	$\bar{x}_6$ vs $\bar{c}_6$: 0.001	$\bar{x}_{12}$ vs $\bar{c}_{12}$: 0.001	$\bar{x}_{21}$ vs $\bar{c}_{21}$: 0.001	$\bar{x}_{27}$ vs $\bar{c}_{27}$: 0.01
6d after exposure		$\bar{x}_{13}$ vs $\bar{c}_{13}$: 0.001		$\bar{x}_{28}$ vs $\bar{c}_{28}$: 0.05
7d after exposure		$\bar{x}_{14}$ vs $\bar{c}_{14}$: 0.001		
11d after exposure				$\bar{x}_{29}$ vs $\bar{c}_{29}$: 0.05

[1]atmospheric Rn-d concentration: 6WL
[2]atmospheric Rn-d concentration: 0.9 WL

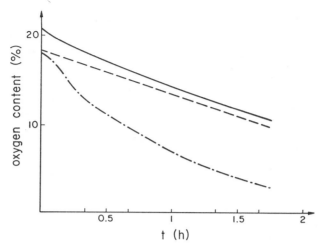

Figure 5. Oxygen consumption measurements of tumor lung tissue before and after exposure to an atmospheric Rn-d concentration of 6 WL: unirradiated control tissue at the beginning of the experiment: ——— ; unirradiated control tissue 24 hours later: — — ; tissue exposed to an atmospheric Rn-d level of 6 WL:—·—.

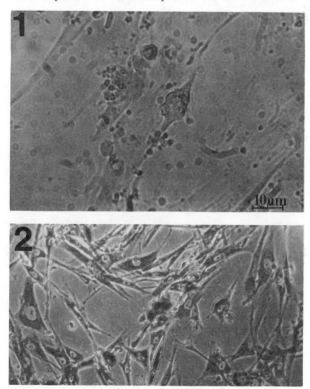

Figure 6. Structural changes of (1) radon-exposed cells (atmospheric Rn-d concentration: 6 WL) compared with (2) control cells.

TABLE III : Oxygen Consumption of Lung Biopsies

Regression Coefficients:Slope (b) and Variation (s)

	1. measurement control tissue	2. measurement control tissue	irradiated tissue[1]
normal lung	b_1 = -0.25		b_2 = -0.04
	s_1 = 0.72		s_2 = 0.15
normal lung	b_3 = -0.04		b_4 = -0.01
	s_3 = 0.05		s_4 = 0.21
tumor center	b_5 = -0.08	b_6 = -0.06	b_7 = -0.13
	s_6 = 0.01	s_6 = 0.02	s_7 = 0.02
tumor center	b_8 = -0.03		b_9 = -0.07
	s_8 = 0.01		s_9 = 0.04

[1]atmospheric Rn-d concentration : 6 WL

TABLE IV: Statistical Comparison of the Oxygen Consumption

Curves; α (two-sided) = Level of Significance

	control vs irradiated[1] tissue α :	oxygen consumption after irradiation
normal lung	0.001	decreasing
normal lung	0.05	decreasing
tumor center	0.0001	increasing
tumor center	0.001	increasing

[1]atmospheric Rn-d concentration : 6 WL

The reduction of the energy metabolism of normal lung tissue as indicated by decreasing oxygen-values, reflects reduced energy requirement due to reduced cellular synthesis during Rn/Rn-d exposure.

This is in contrast to the stimulation of the energy metabolism, following Rn/Rn-d exposure of lung tumor center, indicating possible tumor growth promotion by Rn/Rn-d exposure.

Structural changes of Rn/Rn-d exposed lung cells similar to heat symptoms can be interpreted as being caused by the similar underlying mechanism, e.g. re-arrangements of membrane protein-lipid components, and by releasing some of the attachment points of the cell to the petridish-surface.

Conclusions

A change of following biophysical parameters after radon-222 exposure could be found as preliminary results:

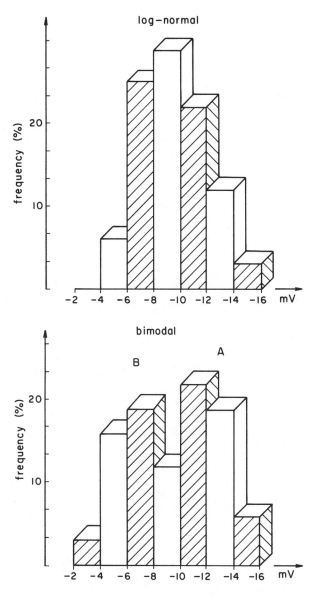

Figure 7. Frequency distributions of TMRP-values after an exposure to an atmospheric Rn-d concentration of 6 WL (bimodal) and of TMRP control values (log-normal).

1) a decrease of the TMRP; 2) a decrease of the oxygen consumption of normal lung tissue; 3) an increase of the oxygen consumption of tumor lung tissue; 4) a deformation of the cell membrane.

Literature Cited

Chameaud, J. and Perraud, R., Cancers Induced by Rn-222 in the Rat, in Proc. of the Specialist Meeting on the Assessment of Radon and Radon Daughter Exposure and Related Biophysical Effects, (Clemente, G. et al, eds) pp.198-209, RD Press, Salt Lake City (1980).

Chameaud, J., Masse, R. and Lafuma, J., Influence of Radon Daughter Exposure at low Doses on Occurrence of Lung Cancer in Rats, in Radiation Protection Dosimetry: Indoor Exposure to Natural Radiation and Associated Risk Assessment, (Clemente, G., F. et al, eds) pp.385-388, Nuclear Technology Publishing, Anacapri (1983).

Chameaud, J., Masse, R., Morin, M. and Lafuma, J., Lung Cancer Induction by Radon Daughters in Rats, in Occupational Radiation Safety in Mining, Vol.1, (Stocker , H., ed) pp.350-353 , Canadian Nuclear Association, Toronto ,(1984).

Cross, F., Palmer, R., F. and Busch, R., H., Influence of Radon Daughter Exposure Rate and Uranium Ore Dust Concentration on Occurrence of Lung Tumors, in Proc. of the Specialist Meeting on the Assessment of Radon and Radon Daughter Exposure and Related Biophysical Effects, (Clemente, G. F. et al, eds) pp.189-197, RD Press, Salt Lake City (1980).

Hofmann, W., Lung Cancer Induction by Inhaled Radon Daughters-What is the Relevant Dose?, in Radiation Protection Dosimetry, Indoor Exposure to Natural Radiation and Associated Risk Assessment, (Clemente, G., F. et al, eds), pp.367-370, Nuclear Technology Publishing , Anacapri (1983).

Jennings, W., A. and Russ, S., in Radon: Its Technique and Use, Murray, J., London (1948).

Steinhäusler, F. and Pohl, E., Lung Cancer Risk for Miners and Atomic Bomb Survivors and its Relevance to Indoor Exposure, in Radiation Protection Dosimetry, Indoor Exposure to Natural Radiation and Associated Risk Assessment, (Clemente, G. F. et al, eds) pp.389-394, Nuclear Technology Publishing, Anacapri (1983).

Steinhäusler, F., Huber, F., Huber, M. and Wutz, W., Korrelation Zelltyp und Membranruhepotential beim Bronchuskarzinom, Wiener Medizinische Wochenschrift, 135. Jg., 18. Jahrestagung der Österr.Ges. f. Lungenerkrankungen und Tuberkulose, (1985).

Steinhäusler, F., On the Validity of Risk Assessment for Radon Daughter Induced Lung Cancer. This volume.

RECEIVED October 1, 1986

Chapter 37

Health Effects Estimation:
Methods and Results for Properties Contaminated
by Uranium Mill Tailings

Dale H. Denham, Fredrick T. Cross, and Joseph K. Soldat

Pacific Northwest Laboratory, Richland, WA 99352

This paper describes methods for estimating potential
health effects from exposure to uranium mill tailings
and presents a summary of risk projections for 57 mill
tailings contaminated properties (residences, schools,
churches, and businesses) in the United States. The
methods provide realistic estimates of cancer risk to
exposed individuals based on property-specific occupancy
and contamination patterns. External exposure to gamma
radiation, inhalation of radon daughters, and consumption
of food products grown in radium-contaminated soil are
considered. Most of the projected risk was from indoor
exposure to radon daughters; however, for some properties
the risk from consumption of locally grown food products
is similar to that from radon daughters. In all cases,
the projected number of lifetime cancer deaths for
specific properties is less than one, but for some
properties the increase in risk over that normally
expected is greater than 100%.

This paper provides information obtained during the performance of
several risk assessment studies conducted at Pacific Northwest
Laboratory for the U.S. Department of Energy's (DOE) remedial action
programs. During these studies we developed a method for estimating
projected health effects at properties in the vicinity of former
nuclear installations. Our emphasis was on properties that had been
or could be affected by uranium mill tailings piles as a result of
deliberately dispersed wind-transported tailings or direct emissions
from the piles themselves. The emphasis on uranium mill tailings
stems from the fact that tailings were used on vicinity properties as
landfill for many years in the United States before the health risk
associated with such use was recognized. Occupants of buildings
constructed with tailings-laden materials or erected on or adjacent
to contaminated land could receive radiation exposures that are
sufficiently high to warrant remedial action. Hence, the estimation
of potential health effects for occupants of properties contaminated

0097-6156/87/0331-0513$06.00/0
© 1987 American Chemical Society

with uranium mill tailings was initiated to support decisions regarding remedial actions on these properties.

A major objective in developing these risk estimation procedures was to provide a method capable of evaluating hundreds of properties in several communities within the DOE Uranium Mill Tailings Remedial Action Program in a timely manner. Therefore, we chose a calculation scheme that could be performed using commercially available database software (dBASE II, a trademark of Ashton-Tate, Culver City, CA), but that at the same time would be flexible enough that assessments for other contaminants could be readily incorporated.

We have previously documented the methodology (Marks et al., 1985a) and presented a summary of the technique (Marks et al., 1985b) at the Maastricht, The Netherlands, Seminar on Exposure to Enhanced Natural Radiation and Its Regulatory Implications. This paper represents a synthesis of the work we have conducted to date on risk assessment at uranium mill tailings vicinity properties.

This approach has also been used for other situations, and could easily be adapted to other contaminated sites or properties by modifying parameters to account for individual site and property characteristics.

Description of Procedures

The objectives in risk estimation are to project the potential health risk to an individual resulting from exposure to a given source and to compare that risk to a comparable, normally occurring risk. The assessment is often expressed as the increased probability of the individual's developing cancer of a specific type. Finally, the risk associated with a given source must be evaluated in terms of the magnitude of the effects that may be caused by the exposure. In the case of tailings piles, this may involve exposure of populations residing or working at a property some distance from the pile. A vicinity property may be commercial or residential, resulting in different exposure patterns to the occupants. Furthermore, the distribution and total number of occupants within a property will influence the total number of projected health effects.

The studies involved determining appropriate environmental pathways that would result in exposure to humans, determining appropriate occupancy factors (number and distribution) within structures, characterizing the source term for each property, selecting an appropriate set of health risk coefficients, calculating health effects, and providing summary reports of potential health effects for each vicinity property.

Because estimates of health risk are based on the levels of radionuclides in or near the vicinity properties, the quality of the potential health risk estimates depends upon the availability of appropriate measurement data. Hence, the first steps involved the determination of the appropriate environmental pathways of exposure and developing the source term for the exposure of persons potentially at risk. For our work, the radiological source-term data was based on measurements made principally by the Oak Ridge National Laboratory and the Mound Laboratory.

The hazards to human health for properties contaminated by tailings surrounding or beneath a structure is principally from the inhalation of Rn-222 and its daughters. In addition, tailings in the soil may contribute to external exposure and, in the case of residence, internal exposure from Ra-226 deposited in the body via the food chain.

The principal problem in estimating potential health effects is one of relating the measured levels of radiation and/or radionuclides in/on a property to the projected cancer deaths if the location of occupants differs from where the measurements were taken. Aberrant high values can also easily distort and invalidate a comparison between properties. For this reason, we have based our estimates on mean or most likely, rather than maximal, values.

The use of gamma-ray measurements to estimate external exposure was relativley straightforward, since the measurements were usually taken at one meter above the surface. The external exposure source-term values were usually the arithmetic means of the gamma-ray measurements taken in the bulidings or in areas of differing exposure rate within a building, and the average of the outdoor measurements. Outdoor gamma levels and soil-sample data provided a perspective on the distribution of contaminants with respect to a structure or its constituent parts. Maps showing gamma-ray exposure rate contours were prepared for individual properties to provide an indication of the respective contamination patterns.

Estimation of exposures from radon gas and/or daughters was a much more difficult task. In constrast to the abundance of reliable gamma measurements, the radon gas and daughter measurements were sparse and, often, unreliable. The reasons are the cost, complexity and inaccuracy of measurement, and the variability of actual levels caused by diurnal and seasonal changes, or other temporary environmental conditions (e.g., opening doors or windows, operating ventilation equipment). Lung cancer risk estimates are based on the working level (WL), defined as the concentration of short-lived radon daughters (i.e., RaA, RaB, RaC, and RaC') totaling 1.3×10^5 MeV of potential alpha energy per liter of air. A working level month (WLM) is an exposure equivalent to 1 WL for 170 hours. The estimated cancer risk for a given exposure in terms of WLM per year varies with the disequilibrium of the daughters. Therefore, the equilibrium factor had to be determined, and risk coefficients appropriate to the different values had to be used to calculate lung cancer risk estimates (Table I).

Radon concentrations obtained from grab air samples are frequently not representative of the individual's exposure throughout either the work or "at-home" day because of the variation in measurement values during the period and the influence of temporary conditions. The use of integrated or continuous measurements is obviously desired to obtain air-concentration values that represent the long-term average conditions to which occupants may be exposed. However, grab-sample values were used when integrated or continuous measurements were not available. When both radon gas and daughter grab-sample values were available, the daughter values were used. Long-term radon gas data are generally preferred over grab-sample data for radon daughters and, certainly, over radon grab-sample measurements. However, the use of long-term radon gas concentration values does

Table I. Risk Coefficient Adjustment Factors as a Function
of Radon Daughter Disequilibrium

Radon and Daughter Ratios	Equilibrium Factor	Adjustment Factor for Radon Daughter Risk Coefficient
1/0.9/0.7/0.7	0.71	1.00
1/0.9/0.6/0.4	0.55	1.05
1/0.6/0.3/0.2	0.29	1.30
1/0.26/0.098/0.084	0.11	2.21

introduce an element of uncertainty, because the equilibrium factor must be estimated. In using radon gas or daughter values, we did not attempt to correct for diurnal or seasonal variations. However, our method provides for such correction when appropriate.

Equilibrium factors were estimated from simultaneous measurements of radon gas and daughters when available. A default equilibrium factor of 0.3 was used when simultaneous gas and daughter values were not available. The default equilibrium factor is based on reported data for Salt Lake City (EPA, 1974) and on data obtained in Edgemont, South Dakota (Jackson et al., 1985), for climatic conditions similar to those in Salt Lake City.

In calculating the annual health effects, an exposure period of 2000 hours (100% indoors) was assigned for workers, and 7000 hours (80% indoors) was assumed for residential properties. We have used a lifetime risk coefficient of one cancer death per 10,000 R of gamma exposure (NAS, 1980), where one R is approximately equivalent to one rem of gamma radiation. The exposure to gamma-rays is assumed to occur over a 50-year period. The increased risk is the percentage above the base lifetime value of 0.164 (i.e., the probability of developing cancer in a lifetime from natural causes). It is based on the age and sex composition of the population of the United States as used in Report 3 of the Committee on the Biological Effects of Ionizing Radiations (NAS, 1980). Projected cancer deaths due to gamma radiation are based on the number of annual-equivalent persons exposed to a specified exposure rate. Use of the linear hypothesis is implicit in this estimation process in that the duration of each individual's occupancy is not used in the risk calculations. The linear hypothesis may be challenged as unduly conservative for gamma radiation, but it is generally accepted as appropriate for radon daughters.

Calculation of lung cancer risk for radon daughter exposure is based on factors developed by the National Council on Radiation Protection and Measurements (NCRP, 1984). The risk coefficients are expressed in terms of lifetime risk from lifetime exposure for a population of mixed ages, comparable to the standardized U.S. population, and range between one and two per 10,000 WLM of exposure. The percent increase in risk is related to a normal lifetime lung cancer risk of 0.041.

At residential sites, ingestion of home-grown garden crops was included as an additional potential exposure pathway. The individuals occupying residential properties were assumed to ingest 10% of the fruit and vegetable segment of their diet from the backyard garden. In the special case of Salt Lake City, the proportion of the fruit and vegetable segment of the diet produced in a home garden is increased to 50% because of local living patterns. The diet itself was assumed to be average rather than maximal, in accordance with the "best-estimate" approach we adopted. The transfer of radionuclides to the garden crops was assumed to arise from the upper (0-15 cm) soil layer. Unless special considerations, such as a fenced garden, indicate selection of a limited area for cultivation, the entire back and side yards were assumed to be equally likely locations for planting. Concentrations of radionuclides in the soil at sample sites, the estimated extent of contaminated areas, and an estimate of the total "backyard" area available for cultivation were used to calculate the weighted-average soil concentration. This concentration was used in the computer code PABLM (Napier et al., 1980) for estimating dose to man from food products grown in contaminated soil.

In most cases, we used the natural background radiation levels reported in the original property surveys. However, in several cases we modified the background value to reflect the influence of building materials when a mean background value could be calculated for parts of a structure that had no apparent relationship to contaminated materials and when such an adjustment was warranted by the nature of the building materials.

The following expressions were employed for health risk as a result of the radiation exposure incurred during occupancy of a property: the cancer risk per individual for gamma and/or for radon daughter exposure; the individual's percent increase in cancer risks relative to the respective, normal cancer risks; and the number of projected excess cancer deaths due to the radiation exposure (external and internal) for the number of occupants at each property.

Results and Discussion

For this paper, we have described our method of calculating health effects at vicinity properties contaminated with uranium mill tailings. In this section, we provide the results of applying the method to estimate health effects for 57 vicinity properties (13 in Canonsburg, Pennsylvania; 3 in Durango, Colorado; and 41 in Salt Lake City). Of these, 38 were residential properties and 19 were commercial properties. The Canonsburg property-specific survey plan only included indoor radon gas and radon daughter measurements, hence only the risk from radon daughter inhalation was estimated for these properties. The survey data available from the other sites included measurements of indoor and outdoor gamma radiation, Ra-226 and U-238 in soil, and radon gas or radon daughters in structures. A summary of the observed range of radiation levels and radionuclide concentrations is included in Table II.

The survey and occupancy data for each property were entered into the data base and a series of results summaries (computer print-outs) were generated for each property. These include an All Surveys Summary Report (the types, numbers, means, and ranges of survey data available per property, including surveys made by groups other than Mound Laboratory or Oak Ridge National Laboratory); a Measurement Summary Report (the gamma, radon, and soil mean values, ranges, and background data used for the health effects calculations); a Health Effects Estimation Report (summary of health effects estimates per property for gamma radiation, radon daughter exposure, and food pathways); a Ranking Report (ranking of all properties, either by type or combined, by total projected cancer deaths). Because the amount of data and the number of significant figures available on the computer printouts is large, only a summary of the information provided on the Health Effects Estimation Report for two Salt Lake City properties, one residential and one commercial, is provided in Table III. These examples are typical of the reports generated for all 57 properties. A summary of the resultant health effects for all the properties considered here is provided in Table IV.

For comparison, we have calculated the health effects for typical residential properties (4 occupants each) based on: 1) the naturally occurring background radiation levels and radionuclide concentrations in four cities across the U.S. (see Table V); and 2) the EPA (CFR, 1981) guideline values (20 μR/h, 0.02 WL, 5 pCi Ra-226/g of soil) for cleanup at inactive uranium processing sites (see Table VI).

Table II. Range of Observed Radiation Levels and Radionuclide Concentrations at Uranium Mill Tailings Vicinity Properties

| | Gamma (μR/h) | | Indoor | | Surface Soil | |
	Indoor	Outdoor	^{222}Rn(pCi/L)	RDC (WL)*	^{226}Ra(pCi/g)	^{238}U(pCi/g)
Canonsburg, PA						
Residential	—	—	0.2–16	0.002–0.09	—	—
Durango, CO						
Residential	12–50	10–50	0.2–2.6	0.003–0.04	1–120	0.5–10
Commercial	20–200	10–30	1–21	0.007–0.2	1–97	0.5–7
Salt Lake City, UT						
Residential	6–70	6–180	0.1–70	0.0005–0.03	1.1–270	0.8–20
Commercial	6–320	10–410	0.1–250	0.001–0.64	0.8–540	0.7–45

* Radon daughter concentration in working levels.

Table III.   Summary of Health Effects for Two Salt Lake City Vicinity Properties

Location	Source	Net Exposure Level	Lifetime Cancer Risk	% Increase in Cancer Risk	Projected Cancer Deaths
A. Residential (4 occupants)					
Outdoor	Gamma	28 µR/h	0.00007	0.04	0.00028
Indoor	Gamma	16 µR/h	0.00052	0.31	0.0021
	Radon	0.018 WL	0.0057	13.9	0.023
Foodstuffs	Ra-226	63 pCi/g	0.0082	5.0	0.033
		Total Project Cancer Deaths for Residential Property:			0.058
B. Commercial (23 occupants)					
Indoor-Bldg 1	Gamma	130 µR/h	0.0013	0.80	0.011
	Radon	0.46 WL	0.053	30.6	0.61
Indoor-Bldg 2	Gamma	110 µR/h	0.0011	0.65	0.0025
	Radon	0.15 WL	0.012	28.0	0.024
Foodstuffs				Not Applicable	
		Total Projected Cancer Deaths for Commercial Property:			0.65

Table IV. Range of Estimated Lifetime Health Effects for 57 Vicinity Properties in Salt Lake City, UT; Canonsburg, PA; and Durgano, CO

| Property Type | Lifetime Health Effects from | | | |
	Gamma	Radon Daughters	^{226}Ra in Foodstuffs	Combined Pathways
Residential	5×10^{-5} to 6×10^{-3}	5×10^{-4} to 1×10^{-1}	3×10^{-4} to 1×10^{-1}	3×10^{-4} to 1×10^{-1}
Commercial	6×10^{-5} to 3×10^{-2}	1×10^{-3} to 6×10^{-1}	not considered	6×10^{-5} to 6×10^{-1}

Table V. Lifetime Health Effects Estimates at Background Radiation Levels for a Typical Residential Property in Several U.S. Cities

Location	Lifetime Health Effects From			
	Gamma	Radon Daughters	^{226}Ra in Foodstuffs	Combined Pathways
Edgemont, SD[a]	1.8×10^{-3}	4.7×10^{-3}	2.6×10^{-3}	9.1×10^{-3}
Middlesex, NJ[b]	1.0×10^{-3}	8.9×10^{-3}	1.8×10^{-3}	1.2×10^{-2}
Durgano, CO	2.5×10^{-3}	1.0×10^{-2}	1.4×10^{-3}	1.4×10^{-2}
Salt Lake City, UT	1.4×10^{-3}	1.3×10^{-2}	1.2×10^{-3}	1.5×10^{-2}

a See Jackson et al., 1985 and Perkins et al., 1981

b See DOE, 1980

Table VI. Lifetime Health Effects Estimates at EPA* Recommended Levels for a Typical Residential Property in the U.S.

Lifetime Health Effects From			
Gamma	Radon Daughters	^{226}Ra in Foodstuffs	Combined Pathways
3.5×10^{-3}	2.0×10^{-2}	5.2×10^{-3}	2.9×10^{-2}

* See CFR, 1981

Summary and Conclusions

We have presented an overview of our approach to the estimation of individual risk and projected health effects for uranium mill tailings vicinity properties in Canonsburg, Pennsylvania; Durango, Colorado; and Salt Lake City, Utah. For the 57 properties evaluated, the greatest number of health effects were attributed to radon daughter exposures in commercial structures, although these represented less than one fatal cancer per lifetime per property. The range of health effects from radon daughters was about an order of magnitude higher than that from gamma radiation for both commercial and residential properties, while the range of health effects from the food pathway was similar to that for radon daughters for residential properties. To refine these estimates would require data on property-specific fruit and vegetable radionuclide concentrations, annual consumption of these foods, and long-term radon gas and radon daughter concentrations, as well as equilibrium factors.

For comparison, the health effects calculated from exposure to natural background radiation levels for typical residential properties were all about 0.01 per property, with radon daughters accounting for more than 50% of the total health effects estimated. On the other hand, the estimated risk for occupants of residential structures cleaned up to the EPA guideline values was about a factor of 3 higher than for residents exposed to natural background levels.

The model presented is considered to be useful for predicting changes in risk associated with changes in exposure (e.g., from remedial actions), if the following remain relatively constant: particle sizes of the carrier aerosols for the radon daughters, the degree of disequilibrium, the degree of unattachment of radon daughters, and the availability (depth) of radionuclides in the soil. We conclude that the change in exposure is accurately modeled, although the absolute values of risk are not well known. This uranium mill tailings methodology can readily be adapted to other remedial action sites by changes in appropriate parameters.

Acknowledgments

This work was supported by the U.S. Department of Energy, Office of Environmental Guidance, under Contract DE-AC06-75RLO 1830. We wish to acknowledge the valuable participation of other Pacific Northwest Laboratory staff members, principally S. Marks, W.E. Kennedy Jr., K.A. Hawley, R.E. Jaquish, D.K. Washburn, and R.D. Stenner, in the development of the work that forms the basis for this paper. B.A. Berven of the Oak Ridge National Laboratory, P.H. Jenkins of Monsanto Research Corporation (Mound Laboratory), and C.D. Young of the Aerospace Corporation generously provided necessary survey reports. We are also grateful for the sponsorship of C.G. Welty Jr. and W.E. Mott of the Office of Environmental Guidance, U.S. Department of Energy, who initiated and encouraged the development of the studies upon which this paper is based.

Literature Cited

CFR, Proposed Cleanup Standards for Inactive Uranium Processing Sites, Report No. 40 CFR 192, Code of Federal Regulations (1981).

Department of Energy, Environmental Assessment of the Properties
Adjacent to and Nearby the Former Middlesex Sampling Plant,
Middlesex, New Jersey, Report No. DOE/EA-0128, U.S. Department of
Energy (1980).

Environmental Protection Agency, Environmental Surveys of the Uranium
Mill Tailings Pile and Surrounding Areas, Salt Lake City, Report No.
EPA-520/6-74-006, U.S. Environmental Protection Agency, Las Vegas, NV
(1974).

Jackson, P.O., V.W. Thomas, and J.A. Young, Radiological Assessment
of the Town of Edgemont, NUREG/CR-4057, Pacific Northwest Laboratory
for U.S. Nuclear Regulatory Commission, Washington, D.C. (1985).

Marks, S., F.T. Cross, D.H. Denham, W.E. Kennedy Jr., and R.D.
Stenner, Risk Assessment in the DOE Assurance Program for Remedial
Action, PNL-5541, Pacific Northwest Laboratory, Richland, WA (1985a).

Marks, S., F.T. Cross, D.H. Denham, and W.E. Kennedy Jr., Estimation
of Health Effects due to Elevated Radiation Exposure Levels in
Structures, Sci. Total Environ. 45:543-550 (1985b).

Napier, B.A., W.E. Kennedy Jr., and J.K. Soldat, PABLM - A Computer
Program to Calculate Accumulated Radiation Doses from Radionuclides
in the Environment, PNL-3209, Pacific Northest Laboratory, Richland,
WA (1980).

NAS, The Effects on Populations of Exposure to Low-Levels of Ionizing
Radiation, (BIER III), National Academy of Sciences (1980).

NCRP, Evaluation of Occupational and Environmental Exposures to
Radon and Radon Daughters in the United States, NCRP Report No. 78,
National Council on Radiation Protection and Measurements, Bethesda,
MD (1984).

Perkins, R.W., J.A. Young, P.O. Jackson, V.W. Thomas, and L.C.
Schwendiman, Workshop on Radiological Surveys in Support of the
Edgemont Clean-up Action Program, NUREG/CP-0021, Pacific Northwest
Laboratory for U.S. Nuclear Regulatory Commission, Washington, D.C.
(1981).

RECEIVED November 12, 1986

MITIGATION METHODS

Chapter 38

Modified Design in New Construction Prevents Infiltration of Soil Gas That Carries Radon

Sven-Olov Ericson[1] and Hannes Schmied[2]

[1]Radon Consultants, Edsviksvägen 33, S-182 33 Danderyd, Sweden
[2]AIB Consulting Engineers, P.O. Box 1315, S-171 25 Solna, Sweden

Dwellings located on permeable soil with strong exhalation of radon often get a contribution to indoor radon
from infiltrating soil gas carrying radon from the ground
into the building. 100 dwellings have been built on
radon dangerous land with different modifications in
design and construction in order to prevent infiltration
of radon. Tight construction, ventilated crawl space,
ventilation/depressurization of the capillary breaking
layer (crushed stone), and mechanical ventilation with
heat recovery by air to air heat exchangers or heat
pumps have been tested. Added building costs and
measured concentration of radon after construction and
3-5 years later are reported. It is concluded that it
is possible to build radon protective and radon safe
dwellings on any land. The added costs have ranged
from zero to 4% of total building costs.

Radon from the soil enters into buildings by convective flow of soil
gas. Transport by diffusion is normally insignificant. In houses
with very high radon concentration, diffusion need not be considered
because it can only provide an insignificant fraction of the source
strength. There are three conditions necessary for infiltration of
soil gas containing radon into the building from the soil:

- there must be an open connection for gas flow from the soil into
  the building,

- there must be a driving force to produce the flow through the
  opening, i.e. a pressure gradient with negative pressure in the
  building relative to the soil, and

- there must be a large enough volume of permeable soil subjacent
  to the building to supply the necessary radon.

They are important aspects to be considered in new construction on
land where protective or radon safe design is recommended.

Recommended Strategy for Cost-Effective Radon Protective and Radon
Safe Construction

Based on our experience with remedial actions and modifications of
design in new construction we recommend the following strategy for
making new buildings (slab on grade, ventilated crawl space or
furnished basement) radon protective or radon safe in a cost-
effective way:

- Avoid open connections to the soil.

- Reinforced concrete of standard quality gives enough resistance
  to prevent convective flow of soil gas provided there are no
  major cracks or other obvious openings. Our results verify
  earlier Canadian conclusions (Central Mortgage & Housing Corp.,
  1979) that a monolithic slab with carefully sealed utility
  entrances will prevent infiltration of radon. Connections with
  basement walls and any joints must be carefully sealed and
  settling should be prevented. Double reinforcement increases
  the resistance to cracking, but is normally not necessary and
  cannot be used as an alternative to prevention of settling.

- Ventilation of the building should provide a reasonable ventilation rate and the negative pressure indoors should be
  minimized. A balanced ventilation system with supply and
  exhaust ventilation is to be preferred to a mechanical exhaust
  ventilation system.

- Minimize the contact between living area and subjacent soil.
  Buildings with slab on grade or ventilated crawl space require
  less modification of the standard design than houses with basements. If basements cannot be avoided on radon dangerous soil,
  it is recommended to keep the basement separated from the living
  space by normally closed doors and if possible, install exhaust
  ventilation from the basement.

- Avoid building blocks with interconnected internal cavities in
  basement walls. Basement walls made from in situ (poured) concrete are recommended. Concrete elements require careful and
  permanent sealing of all joints. Designs similar to swimming
  pools, manure tanks in agriculture and bomb shelters will
  provide a radon safe construction.

- If possible, make the construction forgiving to later appearing
  cracks in the concrete. If a slab is covered with a permanently
  elastic material, like rubberized asphalt or bitumen, future
  cracks in the concrete will not automatically result in a route
  of entry for soil radon.

- Prepare in the design stage for an easy remedy in case the
  passive precautions fail. This is an important component in a
  cost-effective strategy. Ambitious precautions giving low radon
  concentration in 100% of the new buildings without any retrofit
  modifications will give a higher total cost. An accessible and

ventilated crawl space between the living space and the soil
will always provide possibilities for various remedies.
Depressurization and/or ventilation of the capillary breaking
layer has been proven to be an effective remedial action
(Ericson and Schmied, 1984). Preinstallation of a duct from the
capillary breaking layer to the atmosphere or by placing drain
pipes for future distribution of negative pressure in the
capillary breaking layer by suction at one point provide
possibilities for effective remedies. Air from crawl space or
subjacent soil can be very rich in radon and should not be mixed
with exhaust air and possibly mixed with supply air in a heat
exchanger.

With modifications according to the above principles it is
possible to develop any site and still keep acceptable concentrations of radon in the dwellings. Probably the same principles are
applicable also in construction on old waste dumps preventing
infiltration of other gases, for example methane or mercury vapor.
Figures 1 and 2 present two examples of radon safe designs which
in most cases can be regarded as overly safe. These two particular
buildings were erected in the village of Varnhem where the soil is
permeable gravel containing alum shale with an elevated activity of
uranium. The radon concentration in the soil gas is between 100,000
and 300,000 Bq/m^3. The first one is provided with a monlithic concrete slab in the crawl space. The slab and foundation are covered
with rubberized asphalt. The crawl space is ventilated by natural
draught separately from the ventilation of the dwelling. The concentration of radon in the capillary breaking layer has been measured
several times with results ranging from 122,000 to 340,000 Bq/m^3.
Four measurements within the crawl space have ranged from 70 to 240
Bq/m^3. The other house has concrete slabs between the footings in
the crawl space and the crawl space is ventilated as an integral part
of the supply exhaust ventilation system of the living space. Three
measurements have given 55 to 75 Bq/m^3 in the living space. The
modifications in design added approximately 4% to the building costs.
As a contradiction to the above examples is a hypothetical house
designed to maximize the risk for infiltration of soil radon. The
house has a furnished basement in open contact with the main living
space. The basement slab is poured between the footings without
sealing the joints between basement walls and slab. The basement
walls are made from building blocks having interconnected interval
cavities. The ventilation is of the mechanical exhaust type without
air inlets, thus creating a strong negative pressure indoors at a
fairly low air exchange rate. These are common characteristics in
the existing building stock. If surrounded by permeable soil such
buildings often face an infiltration of soil gas equal to 0.1 to 5%
of the total air exchange rate. This means that the infiltrating
soil gas is diluted 20 to 1000 times with outdoor air infiltrating
through leaks above ground level. If the concentration of radon in
the soil gas is 100,000 Bq/m^3 the contribution to indoor radon often
is 100 to 5000 Bq/m^3.
There is no conflict between energy conservation and radon
protective or safe design. Radon safe construction can be made as
energy efficient as can conventional design. Radon safe design only

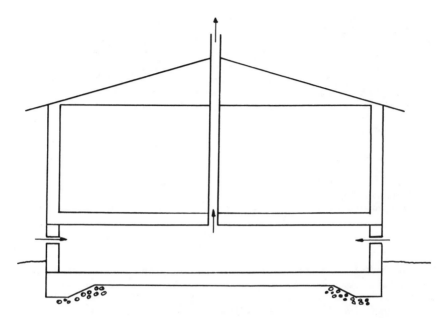

Figure 1. Radon safe dwelling; monolithic slab in a crawl space with separate ventilation, mechanically ventilated living space.

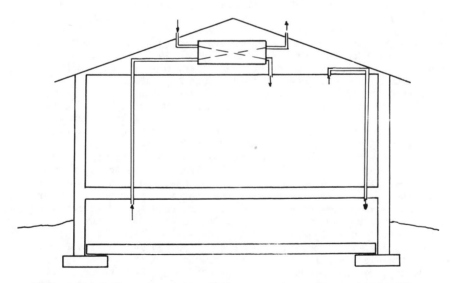

Figure 2. Radon protective design; crawl space provided with a slab, mechanical supply exhaust ventilation of living space and crawl space.

introduces some restrictions on how to conserve energy. Energy conservation by increased tightness of the building and very low air exchange rates must be discouraged due to its effect on indoor air quality and humidity even without regard to radon. With mechanical ventilation systems installed, heat recovery will pay its own costs in temperate climates. In order to avoid strong negative pressure indoors, heat exchangers with balanced ventilation are to be preferred for exhaust ventilation with a heat pump for recovering the heat.

Every modification in design, as well as remedial actions, must be made after due consideration of building physics. Otherwise negative side effects can occur. Examples are locally high relative humidity in air, condensation of moisture and development of mold. With due consideration to the constraints laid by building physics, however, modifications to radon safe design can give positive trade-offs as reduced possibility for moisture transport and development of mold.

Monolithic Slabs

In the same village as the previous two examples, a group of 21 two-story, wood frame dwellings were built in 1979-80. The houses were founded with monolithic slabs (Figure 3). A layer of fine grained moraine was placed under the capillary breaking layer in order to restrict the flow of soil gas to the building from the subjacent gravel. The ventilation systems are of the balanced supply exhaust type with variable speed and heat exchanger. As an attempt to increase the tightness of the foundation, aluminum foil laminated with polyethylene and polyester was applied above the slab in 11 houses. In three houses, the foil was joined by welding the polyethylene and in the other seven by overlapping joints. Measurements have been performed four times from 1980 to 1985. It has not, however, been possible to detect any effect of the aluminum foil. Thus, the slab itself has provided enough tightness. The results are presented in Table I. In most of the houses, the radon concentration is less than 200 Bq/m^3. A few of the houses have higher concentrations indicating some infiltration of soil gas. A contribution of 100 to 500 Bq/m^3 indicate that the infiltrating soil gas, with 200,000 to 300,000 Bq/m^3, is diluted 500-2000 times. If the mechanical ventilation systems are operated at a capacity of 200 m^3/h, the infiltration in these houses would be 0.1 to 0.4 m^3/h of soil gas. The elevated levels have been measured in the same houses every time and in the majority of the houses nothing but low concentrations have been measured. The modifications in design added not more than 1% to the building costs. As an alternative the slabs could have been covered with rubberized asphalt, especially near entering utilities, and the structures prepared for an easy retrofit with mechahnical ventilation or depressurization of the capillary breaking layer to be used in the few buildings where radon infiltrates. With this strategy it is assumed that all houses could have been under 200 Bq/m^3 at an added building cost of approximately 0.5%.

In the town of Ljungsbro 27, 1 1/2-story, wood frame dwellings without basements were built on monolithic slabs. Ventilation is provided by a duct system connected to an exhaust fan. In the sub-

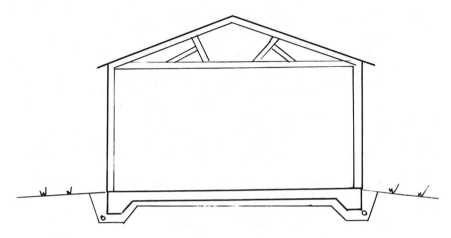

Figure 3. Mechanically ventilated wood frame dwelling with monolithic slab.

Table I. Distribution of Measured Radon Concentration, Bq/m^3, in
21 Detached Houses Designed According to Figure 3

Radon Concentration	Dec '80 – Jan '81*	Feb '81**	Summer '82*	Spring '85***
0-50	1	2		1
50-100	5	5	8	3
100-150	5	2	3	4
150-200	3	3	1	4
200-250	1	1	2	2
250-300	1	1		1
300-350	2	2		1
350-400		1		1
400-450	1			
450-500				1
500-550				
550-600				
600-650		1		
–				
900-950		1		
1200-1250				1

  * 30 days average
 ** grab samples
*** 3 months average

jacent soil, radon concentrations ranging from 10,000 to 211,000
Bq/m^3 were measured. Together with the upper layer of the ground
being permeable gravel and sand, radon protective design was
indicated. In order to get a passive ventilation of the subjacent
soil, perforated ducts were installed in the capillary breaking layer
with contact to the atmosphere through the edge beam on both sides of
the houses (Figure 4). Results from measurements made in 1981, 1982
and 1985 are presented in Table II. In the majority of the houses,
there is no sign of infiltration of soil gas. During the 1982 measurements, the ducts in the capillary breaking layer were sealed in 10
and open in 17 houses. The houses with sealed ducts did not have
higher radon concentrations, indicating that the monolithic slab
prevents infiltration of soil gas, though the mechanical exhaust
ventilation provides a constant negative pressure in the buildings.

Table II. Distribution of Measured Radon Concentration, Bq/m^3,
in 27 Detached Houses Designed According to Figure 4

Radon, Bq/m^3	1981	1982	1985
0-50	5	24	9
50-100	2	1	8
100-150		1	2
150-200			3
200-250			
250-300			1
300-350	1		

Ventilated Crawl Space

Buildings with ventilated crawl space are inherently radon protective. In one widely used design, the crawl space is mechanically
ventilated by the exhaust air from the living space (Figure 5). This
design presents the advantage that the heat in the exhaust air is
used to provide a warm floor in the living space. Recently more heat
is often extracted from the exhaust air after it has passed through
the crawl space. If the air is contaminated with radon in the crawl
space, this is preferentially done by a heat pump. If heat is
transferred to the supply air in an air to air heat exchanger
contamination of the supply air should be avoided. Thus, the
pressure differential between the supply and exhaust air and the
tightness of the heat exchanger has to be carefully considered. We
have seen an example where radon in the exhaust air from the crawl
space contaminated the supply air with 600 to 800 Bq/m^3 of radon. If
the pressure in the crawl space is lower than in the living space,
the radon concentration in the crawl space can be very high without
any substantial transport into the living space. If the pressure in
the crawl space is higher than outdoors, infiltration of soil radon

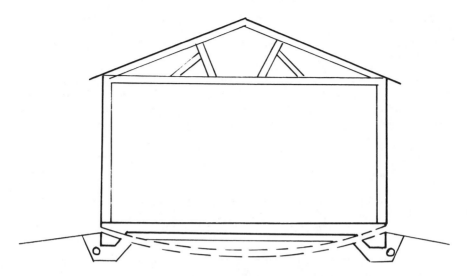

Figure 4. Mechanically ventilated wood frame dwelling on a monolithic slab, natural draught ventilation of the capillary breaking
layer by ducts through the edge beam.

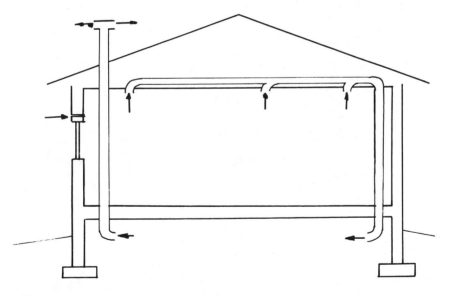

Figure 5. Mechanical exhaust ventilation through the crawl space.

into the crawl space is counteracted efficiently and the radon
concentration in the air from the crawl space can then be
surprisingly low, even on land where radon often infiltrates into
conventional buildings.

In 24 dwellings with exhaust air ventilated crawl space, we have
measured the radon concentration in 1982 and 1985. The results are
presented in Table III. With few exceptions the radon concentrations
have been low. The anomalously high value from 1982 has been
explained by contamination of the supply air in a heat exchanger.
The other high value was measured in a house without a heat exchanger
and in which we measured only 25 Bq/m^3 in 1982. The concentration in
the crawl space was high, 4600 Bq/m^3.

In Table IV, measurements in six dwellings with asphalt painted

Table III. Distribution of Measured Radon Concentration in
24 Detached Houses Designed with Mechanical Exhaust
Ventilation through the Crawl Space

Radon, Bq/m^3	1982*	1985
0-50	13	7
50-100	5	7
100-150	1	7
150-200		1
200-250		1
250-300		
300-350	1	
-		
900-950		1

* Measurements performed in 20 houses.

Table IV. Concentration of Radon in Indoor Air in 6 Detached
Houses Provided with Asphalt Coated Slab in the Crawl Space

House	Measured Radon Concentration, Bq/m^3	
	1982	1985
1	10	50
2	10	100
3	10	100
4	10	140
5	10	50
6	15	20

slab in the crawl space are presented. These houses have natural draught ventilation through air inlets in the foundation walls only or through air inlets and a duct over the roof. The radon concentration in the soil gas under these houses range from 40,000 to 100,000 Bq/m^3. The increase in values from 1982 to 1985 can possibly be attributable to inaccuracy in the detectors used.

Literature Cited

Central Management and Housing Corporation, Canada, Technical Memorandum 33/79 (1979).

Ericson, S.-O. and H. Schmied, Modified Technology in New Constructions and Cost-Effective Remedial Action in Existing Structures to Prevent Infiltration of Soil Gas Carrying Radon. In Proceedings of the 3rd International Conference on Indoor Air Quality and Climate, Stockholm, 5:153-158 (1984).

RECEIVED October 27, 1986

Chapter 39

Remedial Measures To Reduce Radon Concentrations in a House with High Radon Levels

K. D. Cliff, A. D. Wrixon, J. C. H. Miles, and P. R. Lomas

National Radiological Protection Board, Chilton, Didcot, Oxfordshire OX11 0RQ, United Kingdom

Measures to reduce radon concentrations have been studied in an old house in which the radon decay-product concentration initially exceeded 0.3 Working Level (WL). Some of the measures were only partially successful. Installation of a concrete floor, designed to prevent ingress of radon in soil gas, reduced the radon decay-product concentration below 0.1 WL, but radon continued to enter the house through pores in an internal wall of primitive construction that descended to the foundations. Radon flow was driven by the small pressure difference between indoor air and soil gas. An under-floor suction system effected a satisfactory remedy and maintained the concentration of radon decay products below 0.03 WL.

Before standards for indoor exposure to radon can be formally established, work is necessary to determine whether remedies are feasible and what is likely to be involved. Meanwhile, the Royal Commission on Environmental Pollution (RCEP) in the UK has considered standards for indoor exposure to radon decay products (RCEP, 1984). For existing dwellings, the RCEP has recommended an action level of 25 mSv in a year and that priority should be given to devising effective remedial measures. An effective dose equivalent of 25 mSv per year is taken to correspond to an average radon concentration of about 900 Bq m^{-3} or an average radon decay-product concentration of about 120 mWL, with the assumption of an equilibrium factor of 0.5 and an occupancy factor of 0.83.

The range of radon decay-product concentrations found in houses in southwest England spans three orders of magnitude. A few dwellings have been found where levels exceed 0.6 WL. It is clear that there are likely to be an appreciable number of houses in this region with indoor concentrations higher than 120 mWL. In 1982, the Department of the Environment identified two local authorities in the county of Cornwall that were willing to lend council property to test

0097-6156/87/0331-0536$07.00/0

various remedial measures. At the end of 1983 a suitable house with a radon decay-product concentration typically exceeding 0.3 WL was selected for detailed study. This report presents the results of a range of remedial measures carried out through 1984 and 1985 in co-operation with the Building Research Establishment of the Department of the Environment.

Our objectives were to identify the main sources of radon in the dwelling, to test the effectiveness of remedial measures not requiring alterations to the fabric of the building, and to provide a permanent means of reducing the radon concentration to the value typical of UK dwellings, i.e. 25 Bq m^{-3}, which corresponds to 3 mWL.

Measurement techniques

The concentrations of the decay products of ^{222}Rn were determined using a Radon Decay Products Monitor, RDPM (Cliff, 1978a). Room air is sampled through a glass fibre filter by means of an external pump. Radon decay-product activity collected on the filter is counted by a diffused junction alpha-particle detector during sampling and twice afterwards. The concentrations of ^{218}Po, ^{214}Pb and ^{214}Bi(^{214}Po) in air can then be calculated (Cliff, 1978a). Two additional gross alpha counts enable the concentrations of the thoron decay-products (^{212}Pb and ^{212}Bi) to be determined from the same air sample.

Continuous measurements of the potential alpha energy concentration of the radon decay products were made with a Continuous Working Level Monitor (WLM-300) (EDA Instruments Inc., Toronto).

Concentrations of radon gas were measured using scintillation cells of 150 ml capacity fitted with two self closing vacuum connectors (EDA Instruments Inc., Toronto). The cells were filled by flushing with filtered room air. Flushing was carried out for several minutes to ensure complete filling. The cells were left for 3 hours to allow radon and its short-lived decay products to approach secular equilibrium before counting with a photomultiplier and scaler assembly. Measurements of the radon gas concentration in room air, in soil gas beneath the floor and in the voids in the internal walls of the dwelling were made.

Ventilation rates were determined with a nitrous oxide tracer. The gas was released into room air. After allowing time for uniform mixing, the decay in concentration was followed using an infra-red gas analyser, type Miran-101 (Foxboro/Wilks Inc. USA).

Characterisation of the dwelling

The dwelling was a detached building some 100 years old with exterior walls constructed of granite blocks and mortar. Internal load bearing walls were of brick and stone, cement rendered, and plastered. The material under the house contained a high proportion of mine spoil arising from local tin mining operations. Although, at present, there are just a few operating tin mines in Cornwall, the number exceeded 200 in the eighteenth century. Spoil tips from derelict mines have provided the construction industry in Cornwall with a plentiful and cheap source of material for hardcore or as an

aggregate in concrete. Mine spoil from these sources often has
elevated uranium concentrations.

One ground floor room of the dwelling had been retained by the
local Council as an office to collect local taxes. The remainder of
the dwelling had been occupied by a family. A floor plan of the
dwelling showing the room usage is given in Figure 1. The office,
sitting room and hall had wooden suspended floors, whereas the floor
was of solid concrete in the scullery (a room for rough kitchen
work), kitchen and behind the stairs. The office and sitting room
each had an open fireplace and associated chimney. Two air grilles
were set below floor level in the front exterior wall of the house;
one was intended to ventilate the space under the office floor, the
other the space under the sitting room floor. This arrangement, with
just two air grilles in one wall and with no ventilation access to
the other three sides of the underfloor spaces, did not provide
adequate underfloor ventilation. Soil gases could accumulate in the
underfloor spaces and reach high concentrations, except when the wind
velocity was very high.

Measurements of gamma dose rates over the ground floor indicated
values in the range 0.15 µGy h^{-1} to 0.24 µGy h^{-1}. These values are
higher than the average (0.12 µGy h^{-1}) found in stone houses in
Cornwall, but are within the normal range (Wrixon et al., 1984).
Localised higher dose rates were detected close to the floors of the
office, sitting room, hall and kitchen. When the wooden suspended
floors were later removed, areas of high dose rate in the office,
hall and sitting room were found to be associated with discrete lumps
of uraniferous material in the oversite fill. The kitchen had the
highest density of hot spots over the floor. Since the floor was
solid, no attempt was made to raise it to investigate the immediate
sources of high gamma dose rate. Gamma dose rates measured close to
the walls within the dwelling were unremarkable. It was concluded
that the building materials did not contribute significantly to the
indoor radon concentrations.

Initial studies of the indoor concentrations of radon and its
decay products were carried out with all exterior doors and windows
closed, but with all internal doors fully opened. The radon
production rate for the whole house was determined by measuring the
^{218}Po concentration in indoor air and in the air outside the dwelling
and by determining the ventilation rate. The radon production rate,
K, is given by (Cliff, 1978b):

$$K = j(1 + 0.0748j)(C_A - C_A') \text{ Bq m}^{-3} \text{ h}^{-1}$$

where j is the ventilation rate, h^{-1}
      C_A,   is the concentration of ^{218}Po in indoor air, Bq m^{-3}
      C_A' is the concentration of ^{218}Po in outside air, Bq m^{-3}.

The radon production rate for a dwelling, or for an individual
room, is not constant with time, as it is affected by meteorological
and other conditions. Average radon production rates based on longer
term integrated measurements of radon gas concentration would have
resulted in slightly different values from those reported here.
However, this parameter is less variable than "grab" sample
determinations of the concentration of radon or its decay products.

It best indicates the effectiveness of different strategies to reduce the concentration of indoor radon. The radon production rate for the whole dwelling was measured as 280 Bq m^{-3} h^{-1}, with all internal doors open.

When internal doors were closed, large differences were found in the concentrations of radon and its decay products in different rooms. All ground floor rooms had much higher concentrations than those upstairs. Radon production rates were determined for each of the ground floor rooms. In the scullery, kitchen, office and sitting room, the production rates were in the ratio 1 : 5 : 65 : 130. The radon production rate for the sitting room was 7800 Bq m^{-3} h^{-1}.

Evidently radon was entering the dwelling predominantly in the sitting room and office. The likely source of indoor radon was accumulation from soil gas in the poorly ventilated underfloor spaces. The ease with which gases could enter the dwelling was demonstrated by injecting nitrous oxide into the underfloor space. This was detected in both the sitting room and the office within minutes of injection under the floor. It appeared earlier and reached higher concentrations in the sitting room, in keeping with the difference in the radon production rates of the two rooms.

Figure 2 shows the variation in the radon decay-product concentration in the sitting room over a 10 day period at the beginning of the study, together with the wind speed. The ventilation rate will increase with increasing wind speed. If the radon production rate is reasonably constant, the concentration of radon decay products in room air should therefore decrease with increasing wind speed. This effect is seen between days 6 and 8 in Figure 2, but for other periods wide fluctuations in the radon decay-product concentration occur for little change in wind speed. The reason for this variable behaviour was later shown to be due to the Venturi effect of the wind on the two chimneys.

Remedial measures

Some of the remedial measures tested in this study were not regarded as likely to form part of a long-term control stategy. For example, the installation of a mechanical ventilation system, with a heat recovery unit, would not be used in a dwelling of this type, because of the very high installation cost. Nevertheless, the availability of the dwelling enabled devices to be tested under real housing conditions, rather than in the laboratory.

Four remedial measures that did not entail altering the building fabric were tested with the following results.

Forced underfloor ventilation. Two attempts were made to forcibly ventilate the spaces below the suspended wooden floors of the office, hall and sitting rooms. At first, a 30 W centrifugal fan was installed in ducting connected between a hole in a floor-board in the sitting room and the chimney. The ducting was sealed through a plate covering the fire place aperture that prevented air discharged to the chimney from entering the room. The hole in the floor-board from which the ducting was connected was only some 0.6 m from the front exterior wall in which the underfloor ventilation grilles were fixed. When operated, the fan produced a reduction in the radon decay

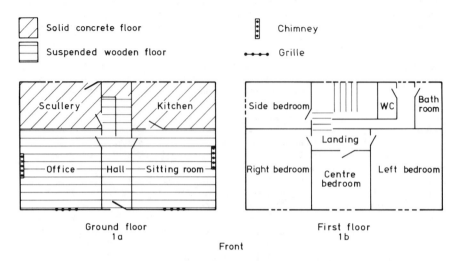

Figure 1.   Floor plans of the study house.

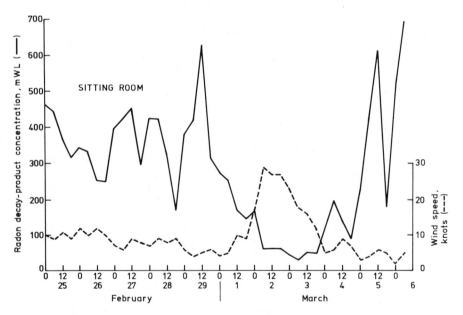

Figure 2. Variation in radon decay-product concentration over a
10 day period and wind speed.

products concentration in the sitting room by a factor of about four. The radon concentration in the exhaust air discharged to the chimney was 2600 Bq m^{-3}. In the second configuration the fan was connected in ducting fixed over the ventilation grille beneath the office floor, such that air was drawn from the underfloor spaces and discharged to the outside. In this case the radon concentration in the exhaust air from the fan was again 2600 Bq m^{-3}, but the radon decay products concentration in the sitting room was reduced by only a factor of about two.

With only two ventilation grilles in the same exterior wall, air movement under the floors was largely restricted to the area close to the front wall. Low resistance air paths existed between the two exterior grilles and only a small proportion of the total underfloor volume was effectively ventilated. This explains the relatively small reductions in the radon decay products concentration achieved in the sitting room. The lower reduction in concentration when the fan was connected directly to the air grille under the office floor was due, in part, to air leakage directly from the office to the underfloor space via gaps behind the skirting boards in that room.

At a later stage, when the suspended wooden floors were removed, it was found that the oversite fill, consisting of coarse sand, gravel and some stones, was very uneven and in places was higher than the bases of the joists supporting the wooden floors. Had the oversite fill been levelled, or excavated, to leave a clear space below the joists, the reduced pressure on the underfloor space resulting from the operation of the fan would have been effective over a larger area. A more effective forced ventilation of the underfloor space could have been effected by having a distributed system of ventilation extract points within the underfloor space, connected by a manifold to a single fan system.

Forced underfloor ventilation is not considered an ideal permanent remedy, and it is preferable to find a passive solution to the problem. A passive method is one requiring no energy source and minimal maintenance by the occupants of the dwelling.

Polymeric barriers. The UK building codes require that when a building is constructed on a concrete raft (slab-on-grade) the concrete base must incorporate a vapour barrier to prevent ingress of moisture. The vapour barrier is usually a 125 μm or 250 μm thick polythene sheet. Provided the polythene sheet is free from holes, it should reduce the diffusion of radon from the underlying soil through the concrete and into the building. An intact sheet of 250 μm polythene in a 10 cm thick concrete base would reduce the flux of radon into a building by about 40% (Cliff, Miles and Brown, 1984). Unfortunately, with normal construction practices in the UK, the polythene vapour barrier is rarely installed free of perforations.

To test the effectiveness of an intact polythene sheet as a radon barrier, two layers of 125 μm sheet were laid across the sitting room floor. The sheets were dressed up the walls to a height of about 60 cm and fixed with adhesive tape. Polythene sheets were also used completely to cover the fireplace.

Figure 3a shows the radon decay-product concentration in the sitting room, together with the wind speed, for the 25 day period

before the polythene sheets were laid. Large excursions in
concentration occurred, particularly between days 13 and 18. These
are not well correlated with wind speed. Figure 3b shows the data
for the period immediately following the laying of the polythene
sheets on day 26. It is seen that quite substantial changes occur in
wind speed, e.g. during day 31, but that the changes in the
concentration of radon decay-products are modest compared with those
of Figure 3a. The peak in radon decay-product concentration during
days 43 and 44 is not related to wind speed.

The mean radon decay-product concentration before laying the
polythene sheets was 240 mWL (Figure 3a), whereas that after laying
was 160 mWL (Figure 3b). This reduction, by a factor of 1.5, was
less than anticipated. The main reasons for the limited success of
polythene sheets were difficulties in achieving a perfect seal
between the polythene and the walls, and also the fact that the pores
in the internal walls of the dwelling provided a path for radon in
soil gas to enter the room above the polythene lining. It was also
difficult to prevent completely air from elsewhere in the house
(notably the office) mixing with that in the sitting room. Spot
measurements while the polythene sheet was in position in the sitting
room showed that the concentration of radon decay products in the
office were typically 10 times higher.

Figure 4 shows the rapid increase in the concentration of radon
decay-products in the sitting room on removal of the polythene
sheets. This was effected in three stages. Initially, the sheets
were removed along the front wall of the room and for a distance of
1.5 m across the floor. A sharp increase occurred immediately. An
hour later the remaining polythene, except that covering the fire
place, was removed. The concentration was still rising at 21:00.
When measurements were resumed at 10:30 the following day, the
concentration had started to fall and continued to fall until 15:15.
The polythene covering the fire place was then removed, producing a
marked increase in the concentration of ^{218}Po, but little change in
the aggregate radon decay-product concentration (mWL). Fresh radon
gas was thus drawn into the room when the fire place was uncovered
because of the Venturi effect on the chimney. This reduced the
pressure slightly in the sitting room and increased the pressure
driven flow of soil gas into the room.

An intact polythene membrane within the concrete base of a
building will prevent pressure driven flow of radon into the building
from the soil, even if the concrete is cracked. Diffusive flow of
radon into the building will also be reduced because of the
comparatively low diffusion coefficient of radon in polythene ($\sim 10^{-7}$
cm^2 s^{-1}). No significant improvement was achieved by substituting a
50 µm sheet of mylar for polythene (mylar diffusion coefficient
$\sim 10^{-11}$ cm^2 s^{-1}). In this case additional difficulties were
experienced in sealing the less flexible material to the walls.

Electrostatic precipitators. Two electrostatic precipitators, of the
type designed for domestic premises or small offices, were operated
in the sitting room. Each precipitator was run at its maximum
nominal air throughput of 270 m^3 h^{-1}. The concentration of radon in
room air was not measured during these tests, but dis-equilibrium

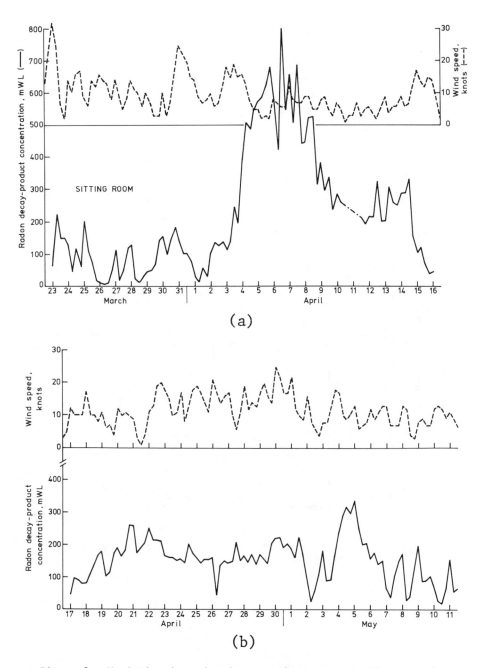

(a)

(b)

Figure 3. Variation in radon decay-product concentration and wind speed, (a) before and (b) after laying polythene.

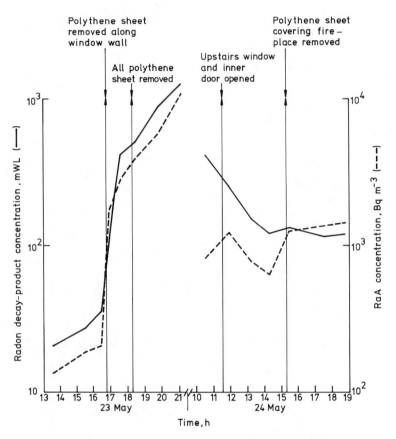

Figure 4. Radon decay-product concentrations on removing polythene sheet from sitting room floor.

between the radon decay-products was measured by the ratio, F',
where:

$$F' = \frac{37 \text{ (mWL value)}}{10 \text{ (}^{218}\text{Po concentration, Bq m}^{-3}\text{)}}$$

F' varies from 0.1, when only ^{218}Po is present, to unity at
equilibrium. Figure 5 shows the changes in the aggregate radon
decay-product concentration, the concentration of ^{218}Po, and the
variation in F' achieved by operating the electrostatic
precipitators. The radon decay-product concentration was reduced by
more than a factor of 10.

The condensation nucleus concentration was measured using a
continuous flow counter (model 3020, TSI Inc. MN, USA). This fell
from 11000 cm^{-3} before operating the electrostatic precipitators to a
minimum of 2000 cm^{-3}, when both devices had been operating for
several hours. The size distribution of the room aerosol was
determined using an 11-stage screen diffusion battery together with a
switching valve (models 3040 and 3042 respectively, TSI Inc. MN, USA)
in conjunction with the condensation nucleus counter. The number
weighted size distributions, before and during the operation of the
electrostatic precipitators, are shown in Figure 6.

The change in aerosol size distribution caused by the
electrostatic precipitators increases the lung dose per unit exposure
to potential alpha energy concentration. The fraction of the total
potential alpha particle energy not attached to aerosol particles was
calculated using Raabe's method (Raabe, 1969). The dose rate to lung
before and after operating the electrostatic precipitators was
calculated using the dosimetric model described earlier by James
(James, 1984) and also in these proceedings. Although the radon
decay-product concentration is reduced by a factor exceeding ten, the
dose per unit exposure is increased by a factor of five, resulting in
a net reduction in lung dose by only a factor two.

Under normal household conditions the effect of electrostatic
precipitators would be less dramatic than shown here, because these
measurements were carried out in a closed room. With normal movement
between rooms, the aerosol concentration would not be reduced by as
much and the increase in unattached fraction of ^{218}Po would be less.

Mechanical ventilation system. A mechanical ventilation system was
installed to supply air to the office, sitting room, hall and stairs
and to extract from the scullery, bathroom and kitchen. Fresh air
was drawn through a cowling set in the window of the front centre
bedroom, which also housed the mechanical ventilation unit and heat
recovery unit. Exhaust air was discharged through a cowling passing
through a window situated half way up the stairs at the back of the
house. This avoided the mixing of inlet and exhaust air. All
bedroom doors were kept closed and these rooms were not regarded as
part of the controlled environment during the operation of the
system. Figure 7 shows diagrammatically the layout of the
ventilation system. During operation of this system the polythene
sheeting was in position on the sitting room floor. Table I gives a

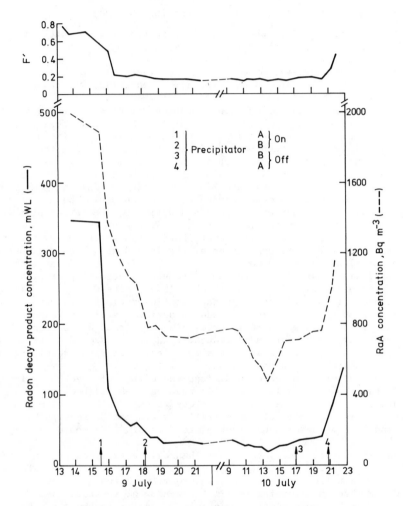

Figure 5.   Radon decay-product and RaA concentrations with
equilibrium factor during the use of electrostatic precipitators.

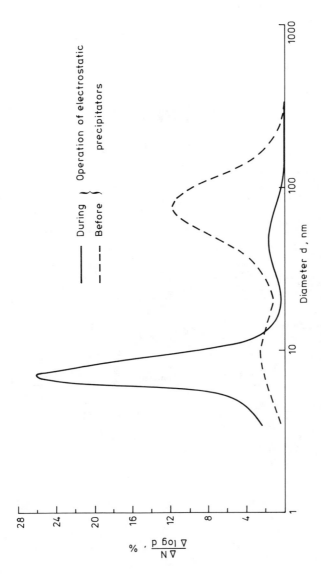

Figure 6.   Change in aerosol number weighted size distributions during operation of electrostatic precipitators.

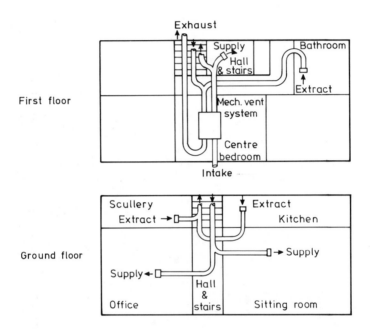

Figure 7.   Mechanical ventilation system.

summary of the effects of mechanical ventilation on radon
decay-product concentrations in the various rooms.

Table I
Results of measurements immediately before and during operation of
the mechanical ventilation system

Mechanical ventilation system	Room or area	Ventilation* rate, h⁻¹	Radon decay-product conc., mWL
Before operation	Sitting room	0.19	150
	Office	--	120
	Kitchen	--	180
	Hall	--	160
	Scullery	--	190
	Bathroom	--	98
During operation (after 17 hours)	Sitting room	0.81	3.8
	Office	--	72
	Kitchen	--	10
	Hall	--	11
	Scullery	--	3.1
	Bathroom	--	6.1

*Ventilation rate was only measured in the sitting room, but the
mechanical ventilation system was balanced to provide approximately
equal ventilation rates for all areas served by the system.

The initial ventilation rate for the whole house was typically
$0.2\ h^{-1}$. When the mechanical ventilation system was running, the
supply and exhaust flow rates were in the range $250-300\ m^3\ h^{-1}$,
providing a ventilation rate of about $0.8\ h^{-1}$ in ventilated rooms.

In all rooms the concentration of the radon decay products was
decreased by mechanical ventilation (Table I). The smallest decrease
occurred in the office, indicating that this room had the highest
radon production rate whereas the polythene sheeting reduced ingress
of radon into the sitting room. Apart from the office, reduction in
radon decay-product concentration was larger than could be explained
by the increased ventilation rate.

The installation of a mechanical ventilation system in an
existing dwelling would not be an economical proposition, even though
such a system does permit greater control over ventilation rate than
ad hoc window opening. Also, the climate in south-west England is
rarely severe enough for the full benefits of a heat recovery system
to be realised. In some new premises, where controlled ventilation
may be desirable to limit the build up of other indoor pollutants,
e.g. kitchen odours, a mechanical ventilation system might be
considered worthwhile.

Structural alterations to the building

The limited effectiveness of some of the above measures strengthened
our determination to investigate a number of more radical remedies.

Replacement of the suspended wooden floors. The preferred remedy for
ingress of radon in soil gas was to replace all the ground floors
with a suspended concrete floor, including multiple underfloor
ventilation points. However, one constraint on this study was to
provide a solution at a reasonable cost and by a method within the
competence of a typical small building firm. Installation of a
suspended concrete floor throughout the building would have entailed
removing the solid floor at the rear of the dwelling and temporarily
removing internal load-bearing walls and installing a support system.
This was deemed to be too complicated. The chosen solution was to
replace the suspended wooden floors by a single concrete floor. This
would incorporate a composite damp proof membrane, to act as the main
barrier to the movement of radon from the soil. Care would be taken
in laying the membrane to ensure that it remained intact. The
concrete floor was designed by the Building Research Establishment
and details of the design are shown in Figure 8. The composite
membrane chosen comprised 1 mm of bitumen compound between two sheets
of 250 μm thick polythene. This permits some movement of the
material above and below it without tearing.

In order that a complete floor could be laid, the partition
walls between the hall, the office and the sitting room were removed.
After the wooden floors had been taken up, samples of soil gas were
taken from a depth of 0.8 m. The radon concentration in these
samples ranged from $1.8 \ 10^5$ to $3.2 \ 10^5$ Bq m^{-3}. These concentrations
in soil gas are above the value of $5 \ 10^4$ Bq m^{-3} suggested as being
likely to cause high concentrations of ^{222}Rn in indoor air (Wilson,
1984). Material from beneath the floors was also analysed for its
radionuclide concentration. The mean values were 790 Bq kg^{-1} and
22 Bq kg^{-1} for ^{226}Ra and ^{232}Th, respectively. These concentrations
of ^{226}Ra far exceed the value of 89 Bq kg^{-1} found in typical granite
(NEA, 1979). The maximum concentration of radium was 3800 Bq kg^{-1}.

A local builder was contracted to instal the floor. Part of the
specification stated that those areas of the internal wall surfaces
that would abut the concrete floor should be cement rendered, with as
flat a finish as possible, before the base concrete was poured. In
the event, the standard of rendering was poor and uneven. The other
stages in the installation of the concrete floor were performed
satisfactorily.

After the concrete floor had been installed, measurements
indicated radon decay-product concentrations of 160 mWL and 140 mWL
on the ground floor and upstairs, respectively. There were
noticeable gaps between the concrete floor and the wall surfaces in a
number of places, some of which extended to the foundations. The
radon production rate was estimated to be 1300 Bq $m^{-3} \ h^{-1}$, more than
four times the value found in the initial study with all internal
doors open. Radon was obviously entering the dwelling with ease,
even though the area of underlying material exposed in gaps between
the floor and the walls was much smaller than that exposed beneath

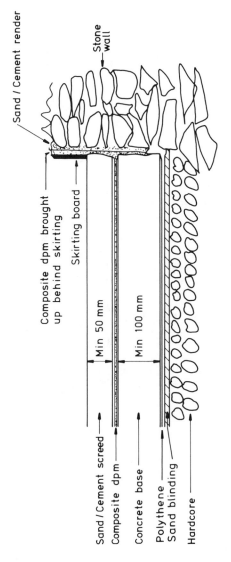

Figure 8.   Section through floor.

the original suspended floor. Measurement of the radon flux from the surface of the new floor demonstrated that little radon was being released from the floor itself.

Further attempts to reduce the ingress of radon. The most obvious paths by which radon could enter the dwelling from the underlying soil were the gaps between the walls and the floor. Attempts were made to seal these gaps with materials which would prevent pressure driven flow of gases and in which radon would have a low diffusion coefficient. Although some gaps were obvious, it was possible that smaller gaps might exist, especially where the new concrete floor joined the old floor at the rear of the dwelling. The larger gaps between the floor and the walls were filled by injecting an expandite rubberised bitumen roof sealant. A rectangular groove, 1 cm square, was chiseled over the joints between the old and the new concrete floors and battens fastened around the periphery of the floor, 1 cm from the walls. A heavy duty pavement sealant was poured between the battens and the wall to cover the wall-floor joints and also into the grooves to cover the floor to floor joints. The average radon decay-product concentrations over a ten day period after the floor-wall joints had been sealed were 72 mWL and 75 mWL on the ground floor and upstairs, respectively.

The effect of covering the fireplaces was investigated and the results are shown in Figure 9. The ventilation was found to decrease in spite of the moderate wind speed. (See the broken bars for the duration of coverage.) The radon decay-product concentration also decreased, showing that the radon production rate was greatly affected by covering the fireplaces. Approximate radon production rates with and without the fireplaces covered were 27 and 230 Bq m^{-3} h^{-1}, respectively. Confirmation that the fireplaces themselves were not significant sources of radon was obtained by blocking off the chimneys and leaving the grates exposed. This behaviour is explained if the major mode of radon entry into the dwelling was by pressure driven flow. Closing the fireplaces would have reduced the pressure difference between room air and the soil gas caused by the Venturi effect of the wind across the two chimney stacks. At this stage of the study, satisfactorily low levels of radon decay-product concentration were achieved simply by eliminating the Venturi effect of the chimney stacks. However, this could not be used as a permanent remedy because of the form of construction of the dwelling. There were no damp proof courses in any of the walls and avoidance of dampness relied on the ventilation afforded by the chimneys.

The relatively high concentrations of radon decay products when the fireplaces were open demonstrated that the seal between the floor and the walls had not significantly improved the situation. However, it seemed very unlikely that the new seal was so poor as to permit the rapid changes in concentration that were being observed. The most likely path by which radon from the soil could enter the dwelling therefore appeared to be the walls, particularly the internal load-bearing walls which were in contact with the subsoil.

To measure the radon concentration in pore spaces of the walls, holes were drilled to a depth of 10 cm to accommodate 12 mm copper

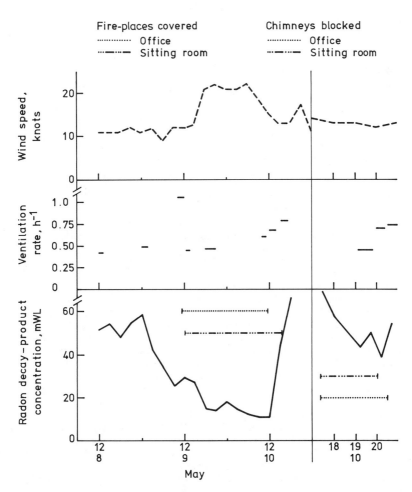

Figure 9. Radon decay-product concentrations and ventilation rates showing the effect of covering the fire-places.

pipes. The pipes were fitted with self closing vacuum connectors and sealed into the walls with epoxy resin. An area of wall of diameter approximately 30 cm, centred on each pipe, was also sealed with epoxy resin, following the method of Jonassen and McLaughlin (Jonassen and McLaughlin, 1980). Scintillation cells were evacuated, connected to the pipes and left in position for at least 18 hours. Internal walls, which are of brick and stone construction, had large voids within them. In all cases the scintillation cells were found to be at atmospheric pressure when they were removed. The highest concentrations of radon were found in the load-bearing walls separating the office from the scullery and the kitchen from the sitting room. The concentration of radon in these walls, at a height of 30 cm above the floor, ranged from $6 \ 10^3$ Bq m^{-3} to $3.9 \ 10^4$ Bq m^{-3}. At 1 m above the floor the concentration of radon was reduced to $5 \ 10^2$ Bq m^{-3}. Thus, at the base of these walls, the radon concentration was a substantial fraction of that in the soil gas. The rapid fall in concentration with height above the floor suggested that radon was rising from the subsoil through the pores of the internal walls and then escaping into room air above floor level.

To prevent the escape of radon from the wall surfaces, all relevant surfaces in the areas of the office, hall and sitting room were coated with an epoxy resin paint. Following application of the paint, the vertical profile of radon concentration in the internal walls was found to be more nearly constant, up to a height of 1 m. The concentration above this height fell quite slowly. Concentrations of radon at all heights were higher after applying the epoxy resin paint than they had been in the initial series of measurements. However, the concentrations of radon in the walls did vary considerably with time, by a factor up to five.

For a period of 17 days after coating the walls with epoxy resin paint, the average radon decay-product concentration on the ground floor was 100 mWL, with a maximum value of 350 mWL. During this period, the concentration of ^{218}Po upstairs was consistently higher than that on the ground floor. These results indicated that, in spite of the change in the profile of the radon concentration in the internal walls, radon was still entering the dwelling and producing unacceptable indoor concentrations. It was shown later that one of the bedrooms now had a significant radon production rate. One wall of this bedroom was an extension of the wall between the kitchen and the sitting room. It is likely that the coat of epoxy resin paint on the downstairs wall was reducing the rate at which radon could escape downstairs, allowing more radon to reach the pore spaces of the bedroom wall and to escape through the untreated surface into the bedroom.

It was now clear that the objective of finding a passive solution to the high radon concentrations in this unusual dwelling was not readily achievable. In the final measure, a system was installed to reduce the pressure under the concrete floor and to remove soil gas containing radon from the soil underlying the house and discharge it to the atmosphere. Such systems have been used successfully in the uranium mining communities of Canada (Chakravatti, 1979; Keith, 1980) and also in Sweden. The system installed was similar to that employed in Sweden (Ericson, Schmied and Clavensjo, 1984).

Under-floor suction system. Holes, 0.1 m in diameter, were bored through the concrete floor in the centre of the office and sitting room areas. Roughly hemispherical cavities, of approximate radius 0.4 m, were excavated under each hole through the floor. Ducting, incorporating a 30 W centrifugal fan, was fixed between each hole and an exhaust outlet through the window of the corresponding room. Micromanometer sensors were set into the ducting, on the low pressure side of each fan, and the speeds of the fans were regulated with variable auto-transformers in the power supply to each fan. For most of the tests of these systems an underpressure of about 100 Pa was maintained in the cavities under the floor. Figure 10 shows the arrangement in diagramatic form.

The variation of the radon decay-product concentration on the ground floor and of the wind speed, for the period prior to installation of the under-floor suction systems through to the completion of the tests, are shown in Figure 11. Opening the two holes through the concrete floor resulted in a large surge in the concentration of radon decay products. This reduced when the window was opened at C to install the ducting. The window was then boarded up. By stage D, all the ducting had been completed and even without the fans operating the substantial reduction in radon decay-product concentration was maintained and a slight further reduction occurred.

The fans were switched on at time E, causing a further decrease in concentration. Low concentrations were maintained for 4 days, at the end of which only one of the fans was kept in operation (office system F to G, sitting room system G to H). The increase in the concentration of radon decay-products after switching off one or both fans is evident. After time I, the radon decay-product concentration decreased, but less rapidly than in the first trial and the final value was not as low. However, a trend to still lower values was apparent when the exercise was concluded on day 19. Measurements of the concentration of radon in the exhausts of the two suction systems were made on three days and the results are given in Table II.

Table II
Radon concentration in exhaust air

| Day | Radon concentration, Bq m^{-3} | |
	Sitting room duct	Office duct
7	$2.5 \ 10^4$	$4.0 \ 10^3$
10	$4 \ 10^{1}$*	$1.0 \ 10^4$
13	$5.7 \ 10^4$	$3.4 \ 10^3$

*Fan off

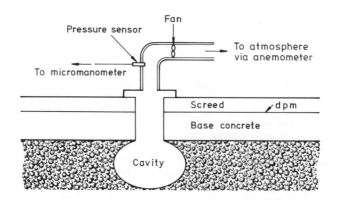

Figure 10. Underfloor under-pressure system.

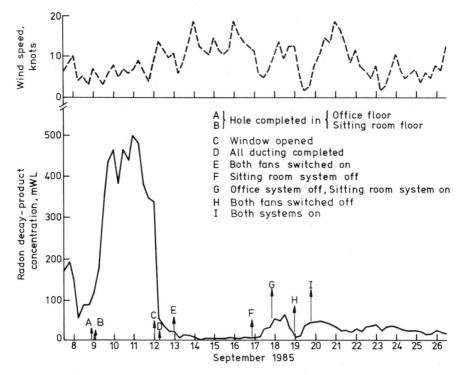

Figure 11. Variation in radon decay-product concentrations before and during the operation of the underfloor under-pressure system.

The radon production rates for the dwelling, whilst the under-floor suction systems were operating varied over the range 4.7 to 18 Bq m^{-3} h^{-1}, confirming the effectiveness of the system.

Whilst the under-floor suction systems were operated, further measurements were made of the radon concentration in the pores of the wall dividing the office area from the scullery. On day 7 (Figure 11), when the concentration of the radon decay-products was low (approximately 7 mWL), the concentration of radon near the base of the wall was about 8800 Bq m^{-3}. This had risen to more than 13000 Bq m^{-3} when measurements were made over-night between days 10 and 11. The radon decay-product concentration on the ground floor on the morning of day 11 was 44 mWL. These concentrations of radon in the wall pores were not significantly different from the values originally found before the walls had been coated with epoxy resin paint. This explains why the bedroom above the sitting room continued to have a significant radon production rate when the suction systems were operated. Radon still migrated within the epoxy resin painted walls and entered the bedroom through the untreated wall surface.

The internal walls between the scullery and the office and between the kitchen and the sitting room descended below the base of the concrete floor to the sub-soil. Clearly, the effectiveness of the suction systems in reducing the soil gas concentration at the base of the walls was limited because of the distance from the suction points.

Conclusions

The subsoil is the principal source of radon in this house. Both the activity concentration of radium-226 in subsoil and of radon in soil gas are above levels for building ground that might result in significant indoor radon concentrations. The radon decay-product concentration in the dwelling before remedial measures were taken was substantially higher than the reference value of 120 mWL.

The highest radon production rates occurred in rooms with suspended wooden floors through which air could pass readily. Simple ventilation of the underfloor space with a 30 W fan was effective in reducing radon decay-product concentrations in these rooms, but not below the reference value in all rooms. More rigorous application of this technique with multiple extraction points and a larger fan could clearly serve as an effective remedy.

Polythene and mylar sheeting laid over the suspended wooden floor reduced the concentrations of radon decay products, but not below the reference level. This technique was only partially successful because of the difficulty of effecting good seals to the walls, indicating the care needed to remedy high radon levels with membrane barriers.

Electrostatic precipitators were effective in reducing the radon decay-product concentration below the reference level, but analysis of the room aerosol indicated that the reduction was largely offset by an increase in lung dose per unit exposure.

Mechanical ventilation of room air reduced the radon decay-product concentration below the reference level by a larger factor than expected from the increased ventilation rate. However, such systems are expensive to instal in existing dwellings.

A new concrete floor incorporating barriers to radon transport from the subsoil appeared to be only partially successful in reducing radon decay-product concentrations. It was shown that the Venturi effect of the wind across two chimney stacks caused pressure-driven flow of radon from the ground. Covering the fireplaces to eliminate this effect resulted in concentrations below the reference level. The floor slab itself was fully effective against radon ingress.

Careful sealing of the floor-to-wall joints and of the ground floor walls was unsuccessful in reducing pressure-driven flow of radon from the subsoil. This merely diverted the flow of radon up through the internal walls of the dwelling and into upstairs rooms. The problem arose in this old dwelling because it has very porous walls and no damp proof course, thus allowing radon to by-pass the sealed floor. Incorporation of a passive radon barrier into the floor of a modern house with less porous walls is likely to be effective.

Suction on cavities created under the concrete floor with low-power fans was highly successful in reducing the radon decay-product concentrltions below the reference level. Such under-floor suction systems are relatively easy to instal and cheap to operate. They appear to hold considerable promise.

Where installation costs are not prohibitive, the ideal remedy would be a suspended concrete floor above a well ventilated space, with the ground below the ventilated space sealed by concrete incorporating a polymeric radon barrier. This would effect a passive remedy requiring little maintenance and no running costs.

Acknowledgments

The work reported here was part-funded, and jointly carried out by the Building Research Establishment of the UK Department of the Environment (Contract No. 7910-2199). The authors would like to thank Dr P R Warren and his colleagues, of the Building Research Establishment, for their interest and advice during this project.

References

Chakravatti, J.L., Control of ^{222}Rn-WL in New Construction at Elliot Lake, presented at Workshop on Radon and Radon Daughters in Urban Communities Associated with Uranium Mining and Processing, Bancroft, Ontario, Atomic Energy Control Board of Canada, AECB-1164 (1979).

Cliff, K.D., The Measurement of Low Concentrations of the Daughters of Radon-222, with Emphasis on RaA Assessment, Phys. Med. Biol. 23: 55-65 (1978a).

Cliff, K.D., Assessment of Airborne Radon Daughter Concentrations in Dwellings in Great Britain, Phys. Med. Biol. 23: 696-711 (1978b).

Cliff, K.D., J.C.H. Miles and K. Brown, The Incidence and Origin of Radon and its Decay Products in Buildings, NRPB-R159, HMSO, London (1984).

Ericson, S-O., H. Schmied and B. Clavensjo, Modified Technology in New Construction, and Cost Effective Remedial Action in Existing Structures, to Prevent Infiltration of Soil Gas Carrying Radon, in Indoor Air 5 (B. Berglund, T. Lindvall and J. Sundell, eds.) pp 153-158, Swedish Council for Building Research, Stockholm (1984).

James, A.C., Dosimetric Approaches to Risk Assessment for Indoor Exposure to Radon Daughters, Radiat. Prot. Dosim. 7: 353-366 (1984).

Jonassen, N., and J.P. McLaughlin, Exhalation of Radon-222 from Building Materials and Walls, in Natural Radiation Environment III (T.F. Gesell and W.M. Lowder, eds.) pp 1211-1224, National Technical Information Center, U.S. Department of Commerce, Springfield, Va. (1980).

Keith Consulting, Summary of Uranium City, Saskatchewan, Remedial Measures for Radiation Reduction with Special Attention to Vent Fan Theory, presented at Workshop on Radon and Radon Daughters in Urban Communities Associated with Uranium Mining and Processing, Port Hope, Ontario (1980).

NEA; Nuclear Energy Agency Group of Experts, Exposure to Radiation from the Natural Radioactivity in Building Materials, OECD Nuclear Energy Agency, Paris (1979).

Raabe, O.G., Concerning the Interactions that Occur between Radon Decay Products and Aerosols, Health Phys. 7: 353-366 (1969).

RCEP; Royal Commission on Environmental Pollution, Tackling Pollution - Experience and Prospects, Tenth Report, HMSO, London (1984).

Wilson, C., Mapping the Radon Risk of Our Environment, in Indoor Air 2 (B. Berglund, T. Lindvall, and J. Sundell, eds.) pp 85-92, Swedish Council for Building Research, Stockholm (1984).

Wrixon, A.D., L. Brown, K.D. Cliff, C.M.H. Driscoll, B.M.R. Green and J.C.H. Miles, Indoor Radiation Surveys in the UK, Rad. Prot. Dosim. 7: 321-325 (1984).

RECEIVED October 31, 1986

Chapter 40

The Feasibility of Using Activated Charcoal To Control Indoor Radon

Rey Bocanegra[1] and Philip K. Hopke[2]

[1]Department of Nuclear Engineering and Institute for Environmental Studies, University of Illinois, Urbana, IL 61801
[2]Departments of Civil Engineering and Nuclear Engineering and Institute for Environmental Studies, University of Illinois, Urbana, IL 61801

Indoor contamination by radon-222 and its decay products has recently been the focus of attention for the Environmental Protection Agency, the news media, and the public in general. This new awareness of the health hazard posed by radon has lead to increased efforts to understand and alleviate the problem. Numerous methods for radon and/or decay product removal from uranium mines and indoor air are found in the literature. A method often cited involves radon adsorption on charcoal. We have studied radon adsorption on activated charcoal as a means of radon mitigation for indoor air. We have identified several important adsorption parameters which need to be investigated. These studies will eventually lead to the design and construction of an apparatus capable of operating within reasonable constraints. The parameters are flow rate, humidity, temperature, and charcoal type. Other interfering organic and inorganic gases will also be examined.

There is an increasing concern regarding the exposure of the general population to increased levels of radon decay products in indoor air. The exposure to the public from increased levels of natural radiation has been the subject of a conference held in Maastricht, The Netherlands, in March 1985 where a number of reports were presented showing high levels of radon in the indoor environment. Other reports in this volume also demonstrate that high levels of radon are not as uncommon as had been previously thought.

Radon in indoor air arises primarily from radium in the soil. The radon in the soil gas flows under a pressure gradient from the soil into the building. In some cases building practices can lead to high radon levels in the living areas of the house. Radon is chemically quite inert and does not pose a significant radiation health hazard in itself because the retained fraction in the body is so low (Mays et al., 1958). It is, however, an excellent vehicle for the dispersion of its short-lived radioactive decay products.

Background

Measurements done in houses in the Reading Prong area of Pennsylvania and in New Jersey have found a number of houses with very
high decay product concentrations. Radon transported through the
ground tends to accumulate in these houses. In one case the activity
was in excess of fifty times the maximum permissible concentration
for workers (0.1 WL) in uranium mines (Guimond, 1985). The
Environmental Protection Agency (EPA) has recommended a limit of
150 Bq/m^3 (4 pCi/l) of radon as an equivalent to 0.02 WL for houses
in areas of high natural radioactivity (Guimond, 1985). Although
radon concentration values for houses in the U.S. have been reported
in the general range of about 4 - 1000 Bq/m^3 (0.1 - 27.0 pCi/l)
(Nero et al., 1983), there have been cases reported showing radon
concentrations an order of magnitude higher. Variations in household
concentrations are due mainly to radium concentration in the soil,
air permeability, and volume of high porosity soil (Sextro et al.,
1986). Sources of radon infiltration are sometimes found to be
very localized (Grimsrud et al., 1983). Sealing cracks, drains,
and other paths for radon can be an effective control measure in
these cases. Small fans can also be used to provide subfloor
ventilation for soil gas dilution and removal before entry into
the house. There may be cases where these measures are not
effective, and larger, more expensive ventilation systems must be
installed. The EPA has installed ventilation systems in houses in
the Reading Prong area in an effort to reduce the radon concentration. In some cases, because of difficult to control pressure
differentials, the ventilation systems actually caused the radon
content to increase (Henschel and Scott, 1986). Thus, alternative
methods to reduce the radon levels could prove useful.

A conceivable way of reducing the exposure to radon and its
decay products is to trap the decay products in a filter and thus
remove them. However, inspection of Figure 1 reveals that within
a matter of minutes equilibrium with polonium-218 is reestablished
and there is a substantial increase in the second and third decay
products as well. Therefore, control of the radon is the key to
exposure control. One method of radon removal is by trapping it
on some adsorber. The ability of charcoal to adsorb noble gases
is well documented. Adams et al. (1959) have shown the superior-
ity of charcoal over other adsorbers including molecular sieve,
silica gel, and alumina. Values for the dynamic adsorption coefficient for radon at room temperature have been reported in the
general range of 2000-7000 cm^3/g. Table I summarizes data available
in the literature for various radon adsorbing charcoals.

Theoretical

The basic mechanism for radon adsorption on charcoal and the effects
of competing processes have been well studied. Several treatments
of the adsorption of radon (and noble gases in general) on charcoal
have appeared in the literature (Adams et al., 1959; Strong and
Levins, 1978; Kapitanov et al., 1967; Siegwarth et al., 1972).
Consider a charcoal bed consisting of N theoretical stages as

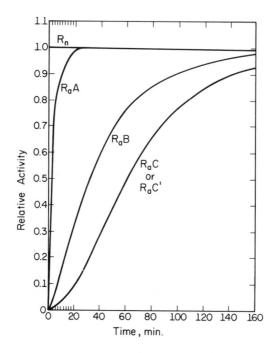

Figure 1. Given a constant radon source, equilibrium is quickly
     established between radon-222 and its short-lived
     daughters.

represented in Figure 2. A mass balance for the adsorbate across each of the N stages is given by

$$\frac{dY}{dt}i = \frac{-FN}{kM} (Y_i - Y_{i+1})$$ (1)

where k = dynamic adsorption coefficient in cm^3/g
   F = flow rate through the charcoal bed in cm^3/min
   Y = volume fraction of adsorbate in the gas phase leaving the ith stage at time t
   M = total mass of the adsorber in grams
   t = time in minutes.

The adsorbate concentration in the Nth stage along the charcoal bed can be found by solving the series of N differential equations. These solutions represent the concentration profile in the Nth stage. For a unit pulse of adsorbate at time t = 0, the solution reduces to

$$Y_N = \frac{N^N F^{N-1} t^{N-1}}{(N-1)! (kM)^N} \exp(-NFt/kM)$$ (2)

The time at which the concentration of the adsorbate is maximum, t_{max}, is found by setting the derivative of Equation (2) equal to zero. The solution is given by

$$t_{max} = \frac{(N-1) k M}{N F}$$ (3)

Since the concentration profile given by Equation (2) is nearly symmetrical, it can be assumed that t_{max} occurs at the midpoint of the curve, therefore, this value is equivalent to the mean residence time, $\langle t \rangle$. For columns consisting of a large number of theoretical stages, the quantity (N-1)/N approaches unity and Equation (3) becomes

$$k = \frac{F \langle t \rangle}{M}$$ (4)

For the case of constant input, the concentration profile for the Nth stage can be found by integrating Equation (2).

$$C_N = \int_0^t Y_N \, dt$$ (5)

At time = tmax (ie. time = $\langle t \rangle$), $Y_N' = 0$, and $Y_N'' = C_N''$. Therefore, the mean residence time for the constant input case is the point of inflection of the concentration profile represented by Equation (5). Figure 3 shows a typical radon concentration profile.

Table 1

Adsorption Coefficient for Various Radon Adsorbing Charcoals

Type	Dynamic Adsorption Coefficient (cm^3 STP Air/g)	Temperature (°C)	Reference
SKT-2M	9650	18	a
No. 2	6300	18	a
SKT-1	9200	18	a
No. 4	7600	18	a
MSKT (5)	7400	18	a
SKT (82)	6000	18	a
No. 7	4500	18	a
SKT (84)	5700	18	a
Ag-2	4250	18	a
No. 12	2000	18	a
Witco AC-337	5660	20	a
Peat	2250	20	b
Sutcliffe-Speakman 207c	3530	25	c
Norit RFL 3	4610	25	c
Norit RFL 111	4660	25	c
Ultrasorb	5000	25	c
Pittsburgh PCB	5690	25	c
Lime Wood	7000	20	d

a. Kapitanov et al., 1967
b. Scheibel et al., 1980
c. Strong and Levin, 1978
d. Gubeli and Stori, 1954

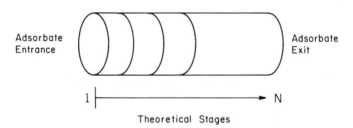

Figure 2. A carbon bed consisting of N theoretical stages.

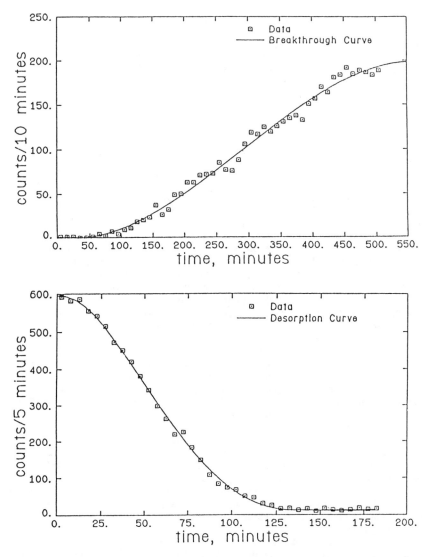

Figure 3. Adsorption and desorption curves for the sorption of radon on coconut based charcoal. (Jebackumar, 1985).

The dynamic adsorption coefficient can be used to gauge the
performance of a charcoal bed under various conditions.
The temperature strongly affects the adsorption process.
At about 100 C, the desorption of radon from charcoal is rapid and
efficient enough to fully regenerate a charcoal bed in a reasonable
amount of time. The desorption concentration profile also features
an inflection point (see Figure 3). Equation (4) can therefore
also be applied to find an optimal temperature for the desorption
process.

Discussion

It is clear from Table I that adsorptive carbon is available for
use in an air cleaning system. A radon removal apparatus would be
required to clean a volume of air large enough to equal the rate
of entrance into the house in order to maintain the radon concentra
tion below the 150 Bq/m^3 (4 pCi/l) limit recommended by the EPA.
Figure 4 shows a possible conceptual design of an indoor radon
removal apparatus based on charcoal adsorption. It consists of
two parallel charcoal beds. The principle of operation is as
follows: Air is pulled into inlet A and passed through Bed 1
while valves V1 and V2 are shut and V3 and V4 are open. Once
breakthrough is detected in detector D1, valves V3, V4, V5, and V6
will close and valves V1, V2, V7, and V8 will open. Heated air will
then be forced into inlet OS1 through Bed 1 and out outlet OS2.
While the radon is being desorbed from Bed 1, Bed 2 will be in the
adsorption cycle. The desorption process takes a much shorter
time than the adsorption thereby insuring that the other bed will
be available to begin the next cycle. There are several engineering
considerations that need to be taken into account. The backflushing
during the desorption stage is achieved using air from outside of
the building to prevent the creation of a pressure gradient that
might enhance the diffusion of radon from the surrounding walls.
The charcoal beds must be thermally isolated from each other. The
air inlets must be positioned far enough apart so as to minimize
feedback of clean air back into the system. To prevent the
accumulation of radon in the house in the event of a valve failure,
all valves should be provided with backups. The volume of air
cleaned per unit mass of carbon increases exponentially with
decreasing temperature (Kapitanov et al., 1967). Thus greatly
increased adsorption capacity can be obtained by cooling the carbon
below ambient temperature. Although this process will require
additional energy input, it may be worthwhile to consider some
form of cooling.
     There may be problems from other adsorbing species in the
house. Carbon-dioxide and water vapor have been found to have an
adverse effect on the adsorption coefficient (Strong and Levins,
1978; Siegwarth et al., 1972). The likeliest place for indoor
radon to accumulate in houses is in the basement or crawl space
where a large surface area is in direct contact with the soil, and
thus the most likely place to put an adsorption system is in these
locations. However, these areas are also commonly used to store
various household chemicals such as painting supplies, etc. These
household items stored in basements can release contaminants that
may be classified into 4 broad categories; aromatics, paraffins,

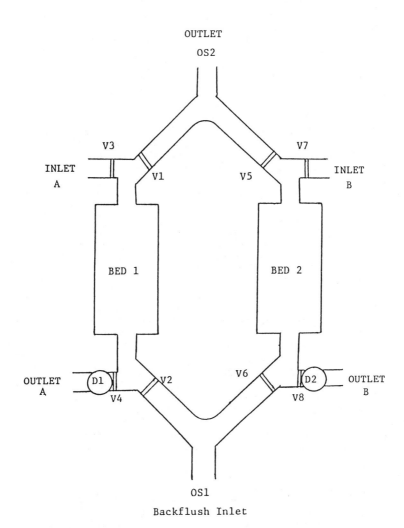

Figure 4. Conceptual design of an apparatus for the removal of indoor radon.

halogenated hydrocarbons, and combustion by-products. Paint, paint
remover, and paint thinner are examples of aromatic compounds.
Paraffins may be released from materials used for hobbies or
recreation. Examples of these are propane lanterns and welding/
brazing equipment. Household cleaning chemicals and garden care
products are the biggest source of halogenated compounds. The
effect of basement airborne contaminants on the adsorption of
radon on charcoal is probably the most important and least studied
area of indoor radon control.

In order to evaluate the practical feasibility of the system
depicted in Figure 4, several experiments will be performed to
find the conditions for optimal performance. The following
parameters will be investigated:

Carbon type: Given a fixed set of conditions (ie. flow rate,
temperature, mass of charcoal, humidity) the type of carbon demonstrating the highest dynamic adsorption coefficient will be
identified.

Flow rate: The system must be capable of processing a moderate
volume of air per unit time to be of practical use. However, the
empirical formulations for the dynamic adsorption coefficient
described in this paper are valid only for a certain range of
conditions. Experiments will be performed to identify the flow
rate/flow channel diameter combination beyond which the formulations
are no longer valid.

Contaminants: Indoor contaminants are expected to compete
for adsorption sites on the charcoal. We will experimentally find
the effect that these contaminants have on the dynamic adsorption
coefficient and on the life-time of the charcoal bed. Since the
number of radon atoms in even the most seriously contaminated
houses is very small, decay product buildup is not expected to
pose a significant problem.

The feasibility of building a radon removal apparatus will
depend on finding the parameters which will give the highest dynamic
adsorption coefficient under operational conditions. However, it
is recognized that real world constraints like initial costs,
operating costs, and physical size may force the ideal parameters
to be compromised. Devices of this type are intended for use in
private homes where these constraints cannot be ignored.

Acknowledgments

Although the information described in this article has been funded
wholly by the United States Environmental Protection Agency under
assistance agreement EPA Cooperative Agreement CR810462 to the
Advanced Environmental Control Technology Research Center, it has
not been subjected to the Agency's required peer and administrative
review, and therefore does not necessarily reflect the views of
the Agency and no official endorsement should be inferred.

Literature Cited

Adams, R.E., W.E. Browning, Jr., and R.D. Ackley, Containment of
Radioactive Fission Gases by Dynamic Adsorption, Industrial and
Engineering Chemistry, 51:1467-1470 (1959).

Grimsrud, D.T., W.W. Nazaroff, K.L. Revzan, and A.V. Nero, Continuous Measurements of Radon Entry into a Single Family House, Proceedings of 76th Annual Meeting of the Air Pollution Control Association, pp. 83-98 (June, 1983) .

Gubeli, O. and M. Stori, Zur Mischadsorption von Radon an Aktivohle mit Verschiedenen Tragergasen, Helvetica Chim. Acta 37:2224-2230 (1954).

Guimond, R.J., Regulation and Guidelines for Enhanced Natural Radiation in the United States, Science of the Total Environment 45:641-646 (1985).

Henschel, D.B. and A.G. Scott, The EPA Program to Demonstrate Mitigation Measures for Indoor Radon: Initial Results, Proceedings of an Air Pollution Control Association International Specialty Conference, p. 110-121, Philadelphia, Pennsylvania (February, 1986).

Jebackumar, R., Feasibility of Radon Control from High Volume Flows and Indoor Air, Master of Science Thesis, Nuclear Engineering Program, University of Illinois at Urbana-Champaign, (1985).

Kapitanov, Y.T., I.V. Pavlov, N.P. Semikin, and A.S. Serdyukova, Adsorption of Radon on Activated Carbon, Internat. Geology Rev. 12:873-878 (1967).

Mays, C.W., M.A. Van Dilla, R.L. Floyd, and J.S. Arnold, Radon Retention in Radium-Injected Beagles, Radiation Research 8:480-489 (1958).

Nero A.V., M.L. Boegel, C.D. Hollowell, J.G. Ingersoll, and W.W. Nazaroff, Radon Concentrations and Infiltration Rates Measured in Conventional and Energy-Efficient Houses, Health Physics 45:410-405 (1983).

Prichard, H.M. and K. Marien, A Passive Diffusion Rn-222 Sampler Based on Activated Carbon Adsorption, Health Physics 48:797-803 (1985).

Sextro, R.G., B.A. Moed, W.W. Nazaroff, K.V. Revzan, and A.V. Nero, Investigation of Soil as a Source of Indoor Radon, this volume.

Siegwarth, D.P., C.K. Newlander, R.T. Pao, and M. Siegler, Measurement of Dynamic Adsorption Coefficients for Noble Gases on Activated Carbon, Proceedings of 12th AEC Air Cleaning Conference, p. 28-46, Oak Ridge, Tennessee (August, 1972).

Strong, K.P. and D.M. Levins, Dynamic Adsorption of Radon on Activated Carbon, Proceedings of the 15th DOE Nuclear Air Cleaning Conference, CONF-780819,627-639 (1978).

RECEIVED August 4, 1986

A LAYMAN'S GUIDE

Chapter 41

The Indoor Radon Problem Explained for the Layman

Philip K. Hopke

Departments of Civil Engineering and Nuclear Engineering and Institute for Environmental Studies, University of Illinois, Urbana, IL 61801

Recently, interest has grown in the discovery of radon as a component of indoor air. This naturally occurring radioactive gas is and always has been present in the environment in which we live. In the previous decade, elevated levels of indoor radon had been found, but only in areas where human activities had been at work to enhance the radon levels. For example, in Grand Junction, Colorado, uranium mill tailings had been found to make excellent concrete and had therefore been used in a number of homes, sidewalks, and public buildings. Subsequently, enhanced radon levels had been found and traced to this building material. In 1977 the general concensus was that the radon problem was primarily related to building materials (UNSCEAR, 1977).

Other areas were found to have high radon-containing houses. However, these geographical regions included disturbed or contaminated land such as in Florida where phosphate mining wastes had been used as fill and houses then built on the land. Another case was in New Jersey where wastes from a radium processing plant had been put into local landfills early in this century and houses were subsequently built on the site after it was fully filled. Thus, the situation was believed to be one where radon in normally constructed houses built on uncontaminated lands were not even considered when thinking about indoor radon. However, recently there have been a number of instances where "high" radon levels have been found in areas that are not affected directly by building materials or wastes in the soil. The impact of natural soils on indoor radon levels is now recognized as potentially substantial and contribute to a large fraction of the average dose of radiation to the general public. It is the purpose of this report to provide an introduction to the radiological aspects of the problem, the general nature of the sources and ingress of radon into buildings, the potential health effects, and the possible methods of mitigating against high radon levels.

Radiological Background. The surface of the earth contains a very heterogeneous distribution of the naturally occurring elements.

0097-6156/87/0331-0572$06.00/0
© 1987 American Chemical Society

Among those elements are a number that are entirely radioactive or
have isotopes that are radioactive. The element potassium, a critical
electrolyte in the blood, includes the radioactive isotope K-40.
This very long half-life (1.25×10^9 years) isotope comprises 0.0117
percent of all potassium. Thus, this isotope is present in all of us
and has always been so. In addition, the materials around us,
including the soil and the building materials, contain both potassium
and the heavy naturally occurring radioactive elements thorium and
uranium that contribute to a level of radiation to which we are all
continuously exposed. Thus, there is always radiation exposure to
the general public and we must understand the exposure due to radon
in this context.   The amount of radioactivity is described in units
of activity. The activity is the number of decay events per unit
time and is calculated as follows

$$A = \lambda N$$

where λ is the probability of decay of the nucleus of a particular
atom in a unit time (a second, for example) and N is the total number
of atoms present in the sample. The units of activity are then disintegrations per second and 1 disintegration per second is called one
Becquerel (1 Bq). However, a substantial amount of the literature in
this field uses the older unit of Curie and fractions thereof. In
the case of radon, we generally deal with picocuries (pCi) or 10^{-12}
Ci. There are 3.710^{10} Bq/Ci or 1 Bq equals 0.037 pCi/l. Since we
are generally interested in the amount of activity per unit volume,
the appropriate units are Becquerels per cubic meter (Bq/m^3). Again,
much of the literature refers to picocuries per liter (pCi/l). A
concentration of 37 Bq/m^3 is equal to 1 pCi/l.

   If we start with a particular number of nuclei at a particular
time which we can call T=0, then the time at which half of the nuclei
originally present are gone is called the half-life. The half-life
and the decay probability are related by

$$\lambda = .693/T_{1/2}$$

Thus, by measuring activity as function of time yields a curve as
shown in Figure 1. However, the pattern of activity is more
complicated if the product of the decay process is also radioactive.
We will return to this problem after introducing the origins of radon.
   There are three heavy naturally occurring radioactive isotopes,
U-235, U-238, and Th-232. Although they are radioactive, their half-
lives are so long that the time since the formation of the universe
and these atoms is not sufficient for them to have decayed to stable
elements. Each of these isotopes decays through a long, complicated
series of shorter lived radioactive atoms until they finally become
nonactive lead isotopes. In this chain, there are a number of
different chemical elements formed and almost all of them are
chemically reactive. Thus, they tend to stay within the material in
which the uranium or thorium atom was originally deposited. However,
in each case, an isotope of radon is formed. Radon is a member of
the chemical group that includes helium, neon, argon, krypton, and
xenon. Although radon is not completely chemically inert as described
in the chapter in this volume by Stein (1977), it only reacts with
unusual reagents that are not found under any known environmental

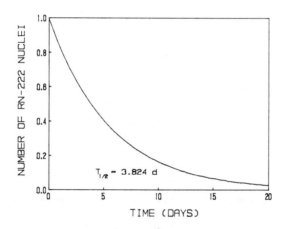

Figure 1.   Simple decay curve for radon-222.

condition. Thus, the "inert" radon can move slowly through solid material such as a soil grain until it reaches an air space in the soil or it decays to a different, chemically active element. The properties of the three radon isotopes are given in Table I along with their names as originally proposed by Marie and Phillippe Curie. Although these names are not the approved scientific designations for these isotopes, they are widely used in the literature. Thus, it is convenient to refer to Rn-222 as radon, Rn-220 as thoron, and Rn-219 as actinon. Because of the very low percentage of uranium that is U-235 and the very short half-life of actinon, it will not be considered further.

The decay sequence for Rn-222 is given in Figure 2. Because of the 4 day half-life of Rn-222, it has the time to penetrate through the soil and building materials into the indoor environment. There is some recent evidence that in spite of its short half-life, 55 seconds, Rn-220 can also penetrate into structures in significant amounts (Schery, 1985). However, the data are limited and the extent of the thoron problem is quite uncertain. It is, therefore, the short-lived decay products of radon that are considered to be particularly important in the exposure of the general public and it is these isotopes on which this report will concentrate.

The short-lived decay products, Po-218 (Radium-A), Pb-214 (Radium-B), Bi-214 (Radium-C), and Po-214 (Radium C'), represent a rapid sequence of decays that result in two alpha decays, two beta decays and several gamma-ray emissions following the beta decay. In alpha decay, a large nucleus breaks up into two pieces; a helium nucleus containing two protons and two neutrons (an alpha particle), and the rest of the nuclear particles. Energy is released as a result of this decay and is partitioned between the alpha particle and the residual nucleus such that energy and momentum are conserved. The nucleus and the alpha thus move away from one another with most of the energy carried by the lighter alpha particle.

The alpha particle interacts with the material through which it is traversing by a large number of electrostatic interactions that transfer a small amount of energy per interaction. It moves in a relatively straight path and moves a relatively well defined distance (range) before slowing to thermal velocity. For example, the range of the 6.0026 MeV alpha from Po-218 decay to Pb-214 has a range in air of about 4.6 cm.

The beta particles emitted by Pb-214 and Bi-214 are the result of the decay process that converts a neutron into a proton in a nucleus. Because of the very small mass of these emitted electrons, there is much less recoil energy available to the residual nucleus. These high energy electrons then interact with matter over a much longer, but more poorly defined range. Because the conversion of the neutron to the proton sometimes occurs preferentially to an excited state of the nucleus, there will also be emission of even more penetrating electromagnetic radiation gamma rays. Thus, there are three different types of emitted radiation with quite different properties in terms of their interaction with the material around them. These differences are reflected in the differences in the amount and location of energy deposition and in ways that the presence of these radionuclides can be detected. To illustrate the behavior of the activity of the radio-active products of the radon decay, the activity of each of the short-lived isotopes is plotted as a function of time for initially pure

Table I. Properties of the Naturally Occurring Radon
Isotopes and Their Major Decay Products

Isotope (Curie Symbol)	Mode of Decay	Half-life	Major Radioactive Emissions Energies (MeV)
Radon-222 (Rn)	α	3.8235 d	5.4897
Polonium-218 (RaA)	α	3.11 m	6.0026
Lead-214 (RaB)	β	26.8 m	.67, .73
	γ		.35192, .29522, .24192
Bismuth-214 (RaC)	β	19.8 m	1.54, 3.27, 1.51
	γ		.60932, 1.7645, 1.12028
Polonium-214 (RaC')	α	163.7 μs	7.6871
Lead-210 (RaD)	β	22.3 y	.017, .061
	γ		.046539
Bismuth-210 (RaE)	β	5.01 d	1.161
	γ		.2656, .3046
Polonium-210 (RaF)	α	138.38 d	5.3044
Lead-206		stable	
- -			
Radon-220 (Tn)	α	55.6 s	6.2883
Polonium-216 (ThA)	α	0.15 s	6.7785
Lead-212 (ThB)	β	10.64 h	.331, .569
	γ		.23863, .30009
Bismuth-212 (RhC)	β	60.6 m	2.251
	γ		.7272
	α	60.6 m	6.0510, 6.0901
Polonium-212 (ThC')	α	.298 μs	8.7844
Thalium-208 (ThC'')	β	3.053 m	1.796, 1.28, 1.52
	γ		2.6146, .5831, .5107
Lead-208		stable	
- -			
Radon-219 (An)	α	3.96 s	6.8193, 6.553, 6.425
	γ		.27120, .4017
Polonium-215 (AcA)	α	1.780 ms	7.386
	γ		.4048, .4270
Lead-211 (AcB)	β	36.1 m	1.38
	γ		.4048, .8318, .4270
Bismuth-211 (AcC)	α	2.14 m	6.623, 6.279
	γ		.3510
Thalium-207 (AcC'')	β	4.77 m	1.43
	γ		.8978

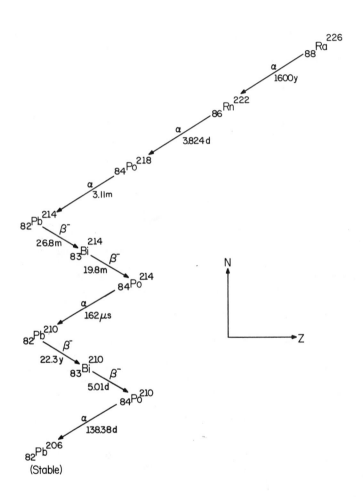

Figure 2. Decay chain for the formation and decay of radon-222.

radon-222 (Figure 3). Note that because radon-222 has a longer half-
life than either of the four short-lived products, they reach the same
activity (number of decays per unit time) as the radon. The mixture
then decays with the 3.8 day half-life of the radon. The total activity would then be the sum of these individual curves.

Radioactivity can be detected by examining the effects on matter
of the deposited energy. The energy deposition results in two possible
interactions; the excitation of an atom or molecule to a higher energy
state or the ionization of an atom or molecule leading to charge
species in the material. The exitation of the molecule is not a
stable condition and the molecule returns to its stable lowest energy
state with the emission of light. The amount of light will depend on
the number of atoms that were excited by the passage of the particle
and that will depend on its energy. The emitted light can then be
used to detect the presence of the activity. For example, zinc sulfide with a little silver added to it, $ZnS(Ag)$, is a very good
material for detecting alpha particles. If the inside of a cylinder
is coated with $ZnS(Ag)$ and the flat edge optically coupled to a photomultiplier tube, then the alpha decay of radon in the detector can be
observed as pulses in the photomultiplier tube.

Alternatively, the ionization of a gas can be used to detect the
passage of an alpha or beta particle. An electrode is strung through
the gas and an electrical potential is applied to the electrode.
Typically a positive charge is applied to attract the electrons that
were pulled from the molecules by the passage of the charged particle.
Often to enhance the output signal, the voltage is set high enough
that the electrons are accelerated toward the electrode and cause
more ionization on their way. Thus, a single event initiates a chain
of events giving rise to a measureable pulse of electrons. The
current flowing is typically time integrated and displayed on a meter
as a rate of activity (counts per second) and each pulse is used to
drive a small audio speaker so that the counter clicks to show the
presence of the radioactivity.

These types of detections are capable of very high sensitivity so
that very small numbers of radioactive atoms can be detected. To
illustrate this idea the number of atoms of Rn-222 for a concentration
of 10 pCi/l can be calculated

$$\lambda N = 10 \text{ pCi/l } (3.7 \times 10^{-2} \text{ disintegrations/s})$$

$$\lambda N \cdot \frac{.693N}{(3.825 \text{ days})(24 \text{ hours})(3600 \text{ s/hr})} = 0.37 \text{ Bq/l}$$

$$N = 1.76 \times 10^5 \text{ atoms/l}$$

This concentration is equivalent to a partial pressure of radon of
6.6×10^{-18} atmospheres. Yet this value is an elevated radon concentration. If all of the decay products formed by the decay of the
radon remain in the air, then there would also be 10 pCi/l of Po-218,
Pb-214, etc. Such a mixture would be said to be in secular equilibrium. From the monitoring of uranium mines, an equilibrium mixture
of these decay products at 100 pCi/l is called a working level (WL).
Thus, a 10 pCi/l equilibrium mixture represents 0.1 working level.

The cumulative exposure to such activity can be expressed as the amount of activity in WL multiplied by time.   In past analyses of occupational exposure by miners, this cumulative exposure has been given in WLM assuming 170 hours in a working month and is calculated as

$$\text{Cumulative Exposure in WLM} = \sum_{i=1}^{n} (WL)_i \left(\frac{t_i}{170}\right)$$

where $(WL)_i$ is the average concentration of radon decay products during exposure interval i expressed in WL and t_i is the number of hours of the exposure i.   The cumulative exposure at a given decay product concentration is four times that for occupational exposure (8766 against 2000 hours on an annual basis).

A better way to express the activity of the radon decay products is as the Potential Alpha Energy Concentration (PAEC).   This quantity incorporates the deposition of energy into the air and is expressed as MeV/m^3

$$PAEC = \frac{(E_1 + E_3)}{\lambda_1} \cdot A_1 + \frac{E_3}{\lambda_2} \cdot A_2 + \frac{E_3}{\lambda_3} \cdot A_3$$

$$= 3.69 \times 10^3 \, A_1 + 4.01 \times 10^4 \, A_2 + 5.82 \times 10^4 \, A_3$$

where A_1, A_2, and A_3 are the activity concentration in Bq/m^3 of Po-218, Pb-214, and Bi-214, respectively, and E_1 and E_3 are the alpha energies emitted by Po-218 and Po-214, respectively.   Working level of 100 pCi/l in equilibrium deposits 1.3×10^5 MeV/l.   Thus, the PAEC can be reported in "working levels" by dividing by 1.3×10^5 or the working level value of the PAEC can be found from

$$PAEC(WL) = 0.10 \, I_1 + 0.52 \, I_2 + 0.38 \, I_3$$

where I_1, I_2, and I_3 are the activities of the three decay products in pCi/l.

In a real atmosphere, there are losses of the decay products from the air by attachment to environmental surfaces such as walls, floors, furniture, and the people in the room.   The decay products also attach to airborne particles.   However, the airborne particles keep the radioactivity in the air.   Typically, the ratio of the decay products to radon termed the equilibrium factor, F, ranges from 0.3 to 0.5.   Thus, for a value of F of 0.5, 10 pCi/l of radon represents 0.05 working levels.

Another concept relating to the decay products is that of the "unattached" fraction.   Although it is now known that the decay product atoms are really attached rapidly to ultrafine particles (0.5 to 3 nm in diameter), there is a long history of an operationally defined quantity called the "unattached" fraction.   These decay products have much higher mobilities in the air and can more effectively deposit in the respiratory system.   Thus, for a long time the "unattached" fraction has been given extra importance in esti- mating the health effects of radon decay products.   Typically most of the "unattached" activity is Po-218 and the value of unattached frac-

tion, f_p, is usually in the range of 0.001 to 0.05 in indoor air
although it could be higher for very low concentrations of particles
in the air.

Sources of Indoor Radon

Soil. As previously discussed, uranium and its decay products are
ubiquitous and present in low concentrations. Thus, soils contain an
average concentration of radium-226 of approximately 1 pCi per gram
of soil. Thus, a kilogram of soil produces 37 atoms of radon-222
every second. Not every radon atom escapes from the soil grains into
the air in the soil. The fraction of radium actually emitting radon
into the soil gas is called the emanating radium. This fraction is
typically 5-30%, but in fine grained materials may be as high as 60%.
This radon leads to concentrations in the soil gas that would be
unacceptable for indoor air. The radon concentration in the soil
will depend on the radium concentration, the fraction of emanation,
the pore volume, the fraction of pore volume filled with water, and
the rate of exchange of air with the open atmosphere. Concentrations
of 10,000 to 25,000 Bq/m^3 (270 to 675 pCi/l) are normal. Values in
excess of 370,000 Bq/m^3 (10,000 pCi/l) have been observed.

The infiltration of these high concentrations into houses or
other structures provides the most important source of indoor radon
under most circumstances. The flow into the structure is generally
due to a pressure differential between the house and the soil. Values
of 8 Pa are quite common. The pressure differential then drives the
soil gas through cracks or other openings between the structure and
the soil.

It is important to note that it is a combination of the soil gas
concentration, the pressure differential, the leakiness of the structure and the total volume of soil gas that can be exhausted into the
structure that governs the indoor radon concentration. Thus, houses
with high radon concentrations can be found not only on high radium
concentration soils, but on soils of average or below average radium
that are extremely porous such as reported in this volume by Sextro
et al. (1987). Thus, the combination of radium concentration, soil
porosity and volume, and building characteristics make it very difficult to predict the radon concentration that will be found in any
given house. However, areas of increased probability of an indoor
radon problem may be identified based on surficial geology. Part of
the national survey of the United States that is being initiated now
will be to test this possibility. Diffusion does also play a role in
the total ingress of radon into a structure. However, it is a relatively short range process and thus generally does not lead to high
indoor radon levels.

Another potential source of indoor radon is from utilities
including water and natural gas. As reported by Hess et al. (1987)
there are wells that can have extremely high radon concentrations per
unit volume of water. Thus, use of such water could be an important
indoor radon source. Radon in water is typically of the order of
several thousand picocuries per liter or less. Surveys of U.S.
drinking water sources indicate that 74% had concentrations below
2,000 pCi/l and only 5% had values above 10,000 pCi/l. The problem
appears to exist primarily in deep drilled wells and a concentration

of 10,000 pCi/l in water will increase the indoor radon concentration by about 1 pCi/l.

Natural gas when it is removed from the ground will have radon admixed in it. However, in many cases gas is stored and transmitted over relatively long pipelines giving a decay period during which the 3.825 day Rn-222 concentration can decrease. Measurements in the Houston, TX, area yielded an average concentration of approximately 50 pCi/l in natural gas (Gesell, 1973). Values in the gas distribution system ranged from 1 to 1000 pCi/l with an average around 20 pCi/l (Johnson, 1973). With typical use rates in stoves, it is estimated that gas combustion will contribute less than 0.1 pCi/l even without venting to outside the structure (Nero, 1983). Thus, natural gas does not appear to be an important indoor radon source.

In the past, it was thought that building materials were the principal sources of indoor radon (UNSCEAR, 1977). However, most recent studies have shown that with the exception of some unusual materials such as Swedish alum-shale concrete, the effect of building materials on indoor radon is small.

Distribution of Measured Radon Concentrations

Before addressing the question of the health impacts of radon and its decay products, the levels that have been found in indoor air will be reviewed. During the past decade, there has been an increasing number of surveys on the concentrations of airborne radioactivity in homes and other structures. A number of systematic national surveys have been conducted during the past five years so that there is substantial information available on radon concentrations in many European countries and Canada. A summary of these studies is presented in Table II.

There are obviously higher values in Scandinavia than in southern Europe and in many local areas such as Cornwall in the United Kingdom (Cliff et al., 1987) there are some substantially higher concentration areas. In Sweden there is a clear effect of alum shale concrete on indoor radon levels both from the perspective of increasing the concentrations indoors, but also because of an increased percentage of the population living in houses of such construction (Swedjemark et al., 1987). Thus, the largest fraction of people live in houses with radon concentrations below 110 Bq/m^3.

In the United States, a national survey is in the planning stage (Magno and Guimond, 1987). There have been a number of local area surveys. Nero et al. (1984) have developed an aggregation of 19 of these surveys representing 552 houses in the U.S. to produce the distribution shown in Figure 4. This distribution is not based on a statistical sampling strategy and therefore its lognormal form should not be the basis of detailed analysis and interpretation. The mean radon concentration is 55 Bq/m^3 and median is 33 Bq/m^3 based on this limited sample. It is clear from all of the measured distributions that it is likely that in any survey there will be some high values found. At the present time, it is not possible to predict on a local basis where those high values will be found.

Health Effects

The principal health effect of airborne radioactivity is the induction

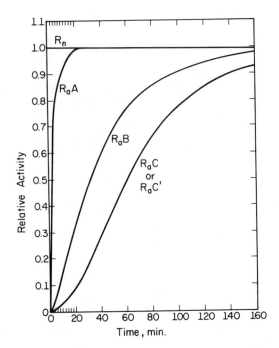

Figure 3.   Growth and decay curves for radon and its short-lived
decay products.

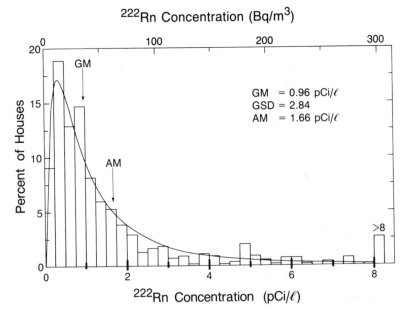

Figure 4.   Frequency distribution of measured radon concentrations in U.S. houses derived from multiple surveys (taken from
Nero et al., 1984, and used with permission).

Table II.   Summary of National Surveys for Radon Concentrations in Homes

Country	Date of Completion	Number of Samples	Median Concentration of Radon (Bq/m^3)	Ref.
Sweden	1982	750	101*	5
West Germany	1983	5970	40	10
United Kingdom	1985	2240	29	3
Ireland	1986	220	61	2
Finland	1986	4500	90*	6
Norway	1985	1500	100 (160*)	4
Japan				
(Honshu)	1986	251	17.7	1
(Hokkaido)	1986	7	35.5	1
Belgium	1983	73	41	8
Italy	1985	1000	25	7
Netherlands	1984	1000	24	9
Canada	1980	13413	33*	11

*   Arithmetic mean
1.   Aoyama et al, this volume.
2.   McLaughlin, this volume.
3.   Cliff et al, this volume.
4.   Stranden, this volume.
5.   Swedjemark et al, this volume.
6.   Castren, this volume.
7.   G. Sciocchetti et al., Sci. Total Environ. 75:327 (1985).
8.   A. Poffijn et al., Sci. Total Environ. 75:335 (1985).
9.   L.W. Put et al., Sci. Total Environ. 75:441 (1985).
10. Urban et al., Report No. KfK 3805, Kernforschungszentrum Karlsruhe
     GmbH, Karlsruhe, West Germany (1985).
11. Letourneau, Rad. Prot. Dos. 7:299 (1983).

of lung cancer. There is clear epidemiological evidence of lung
cancer in miners at relatively high levels of airborne radioactivity
from radon decay products. However, there are substantial uncertainties as to how to extrapolate these high exposure doses to the lower
levels typical of indoor air in order to decide how low the concentration must be to be "safe". There always will be exposure to radon
and its decay products. This exposure cannot be eliminated and thus
the questions are the level of risk that occurs for long term exposure
to low levels and what that maximum level should be.

The health effects of radon on the general population have not
been directly demonstrated. There are on-going epidemiology programs
that hope to obtain some direct determinations of health effects.
However, at this time the health effects are predicted through a
combination of models of the deposition of the radioactivity in the
human lung, the dose that deposited radiation imparts to the surrounding tissue, and the probability that such a dose will induce lung
cancer. There are several different dosimetric models that arrive at
different results for the same concentration of radon decay products
and unattached fraction. Some of these differences will thus lead to
different estimates of the effects of indoor radon and its decay products.

In the United States, the National Council on Radiological
Protection and Measurements (NCRP) has recently issued two reports on
this subject (NCRP, 1984a,b). The lifetime risks have been estimated
and it is predicted that exposures of 2 WLM/y would reduce the probability of living to age 70 by 1.5% and an exposure level of 5 WLM/y
would reduce that probability by 4% (NCRP, 1984a). For an average
dose to the population of 0.2 WLM/y, the model predicts about 9000
deaths from lung cancer per year in the U.S. population (NCRP, 1984b)
for nonsmokers.

Based on these results, the NCRP (1984a) recommends that remedial
action should be taken for exposure above 2WLM/y. The value of 2
WLM/y can be converted into an average working level of 0.04.

$$WL = \frac{WLM}{12} \cdot \frac{170}{720}$$

The factor of 170/270 accounts for the difference in exposure time.
A working level month assumes exposure only for the working period of
a month, 170 hours, whereas exposure may occur for all of the month.
If the equilibrium factor is 0.5, then the NCRP recommended action
level is equivalent to 8 pCi/l or twice the current EPA guidline of 4
pCi/l. It is important to note that NCRP has made its recommendation
on the decay product not the radon concentration. There is currently
some uncertainty as to the precise level at which action should be
taken. There is general agreement that remedial action is clearly
needed above 20 pCi/l and at levels of 100 pCi/l and higher, immediate
action is required.

<u>Mitigation Against Indoor Radon</u>. The focus of this volume has not
been on remedial measures. There is a review of construction practices that can be used to minimize radon infiltration or provide
inexpensive remedial action if a problem is discovered after construction (Ericson, 1987). A review of a variety of commonly used measures
and their implementation in a particular house has been given by

Cliff et al. (1987). Basically, there are two ways to mitigate
against high radon levels: prevent its infiltration or dilute it
inside the house. Crack sealing and subfloor ventilation have been
employed to reduce radon infiltration. Currently a variety of air-
to-air heat exchange ventilation systems are available to provide
dilution without excessive energy costs by recoving most of the heat
in the air before venting it to the outside. More complete descriptions of remedial measures are given in NCRP (1984a), and in the
proceedings of the recent APCA Specialty Conference on Indoor Radon
(APCA, 1986).

It is possible to remove radon decay products from indoor air by
filtration. The effects of air cleaning on dose levels are described
by Jonassen (1987). However, there are major uncertainties in the
effectiveness of air cleaning to remove the decay products because
the particles are also removed. When the particles are removed, the
"unattached" fraction increases and although there are fewer decay
products, they are more effective in depositing their dose of radiation to the lung tissue. Thus, there will be much lower dose reduction than there is radioactivity reduction. It, therefore, may be
more protective of health to control the radon rather than its decay
products.

Summary

Indoor radon in most houses come primarily from soil gas infiltrating
into the house because of pressure-driven flow. The radon decays
into a series of decay products to which most of the health effects
are attributed. These decay products begin attached to ultrafine
particles that either plateout on surfaces such as walls, furniture,
etc., or become attached to larger particles that are present in the
indoor air. The nature of those particles depends on the kinds of
sources that exist in the house such as smokers, gas stoves, etc.
The amount of particles determine the quantity of decay products that
stay in the air (equilibrium fraction, F) and the fraction of activity
associated with the "unattached" or ultrafine mode of the size
distribution (f_{pot}). These decay products are certainly harmful at
high concentrations but we cannot yet detect the effects at normal
levels because the vast majority of lung cancer death are due to
smoking. Models predict that potentially 9000 lung cancer deaths per
year in the United States are due to indoor radon. Methods are
currently available and new methods are being developed and tested
for lowering the levels of radon in indoor air.

Literature Cited

APCA, Indoor Radon, Air Pollution Control Association, Pittsburgh, PA
(1986).

Ericson, S.-O. and H. Schmied, Modified Design in New Construction
Prevents Infiltration of Soil Gas Carrying Radon, this volume (1987).

Cliff, K.D., A.D. Wrixon, B.M.R. Green, and J.C.H. Miles, Radon and
Its Decay-Product Concentrations in UK Dwellings, this volume (1987).

Gesell, T.F., Some Radiological Aspects of Radon-222 in Liquified Petroleum Gas, in Noble Gases, R.E. Stanley and A.A. Moghissi, eds., U.S. Energy Research and Development Administration Report, CONF-730915, pp. 612-629 (1973).

Hess, C.T., J.K. Korsah, and C.J. Einloth, Radon in Houses Due to Radon in Potable Water, this volume (1987).

Jonassen, N., The Effect of Filtration and Exposure to Electric Fields on Airborne Radon Progeny, this volume (1987).

Johnson, R.H., D.E. Bernhardt, N.S. Nelson, and H.W. Galley, Radio-logical Health Significance of Radon in Natural Gas, in Noble Gases, R.E. Stanley and A.A. Moghissi, eds., U.S. Energy Research and Development Administration Report, CONF-730915, pp. 532-539 (1973).

Magno, P.J. and R.J. Guimond, Assessing Exposure to Radon in the United States: An EPA Perspective, this volume (1987).

NCRP, Exposures from the Uranium Series with Emphasis on Radon and Its Daughters, NCRP Report No. 77, National Council on Radiation Protection and Measurements, Bethesda, MD (1984a).

NCRP, Evaluation of Occupational and Environmental Exposures to Radon and Radon Daughters in the United States, NCRP Report No. 78, National Council on Radiation Protection and Measurements, Bethesda, MD (1984b).

Nero, A.V., Airborne Radionuclides and Radiation in Buildings: A Review, Health Phys. 45:303-322 (1983).

Nero, A.V., M.B. Schwehr, W.W. Nazaroff, and K.L. Revzan, Distribution of Airborne ^{222}Radon Concentrations in U.S. Homes, LBL Report No. 18274, EEB-Vent 84-33, Lawrence Berkeley Laboratory, Berkeley, CA (1984).

Schery, S.D., Measurements of Airborne ^{212}Pb and ^{220}Rn at Varied Indoor Locations within the United States, Health Phys. 49:1061-1067 (1985).

Stein, L., Chemical Properties of Radon, this volume (1987).

Swedjemark, G.A., A. Buren, and L. Mjones, A Comparison of Radon Levels in Swedish Homes in the 1980s and Thirty Years Ago, this volume (1987).

United Nations Scientific Committee on the Effects of Atomic Radiation (UNSCEAR), Sources and Effects of Ionizing Radiation, UNIPUB, New York (1977).

RECEIVED August 20, 1986

INDEXES

Author Index

Subject Index

Production by Cara Aldridge Young
Indexing by Keith B. Belton
Jacket design by Pamela Lewis

Elements typeset by Hot Type Ltd., Washington, DC
Printed and bound by Maple Press Co., York, PA

Recent ACS Books

Writing the Laboratory Notebook
By Howard M. Kanare
145 pages; clothbound ISBN 0-8412-0906-5

Nuclei Off the Line of Stability
Edited by Richard A. Meyer and Daeg S. Brenner
ACS Symposium Series 324; 532 pp; ISBN 0-8412-1005-5

Geochemical Processes at Mineral Surfaces
Edited by James A. Davis and Kim F. Hayes
ACS Symposium Series 323; 684 pp; ISBN 0-8412-1004-7

Polymeric Materials for Corrosion Control
Edited by Ray A. Dickie and F. Louis Floyd
ACS Symposium Series 322; 384 pp; ISBN 0-8412-0998-7

Porphyrins: Excited States and Dynamics
Edited by Martin Gouterman, Peter M. Rentzepis, and Karl D. Straub
ACS Symposium Series 321; 384 pp; ISBN 0-8412-0997-9

Agricultural Uses of Antibiotics
Edited by William A. Moats
ACS Symposium Series 320; 189 pp; ISBN 0-8412-0996-0

Fossil Fuels Utilization
Edited by Richard Markuszewski and Bernard D. Blaustein
ACS Symposium Series 319; 381 pp; ISBN 0-8412-0990-1

Materials Degradation Caused by Acid Rain
Edited by Robert Baboian
ACS Symposium Series 318; 449 pp; ISBN 0-8412-0988-X

Biogeneration of Aromas
Edited by Thomas H. Parliment and Rodney Croteau
ACS Symposium Series 317; 397 pp; ISBN 0-8412-0987-1

Water-Soluble Polymers: Beauty with Performance
Edited by J. E. Glass
Advances in Chemistry Series 213; 449 pp; ISBN 0-8412-0931-6

Historic Textile and Paper Materials: Conservation and Characterization
Edited by Howard L. Needles and S. Haig Zeronian
Advances in Chemistry Series 212; 464 pp; ISBN 0-8412-0900-6

For further information and a free catalog of ACS books, contact:
American Chemical Society, Sales Office
1155 16th Street, NW, Washington, DC 20036
Telephone 800-424-6747